Environmental Ethics
for Canadians

Second Edition

Environmental Ethics
for Canadians

Edited by Byron Williston

OXFORD
UNIVERSITY PRESS

OXFORD
UNIVERSITY PRESS

Oxford University Press is a department of the University of Oxford.
It furthers the University's objective of excellence in research, scholarship,
and education by publishing worldwide. Oxford is a registered trade mark of
Oxford University Press in the UK and in certain other countries.

Published in Canada by
Oxford University Press
8 Sampson Mews, Suite 204,
Don Mills, Ontario M3C 0H5 Canada

www.oupcanada.com

Library and Archives Canada Cataloguing in Publication

Environmental ethics for Canadians / edited by Byron
Williston. -- Second edition.

Includes bibliographical references and index.
ISBN 978-0-19-901449-1 (paperback)

1. Environmental ethics--Canada. 2. Environmental
ethics. I. Williston, Byron, 1965-, editor

GE42.E585 2015   179.1   C2015-903452-3

Printed and bound in Canada

3 4 5 — 21 20 19

Contents

Preface

I grew up in Whistler, British Columbia. When I was a kid, Whistler was nothing like the tourist colossus it has since become. Rather than the current slick four-lane highway connecting Whistler and Vancouver, at that time there was just a pothole-filled two-lane road etched precariously into the narrow shelf between mountain and sea. The road often washed away, and driving it was always an adventure. Whistler itself was a sleepy little town populated mostly by hippies and hard-core ski bums. There was no Blackcomb Mountain to ski on, no Whistler Village, and you could access the upper runs only by the two lifts—one of which was fetchingly called "the Olive Chair"—rising up into the airy heights from the mountain's south side. The skiing, naturally, was sensational. Whistler has fierce fall lines and big, beautiful bowls. And it *had* fabulous tree skiing: long, steep fields of loosely packed evergreens. Entering one of these fields after a fresh snowfall was a stunning experience. The trees blocked all sound except that of the wind, the air was suffused with the smell of pine, and the light bent rakishly, providing uncanny depth perception. Back then, the human presence on this vast natural place was so insignificant that even on weekends you could feel as though you were lost in a wild space. Because there were no lifts going high into the glacier, as there are now, you could hike up there and see nobody at all, all day.

Whistler isn't like that anymore. There are now two mountains to ski on, dozens of lifts, hotels creeping up the hills, often deafening noise from the bars and restaurants, and lots of cars and trucks. More important, it is nearly impossible to find the sort of serenity at the top of the mountain that used to be so readily available. There's still some great skiing, but not nearly as much tree skiing, and many more people to navigate around. In this place, our relentless encroachment into nature seems to have deprived us of something important. I mention Whistler because I know it well and because it symbolizes the way many have come to think and feel about wild nature more generally. The disenchantment goes way beyond spoiled recreational possibilities, of course. We are now legitimately worried about the extent to which we are not just pushing into but also degrading the whole natural world, in some cases dramatically. In his book *The End of Nature*, Bill McKibben argues that we are now in an age in which "nature" is no more. What he means by this is that if we understand nature as something that is untouched by our activities and schemes, then there really is nothing natural any more. For instance, with the phenomenon of climate change, our activities—chiefly massive deforestation and the burning of fossil fuels—may already have altered every square centimetre of the planet. Because of this, according to McKibben, nature is now a kind of *artifact*. That is an inescapably sad thought.

This reality can make us think that we are now in a position where we need to constrain our activities so as to establish a better relationship with nature. And this is a multi-faceted task. It involves thinking seriously about population stabilization, reduction in consumption, altering our patterns of land use, learning how to build sustainable cities, learning more about ecology generally, seeing *ourselves* as ecologically constrained beings, and much more. But even more fundamentally, it involves rethinking the nature of our duties. In the face of an ecological crisis, what exactly is the scope of our obligations? Asking this question brings us into the sphere of environmental ethics.

We can begin to understand what environmental ethicists are trying to do by examining a distinction that has been a part of philosophical ethics for most of its history. This is the distinction between agents and patients. An agent is a moral actor, someone capable of performing actions for specifically moral reasons. Anyone who can perform an action and explain it to herself and others by saying that she did it because it was to her advantage, or was something a kind person would do, or was something required by the categorical imperative, or, simply, was the right thing to do, is a moral agent in this sense. A moral patient, by contrast, is someone or something toward whom or which moral agents may have duties. All moral agents are also moral patients but not vice versa. That is, moral agents have duties toward all other moral agents, but some moral patients have no duties toward anyone or anything else.

Traditionally, it has been assumed that the only moral agents in the world are normally functioning adult humans. This definition excludes, in the first place, some severely impaired adult humans and all non-adult humans and, in the second place, all non-humans. In defining our duties, we are asking what agents owe to patients. There are therefore two immediate tasks for moral philosophy at the level both of normative ethics and applied ethics. The first is to define who is and who is not a member of the class of moral agents; the second is to define who or what is and is not a member of the class of moral patients. As we have said, on the traditional picture the first task is relatively straightforward: since agency requires the ability to act for reasons, only those entities that are capable of doing this are genuine moral agents, and only normally functioning adult humans fit this description. There may be reasons to reject this conception of moral agency but we will not linger on them here.

The really hard question is what things are in the class of moral patients. It is crucial to know this, because with respect to anything in this class, we are not permitted to do whatever we like. Our actions toward these beings will be subject to moral constraint. Most of us think that other moral agents, severely impaired adult humans, and children are moral patients. But what about the non-human world? Environmental ethics begins when we take seriously the task of widening the circle of moral considerability beyond the human sphere. Many philosophers believe that *this* task—finding a way to think of individual animals, plants, or even ecosystems as moral patients—is fundamental if we are to solve our most pressing environmental problems. This book is meant to aid Canadians in thinking through these problems clearly and systematically.

Why the national focus? Why environmental ethics *for Canadians?* The first and most obvious answer is that most of the work done in environmental ethics, at least judging by the contents of most anthologies for university courses, uses case studies and examples from the American and global context. This perspective can give Canadian students studying such material the mistaken impression that there are no environmental issues in Canada, that the environment is someone else's problem. But it is not: Canada has immense and complex environmental issues to resolve. To illustrate, consider Canada's record on environmental protection as documented by the Washington-based Centre for Global Development. In its 2013 survey, Canada came in thirteenth place among 27 of the world's wealthiest countries. This middle-of-the-pack result hides a very poor environmental performance because the ranking is based on a "commitment to development index" (CDI) that includes performances in foreign aid, openness to trade, openness to migration, technology creation, and more. On the criterion of environmental protection, Canada ranked twenty-seventh, dead last. According to the report, Canada "has the dubious honor of being the only CDI country with an environment score which has gone down since we first calculated the CDI [in 2003]. This reflects rising fossil fuel production and its withdrawal from the Kyoto Protocol, the world's only treaty governing the emissions of heat-trapping gasses. Canada has dropped below the U.S. into bottom place on the environment component."[1]

Given Canada's traditionally progressive values, its internationalism, its wealth and sophistication, this is frankly embarrassing. My hunch is that this very poor record would come as a profound shock to many Canadians were they to learn of it, which is all the more reason for us to come to grips with what we are currently doing to our environment in the pursuit of unconstrained economic growth and enhanced resource extraction. This book was written in part as a response to this crisis. It is a small way of pushing back against our government's attempts to deceive Canadians about the mess we are in. The moral failure and the deception are evident in the way environmentally destructive policies are couched in the bogus rhetoric of striking a "balance between the environment and the economy," in the muzzling of government scientists, in the gutting of Canadian environmental protection laws (especially with Bill C-38, introduced in 2012), in the abject failure to consult meaningfully with First Nations communities about industrial projects that will affect lands to which they have or claim title, and in the closure of internationally renowned environmental research facilities like the Experimental Lakes Area (ELA) in northern Ontario on the transparently fabricated grounds that they are too costly to maintain (the government of Ontario has, happily, saved the ELA from the federal government's shameful abandonment of it). If we are to confront the damage to the environment set in motion by our governments—not to mention their all-out assault on the *truth* about what is going on—we need to enhance our awareness of the facts.

To that end this text contains a multitude of case studies and examples from the Canadian context: issues concerning our biodiversity crisis, the way we have failed to take seriously the ecological wisdom of our First Nations peoples, the way our lifestyles are crowding out megafauna like the grizzly bear, the sustainability of our agricultural practices, the "insider" manner in which environmental issues are dealt with at the political level, the problem of dioxins in the breast milk of Canadian mothers, the question of whether or not our fresh water ought to be sold in bulk to other countries, the current effects of climate change on our ecosystems, the advisability of developing more nuclear power plants or of mining the tar sands in Alberta, our federal government's generally obstructionist approach to international negotiations on climate change, the legacy of the tar ponds in Nova Scotia, the alarming decline in key pollinators like honeybees, and much more.

This list only scratches the surface of the environmental challenges we face in Canada. It is my hope that this book will both help us understand where our main problems lie and give us the philosophical tools we need to meet the challenges in a principled and rationally defensible way.

A word of thanks is in order to the good and diligent people at Oxford University Press. Thanks first to Ryan Chynces for originally suggesting I take this project on. At the time, I doubted that I could fit it into a busy schedule, but he convinced me that it was a timely and worthwhile thing to do. Thanks as well to my developmental editors, Kathryn West, Patricia Simoes, and (for the second edition) Meg Patterson and Judith Turnbull. They are all exemplary editors, always encouraging, with an ever-ready supply of incisive suggestions for improving the manuscript. They unfailingly helped me see the shape of the larger work through the often jumbled geometry of its various parts. I also thank all my students and colleagues at Wilfrid Laurier University for talking to me about these issues over the years and Martin Schönfeld, the self-styled Mad Hun, for getting me interested in environmental ethics in the first place. Finally, thanks to Shanna Braden for her inspiration. Well before just about anyone had even heard about climate change, and while still a teenager, Shanna canvassed for Greenpeace on this issue, knocking tirelessly on the doors of Ottawa's (mostly) skeptical citizens. People like Shanna embody the green virtues we so desperately need.

Byron Williston, Waterloo, Ontario

List of Contributors

Nir Barak is a graduate student in Political Science, Hebrew University of Jerusalem, Israel.

William F. Baxter (1929–1998) was Professor of Law at Stanford University and Assistant Attorney General in the Antitrust Division of the United States Department of Justice during the Reagan administration.

Allen Carlson is Professor Emeritus of Philosophy, University of Alberta, Edmonton, AB.

Avner de-Shalit is Professor of Political Science, Hebrew University of Jerusalem, Israel.

Sue Donaldson is an independent writer and researcher living in Kingston, ON.

David Ehrenfeld is Professor of Ecology at Rutgers University, New Brunswick, NJ.

Stephen M. Gardiner is Associate Professor of Philosophy, University of Washington, Seattle, WA.

Trish Glazebrook is Professor of Philosophy, University of North Texas, Denton, TX.

Thomas E. Hill, Jr is Kenan Professor of Philosophy, University of North Carolina, Chapel Hill, NC.

Dale Jamieson is Professor of Environmental Studies and Professor of Philosophy at NYU, New York, NY.

Will Kymlicka is Canada Research Chair in Political Philosophy at Queen's University, Kingston, ON.

Aldo Leopold (1887–1948) was an ecologist, forester, naturalist, and environmentalist. He was a professor at the University of Wisconsin and founder of the "land ethic."

Robert F. Litke (retired) was Professor of Philosophy, Wilfrid Laurier University, Waterloo, ON.

Reverend Thomas Robert Malthus (1766–1834) was an English scholar working in the fields of political economy and demographics.

Ronald Moore is Professor of Philosophy, University of Washington, Seattle, WA.

Bruce Morito is Associate Professor of Philosophy in the Centre for Global and Social Analysis, Athabasca University, Athabasca, AB.

Arne Naess (1912–2009) was a Norwegian philosopher who taught at the University of Oslo. He is the founder of "deep ecology."

Glenn Parsons is Associate Professor of Philosophy, Ryerson University, Toronto, ON.

Ronald L. Sandler is Associate Professor of Philosophy, Northeastern University, Boston, MA.

Martin Schönfeld is Professor of Philosophy, University of South Florida, Tampa, FL.

Julian Simon (1932–1998) was Professor of Business Administration at the University of Maryland and a senior fellow at the Cato Institute.

Peter Singer is Ira W. DeCamp Professor of Bioethics at Princeton University, Princeton, NJ, and Laureate Professor at the Centre for Applied Philosophy and Public Ethics at the University of Melbourne, Australia.

Georges Sioui is Associate Professor of Philosophy and Coordinator of the Aboriginal Studies Program, University of Ottawa, Ottawa, ON.

Ingrid Leman Stefanovic is Dean of the Faculty of Environment, Simon Fraser University, Vancouver, BC.

Christopher D. Stone is J. Thomas McCarthy Trustee Chair in Law, University of Southern California, Los Angeles, CA.

David Suzuki is a geneticist, broadcaster, environmental activist, and co-founder of the Suzuki Foundation.

Paul W. Taylor is Professor Emeritus of Philosophy, Brooklyn College, City University of New York, Brooklyn, NY.

Karen J. Warren is Professor of Philosophy, Macalester College, St Paul, MN.

Jennifer Welchman is Professor of Philosophy, University of Alberta, Edmonton, AB.

Anthony Weston teaches in the Department of Philosophy, Elon College, Elon, NC.

Laura Westra is Professor Emerita in Philosophy, University of Windsor, Windsor, ON.

Byron Williston is Associate Professor of Philosophy, Wilfrid Laurier University, Waterloo, ON.

Introduction

Philosophical ethics gives us tools to help us to know both what to do and what sort of people to be and how to defend our views and actions rationally. In this introduction, we will look at two of the three major domains of ethics: meta-ethics and normative ethics. In sections A–C, we will examine some meta-ethical questions before going on, in section D, to do some normative ethics. The third domain of ethics is applied ethics, the effort to apply the insights of the other two domains to concrete questions in our business, environmental, medical, or other professional practices. Since the rest of this book is an extended exercise in applied ethics, I won't say much about the nature of this domain in the introduction. What follows in this introduction is by no means meant to be a comprehensive description of philosophical ethics. As you can imagine, the various positions staked out here have undergone considerable refinement over the years, and in the present context we can only scratch the surface of this rich history of thinking. The arguments, counter-arguments, and critiques I present have all been made numerous times by other philosophers. The goal here is not to present new arguments or to provide all the philosophical details but to give you a sense of the main questions and concerns that have come to shape the discipline in the roughly 2500 years of its development. Interested readers are encouraged to consult the texts listed at the end of this chapter for elaboration of any of the themes explored here.

Ⓐ The Nature of Moral Assessment

Meta-ethics is the study of abstract questions concerning the nature of moral justification or assessment, the meaning of fundamental moral terms like "good," "right," and "impermissible," and whether or not we have a reason to be moral. Morality is a fundamentally social phenomenon. We all need certain things to live and flourish—food, shelter, a sound education, good medical care, and a thriving social milieu that produces the cultural goods we want, allows for the development of friendship among people, gives rise to opportunities for play, recreation, and meaningful work, and so on. Such goods do not fall into our lives like manna, however. Nature is stingy in supplying them. They can only be obtained through cooperation and mutual restraint. How do we achieve this? A necessary condition is that each of us conducts himself or herself in ways that do not unduly infringe on or impede the ability of others to live their lives as they see fit. That is, the moral life requires that the projects we each pursue are robustly *constrained by* consideration of the needs and interests of others.

However, this leaves open the possibility that we sometimes *can* infringe on the freedoms of others (and they can do so to us) as long as the infringement is not "undue" or inappropriate. Sometimes others will get in the way as we go about pursuing our various projects, and the question then is whether or not they are *justified* in doing so. As soon as we raise the question of justification, we admit, at least tacitly, that the constraints that define moral behaviour cannot be applied or followed in a haphazard or arbitrary fashion. If Jones constrains himself from stealing the cow of his neighbour to the right but does not do so with his neighbour to the left, we naturally want to know why he is behaving this way. Quite apart from the morality of stealing from either of his neighbours, we want to know how he can justify treating his two neighbours differently. In response, he might say that the neighbour to his left stole *his* cow last year, or that he needs this cow to feed his starving family and this neighbour, in contrast to the other, has a surplus of cows, or that this neighbour poses less of a threat to him than the other does, or that one should only steal from people to one's left and never from those to one's right. Jones is trying, however clumsily, to justify his behaviour. As such, he is engaged in the social institution that is morality.

Morality is about making certain kinds of judgments or assessments, and the judgments that comprise morality are organized in a systematic way. But what are they *about*? Although it is possible to say that we can make moral assessments of agents with no reference whatever to the things they do, this idea is unattractive in view of the fundamentally social and cooperative nature of morality.[1] A better answer is that moral assessments are focused on *actions* (understood broadly to include omissions)—what we do (or fail to do). And the most basic distinction among our moral judgments is between those actions that are **permissible** and those that are **impermissible**. Another way to put this distinction is to say that some actions are *right* and some are *wrong*. The category of wrong or impermissible actions is relatively unproblematic. Most people think that murder, which is defined as the unjust killing of someone, is wrong, and nobody is therefore permitted to do it. Right actions/omissions are more complicated. They can be divided into those that are obligatory and those that are optional. An obligatory action is one that is required. One way to understand a requirement is to say that if you are required to perform a certain action—say, giving 5 per cent of your income to charity—then you can be justly blamed, and perhaps punished, for failing to do it.

Optional actions come in two kinds. The first comprises morally neutral actions. Deciding to become a firefighter is obviously permissible, but it is just as clearly not a moral requirement. You could not, other things being equal, be justly blamed for deciding to become a teacher rather than a firefighter. Nor, however, would we say that you should be morally praised for making this decision. On the other hand, there are actions that are **supererogatory**: they go "above and beyond the call of duty." Rushing into a burning building to save a cat (to simplify, suppose you are not a firefighter) is not a moral requirement. However, if you do this (a) you *can* be morally praised for doing so (the action is not merely optional), but (b) you cannot be blamed for failing to do so (the action is not a requirement).

So morality is fundamentally about actions, about what we do. But the question remains: how do we decide where any particular action belongs in the taxonomy just sketched? The answer is that our judgment of actions—their rightness or wrongness—is itself dependent either on their consequences or on the internal states of the agents performing them.

Let's look at consequences first. Suppose Beckett proposes to refuse to pay his taxes. When asked to justify this action, he might cite any number of reasons. He might say that if he pays his taxes he won't be able to install a swimming pool in his backyard this summer, or that he didn't vote for the government so he should not be forced to pay taxes to them, or that the government is using his tax money for immoral purposes, funding a covert and unjust war against the peasantry in Bolivia, for example, or, finally, that all taxation is robbery and nobody therefore should be compelled to pay taxes. This is a motley collection of reasons, but if pushed on any of them, Beckett could respond by pointing to the consequences of his actions. He could say that he will despair if his desire for a pool goes unfulfilled, or that governments one does not vote for will have more money than they strictly deserve if those who didn't vote for them pay their taxes and they will get up to no good with this extra revenue, or that his tax dollars will aid in the unjust treatment of the Bolivian peasants, or, finally, that the robbery that is taxation will encourage more widespread robbery among citizens, which will lead to total social chaos.

Whichever argument he makes, Beckett will say that the consequences of his paying taxes are *bad*, so withholding them is right or permissible for him. If we want to oppose him and stick to the assessment of consequences, we would need to show that he is wrong about this. For example, although we do not need to deny that his falling into a state of despair because he can't get a pool is a bad thing, we can say that its badness is outweighed by the badness of his refusing to pay his taxes, since if everyone acted this way, the result would be the loss or severe impairment of key governmental services. On the other hand, we might be inclined to agree with the justification springing from a consideration of the plight of the Bolivian peasants. Here, the best consequences—forcing the government to cease funding an unjust war—might be achieved by an act of civil disobedience like Beckett's.

What about internal states? Here, we have in mind things like motives, character (dispositions), intentions, and even principles. As with consequences, we do not abandon the basic distinction between right and wrong actions. We merely say that our moral assessment of the action is dependent on a consideration of agents' internal states. For example, suppose the government has imposed conscription on its citizens to help fight a just war. Smith lies about his birth date or the severity of a medical condition in order to avoid going off to fight. If he had done this simply because he lacked courage, we would be inclined to blame him. In this case, we would say that his dispositions were vicious. If he had done it because he supports the other side in the combat, and since we are supposing this is a just war, we would say that his motives or intentions were evil. On the other hand, he might have done it because he is a religiously inspired pacifist, in which case we might be reluctant to blame him at all, although we could not quite bring ourselves to praise him. Or he might have done it because his aging grandmother, who desperately needs him to care for her, would suffer horribly if he were to go to war, in which case we *might* be inclined to praise him.

With all of these permutations, our moral assessment is still focused primarily on the action, the lie. But we allow our judgment of the rightness or wrongness of that action to alter depending on what we learn about the agent's internal state. In the first two examples, we say that the lie was wrong because it was the product of a vicious or evil internal state. In the second two examples, we say that the lie was right because it was the product of a conscientious or compassionate internal state. Of course, we might think that some actions, like lies, are *always* wrong. Looking to agents' internal states does not by itself force us to the view that sometimes actions are wrong, sometimes right. Kant, for example, thought that lying was always wrong. But the crucial point is that for him the action was wrong in all circumstances because of some feature of the principle (Kant called it a "maxim") on the basis of which the agent proposes to act—namely, its non-universalizability (we'll get to this concept below). A principle is not a consequence of the action, nor is it simply the action itself, so it makes sense to include it in the category of internal states (even if this stretches the latter notion somewhat).

Let's return to Jones and his neighbours' cows. At the most general level, what we want to say is that Jones's action was either right or wrong and that to decide this issue we must look either to his internal states or to the consequences of his action (or perhaps some combination of the two). Once we do all this, we will be in a position to decide whether or not Jones is culpable for his actions and, if he is, what punishment, if any, ought to be meted out to him.

Ⓑ Challenges to Morality I: Ethical Egoism

However, we might wonder how this whole enterprise of assessing others' actions morally—and the associated practices of praising them, blaming them, and punishing them—can itself be justified. Moreover, quite apart from whether or not we should be morally assessing the actions of other people, why should we ourselves *be* moral? Let's begin with this second question, and let's call it the challenge of ethical egoism. We will then address the first question in the following section.

In Book II of Plato's *Republic*, Glaucon, a young Athenian aristocrat, puts forward a theory about the nature of justice that goes to the heart of the enterprise of philosophical ethics. "Justice" here can be thought of as referring to morality in general insofar as it places constraints on the otherwise unbridled pursuit of self-interest. Here is Glaucon:

> This, they say, is the origin and essence of justice. It is intermediate between the best and the worst. The best is to do injustice without paying the penalty; the worst is to suffer it without being able to take revenge. Justice is a mean between these two extremes. People value it not as a good but because they are too weak to do injustice with impunity.[2]

The strong person, by contrast, considers it "madness" to be bound by the terms of moral agreement. This person simply does what he wants. Even the weak wish they could act as they desire, but they are too afraid of the consequences of doing so. To test this theory, Glaucon tells an arresting little tale. Imagine, he says, a powerful ring that allowed you to become invisible if its setting is turned inward. The ring thus gives you the power to do whatever you want with no fear of the consequences. Would you not use this new power to fulfill all your desires no matter how "immoral"? Glaucon is convinced that we would answer this question in the affirmative.

If we do answer yes, we have committed ourselves to the view that for each of us, it is best to perform those actions that fulfill our own desires and interests regardless of the effect such actions have on others. This is **ethical egoism**. It may not be the way we in fact act—because we are too afraid to, given the existence of a moral code enforced by other members of society—but it is the ideal. What response can be given to Glaucon? Over the years, a number of arguments against ethical egoism have been put forward by philosophers. Let's focus on three of them.

The first is the so-called **publicity argument**.[3] One feature of a moral system is that if it is deemed to be correct, we would appear to have a duty to make its principles public. It is difficult to imagine people being held to the demands of a moral doctrine they know nothing about because it is not made accessible to them. The ethical egoist, however, cannot make his moral theory public. The reason is contained, implicitly, in Glaucon's myth of the ring. I might think it is best to pursue my interests at the expense of others but also judge that the best way to achieve this is to fool others into believing that I am committed to the terms and constraints of their code (my invisibility while wearing the ring expresses this). If I did not do this, they would presumably take measures to protect themselves against me, which would result in making it more difficult for me to exploit them for my own ends. So the ethical egoist must live a morally schizophrenic life, something that might be difficult to accomplish psychologically. More important, if a necessary feature of a moral theory is that it can be publicized successfully, then ethical egoism is not a moral theory.

Second, the ethical egoist is barred from achieving some important goods like friendship. Assuming the ethical egoist wants such goods, this is a problem. Here's how the argument goes:

1. The ethical egoist seeks the goods of friendship.
2. The goods of friendship demand that one tend to the interests of the friend.
3. Doing this often goes against self-interest.
4. Therefore, the ethical egoist must abandon her egoism in order to fulfill her own desire for friendship.

Of course, the ethical egoist could respond by claiming, as was the case with the publicity requirement, that her egoism just needs to be covert in such cases. But this is probably an unstable way to live, because such a ruse would likely be detected by even a minimally perceptive person. If Sally discovers that the *only* reason her alleged friend Betty drives her to work when her car breaks down is because she, Betty, wants to maintain access to Sally's social circle (and she probably *would* discover this), then Sally has good reason to tell Betty to take a hike.

The final criticism is directly relevant to environmental ethics and concerns our duties to future generations and to the rest of nature. Here, we focus on the counterintuitive nature of the ethical egoist's principles. Imagine a person who claimed that his desire to drive a gas-guzzling vehicle is more important than the damage to the interests of future generations such behaviour might cause. He does not need to deny that his actions will have this effect, just that the interests of posterity matter as much as his interests do (or that they matter at all). Suppose further that the interests that are being damaged are basic or vital interests. If these kinds of activities cause climate change that leads to massive droughts in 50 years and this causes a partial collapse of the global food system, then they threaten the basic interests of anyone dependent on that system. In contrast, the desire to drive is only about pleasure: our ethical egoist likes to spend his spare time driving fast down country roads. Most of us would see this as a clear case of the driver having to curb his desire. The ethical egoist thinks not, but when a moral theory—assuming ethical egoism is one—is counterintuitive to such an extent, it is probably false.

Ⓒ Challenges to Morality II: Ethical Relativism

Ethical relativism is the view that moral codes and the practices associated with them arise from particular cultures and that there is no set of transcultural moral standards against which any particular code or practice can be assessed. Therefore, we should refrain from judging the moral practices of other cultures. Rather, we should be tolerant of their differences. This line of thinking is a challenge to morality because morality itself very often seems to force us into condemning the practices of other cultures. For example, many people think it is appropriate to criticize cultures that practice female genital mutilation. But if ethical relativism is the correct position, such criticism is misguided. Formally, the argument for ethical relativism looks like this:[4]

1. There is a large diversity of moral practices and beliefs in the world.
2. Judgments about right and wrong are dependent on, or arise from, the particular cultural milieu in which they are generally expressed and acted on.
3. There are no moral standards that transcend all cultures and that can be used to assess the moral practices and beliefs of any particular culture.
4. Therefore, we ought to be tolerant of other such practices and beliefs.

What should we make of this argument? Premise (1) is a **descriptive claim**. That is, it tells us something about the way the world is, not about how it should be. To verify it, you really just need to look around the world a little. Also, the work of anthropologists and historians provides a rich and readily accessible cache of examples. Here is one from the *History* of Herodotus, one of the great early Greek historians. He is speaking of a people called the Massagetae:

> The following are some of their customs. Each man has but one wife, yet all the wives are held in common. Human life does not come to its natural close with these people; but when a man grows very old, all his kinsfolk collect together and

offer him up in sacrifice. . . . After the sacrifice they boil the flesh and feast on it; and those who thus end their days are reckoned the happiest. If a man dies of disease they do not eat him but bury him in the ground, bewailing his ill-fortune that he did not come to be sacrificed.[5]

Premise (2) is not about the fact of diversity but about another (alleged) fact, one concerning the origins of moral practices and beliefs. Here is a statement that expresses this view well:

The "right" way is the way which the ancestors used and which has been handed down. The tradition is its own warrant. It is not held subject to verification by experience. The notion of right is in the folkways. It is not outside of them, of independent origin, and brought to test them. In the folkways, whatever is, is right. This is because they are traditional, and therefore contain in themselves the authority of the ancestral ghosts. When we come to the folkways, we are at the end of our analysis.[6]

If we think of morality as a set of practices and beliefs that together comprise something like a "code" that tells people which actions are right and which are wrong, premise (2) tells us that the culture itself is always the author of this code. This statement is much more controversial than premise (1). To take just one counter-example, many religious people would argue that the moral code with which they operate is distinct from, and capable of standing in judgment on, the code of the larger culture they inhabit.

Premise (3) is meant to follow logically from premises (1) and (2). And—setting aside the question of the soundness of premises (1) and (2)—it surely does. For if it is the case that there are sometimes radically different sets of moral practices and beliefs, and if the explanation for this diversity is that each set arises from a particular cultural milieu, then there cannot be a set of moral practices and beliefs that is independent of all culture and able to sit in judgment on any culture. If we thought such a set existed, we would be mistaken. The set we picked out for this role would *in fact* be just one more emanation of a particular culture, even if it were dressed up as an absolute, universal, or transcultural moral code.

The argument's conclusion, (4), is an explicitly **normative claim**, a statement about the way the world ought to be. It states that because of what the argument has so far established, it is impermissible to be intolerant of cultures whose moral practices and beliefs are different from one's own. When you contemplate some of these practices—child labour, racial discrimination, environmental destruction, the oppression of women, and so on—you may find it difficult to avoid a feeling of disgust or outrage, but you must not express these feelings in the language of moral condemnation. In fact, you should probably work on eliminating such negative reactions altogether lest they lead you down the path of moral condemnation.

Before looking at what is wrong with ethical relativism, we should say something about what the theory is not saying. It is not the same as **moral scepticism**. The moral sceptic thinks that there are no moral truths. That is, she asserts that there is nothing anywhere in the world to make a sentence of the form "*x* is right" true. Compare this with a simple descriptive statement. The sentence "snow is white" has truth conditions. Whether or not snow is in fact white makes the sentence either true or false. But moral sentences are not like this. Just what they *are* has been the subject of much debate among moral philosophers. Perhaps, to take just one prominent possibility, they are merely expressions of approval. The sentence "*x* is right," on this construal, just means "I approve of *x*." But the ethical relativist is not a moral sceptic, because she *does* believe there are truth conditions for moral sentences. For the ethical relativist, the sentence "*x* is right" is true if and only if it is an element in the moral code of the culture of the person asserting it. Otherwise it is false. For example, if society *Y* condones female genital mutilation and a member of *Y* says, "Female genital mutilation is wrong," she would be making a false claim, not just expressing disapproval of the practice (though she might be doing that as well).

This observation is a good starting point for an examination of some criticisms that have been made of ethical relativism by philosophers over the years. We will consider three problems with ethical relativism, beginning with the problem of moral error. The person living in a society that condones female genital mutilation might think that in asserting "female genital mutilation is wrong," she is neither making a false claim nor merely expressing her disapproval of the practice. Rather, she is claiming that other members of her society are in *error* in allowing this practice to continue. She might argue as follows:[7]

1. Elements of a society's moral code may be in error. Think for example of the belief in Nazi Germany that it is right to kill Jews.
2. Ethical relativism allows for the claim that a moral practice is erroneous only if that practice is contrary to the society's moral code.
3. When elements of the moral code are themselves suspect, this is not sufficient. Think again of the "Nazi code," which condoned the killing of Jews.
4. In cases like this, we are justified in seeking a critical standpoint from outside the society's code.
5. But ethical relativism thinks there is no such critical standpoint.
6. Therefore, ethical relativism is false.

The key to this argument is the very strong intuition we have that some moral practices and beliefs are so bad that we cannot rely on the culture that has them to supply us with the standpoint we require to think intelligently and responsibly—that is, critically—about them. If ethical relativism does not allow us to take this transcultural step—the one that supplies the claim about an element of the code being in error—then it is a false theory.

The second criticism is related to this one but concerns the nature of moral disagreement. Suppose Bob, a member of society *S*, says to John, a member of society *Q*, "Rounding up and torturing citizens who speak out against their government is wrong." John, on the other hand, says, "Actually, Bob, there is nothing wrong with that sort of thing. Such people are a nuisance, and the less we see and hear of them the better." To all appearances, John and Bob are disagreeing about something. But can ethical relativism make sense of this idea? Let's suppose that both John and Bob are correctly stating an element of the moral codes of their respective societies, *S* and *Q*. Then what the ethical relativist has to claim is that Bob is really saying, "In my society, rounding up and torturing citizens who speak out against the government is not condoned." John, on the other hand, is really saying, "In my society, doing so is condoned."

However, once they translate their claims into this form—the form approved of by the ethical relativist—they should shake hands and get on to the next topic because *they no longer disagree about anything*. They are each making a descriptive claim about what goes on in their own society.[8] As long as neither suspects the other of dissembling, each of them should agree with the other. The problem is that this is not at all how we understood the original conversation they were having. We were right in thinking that they were having a genuine moral disagreement. Here's how the criticism looks formally:

1. There are genuine moral disagreements.
2. Ethical relativism denies this.
3. Therefore, ethical relativism is false.

The final criticism of ethical relativism has to do with our alleged duty to be tolerant of other cultures. Doesn't the term "duty" sound a bit weird in the mouth of a relativist? It should, because it contradicts the theory. Suppose James finds ethical relativism attractive but lives in a society that strongly disapproves of the practices and beliefs of all other cultures. In fact, James's society is a Christian theocracy that believes that all infidels (no matter the culture they come from) should be persecuted and, if they refuse to declare allegiance to the state religion, summarily executed. If James is inclined to be tolerant of infidels, he will find no support for his views in the doctrine of ethical relativism. His claim that infidels should be tolerated is, from the standpoint of this theory, simply false. In other words, if ethical relativism is correct, then we have a duty to be tolerant of others' moral practices and beliefs *only* if our culture happens to approve of tolerance. That, we might think, is a pretty shaky foundation for a duty of tolerance. More important, it shows that if the claim about tolerance is meant to be an absolute, transcultural moral standard, then, of course, the ethical relativist is in no position to assert it.

The two major challenges to morality both fail. It is not the case that whatever is in the interest of an agent is by that fact right (ethical egoism), nor is it the case that we are necessarily misguided in seeking transcultural or universal moral standards (ethical relativism). These results open the door to an investigation into the nature of our moral duties, the normative constraints we are justified in placing on both our own behaviour and that of others. This brings us to the sphere of **normative ethics**, the attempt by various philosophers to spell out the nature of these duties and to provide accounts of how such duties are philosophically grounded.

Ⓓ Four Normative Theories

1. Utilitarianism

The first normative theory is a form of **consequentialism**, the view that what matters morally are the consequences of our actions. **Utilitarianism** is a consequentialist view according to which we should strive, with respect to each action we perform, to maximize welfare among all those affected by the action (more precisely, this is **act-utilitarianism**). Among utilitarianism's earliest exponents was Jeremy Bentham (1748–1832). The doctrine was then substantially revised by John Stuart Mill (1806–1873). Bentham grounded his account in three fundamental claims: (a) that sentience is an entity's most morally important attribute; (b) that evidence for sentience is provided by the ability an entity has to feel pleasure and pain; and (c) that pleasures and pains can be quantified. On the basis of these claims, Bentham thought it was possible to be quite precise when deliberating about how we should act.

Suppose we rate pains on a scale from 0 to −10, with −10 being the worst, and we rate pleasures on a scale of 0 to +10, with +10 being the best. For example, if Jane is contemplating punching Fred in the nose because he has done something morally wrong (though not terrible), as a good act-utilitarian she must reason as follows. Fred, who is quite sensitive to physical pain, would derive −5 units from being punched. Jane herself, who enjoys punishing the wicked but realizes that Fred has been only mildly wicked, would derive +1 units of pleasure from hitting him. The total score here is −4. So long as Jane's refraining from the punch would produce a higher score than this—and it's hard to see how it could fail to do so given the scant information about our two characters provided here—the action would produce less pleasure or welfare than some feasible alternative, and she should therefore not do it.

This is, of course, hyper-simplified. The calculations will get extremely difficult when we factor in all the people affected by the proposed course of action, including those (perhaps) in future generations. Also, we don't always know what consequences our actions will bring about, so we need to factor probabilities into our calculations. But in principle this can be done with any proposed course of action. In spite of, or perhaps because of, its simplicity, the idea is not unattractive, and here are three reasons why this is so.

First, it seems to conform to our ordinary way of making moral judgments. For example, on 9/11 George W. Bush, on hearing about the planes that hit the World Trade Center in New York, ordered Air Force jets to shoot down the remaining passenger jet (the one that eventually crashed in Pennsylvania). He was prepared to order that plane shot down, thereby killing some 200 innocent people, rather than allow it to cause a potentially greater loss of life by, for instance, being flown into another building. Many people think this was the right choice, and it would have been fairly easy to justify on Benthamite principles. If this is the way we react to the decision, chances are we are utilitarians at heart.

Second, the doctrine can be a potent tool for social reform. Bentham himself was a tireless social reformer, an outspoken critic of the horrors of early industrial capitalism in Britain. At that time, as the novels of Charles Dickens illustrate, child labour was the norm, and children were often forced to work long hours in appalling conditions. One way to justify such practices is to insist that the pleasures and pains of children are not worth as much as the pleasures and pains of adults, if they are worth anything at all. One of Bentham's key claims, however, is that everyone whose pleasures and pains are affected by a course of action gets a "vote" on that decision.

Third, because of the latter claim, Benthamite utilitarianism has become instrumental in the moral struggle to end discrimination against non-human animals. As we will see in

Chapter 1 of this volume, Peter Singer, the most prominent figure in the animal liberation movement, draws explicitly on Bentham, especially on the idea that the key question for ethics is not whether a being can reason but whether it can suffer.

In spite of these points, the doctrine has received extensive criticism from philosophers. Let's focus on four ubiquitous criticisms.

First, utilitarianism runs counter to some of our most deeply entrenched intuitions. For example, most of us think that what it means to issue a promise is that the promisor will follow through on the terms of the promise *no matter what*. Or, if this is too extreme, then at least the promisor will not break the promise unless it is absolutely necessary to do so. If I promise to meet you for lunch but on the way am detained by a roadside accident and must stop to give assistance to injured people (suppose I'm a doctor), then I'm justified in breaking the promise. But the strict utilitarian might have to break virtually every promise she makes, because a fresh calculation of consequences might indicate that *just a little more* welfare would be produced by doing so. No matter how small the increase in welfare, the act-utilitarian is duty-bound to pursue this course.

But even when the utility gains of promise-breaking are quite large, we might balk at the idea that the agent ought to break the promise. For example, a married person might calculate that he should break his promise of fidelity to his wife because by having an affair, which he is reasonably certain she would not learn about, he could bring an enormous amount of pleasure both to himself and to his would-be paramour. In fact, if this is the way the calculation were to come out, he would be duty-bound to have the affair. But promises, so goes the counter-argument, are in place precisely to guard against outcomes like this. They are meant to bind the promisor's will into the future. Would it be rational *ever* to believe the promises of a strict act-utilitarian?

Utilitarians have responded to worries like this by arguing that the best consequences will be realized if everyone follows certain rules in the appropriate circumstances. So even if on a particular occasion it seemed that utility could be maximized by breaking a promise, the promise-keeping rule, if adhered to by everyone, would achieve this result more reliably. This is **rule-utilitarianism**. More technically, it is the claim that an act is right if and only if it conforms to a rule that, if followed by everyone, will produce the best consequences. However, this revision to act-utilitarianism, though it looks attractive, is not plausible. It works if, but only if, following the appropriate rule in every situation would maximize utility. But in our example of the broken promise, we have supposed that the married man has correctly surmised that his wife will not discover his affair and that, partly because of this, more utility will be produced by breaking the promise than by keeping it. If this is correct, then he has a duty to violate the rule. We do not have to deny that sometimes following a rule will maximize utility; it's just that whether or not this is the case must be assessed on an act-by-act basis. In other words, it appears that rule-utilitarianism, if it is to avoid becoming a form of rule-worship, reduces to act-utilitarianism.

The second criticism of act-utilitarianism is that it is too demanding, as it may require us to perform actions that are extremely difficult—or even psychologically impossible. Suppose Frank is walking along the bank of a river and he sees two people drowning. He notices to his horror that one of them is his wife, and someone next to him (who can't swim) informs him that the other person is a famous doctor on the cusp of discovering a cure for cancer. Frank, himself not a very strong swimmer, knows that he cannot save both people but can probably save one. What should he do? An immediate calculation of the consequences of the two decisions shows him unequivocally that more good will come to more people if he saves the doctor. So his duty is clear, but, of course, he finds the task impossible and saves his wife instead. According to the utilitarian, the action Frank performed was wrong. His emotional attachment to his wife was irrelevant from the standpoint of making the correct moral judgment. Most of us can sympathize strongly with Frank, because we believe that particular attachments—like those we have to family members—place special moral burdens on us. We believe that the sort of *impartiality* utilitarianism asks us to display in cases such as this is out of place.

Third, utilitarianism eliminates supererogation. Think again of the person who rushes into a burning building to save a cat. In doing so, he brings an enormous amount of pleasure to the cat's owner, not to mention to the cat. To say that the act was supererogatory is to claim that it is above and beyond the call of duty. But for the utilitarian, in any situation there is just one right thing to do. If you do it, you may be praised, but if you don't, you can rightly be blamed. It all depends on how the figures add up. Our reaction to the individual's act is not just about whether or not we have recourse to a word, "supererogatory." It has to do with our whole moral practice. Were the person to refuse to go into the building, the utilitarian would expect us to criticize him morally, and criticizing is a meaningful social act with real consequences. We will think of him in a less favourable light, perhaps subject him to a certain amount or kind of social ostracism, and so on. The concept of the supererogatory is meant to give our moral practices some subtlety and flexibility, qualities that allow us to better cope with a moral reality that is often messy and complex. So the critique of utilitarianism here is that in asking us to eliminate the supererogatory from our moral repertoire, it is oversimplifying the practice of morality.

Fourth, utilitarianism has generally been thought to conflict with the requirements of *justice*. One aspect of this criticism has to do with how the benefits and burdens of decisions are distributed across a population, a question central to environmental ethics. Questions of environmental justice arise often for low-income people and non-white or Aboriginal people. Toxic waste sites are disproportionately located in their neighbourhoods. Exploitation of the resources on Aboriginal peoples' traditional lands tends to proceed without adequate consultation with them. Or, their land rights are encroached upon in more indirect ways. For example, the Innu of Labrador have had to endure years of low-level test flights by NATO fighter jets over their territory. The flights have caused health problems for the people and disturbed the migratory patterns of many animals on which the people depend. Attempts to justify actions like this sometimes proceed on utilitarian grounds. The actions, so goes the argument, maximize aggregate welfare because NATO pilots use the flights to hone their skills just in case they are called upon to protect us from our enemies, whereas the people who are disadvantaged—the Innu—are not particularly numerous. But if we are inclined to oppose such actions, we will likely want to say that people are as a *matter of right or justice* entitled to not be treated or interfered with in this sort of way, no matter what the utilitarian calculation suggests. Rights cannot be abrogated in the interests of promoting aggregate welfare.

2. Deontology

The question of normative ethics is, "What makes an act right?" Utilitarians answer by pointing to consequences, but we have seen that there are problems with that approach. Partly in response to these problems, we might be inclined to suggest that what really matter are the rules or principles on the basis of which agents act. This is the standpoint of **deontology**, which means "duty-based." It holds that we should focus on these rules or principles and try to determine which of them are right and which are wrong, quite apart from the consequences of acting on them. The most famous exponent of this approach to morality is Immanuel Kant (1724–1804). For Kant, sound morality was grounded in the **categorical imperative**. There are three versions of this imperative, but we will focus here on just two. First, let's analyze the concept itself. An "imperative" is a command. To say that a command is "categorical" is to contrast it with a **hypothetical imperative**. What is a hypothetical command? Here are some simple examples:

- If you are hungry, eat some food.
- You can only get to heaven by following the Ten Commandments. Do so.

What unites these examples is the idea that there is a command to pursue a certain course of action, but the command "binds" only if something else is true. So you should eat some food but only if you are hungry. In the second example, the structure is the same even though the if-clause is merely implicit: we could have said, "If you want to get to heaven, follow the Ten Commandments." For any hypothetical imperative whose if-clause is not satisfied or is false, the command is cancelled. By contrast, no such if-clause can cancel a categorical imperative. Here is what categorical imperatives look like:

- Follow the Ten Commandments.
- Get out of bed.

For Kant, all genuinely moral duties are categorical imperatives. They apply to us, if they do at all, whatever our other inclinations, desires, sentiments, and so on. If, for example, we have a duty to keep a promise, then this applies to us in the sense that it ought to motivate us to act in accordance with it, even if we do not feel like keeping the promise or believe that utility will be maximized if we break it. The only circumstance in which the duty would cease to apply is if we had been prevented from keeping it through no fault of our own. If Bill has promised to visit Mary in the hospital but before leaving his house is waylaid by thieves who tie him to a chair, preventing him from moving while they ransack his house, then his duty to Mary is cancelled. As Kant famously put it, "ought implies can." Further, only if our actions are motivated by categorical imperatives are they genuinely moral actions. Now let's look at the two versions of the categorical imperative.

The first version (CI-1) says that we should "act only according to that maxim by which you can at the same time will that it should become a universal law."[9] A "maxim" is a general principle of action. We are constantly faced with choices between two or more courses of action among which we must deliberate, since we can very often only choose one thing. Think about Canada's decision to allow the extraction of bitumen from Alberta's tar sands, an egregiously polluting activity. Should we do this or not? CI-1 offers a way to conduct our deliberations about such choices systematically. Let's suppose that each one of us is required to make a decision about the morality of this proposal so that we know which way to vote in the next federal or provincial election, for example. Heather is one such agent. We can break her deliberation down into three steps. First, she describes a course of action: "Since it makes our economy strong, we should exploit the tar sands with no regard for the environmental consequences of doing so." Next, she generalizes the proposal: "In the interest of strengthening its economy, any country with a valuable natural resource should exploit it with no regard for the environmental consequences of doing so." This is Heather's **maxim**. Finally, she should test her maxim for **universalizability**. That is, she should ask what would happen if everyone actually did this.

Kant's claim is that it is permissible to act on the maxim if, and only if, doing so would not result in a contradiction, which is the same as saying that it is universalizable. Let's suppose that if everyone acted on Heather's maxim, the result would be environmental catastrophe on a global scale. But since the economy cannot function except on the basis of a healthy environment, the widespread degradation of our natural capital would severely impair the global economy. What has happened to Heather's deliberations? Despite appearances, the Kantian is not arguing like a consequentialist here. He is not saying that we should refrain from exploiting the tar sands because such exploitation would produce a bad outcome (it would fail to maximize utility, say). Rather, he is saying that Heather is *contradicting* herself and that it is therefore logically impossible for the maxim to be universalized. Here's how the contradiction looks formally:

- Heather: "I am committed to acting in ways that strengthen the economy."
- Heather: "I am committed to acting in ways that weaken the economy."

Because the maxim generates a contradiction, it is immoral for Heather to act on it. The key to this theory is the idea that if an action is permissible (or impermissible) for one agent, then it should be permissible (or impermissible) for all agents in relevantly similar circumstances. That is, Kant is attempting to weed out of our moral deliberations factors that, he believes, are irrelevant to sound morality. Most of these factors play a large part in how we do generally act, factors like our particular attachments, our emotions, our desires, and so on. We may recognize that if everyone did what we are proposing to do, the results would be less than optimal, but we insist that there is something special about us that gives us permission to perform the action.

So, to pursue the example one step further, we might ask, "What is special about Canada that gives it moral permission to develop this resource unsustainably?" Put another way, suppose every country had its own cache of tar sands (or something analogous). Since we could not permit everyone to develop their resource (on pain of environmental catastrophe), would we argue that Canada alone has the right to develop its resource? On what basis could we make this claim? CI-1 is important precisely because it defuses the attempt, endemic to moral decision-making, to make exceptions of ourselves. You act this way every time you jump a queue at the coffee shop or support the unsustainable development of our resources, both of which actions are unjustifiable on Kantian grounds.

The second version of the categorical imperative (CI-2) says, "Act so that you treat humanity, whether in your own person or in that of another, always as an end and never as a means only."[10] Kant elaborates on the idea this way:

> Now, I say, man and, in general, every rational being exists as an end in himself and not merely as a means to be arbitrarily used by this or that will. In all his actions, whether they are directed toward himself or toward other rational beings, he must always be regarded at the same time as an end. . . . [R]ational beings are designated "persons," because their nature indicates that they are ends in themselves, i.e., things which may not be used merely as means.[11]

Actions are right to the extent that they conform to this principle, but what does it mean to treat something as an "end in itself"? Let's get at this by looking at the contrasting notion, that of treating something as a mere means. Here are some examples:

- I use my car to get to work.
- I use the pharmacist to get my medication for me.

In each case, the "thing"—car, pharmacist—is a mere instrument for my purposes. There is, other things being equal, nothing wrong with treating these things this way. This is unproblematic in the case of the car. But reference to the pharmacist indicates that there is another class of things in the world that we cannot treat as a *mere* means to our ends—namely, persons. What is it about persons that gives them this special status? For Kant, it is the fact that they are rational natures—that is, autonomous agents. **Autonomy** refers to the ability to act on the basis of self-legislated reasons (the word derives from the Greek words "auto" or "self" and "nomos" or "law"; so: "giving the law to oneself"). We are all assailed by desires, emotions, and inclinations. But most of us think we should not act on all of them. On what basis do we decide which ones to act on and which ones to suppress or ignore? Usually we do this by appealing to a higher-order set of ideals or projects that more deeply define who we are. Isaac, a reformed smoker, might still like smoking and want to smoke, but he also wants to be healthy, and so, in the name of this ideal, he suppresses the desire for cigarettes. Here, he acts for reasons that he has, we suppose, autonomously generated.

Kant makes three further claims. First, that each of us inevitably conducts *himself or herself* this way. Second, that we should also notice that every *other* "rational nature" does so as well. And finally, that this should cause us to act in a way that respects this fact about ourselves and others. This is what it means to respect rational natures wherever we find them. On what grounds could we respect this fact about ourselves but not when it comes to others if what we respect is precisely *the fact itself*? Morally, it does not matter where we find autonomy. This is a moral doctrine with substance: it requires us to conduct ourselves in such a way that we do not undermine or bypass the autonomy of other people. We cannot lie to them, for example, because in order to act autonomously a person must have full access to all relevant information, but liars conceal this information. A whole host of other actions—coercion, for example—will be impermissible on similar grounds.

Let's consider three standard criticisms of deontology, two directed at CI-1 and one at CI-2. The first criticism of CI-1 is that it allows us to generate "duties" that are trivial or immoral. Here's an example of a trivial maxim: "Always brush your teeth one hour before you go to bed." Obviously, the maxim is universalizable. No contradiction would result from it, but it is clearly too trivial to count as a moral rule. True, replies Kant, but what it does do is pick out a permissible action. Because it is trivial, it cannot be obligatory, so it is better to think of it as a non-supererogatory optional action (see section A, above). The other point is that immoral actions can be based on universalizable maxims. An example is: "Torture all those whose last name has nine letters in it." Unfortunately for people with names like "Williston," it is difficult to see what sort of contradiction could be generated out of this maxim. Is it therefore permissible to act on it? Here, the best response is to insist that CI-1 and CI-2 come as a package. And since we have clear grounds, on the basis of CI-2, for excluding torture from the list of permissible actions, the maxim is not morally sound (phew!).

The second criticism of CI-1 is that it is absolutist. That is, whereas most of us think that there are legitimate exceptions to any moral rule, Kant denies this. There is an undeniably absolutist element in Kant's thinking. To deny it would be to cancel the defining feature of his moral philosophy. Many critics point to his stance against lying—about which he was a staunch absolutist—to argue that the theory as a whole is flawed. That is probably too quick. One way to get around the problem is to note that much depends on how we frame our maxims. Although we may not want to say that a given action is simply wrong, wherever and whenever it occurs, we can formulate the maxim so that it defines circumstances under which we can all agree that it is wrong. So, for example, we may want to say that the claim "theft is wrong" is false—because, say, theft may be the only means available to feed one's starving family—which leaves open the possibility that theft for pure profit is impermissible. In this fashion, we can build reasonable exceptions into a system of rules that is otherwise absolutist in orientation.

The third criticism of CI-2 focuses on those beings whom it seems to leave outside the sphere of moral concern. The connection Kant draws between a person's capacity for rationality and his or her moral standing should make us wonder how we are to treat beings that are not rational in this sense. These beings would include fetuses, small children, mentally impaired people, non-human animals, and other living things. Are they mere "things"? If so, are we permitted to treat them as mere means to our ends? The best way to answer may be to divide the list of beings just mentioned into those that are potentially rational and those that are not. So fetuses and small children are worthy of respect just because they will eventually become fully rational adults. But the rest of the beings on the list may not fare as well. What is the Kantian basis for treating mentally impaired humans, non-human animals, and other living things morally? As we will see in Chapter 2, Paul Taylor employs Kantian premises to argue that living things as such are worthy of respect, but not everyone agrees with this use of Kant. If we balk at it, it may be because we see a real limitation in the theory here, especially for environmental ethics.

3. Social Contract Theory and Contractualism

If we find the idea that morality is a fundamentally social phenomenon attractive, we may be drawn to social contract theory or contractualism, both of which claim that actions are right by virtue of having arisen from a certain kind of agreement among people. **Social contract theory** originated with Thomas Hobbes (1588–1679), who argued that in a state of nature we all have good reason to "sign a contract" bringing civil society, complete with a nearly all-powerful sovereign, into being. The reason is that in the state of nature preceding the establishment of civil society, life is "solitary, poor, nasty, brutish and short."[12] It is that way because everyone is self-interested, there are no external constraints on anyone's actions, and people are therefore out to get as much as they can to secure their own interests. This is the infamous "war of all against all." For Hobbes, we are all fundamentally equal in this state, not because we are all deserving of respect but because any person can kill or dispossess any other person. To get out of this horrible state, we agree to establish an authority figure whose function is to limit our natural liberty in the interest of maintaining civil peace.

For moral philosophy, this is the key claim. The fact that we have signed an agreement acts as a constraint on our future relations with other signatories, whereas in the state of nature there were no constraints on our actions. This is a powerful idea. For Hobbes, the sovereign was required to back up the demands of the contract. In setting up the sovereign, the contracting parties are clear that they are relinquishing their own authority over crucial matters like dispute resolution:

> And, therefore, as when there is a controversy in an account, the parties must by their own accord set up for right reason the reason of some arbitrator or judge to whose sentence they will both stand, or their controversy must either come to blows or be undecided, for want of a right reason constituted by nature, so it is also in all debates of what kind soever.[13]

But nobody gives up the perspective of self-interest in making this move. Instead, each of us judges that that interest will be best served by allowing an independent authority to decide our quarrels for us. This makes sense: if you and I disagree about who gets what, I might not like what the judge rules (because it is less than I wanted), but it is better than the possible alternative—namely, that you would have gotten everything because you were able to overpower me. The judge then has two functions. First, to define what is right and wrong in particular cases. This effectively gives the judge the power to fix the content of moral principles. Second, to apply sanctions to those who contravene the rules. So although morality is reduced to self-interest on this model the cooperative enterprise does not seem to be threatened.

It is worth emphasizing the role given to *correct procedure* in deciding how to resolve conflicts among citizens. Hobbes is saying that each of us will sign the social contract, and consider ourselves bound by its terms, if we are sure that every other member of the group will do likewise. Again, we might not like any particular decision the judge makes, but we agree to abide by the results of the procedure. In other words, its origin in an agreement among signatories *legitimizes* the procedure even when that "procedure" is simply the will of a sovereign power.

A more recent version of the theory, labelled **contractualism**, was provided by John Rawls (1921–2002). Like Hobbes, Rawls was in the business of providing a justification for specific moral and political principles governing people's relations with one another. But he goes about it differently, partly because he was inspired as much by Kant as by Hobbes. Rawls constructs a thought-experiment. Imagine you were a party to the original agreement but that you and your co-signatories negotiated the terms of the contract from behind a **veil of ignorance**. That is, you were not aware of specific features that define who you are: your place on the socio-economic ladder, your race, your gender, and so on. Rawls describes this situation as the **original position**. Now, Rawls does not think there ever was such an agreement among us. His theory is explicitly hypothetical. What he is doing is asking what we would agree to—given some of our deepest values and ideals—if we had been called upon to shape the principles that govern the institutions of justice.

Suppose someone suggests that those with exceptional physical "talents," such as great strength or beauty, ought to receive 95 per cent of society's wealth. Everyone in the original position would be likely to reject this principle for the very good reason that they are, given the relative scarcity of very strong and beautiful people in the world, unlikely to be among the charmed elite. More seriously, Rawls thinks that utilitarian thinking generally would be rejected by these people. If someone proposed to abolish individual rights and liberties—the right to free speech, habeas corpus, and the like—whenever doing so would maximize utility, everyone would, given a moderate degree of risk aversion, object. After all, *your* rights and freedoms might be the first to go. You might be the victim of the utilitarian calculus. What *would* we agree to in the original position? First, that there be an extensive catalogue of basic individual rights and liberties. Second, that if there is to be economic inequality, it must be to the benefit of the worst-off in society. That is, such people must be better off than they would be in a system of strict economic equality. Finally, everyone must have equal access to positions of power and prestige in society.

These principles aside, what is the key difference between the Hobbesian and Rawlsian versions of the theory, between social contract theory and contractualism? Both theories, suitably expanded, claim that the content of morality comes from an agreement among rationally informed agents. The main difference has to do with what motivates agents to follow the rules, to be moral. For Hobbes, a "contract without the sword is but words." That is, he does not suppose that agents will be motivated to obey the pronouncements of the judge in conflict resolutions unless that judge has the power and the will to enforce his decisions. This way of seeing things has the following implication. Suppose I discover on some particular occasion that I will gain by reneging on my contract with you. And I also discover that I will get away with it. That is, neither you nor the judge will know that I breached the contract. It would appear in this case that I have no incentive to abide by the contract. I have no motivation to be moral.

Rawls's version of the theory seeks to avoid this outcome. It says that we are all equal, not because we have the power to damage one another but because everybody's interests matter in an equal moral sense. This is the sense in which he is deeply indebted to Kant. In other words, the parties to a contract are defined as having a fundamental respect for each other. That is why they agree to consider factors like race and gender irrelevant in the original position. And because we go into the agreement with a built-in sense of respect for each other, we also do not renege on our agreements, because to do so would violate that fundamental value or commitment. We are motivated to accept the principles we devise because we think that is the right thing to do, and we always act in accordance with what is right.

So the basic difference between Hobbes and Rawls is that Hobbes thinks people are motivated only by self-interest whereas Rawls thinks that people can be motivated by the consideration that acting a certain way is, simply, right. Morality itself is motivating. Another way to put the same point is that although Hobbes and Rawls both think there is a fundamental equality among us, they conceive of it differently. Equality, for Hobbes, means that we are equally powerful in the state of nature. Even weaklings can band together to defeat powerful individuals. Rawls, however, is talking about moral equality, the idea that we all ought to treat each other, or refrain from treating each other, in a certain way. One understanding of equality is descriptive, the other is normative.

Two further points are worth emphasizing. First, contractualism reveals its debt to Kant in another, related way. The original position is a way of allowing the moral ideal of *impartiality* to generate concrete principles to structure our institutions. The corollary is that it is precisely partiality that gets in the way of our doing this. When Kantians argue that we should view our particular identities—my status as *this* person, a member of *this* economic class or gender, race, tribe, and so on—as irrelevant to the process of justifying the decisions we make, they are trying to move beyond the perspective of partiality. That is the whole point of CI-1: if it is right for you do *x* here and now, then it is right for anyone similarly placed. There's nothing morally special about you. In applying the test of universalizability to our maxims, Kant is, in a sense, forcing us to *erase* our particular identities, just as Rawls is explicitly doing with the device of the veil of ignorance. In both cases, we can attain the status of properly moral agents only if we make this move. So all three normative theories considered so far—utilitarianism, deontology, and contractualism—are impartialist.

Second, the theory is attractive partly because everyone counts in a radical way. Utilitarians think everyone counts too. If your interests are going to be affected by someone else's decision, then your vote must be counted. But for utilitarianism, your vote may get swamped by the tide of contrary votes, and crucially, you have no further *moral grounds* on which to complain about this. You are therefore morally bound to abide by the terms of the decision. For contractualism, however, every person effectively has a *veto* on any decision. The difference between the two theories springs from how they define right action. For utilitarianism, actions are made right by their consequences. But for contractualism, what makes an action right is that it is the product of a properly arrived-at *agreement*. Therefore, if you do not sign the contract, then you are effectively saying that you do not agree to its terms and you cannot be bound by it. Of course, this might mean that things will go very badly for you, since there are advantages to being inside the circle of the contract's signatories. You have effectively placed yourself outside the bounds of moral consideration. But for what it's worth, you can still protest while those inside the circle ignore you as they construct the institutions of civil society.

Now let's focus on one very important criticism that philosophers have made of the theory. It applies to both versions and is similar to one of the criticisms of Kant considered above. It has to do with the exclusion of certain beings from the circle of moral concern. Again, a person needs to possess certain properties or attributes to be able to enter into the negotiations culminating in a contract. The metaphor of a contract would appear to exclude anyone incapable of understanding its "terms." This applies clearly to children and the mentally impaired, but does it also apply to members of future generations or members of foreign cultures? For that matter, will there not be a tendency to exclude from negotiation anyone deemed already to lack sufficient *power* to affect the interests of the signatories? Historically, for example, the theory was invoked explicitly to exclude anyone who did not own property. Nor did women fare very well under the terms of the contract. From the standpoint of environmental ethics, the most troubling exclusion has to do with members of future generations. Many of our environmental decisions have implications that will directly affect those people, yet how can they be genuine parties to our contracts?

4. Virtue Ethics

Virtue ethics is the oldest of the normative theories we are considering here. It began with the work of Aristotle (384–322 BCE), whose articulation of the basic tenets of the theory are still canonical. For Aristotle, actions are made right by being the product of the correct character or disposition of the agent performing them. Aristotle begins his analysis by considering what the concept "good" means. And here he makes a basic distinction between kinds of goods:

> If, then, there is some end of the things we do, which we desire for its own sake (everything else being desired for the sake of this), and if we do not choose everything for the sake of something else (for at that rate the process would go on to infinity, so that our desire would be empty and vain), clearly this must be the good and the chief good. Will not the knowledge of it then have a great influence on life? Shall we not, like archers who have a mark to aim at, be more likely to hit what is right? If so we must try, in outline at least, to determine what it is . . .[14]

The distinction is between instrumental and intrinsic goods. Instrumental goods are those that are desired entirely for the sake of something else (e.g., medicine, physical exercise); intrinsic goods are those that are desired entirely for their own sakes. There is only one intrinsic good, and that is happiness. The Greek word for it is *eudaimonia*, but "happiness," to the extent that it connotes a psychological state, is a misleading translation of it. Better is "flourishing" or "excellence." Obviously, it is possible to flourish or be excellent at something in the absence of happiness (if we think of happiness as, say, cheerfulness).

To unpack this key notion, we need to look at Aristotle's conception of *function*. Everything has a function peculiar to it by virtue of the kind of thing it is. For example, the function of a knife is to cut well, and the function of a carpenter is to build sound structures. Notice that these descriptions presuppose that the thing can perform its function well or badly, and this gives us room to criticize or praise it. We might say that the knife is bad because it does not perform its cutting function properly or that Jones is a good carpenter because he builds sound structures. So if the broad claim about function is true of everything there is, what is the human function? The ancient Greek philosophers were fond of dividing the mind up into various parts, then assigning a specific function to each part. Plato thinks that the human soul is divided into three distinct parts: the appetitive (home of desire), the spirited (home of the emotions), and the rational (home of the intellect). The job of the rational part of the soul is to control the "lower" powers, sometimes by violently suppressing them. Aristotle works with the same general model, although he thinks that desire and emotion are more amenable to rational training than Plato does. For this reason, one sees the metaphors of conflict and even "civil war" among the parts of the soul far less often in Aristotle than in Plato.

We humans are living up to our potential when reason is in charge of the whole person. Aristotle thinks that one way in which we can exercise our rational capacity to its utmost extent is through contemplation of the heavens. But arguably just as important is to *live* in a rational manner, and this means acting in accordance with the virtues. As Aristotle puts it, "happiness is an activity of the soul in accordance with virtue."[15] A virtue is a state of character that causes or allows us to act in a specific way as the occasion demands. The key virtues are courage, liberality, justice, temperance, pride, wit, and friendliness. The idea, therefore, is to locate the rightness of actions in the dispositions of agents to behave justly, temperately, and so on. Aristotle's famous "doctrine of the mean" helps clarify what it means to be disposed to be virtuous. Each virtue is said to be a mean between two extremes, one of which is a defect or deficiency in the motivation that defines the virtue, the other an excess of it. So, for example, liberality (generosity) is midway between stinginess (defect) and prodigality (excess), and proper pride is midway between undue humility (defect) and vanity (excess). In principle, every virtue admits of this sort of analysis.

Even with the doctrine of the mean, however, our analysis is not complete. This is because we can imagine someone who seems not to be either defective or excessive with respect to a particular virtue but is nevertheless not properly virtuous. Suppose James, a Greenpeace activist, has just powered his tiny rubber dinghy toward an offshore oil rig in the North Sea, in very heavy seas and under the threatening watch of the Danish navy, whose state oil company owns the rig. Whatever we think of his politics, we might see his action as a clear example of courage. It certainly does not look like either cowardice (defect) or recklessness (excess). James executes the operation masterfully, neither hesitating at an inappropriate time (as a timid person would) nor rushing headlong when it might be better to wait a bit (as a reckless person would). But what if we learned that James was able to perform the action only because he had taken a fear-inhibiting drug or because he had learned the night before that he had just two months to live and had as a result become unhinged? In such cases, we might be inclined to deny that he acted courageously. To get our bearings on this issue, let's look at what Aristotle had to say about justice and temperance:

> Actions, then, are called just and temperate when they are such as the just or temperate man would do; but it is not the man who does these that is just and temperate, but the man who also does them as just and temperate men do them.[16]

And how do just and temperate (and also courageous) people perform actions?

> [I]f the acts that are in accordance with the virtues have themselves a certain character it does not follow that they are done justly or temperately. The agent also must be in a certain condition when he does them; in the first place he must have knowledge; secondly, he must choose the acts and choose them for their own sakes; and thirdly, his actions must proceed from a firm and unchangeable character.[17]

James, then, may not have performed the courageous action *as a courageous person would*. There is thus a key distinction between those who perform actions that are merely in accordance with virtue and those that are done from the virtues. The latter requires knowing what one is doing, choosing the action because it is the right thing to do, and acting from "a firm and unchangeable character." James may satisfy the first condition, though perhaps not: his mind may be too altered by drugs or emotions for him to fully appreciate what he is doing. But he almost certainly does not meet the second and third conditions. He is not acting *because of* the rightness of what he is doing. Instead, the drug or his despair is the proximate cause of his actions. And his character is not firm and unchanging. We may suppose that in the absence of the drug or the bad news, he would not be doing what he is doing or any other "courageous" thing for that matter.

What we might want to say about James is that he lacks **phronesis**, or "practical wisdom." For Aristotle, this is the master virtue because it allows one to deal in just the right way with the emotions that fuel our actions:

> For instance, both fear and confidence and appetite and anger and pity and in general pleasure and pain may be felt both too much and too little, and in both cases not well; but to feel them at the right times, with reference to the right objects, toward the right people, with the right motive, and in the right way, is what is both intermediate and best, and this is characteristic of virtue.[18]

Phronesis is skill at perceiving what situations require and then marshalling the emotions—in just the right degree and quality—to get the job done. How do we learn this kind of wisdom? Aristotle famously says that you become just or courageous or moderate by doing just or courageous or moderate things. What this means is that you gain the appropriate character by living in a society with the right kind of people, those who can show you the moral ropes. You do not first learn a set of rules, then decide how to apply them on a case-by-case basis.

Think of how you might learn to cook well. You can buy recipe books and teach yourself, or better still, you can apprentice with a master chef. In the latter case, you will learn to cook well by being told what to do at first, watching and learning from the way the chef does things, and incrementally incorporating into your own behaviour what you learn from these observations. In other words, you begin simply by doing what you are told so that you may come, in time, to do those same things yourself at the right time, for the right reasons. Similarly, if you want to be courageous, your best course is to work closely with someone who is genuinely courageous. At first, you let this person tell you what to do, and you may not understand exactly why you should be performing this action rather than another one. But you will, one would hope, become increasingly *self-sufficient*. And this means that you will come to know how to see and feel in exactly the right way to produce the appropriate action.

Let's briefly consider two criticisms of virtue ethics, criticisms that have resurfaced in one form or another among moral philosophers for over 2000 years. The first is a challenge from relativism: that the list of virtues required for morality differs radically from one culture to the next. One example is the inclusion of pride on the list of virtues put forward by the ancient Greeks. Yet Christian writers considered pride a sin. So, to generalize, if one society or culture operates with a catalogue of the virtues that is totally different from another society's catalogue, then there is no possibility of genuine moral dialogue between the two cultures. To return to a problem examined above (section C), there is therefore no possibility of moral disagreement between these two cultures, and this is an unwelcome result (as we have seen). The virtue theorist could respond to this charge by denying that there is *radical* disagreement among cultures as to what makes up the list of virtues. Although the place of a virtue like pride comes and goes, every culture values virtues like courage, justice, and temperance. This claim would need to be backed up by anthropological evidence, and it would be necessary to show that distinct cultures mean the same thing by the same virtue terms. Nonetheless, assuming we could fulfill these tasks, such a response might be adequate.

The second criticism has to do with the possibility of evil virtues, or better, of virtues being used in the cause of evil. Think for example of the "courageous terrorist." This person's job is to kill large numbers of innocent people through suicide bombing, something which certainly seems to demand a good deal of courage. Are we therefore required to praise the virtue of the suicide bomber? This seems absurd. If it is, we may have good reason to reject the theory. Here's how the argument in support of the rejection would go formally:

1. For virtue ethics, an action is right if it is the product of a virtuous disposition.
2. But a person can practise the virtues for evil purposes (our terrorist).
3. Therefore, for virtue ethics it can be right to act in an evil manner.
4. This is absurd, and virtue ethics is therefore an inadequate normative theory.

This is a powerful argument, but we may be able to avoid the conclusion by denying premise (2). That is, we could say that the terrorist is not *really* courageous, appearances to the contrary notwithstanding. One way to cash this idea out is to say that you have either all the virtues or none at all. So you cannot be genuinely courageous unless you are also just, for instance. And since he is in the process of treating others unjustly, the terrorist's courage is a sham. This is the **unity of the virtues thesis**, which Aristotle himself advocates. Not every philosopher agrees with it, but it may be the only way to escape the conclusion of the argument from evil virtues.

Environmental ethics is a form of **applied ethics**. It employs the concepts and vocabulary of meta-ethics and, especially, normative ethics in an effort to rationally ground our duties to nature (just as other forms of applied ethics do in such fields as business and medicine). To see how it works, read on.

Further Reading

Fred Feldman. 1978. *Introductory Ethics*. Englewood Cliffs, NJ: Prentice Hall.

William Frankena. 1973. *Ethics*. Englewood Cliffs, NJ: Prentice Hall.

J.L. Mackie. 1977. *Ethics: Inventing Right and Wrong*. New York: Penguin.

Louis Pojman. 2002. *Ethics: Discovering Right and Wrong*. Belmont CA: Wadsworth.

James Rachels. 1993. *Elements of Moral Philosophy*. New York: McGraw-Hill.

Russ Schafer-Landau. 2003. *What Ever Happened to Good and Evil?* Oxford: Oxford University Press.

Russ Schafer-Landau. 2015. *The Fundamentals of Ethics,* 3rd edn. Oxford: Oxford University Press.

Bernard Williams. 2012. *Morality*. Cambridge: Cambridge University Press.

Part I

Moral Standing

In Part I, Moral Standing, we examine what philosophers have had to say about who or what a moral patient is. To whom or to what do we owe duties? What do we do when our duties to one thing, or one kind of thing, conflict with our duties to another? The first three chapters of the book describe the traditional approach to environmental ethics over the past 40 years or so in light of these fundamental questions about value, obligation, and principled conflict resolution. Environmental ethics begins the moment we reject the view that only humans can be moral patients, but this by itself does not tell us how far to widen the circle of moral standing. So we begin (Chapter 1) with what is perhaps the easiest case—that of sentient non-human animals. Because these beings are sentient, many philosophers believe that they deserve moral consideration. After all, surely the most important fact about moral patients is that their interests, expressed in their pains and pleasures, can be affected by what we do to them. If so, then many non-human animals deserve moral consideration. However, some think that this is not enough (Chapter 2): all living things, not just sentient non-human animals, can be said meaningfully to have interests, so we owe them all moral consideration. But not all living things can experience pain and pleasure, so there must be some other way to define what it means to have an "interest." Finally, some philosophers (Chapter 3) think that we should move beyond taking only individual things—animals or plants—as morally considerable and widen the circle of moral standing so that it includes systems of living things—ecosystems and the biosphere as a whole. However, this will work only if such things are enough like living things that we can attribute interests to them.

Animal Welfarism

Ecological Intuition Pump

Imagine living in a society where the economy is based on the slavery of non-whites. If someone were to stand up and say that enslaving non-whites was wrong, you might wonder what grounds she had for this claim. How do you think you would react if she had gone on to explain that there were no morally relevant differences between whites and non-whites that could justify the sort of radically different treatment of them we find in a slave economy? Or imagine the same sort of political radical arguing that women ought to be given the vote at a time when they were excluded from participation in the public sphere. The claim would be similar: there is no morally relevant difference between women and men that would license this sort of discrimination. If you think you might have found these two challenges compelling, try generalizing them even further so that they apply to the way in which we treat many non-human animals today. We routinely torture and kill them, but what justifies this behaviour? Is there some property we have and they lack that makes the difference? What could it be? Are you sure that whatever property you isolate is not just some value-neutral difference between us and the rest of the animal world? Can we consistently both side with our abolitionist/suffragette and continue treating animals as we do?

Ⓐ Introduction

To possess **moral standing** is to be an entity whose existence and fundamental interests have "positive moral weight."[1] To have positive moral weight means that *other* agents have duties to constrain themselves in specific ways when dealing with such entities. For example, we say that humans have moral standing, and this means, in large part, that other humans may not treat them in certain ways (torture them, say). **Anthropocentrism** is the view that only humans have genuine interests of this sort, so we need only look to what they value to discover what all our duties are. Environmental ethics begins the moment we challenge this view. But what things beyond humans might have moral standing? The most obvious place to look is at certain non-human animals.

Philosopher Mark Rowlands lived with a wolf for 11 years. The wolf taught him a lot about love, death, and happiness, but perhaps the deepest lesson he learned from it had to do with how we draw the line between those who count morally and those who do not. The idea of spending significant time with a wild animal might remind you that almost all of the animals we humans deal with have been rendered fairly powerless through the long historical process of domestication. This notion caused Rowlands to re-evaluate the contractarian view of civilization. The view that moral or political communities are held together by a contract among their members is a very powerful one. In early modern philosophy, it originates with Hobbes but also finds potent expression in the political thought of the late-twentieth-century philosopher John Rawls. Its central claim is that the norms governing the community, and the power required to enforce those norms, are justified by the fact that the community's members have agreed to this state of affairs, even if only tacitly. In the same way that you might sign a contract with a plumber that clearly lays out what each of you is required to do for the other—he fixes your pipes and you pay him—we have all signed a contract with one another to refrain from interfering in each other's lives in specified ways. For example, we agree to refrain from seizing one another's property by force or fraud. A political authority—what Hobbes calls "The Leviathan"—is then set up, by us, to enforce these agreements.

Why "sign" such a contract? Because you reckon that you can be either helped or harmed by other would-be members of the community, but that they will help you or refrain from harming you only if you reciprocate. Each person therefore agrees to forswear attacks on the person or property of others on the condition that those others adopt the same stance toward him or her. Key to this arrangement is the recognition by all concerned that there is a rough equality of power among them. I may be able to attack someone weaker than I am, but that person can join forces with others and either resist me or attack me in turn. It is probably best to play it safe and effectively "disarm" everyone. The key aspect of social contract theory for our purposes is that being a party to this agreement is necessary and sufficient to confer moral standing on each of us. It draws us into the circle of moral concern, isolating us from the wild, natural, and essentially amoral forces outside. To have moral standing means that one's interests must be taken into account by other members of the moral community, a process that gives such interests positive moral weight.

This move into the circle of moral considerability *empowers* us relative to those who do not sign up. As Rowlands puts it:

> Those who fall outside the scope of the contract fall outside the scope of civilization. They lie outside the boundaries of morality. You have no moral obligations to those who are significantly weaker than you. That is the consequence of the contractual view of civilization.[2]

Ever since we began domesticating non-human animals some 10,000 years ago, this has been our attitude toward them: they have been placed outside the moral circle. We have tamed them so that they may be used as resources: pets, food, beasts of burden, fodder for experimental research, sources of entertainment, and so on. Even wild species have been rendered mostly powerless against us. Technology and the rise of urban civilization have effectively separated most of us from the really dangerous animals, a separation that is only reinforced by the phenomenon of zoos. At least as far as non-human animals are concerned, Rowlands's equation seems exactly right: to be without power is to be without moral standing and vice versa. It does not follow, however, that we may do what we want with those lacking moral standing. Let's distinguish two ways in which a thing can have value: (1) To have **intrinsic value** is to be valuable independently of other entities or their interests. (2) To have **extrinsic value** is to be valuable relative only to the values or interests of some other entity. Until quite recently, most philosophers thought that non-human animals have, at best, extrinsic value.

But this *can* place constraints on what we may do with such entities. Kant, for instance, thought that even though we have no duties *to* non-human animals, we have duties *regarding* them. He argued that there are two sources of the moral duties we may have regarding things, like non-human animals, that have merely extrinsic value. The first arises from the fact that the entities in question might be the property of some person. If so, then the constraint is rooted in my duty—if I have one—not to interfere with another's property. Second, it may be thought that to be cruel, say, to non-human animals is likely to lead to cruelty toward humans. Again, the duty not to be cruel to non-human animals is entirely derivative from some other moral concern, in this case the obligation to reduce our chances of being cruel to humans. According to the idea that animals have only extrinsic value, and without at least one of these two constraints in place, we can treat non-human animals any way we like.

Ⓑ Moral Standing and Speciesism

The view that all—and only—members of our species have moral standing is intimately linked to the claim that only humans have intrinsic value. This idea, in turn, can be justified by pointing to features or properties of humans that confer on them this special status. Historically, many properties have been proposed for this role, the most prominent being rationality, personhood, the possession of a soul, the capacity for moral agency, and linguistic capacity.

There are two problems with any such appeal to a special property. First, even if we accept the idea that only humans possess the relevant feature, we need to ask why this should matter morally. How, for example, does rationality or linguistic capacity confer specifically *moral* standing on its bearer rather than simply marking some value-neutral natural difference between the members of various species? Would it not be just as dubious to argue that a fish's unique capacity to breathe underwater gives it, and it alone, moral standing? Presumably so, but what is the difference between this claim and the claim that linguistic capacity or the possession of a soul confers such status? Second, every criterion proposed fails to apply to all humans. Fetuses and infants lack both personhood (at least any very psychologically complex version of it) and rationality. If to lack the relevant criterion is to be deprived of moral standing, do we say that these humans lack moral standing? Presumably not.

For Peter Singer, appeal to these special properties—any of them—to ground the claim that their bearers alone have moral standing is a form of **speciesism**. To be racist is to believe that being a member of the approved racial group is necessary for full moral standing. To be sexist is to believe that having the approved gender is necessary for full moral standing. Similarly, to believe that *being human* is necessary for full moral standing is to be speciesist. In all three cases, the claim will be "grounded" by appeal to properties that are deemed necessary for moral standing, properties that the disfavoured group—people of the wrong colour or gender or species—lack. But the properties are morally neutral or irrelevant in all three cases. This might make us think that the search for a suitable criterion of moral standing is doomed. Rather than abandoning the search altogether, however, one might instead respond to these worries by casting a wider net. Perhaps the relevant criterion for moral standing is simply sentience. This, at least, is Singer's provocative suggestion.

ALL ANIMALS ARE EQUAL

Peter Singer

In recent years, a number of oppressed groups have campaigned vigorously for equality. The classic instance is the black liberation movement, which demands an end to the prejudice and discrimination that has made blacks second-class citizens. The immediate appeal of the black liberation movement and its initial, if limited, success made it a model for other oppressed groups to follow. We became familiar with liberation movements for Spanish-Americans, gay people, and a variety of other minorities. When a majority group—women—began their campaign, some thought we had come to the end of the road. Discrimination on the basis of sex, it has been said, is the last universally accepted form of discrimination, practised without secrecy or pretence even in those liberal circles that have long prided themselves on their freedom from prejudice against racial minorities.

One should always be wary of talking of "the last remaining form of discrimination." If we have learnt anything from the liberation movements, we should have learnt how difficult it is to be aware of latent prejudice in our attitudes to particular groups until this prejudice is forcefully pointed out.

A liberation movement demands an expansion of our moral horizons and an extension or reinterpretation of the basic moral principle of equality. Practices that were previously regarded as natural and inevitable come to be seen as the result of an unjustifiable prejudice. Who can say with confidence that all his or her attitudes and practices are beyond criticism? If we wish to avoid being numbered amongst the oppressors, we must be prepared to rethink even our most fundamental attitudes. We need to consider them from the point of view of those most disadvantaged by our attitudes and the practices that follow from these attitudes. If we can make this unaccustomed mental switch, we may discover a pattern in our attitudes and practices that consistently operates so as to benefit one group—usually the one to which we ourselves belong—at the expense of another. In this way, we may come to see that there is a case for a new liberation movement. My aim is to advocate that we make this mental switch in respect of our attitudes and practices toward a very large group of beings: members of species other than our own—or, as we popularly though misleadingly call them, animals.

In other words, I am urging that we extend to other species the basic principle of equality that most of us recognize should be extended to all members of our own species.

All this may sound a little far-fetched, more like a parody of other liberation movements than a serious objective. In fact, in the past the idea of "the Rights of Animals" really has been used to parody the case for women's rights. When Mary Wollstonecraft, a forerunner of later feminists, published her *Vindication of the Rights of Women* in 1792, her ideas were widely regarded as absurd, and they were satirized in an anonymous publication entitled *A Vindication of the Rights of Brutes*. The author of this satire (actually Thomas Taylor, a distinguished Cambridge philosopher) tried to refute Wollstonecraft's reasonings by showing that they could be carried one stage further. If sound when applied to women, why should the arguments not be applied to dogs, cats, and horses? They seemed to hold equally well for these "brutes," yet to hold that brutes had rights was manifestly absurd; therefore, the reasoning by which this conclusion had been reached must be unsound, and if unsound when applied to brutes, it must also be unsound when applied to women, since the very same arguments had been used in each case.

One way in which we might reply to this argument is by saying that the case for equality between men and women cannot validly be extended to non-human animals. Women have a right to vote, for instance, because they are just as capable of making rational decisions as men are; dogs, on the other hand, are incapable of understanding the significance of voting, so they cannot have the right to vote. There are many other obvious ways in which men and women resemble each other closely, while humans and other animals differ greatly. So, it might be said, men and women are similar beings and should have equal rights, while

humans and non-humans are different and should not have equal rights.

The thought behind this reply to Taylor's analogy is correct up to a point, but it does not go far enough. There are important differences between humans and other animals, and these differences must give rise to some differences in the rights that each have. Recognizing this obvious fact, however, is no barrier to the case for extending the basic principle of equality to non-human animals. The differences that exist between men and women are equally undeniable, and the supporters of women's liberation are aware that these differences may give rise to different rights. Many feminists hold that women have the right to an abortion on request. It does not follow that since these same people are campaigning for equality between men and women, they must support the right of men to have abortions too. Since a man cannot have an abortion, it is meaningless to talk of his right to have one. Since a pig can't vote, it is meaningless to talk of its right to vote. There is no reason why either women's liberation or animal liberation should get involved in such nonsense. The extension of the basic principle of equality from one group to another does not imply that we must treat both groups in exactly the same way or grant exactly the same rights to both groups. Whether we should do so will depend on the nature of the members of the two groups. The basic principle of equality, I shall argue, is equality of consideration, and equal consideration for different beings may lead to different treatment and different rights.

So there is a different way of replying to Taylor's attempt to parody Wollstonecraft's arguments, a way that does not deny the differences between humans and non-humans but goes more deeply into the question of equality and concludes by finding nothing absurd in the idea that the basic principle of equality applies to so-called "brutes." I believe that we reach this conclusion if we examine the basis on which our opposition to discrimination on grounds of race or sex ultimately rests. We will then see that we would be on shaky ground if we were to demand equality for blacks, women, and other groups of oppressed humans while denying equal consideration to non-humans.

When we say that all human beings, whatever their race, creed, or sex, are equal, what is it that we are asserting? Those who wish to defend a hierarchical, inegalitarian society have often pointed out that by whatever test we choose, it simply is not true that all humans are equal. Like it or not, we must face the fact that humans come in different shapes and sizes; they come with differing moral capacities, differing intellectual abilities, differing amounts of benevolent feeling and sensitivity to the needs of others, differing abilities to communicate effectively, and differing capacities to experience pleasure and pain. In short, if the demand for equality were based on the actual equality of all human beings, we would have to stop demanding equality. It would be an unjustifiable demand.

Still, one might cling to the view that the demand for equality among human beings is based on the actual equality of the different races and sexes. Although humans differ as individuals in various ways, there are no differences between the races and sexes as such. From the mere fact that a person is black or a woman, we cannot infer anything else about that person. This, it may be said, is what is wrong with racism and sexism. The white racist claims that whites are superior to blacks, but this is false: although there are differences between individuals, some blacks are superior to some whites in all of the capacities and abilities that could conceivably be relevant. The opponent of sexism would say the same: a person's sex is no guide to his or her abilities, and this is why it is unjustifiable to discriminate on the basis of sex.

This is a possible line of objection to racial and sexual discrimination. It is not, however, the way that someone really concerned about equality would choose, because taking this line could, in some circumstances, force one to accept a most inegalitarian society. The fact that humans differ as individuals, rather than as races or sexes, is a valid reply to someone who defends a hierarchical society like, say, South Africa, in which all whites are superior in status to all blacks. The existence of individual variations that cut across the lines of race or sex, however, provides us with no defence at all against a more sophisticated opponent of equality, one who proposes that, say, the interests of those with IQ ratings above 100 be preferred to the interests of those with IQs below 100. Would a hierarchical society of this sort really be so much better than one based on race or sex? I think not. But if we tie the moral principle of equality to the factual equality of the different races or sexes, taken as a whole, our opposition to racism and sexism does not provide us with any basis for objecting to this kind of inegalitarianism.

There is a second important reason why we ought not to base our opposition to racism and sexism on any kind of factual equality, even the limited kind that asserts that variations in capacities and abilities are spread evenly between the different races and sexes: we can have no absolute guarantee that these abilities and capacities really are distributed evenly, without regard to race or sex, among human beings. So far as actual abilities are concerned, there do seem to be certain measurable differences between both races and sexes. These differences do not, of course, appear in each case, but only when averages are taken. More important still, we do not yet know how much of these differences is really due to the different genetic endowments of the various races and sexes and how much is due to environmental differences that are the result of past and continuing discrimination. Perhaps all of the important differences will eventually prove to be environmental rather than genetic. Anyone opposed to racism and sexism will certainly hope that this will be so, for it will make the task of ending discrimination a lot easier; nevertheless, it would be dangerous to rest the case against racism and sexism on the belief that all significant differences are environmental in origin. The opponent of, say, racism who takes this line will be unable to avoid conceding that if differences in ability did after all prove to have some genetic connection with race, racism would in some way be defensible.

It would be folly for the opponent of racism to stake his whole case on a dogmatic commitment to one particular outcome of a difficult scientific issue that is still a long way from being settled. While attempts to prove that differences in certain selected abilities between races and sexes are primarily genetic in origin have certainly not been conclusive, the same must be said of attempts to prove that these differences are largely the result of environment. At this stage of the investigation, we cannot be certain which view is correct, however much we may hope it is the latter.

Fortunately, there is no need to pin the case for equality to one particular outcome of this scientific investigation. The appropriate response to those who claim to have found evidence of genetically based differences in ability between the races or sexes is not to stick to the belief that the genetic explanation must be wrong, whatever evidence to the contrary may turn up: instead, we should make it quite clear that the claim to equality does not depend on intelligence, moral capacity, physical strength, or similar matters of fact. Equality is a moral ideal, not a simple assertion of fact. There is no logically compelling reason for assuming that a factual difference in ability between two people justifies any difference in the amount of consideration we give to satisfying their needs and interests. The principle of the equality of human beings is not a description of an alleged actual equality among humans: it is a prescription of how we should treat humans.

Jeremy Bentham incorporated the essential basis of moral equality into his utilitarian system of ethics in the formula: "Each to count for one and none for more than one." In other words, the interests of every being affected by an action are to be taken into account and given the same weight as the like interests of any other being. It is an implication of this principle of equality that our concern for others ought not to depend on what they are like or what abilities they possess—although precisely what this concern requires us to do may vary according to the characteristics of those affected by what we do. It is on this basis that the case against racism and the case against sexism must both ultimately rest, and it is in accordance with this principle that speciesism is also to be condemned. If possessing a higher degree of intelligence does not entitle one human to use another for his own ends, how can it entitle humans to exploit non-humans?

Many philosophers have proposed the principle of equal consideration of interests, in some form or other, as a basic moral principle, but as we shall see in more detail shortly, not many of them have recognized that this principle applies to members of other species as well as to our own. Bentham was one of the few who did realize this. In a forward-looking passage, written at a time when black slaves in British dominions were still being treated much as we now treat non-human animals, Bentham wrote:

The day may come when the rest of the animal creation may acquire those rights which never could have been witholden from them but by the hand of tyranny. The French have already discovered that the blackness of the skin is no reason why a human being should be abandoned without redress to the caprice of a tormentor. It may one day come to be recognized that the number of the legs, the villoscity of the skin, or the termination of the os sacrum, are reasons equally insufficient for abandoning a sensitive being to the same fate. What else is it that should trace the insuperable line? Is it the faculty of reason, or perhaps the faculty of discourse? But a full grown horse or dog is beyond comparison a more rational, as well as a more conversable animal, than an infant of a day, or a week, or even a month, old. But suppose they were otherwise, what would it avail? The question is not, Can they reason? nor, Can they talk? but, Can they suffer?[3]

In this passage, Bentham points to the capacity for suffering as the vital characteristic that gives a being the right to equal consideration. The capacity for suffering—or more strictly, for suffering and/or enjoyment or happiness—is not just another characteristic like the capacity for language or for higher mathematics. Bentham is not saying that those who try to mark the "insuperable line" that determines whether the interests of a being should be considered happen to have selected the wrong characteristic. The capacity for suffering and enjoying things is a prerequisite for having interests at all, a condition that must be satisfied before we can speak of interests in any meaningful way. It would be nonsense to say that it was not in the interests of a stone to be kicked along the road by a schoolboy. A stone does not have interests, because it cannot suffer. Nothing that we can do to it could possibly make any difference to its welfare. A mouse, on the other hand, does have an interest in not being tormented, because it will suffer if it is. . . .

The racist violates the principle of equality by giving greater weight to the interests of members of his own race when there is a clash between their interests and the interests of those of another race. Similarly, the speciesist allows the interests of his own species to override the greater interests of members of other species.[4] The pattern is the same in each case. Most human beings are speciesists. I shall now very briefly describe some of the practices that show this. . . .

It is not merely the act of killing that indicates what we are ready to do to other species in order to gratify our tastes. The suffering we inflict on the animals while they are alive is perhaps an even clearer indication of our speciesism than the fact that we are prepared to kill them. In order to have meat on the table at a price that people can afford, our society tolerates methods of meat production that confine sentient animals in cramped, unsuitable conditions for the entire durations of their lives. Animals are treated like machines that convert fodder into flesh, and any innovation that results in a higher "conversion ratio" is liable to be adopted. As one authority on the subject has said, "cruelty is acknowledged only when profitability ceases."[5] . . .

Since, as I have said, none of these practices cater for anything more than our pleasures of taste, our practice of rearing and killing other animals in order to eat them is a clear instance of the sacrifice of the most important interests of other beings in order to satisfy trivial interests of our own. To avoid speciesism, we must stop this practice, and each of us has a moral obligation to cease supporting the practice. Our custom is all the support that the meat industry needs. The decision to cease giving it that support may be difficult, but it is no more difficult than it would have been for a white Southerner to go against the traditions of his society and free his slaves: if we do not change our dietary habits, how can we censure those slaveholders who would not change their own way of living?

The same form of discrimination may be observed in the widespread practice of experimenting on other species in order to see if certain substances are safe for human beings, or to test some psychological theory about the effect of severe punishment on learning, or to try out various new compounds just in case something turns up. . . .

In the past, argument about vivisection has often missed this point, because it has been put in absolutist terms: Would the abolitionist be prepared to let thousands die if they could be saved by experimenting on a single animal? The way to reply to this purely hypothetical question is to pose another: Would the experimenter be prepared to perform his experiment on an orphaned human infant if that were the only way to save many lives? (I say "orphan" to avoid the complication of parental feelings, although in doing so I am being over-fair to the experimenter, since the non-human subjects of experiments are not orphans.) If the experimenter is not prepared to use an orphaned human infant, then his readiness to use non-humans is simple discrimination, since adult apes, cats, mice, and other mammals are more aware of what is happening to them, more self-directing, and, so far as we can tell, at least as sensitive to pain as any human infant. There seems to be no relevant characteristic that human infants possess that adult mammals do not have to the same or a higher degree. (Someone might try to argue that

what makes it wrong to experiment on a human infant is that the infant will, in time and if left alone, develop into more than the non-human, but one would then, to be consistent, have to oppose abortion, since the fetus has the same potential as the infant—indeed, even contraception and abstinence might be wrong on this ground, since the egg and sperm, considered jointly, also have the same potential. In any case, this argument still gives us no reason for selecting a non-human, rather than a human with severe and irreversible brain damage, as the subject for our experiments.)

The experimenter, then, shows a bias in favour of his own species whenever he carries out an experiment on a non-human for a purpose that he would not think justified him in using a human being at an equal or lower level of sentience, awareness, ability to be self-directing, etc. No one familiar with the kind of results yielded by most experiments on animals can have the slightest doubt that if this bias were eliminated, the number of experiments performed would be a minute fraction of the number performed today. . . .

It is significant that the problem of equality, in moral and political philosophy, is invariably formulated in terms of human equality. The effect of this is that the question of the equality of other animals does not confront the philosopher, or student, as an issue itself—and this is already an indication of the failure of philosophy to challenge accepted beliefs. Still, philosophers have found it difficult to discuss the issue of human equality without raising, in a paragraph or two, the question of the status of other animals. The reason for this, which should be apparent from what I have said already, is that if humans are to be regarded as equal to one another, we need some sense of "equal" that does not require any actual, descriptive equality of capacities, talents, or other qualities. If equality is to be related to any actual characteristics of humans, these characteristics must be some lowest common denominator, pitched so low that no human lacks them—but then the philosopher comes up against the catch that any such set of characteristics that covers all humans will not be possessed only by humans. In other words, it turns out that in the only sense in which we can truly say, as an assertion of fact, that all humans are equal, at least some members of other species are also equal—equal, that is, to each other and to humans. If, on the other hand, we regard the statement "All humans are equal" in some non-factual way, perhaps as a prescription, then, as I have already argued, it is even more difficult to exclude non-humans from the sphere of equality.

This result is not what the egalitarian philosopher originally intended to assert. Instead of accepting the radical outcome to which their own reasonings naturally point, however, most philosophers try to reconcile their beliefs in human equality and animal inequality by arguments that can only be described as devious. . . .

What else remains? . . . Stanley Benn, after noting the usual "evident human inequalities," argues, correctly I think, for equality of consideration as the only possible basis for egalitarianism. Yet Benn, like other writers, is thinking only of "equal consideration of human interests." . . . Benn's statement of the basis of the consideration we should have for imbeciles seems to me correct, but why should there be any fundamental inequality of claims between a dog and a human imbecile? Benn sees that if equal consideration depended on rationality, no reason could be given against using imbeciles for research purposes, as we now use dogs and guinea pigs. This will not do: "But of course we do distinguish imbeciles from animals in this regard," he says. That the common distinction is justifiable is something Benn does not

question; his problem is how it is to be justified. The answer he gives is this:

[W]e respect the interests of men and give them priority over dogs not insofar as they are rational, but because rationality is the human norm. We say it is unfair to exploit the deficiencies of the imbecile who falls short of the norm, just as it would be unfair, and not just ordinarily dishonest, to steal from a blind man. If we do not think in this way about dogs, it is because we do not see the irrationality of the dog as a deficiency or a handicap, but as normal for the species. The characteristics, therefore, that distinguish the normal man from the normal dog make it intelligible for us to talk of other men having interests and capacities, and therefore claims, of precisely the same kind as we make on our own behalf. But although these characteristics may provide the point of the distinction between men and other species, they are not in fact the qualifying conditions for membership, or the distinguishing criteria of the class of morally considerable persons; and this is precisely because a man does not become a member of a different species, with its own standards of normality, by reason of not possessing these characteristics.

The final sentence of this passage gives the argument away. An imbecile, Benn concedes, may have no characteristics superior to those of a dog; nevertheless, this does not make the imbecile a member of "a different species" as the dog is. Therefore, it would be "unfair" to use the imbecile for medical research as we use the dog. But why? That the imbecile is not rational is just the way things have worked out, and the same is true of the dog—neither is any more responsible for their mental level. If it is unfair to take advantage of an isolated defect, why is it fair to take advantage of a more general limitation? I find it hard to see anything in this argument except a defence of preferring the interests of members of our own species because they are members of our own species.

To those who think there might be more to it, I suggest the following mental exercise. Assume that it has been proven that there is a difference in the average, or normal, intelligence quotient for two different races, say whites and blacks. Then substitute the term "white" for every occurrence of "men" and "black" for every occurrence of "dog" in the passage quoted, and substitute "high IQ" for "rationality," and when Benn talks of "imbeciles," replace this term by "dumb whites"—that is, whites who fall well below the normal white IQ score. Finally, change "species" to "race." Now reread the passage. It has become a defence of a rigid, no-exceptions division between whites and blacks, based on IQ scores, notwithstanding an admitted overlap between whites and blacks in this respect. The revised passage is, of course, outrageous, and this is not only because we have made fictitious assumptions in our substitutions. The point is that in the original passage, Benn was defending a rigid division in the amount of consideration due to members of different species, despite admitted cases of overlap. If the original did not, at first reading, strike us as being as outrageous as the revised version does, this is largely because although we are not racists ourselves, most of us are speciesists. Like the other articles, Benn's stands as a warning of the ease with which the best minds can fall victim to a prevailing ideology.

Drawing inspiration from the nineteenth-century philosopher Jeremy Bentham, Singer claims that a necessary and sufficient condition for moral standing is sentience. Sentience is the capacity to experience pain and pleasure. For Bentham, the key question to ask is not whether a being can reason but, more basically, whether it can suffer. If the answer is yes, the being has moral standing. But by endorsing this view, hasn't Singer made the same error as the people he criticizes? That error, recall, is to isolate a special natural property and claim that possession of it by any entity confers moral standing on that entity. If this move is illegitimate with respect to language, rationality, personhood, and so on, why is it legitimate with respect to sentience? Singer's answer comes in two parts. First, to have moral standing requires having interests. Second, to have interests requires having the capacity to suffer, or be sentient. Many non-human animals are clearly sentient, which means they have interests. This fact then confers moral standing on them.

Following on this, Singer's key claim is that we are required to weigh the interests of different species *impartially* when they conflict. The reason is straightforward: every sentient being has an equal interest in avoiding pain. This does not mean that suffering is the same across species. As Christopher Belshaw puts it, "slapping a small child may be morally wrong while slapping an elephant with the same degree of force is something to which we, like the elephant, may well be indifferent."[6] But it does mean that every species has an interest in avoiding pain relative to its peculiar characteristics. Singer is a utilitarian, so he thinks that once we accept the claims just described, it is incumbent on us to act so as to maximize welfare across sentient species. The view has clear implications for how we treat non-human animals. Using them for food, at least when this involves the cruel practices employed in factory farms, is out. So too, however, is using them as fodder for experimental research and in certain sports like bullfighting and fox hunting, as well as breeding them either for their pelts or, in some cases, as pets.

ⓒ Beyond Utilitarianism

Act-utilitarianism is the view that with respect to every particular act we undertake, or propose to undertake, we must ask the question about whether or not it will maximize welfare. If it will, then we have a duty to perform it. If it will not, then we are not permitted to perform it. Act-utilitarianism has often been criticized for failing to place absolute prohibitions on specific acts. At most, the utilitarian can claim that acts are only conditionally wrong, the wrongness being conditional on whether or not their performance would maximize welfare. One explanation for the power of rights discourse in our moral culture is that appeals to rights can block this way of thinking. If an entity has a *right* not to be interfered with in specific ways, then this claim cannot be overridden by appeals to general welfare. How does all of this affect how we treat non-human animals?

Tom Regan thinks that utilitarians like Singer go wrong in emphasizing the *interests* of non-human animals rather than the entity that *has* such interests. More specifically, and technically, Regan holds that non-human animals have intrinsic value in virtue of being "subjects-of-a-life." Here is how Regan describes this crucial concept:

> To be the subject-of-a-life involves more than being merely alive and more than being merely conscious. [It is to] have beliefs and desires, perception, memory and a sense of the future, including their own future; an emotional life together with feelings of pleasure and pain; preference and welfare interests; the ability to initiate action in pursuit of desires and goals; a psychological identity over time; and an individual welfare in the sense that their experiential life fares well or ill for them, independently of their utility for others.[7]

This is a marvellously rich articulation of an intuitively powerful idea. To the extent that we agree that some entities have intrinsic value, it is surely because they have this complex list of capacities. For Kant, who provides the philosophical inspiration for Regan's views, it is impermissible to treat such beings as mere means to others' ends. Rather, they are ends in themselves, to be treated with respect. And insofar as we describe some non-human animals this way, the position is clearly stronger than the utilitarian one, for it entails that there are absolute, rather than merely conditional, prohibitions on treating non-human animals in certain ways.

As a brief illustration of the difference between the two positions, take the issue of eating animals. The position Singer defends does not strictly entail vegetarianism (though *he* thinks it does). From the fact that any non-human animal has an interest in avoiding pain, it does not follow that it has an interest in avoiding premature death.[8] As long as it is treated humanely while alive and killed painlessly, there appears to be no strictly utilitarian objection to eating it. Not so for Regan, because any subject-of-a-life is defined, partly, as having a sense of its interests unfolding into the future. To have these projects thwarted—through premature death, for instance—is to be harmed in a way that goes beyond the capacity to suffer sensibly.

Imagine that some humans were being bred so that their organs might be harvested for the benefit of other humans whose own organs have failed. This is the premise of Kazuo Ishiguro's marvellous novel *Never Let Me Go*. Even if these people were kept entirely unaware of what was in store for them and were killed painlessly, we would still think it was wrong to treat them this way. Why? Because people are not just organ resources. For Regan, the "fundamental wrong" with eating non-human animals is that to do so is to treat them as mere resources: here essentially for us and not for themselves. We would simply not countenance treating other humans in this fashion, but there is no significant distinction between human and non-human animals with respect to the capacities Regan has isolated.

Is the claim that there is no difference between human and non-human animals on this score correct? One of the main criticisms of Regan's animal rights view is that the definition of subject-of-a-life he offers is too narrow to capture the vast majority of non-human animals. It surely goes without saying that mussels have no sense of the future, but is it any less absurd to attribute such a sense to a cow? Regan has tried to avoid this problem by broadening the definition of subject-of-a-life so that it includes only consciousness and the possession of an individual welfare. But what is so important about these things? Is it that the animal is somehow *aware* in a way that must be respected? If so, the difference between this view and that of the utilitarian would seem to have faded somewhat. As Belshaw has shown, Regan confronts an inescapable dilemma here. If he chooses the narrow definition of subject-of-a-life, it does not appear to apply to very many non-human animals, whereas if he chooses the broad definition, then it is difficult to see why premature death is *necessarily* a bad thing for the entity in question.[9] This, we have seen, is exactly the problem the would-be utilitarian vegetarian has. Either way, we can apparently go on eating most of the animals we already do as long as they are treated humanely while alive and killed painlessly.

Ⓓ Animal Welfarism and Animal Citizenship

Philosophers like Singer and Regan have done much to raise our awareness of the sorry condition in which many non-human animals still live and die. Much of the meat we eat is still produced in a manner that is morally unjustifiable. Further, the work done by the animal rights movement is important because it represents the first step in a broader environmental awareness. By defining a certain stance toward the non-human world as speciesist, Singer forces us to expand our moral vision in a way not systematically contemplated before he arrived on the philosophical scene. This has had many practical consequences. For example, one of the most vociferous animal rights campaigns in recent times is focused on the effort to ban the importation of Canadian seal products—furs, meat, blubber—into the EU. This effort has been a great political success, so much so that Canada is now looking away from Europe and seeking to penetrate the Chinese market for these goods. The EU ban on Canadian seal products is based on the claim that the seal hunt is cruel. It is crucial to note what this claim means: individual seals are treated in morally unjustifiable ways through the practices employed in the hunt. Cashed out, this claim must rest on the notion that such individuals are intrinsically valuable.

We will return to the position of the EU on the Canadian seal hunt (section E, below). For the moment, it is important to note that some critics of Singer and Regan suggest that the main problem with the animal rights doctrine is precisely its *individualism*, the idea that the primary bearers of value are individual animals rather than species or ecosystems. Call this position **animal welfarism**, the view that some individual non-human animals have intrinsic value and that this imposes on us more or less stringent duties to them. While this is a laudable first step beyond narrow anthropocentrism, it still conceptualizes the issues too narrowly. One question it leaves entirely untouched has to do with what "political" relation we ought to have with the various kinds of animals with which we interact. This is a new area of research in animal ethics, nicely articulated by Donaldson and Kymlicka in our next article.

ANIMAL CITIZENSHIP

Sue Donaldson and Will Kymlicka

Political philosophy has been largely invisible in the animal rights debate. Political philosophers have seen the animal question as an issue for ethicists, irrelevant to the core topics of political philosophy, such as theories of democracy, citizenship, sovereignty, constitutionalism, and so on. And this indifference is largely reciprocated: the vast bulk of work done in animal ethics has not drawn upon the core concepts of political philosophy.

Recently, however, various authors have argued that the governing of human-animal relations is a fundamental issue of political theory, situating animals within our theories of citizenship, democracy, and sovereignty. This new approach reflects dissatisfaction with

two key features of the traditional approach to animal ethics: (1) the narrow focus on intrinsic moral status; and (2) the "laissez-faire" intuition.

Until recently, many philosophers working on animal ethics have focused exclusively on one question: namely, the *intrinsic moral status of animals* (typically grounded in the possession of sentience) and the rights that flow from this intrinsic status. This is an important question, but it ignores equally important issues that arise from the different types of relationships humans have with different animals. For example, domesticated dogs and wild wolves possess similar forms of sentience. If this were all that mattered ethically, we would have the same obligations to dogs as to wolves. But our relationship to dogs is historically very different, and we have specific obligations to them in virtue of the ways in which we have brought them into our society and bred them to be dependent on us. Dogs, we might say, are members of a shared human-animal society in a way that wolves are not, and they therefore should have membership rights, as well as the rights owed to all sentient beings. We need a theory of membership, as well as a theory of intrinsic status.

Describing domesticated animals as members of a shared society may seem naive given that they have been brought into society as a caste group to serve us. The history of human-animal relations is one of domination and exploitation, and the goal of animal advocacy must be to end the subordination and sacrifice of non-human animals to serve human purposes. And here we face an important choice: do we end subordination by fully recognizing the membership rights of domesticated animals or by severing our connections with them?

For many theorists, particularly "abolitionists," the way to end domination is *for us to stop exercising power over animals*. We should empty the cages, liberate captive animals (through re-wilding or gradual extinction), and cease ownership and exploitation.[10] As for free-living animals we should "let them be."[11] The goal is not to exercise power more responsibly, or more justly, but to renounce power entirely. Palmer calls this the "laissez-faire intuition" in animal rights theory.[12] This approach is prevalent because many animal rights theorists doubt that power can ever be exercised justly. Given animals' vulnerability and dependence, and human self-interest in using animals for our own ends, relationships with them will inevitably degenerate into tyranny. Justice, on this view, requires severing all relationships with

animals in what Acampora calls "species apartheid": humans living amongst humans; animals living amongst animals.[13]

But this image of a world without human-animal relations is untenable. It fails even a minimal test of empirical feasibility: there is no possible world in which animals are not constantly affected (for better or worse) by us. Humans will inevitably be making decisions that affect animals' environments, their mobility (on land, sea, and air), their food and water sources, and the risks they face. Any plausible theory of animal rights, therefore, must consider how the exercise of power can be rendered just, and how it can be responsive to the good of those who are subject to that power. Species apartheid is also morally untenable. It ignores the fact that we have acquired moral responsibilities toward particular groups of animals due to our own past actions, including responsibilities to attend to needs arising through domestication or through human-induced changes to habitat. Our previous actions have made some animals vulnerable to new types of harm and risk. To simply "let them be" at this point in history is to wash our hands of moral responsibilities we have inherited.

In short, it is increasingly recognized (1) that animals not only have intrinsic moral status but also morally significant relationships and memberships; and (2) that we cannot avoid the exercise of power by "letting them be," but need to acknowledge the inevitability of asymmetric power and hold that power accountable. Taken together, these two insights push us in the direction of a distinctly political theory of animal rights. The first point is recognized within other "relational" approaches to animal ethics, such as care ethics,[14] capability ethics,[15] or posthumanist ethics.[16] But the evidence to date suggests that these approaches underestimate the risks associated with the exercise of power and lack safeguards for protecting basic rights and interests. They rely overly on individual goodwill, neglecting the legal, social, and political dimensions of animal justice.

A distinctly political theory of animal rights, by contrast, would start by locating animals squarely within the fundamental principles of the constitutional order. Different political theories have different conceptions of what this constitutional order is or should be, and it would be interesting to speculate about how animals might fit within, say, a communist, anarchist, or theocratic political order.[17] But today, as for the past 200 years, the most influential conception of political order is that of the territorial nation-state, and so this is the starting point for most current work on political theories of animal rights.

This prevailing political order is defined by territorially-bounded states, each of which derives its legitimacy from ideas of popular sovereignty and/or national self-determination. According to this conception, the world is composed of "peoples" who inhabit different territories. They have the right to govern themselves and their territory, and the nation-state is the vehicle by which peoples enact these rights of self-government. This conception of the nation-state emerged first in Europe, but with decolonization has replicated itself around the world.

How can we locate animals within this conception of political order? In our recent book,[18] we argue that animals, like humans, occupy three distinct statuses:

1. some animals should be recognized as members of "the people" in whose name nation states govern, and hence as our co-citizens. We argue this is true of domesticated animals;

2. other animals should be recognized as having the right to live autonomously on their own territories, and hence as exercising their own sovereignty. We argue this is true of many free-living animals;

3. yet other animals occupy an in-between status, living amongst us but not as members of a shared society, and hence as "denizens" rather than citizens. We argue this true of the non-domesticated free-living animals residing in urban and suburban regions.

We will briefly say a word about each of these categories.

Domesticated animals have been brought into human societies through confinement and selective breeding. We've made them dependent on our care, foreclosing any (immediate) option of a more independent existence. We have coerced their participation in our schemes of social cooperation, exploiting them for food and labour. They are members of a shared society with us, but as a subordinated class intended to serve us. Every dimension of their lives is governed and regulated by a human political order, an order which ruthlessly ignores their interests.[19] They are tyrannized, in short. How do we transform caste hierarchy into relations of justice? As in human cases of caste hierarchy, justice requires recognizing the full and equal membership of subordinated groups, and citizenship is the tool we use to convert relations of caste hierarchy into relations of equal membership. Domesticated animals should be recognized as full members and co-citizens of society, sharing in the same rights to protection (basic rights to life and liberty), provision (social rights), and participation (the right to have a say in how society is structured) as human citizens. Under these conditions, the exercise of power entailed in governing a shared human-animal society can be legitimate, not tyrannical, because society is dedicated to the flourishing of all of its members.

How to enable domesticated animals to "have a say" is a challenge, as it is for many groups of humans (children, the cognitively disabled, future generations). But various models of guardianship, trusteeship, and "dependent agency" have been developed.[20] It is important not to underestimate domesticated animals' capacities for communication, cooperation, and agency. Domestication is only possible for animals capable of entering into relations of trust, reflexive communication, and norm sensitivity with humans. We cannot have this sort of shared sociability with many animals on the planet, but we can with domesticated animals. Thus, domestication not only makes the extension of co-citizenship morally necessary, but also possible.

Wilderness animals are subject to human invasion, colonization, displacement, and habitat destruction. We treat their lands as *terra nullius* that we can pollute, denude, degrade, and occupy without justification. How can we prevent this injustice? In intra-human politics, we attempt to block this kind of aggression through recognition of the rights of "peoples" to their own territory, and to autonomy on that territory. Principles of sovereignty regulate relations between different peoples or states, according bounded political communities a right to maintain themselves as viable, self-governing societies on traditional territories or homelands. Sovereignty offers protection from outsiders who would expel peoples, steal their land and resources, turn them into client states, or impose unfair burdens on them (such as cross-border pollution). Sovereignty also provides a secure basis from which to negotiate fair terms of cooperation (e.g., trade and mobil-

ity rights) and forms of assistance or intervention that do not undermine autonomy. We argue that the same principles should apply to wilderness animals. Wild animal communities should be seen as having a sovereign "right to place" that blocks human aggression, and our relations with these communities should be governed by norms of international justice—a true "law of peoples" between human and animal communities—rather than by brute force.

As with citizenship for domesticated animals, enacting sovereignty for wilderness animals requires creative adaptation. While some wilderness animals live in discrete habitats, other free-living animals migrate over extensive areas of land, water, or air. But various models of partial, overlapping, interstitial, and sub-state sovereignty rights have been developed, along with ideas of mobility corridors and international commons that can address these complexities in ways that uphold underlying rights to territory and autonomy.[21]

Liminal Animals: Not all animals can be neatly categorized as either our co-citizens (full members of a shared cooperative scheme) or as members of some other sovereign society (other nations occupying distinct territories). Countless animals (rats, mice, squirrels, crows, raccoons, pigeons, and many others) live amongst us, but not as part of a shared cooperative scheme like domesticated animals. Currently, these "liminal" animals have no protection from human violence. They are treated as pests, invaders, and aliens, and are ruthlessly killed or expelled. In this respect, they share similar vulnerabilities to groups of liminal humans who live amongst us without participating in a common citizenship, including migrant workers, foreign visitors, or isolationist religious groups such as the Amish. They are all vulnerable to being stigmatized and exploited. One way to limit this vulnerability is to ensure that "denizens" have the option of becoming citizens. However, many denizens do not wish to become citizens, but prefer a looser relationship of tolerant co-existence involving fewer mutual obligations. And this arrangement makes sense for many liminal animals. It is doubtful that they would benefit from (or be capable of) being incorporated into the kinds of cooperative citizenship relations we can have with domesticated animals. Co-citizenship provides robust rights of provision and protection, but it also imposes robust obligations to adhere to citizenship norms, such as refraining from killing, theft, property invasion, and destruction, and otherwise imposing undue burdens on other citizens. To incorporate liminal ani-

mals into these civic norms would require massive coercion and interference in their ways of life. What non-domesticated animals living amongst us need is secure denizenship: they need to be protected from our violence, our negligence of their interests, and our refusal to recognize their secure rights of residency. They need tolerant co-existence or conviviality rather than intimate cooperation.

This is just a sketch of how one might develop a distinctly political theory of animal rights, and much work remains to be done, not only in developing liberal conceptions of animal citizenship and sovereignty, but also in developing alternative political theories of animal rights. We expect this to be one of the most exciting areas of future work in the field.

One way to connect these thoughts about animal citizenship to the more traditional focus on intrinsic value is through the claim that non-human animals have rights. For example, philosopher Mark Sagoff asks what we should do once we recognize our "obligation to value the basic rights of animals equally with that of human beings," and this question is, implicitly, also at the heart of Donaldson's and Kymlicka's analysis.[22] For Sagoff, the answer that follows is a direct attack on the positions of Singer and Regan. Key to Sagoff's argument is the idea that humans have both positive rights and negative rights. A **negative right** is a right not to be interfered with in certain ways. My right not to be prevented from speaking freely when this activity poses no clear and imminent danger to others is an example. Here, others have a duty to forbear from interfering with me. A **positive right** is a right to be provided with certain conditions for flourishing. Sagoff calls this a "basic right." To the extent that I can be said to have a right to health care, this is a positive or basic right because it places a duty on government to provide this service to me.

Sagoff argues that there *are* positive rights: society has a duty to "rescue individuals from starvation," for instance. If we grant this claim, the rest of his argument seems to go through quite swiftly. It has the following formal structure:

1. To possess intrinsic value is to have both positive (or basic) and negative rights.
2. Both human and non-human animals possess intrinsic value.
3. So both human and non-human animals have positive and negative rights.
4. We should value the basic rights of human and non-human animals equally.

5. To be rescued from unnecessary death and misery is a key basic right.
6. For non-human animals, this right is most threatened in nature.
7. Therefore, humans are morally required to reduce the misery and unnecessary death of animals in nature, as much as possible.[23]

What are we to make of this argument? Sagoff offers us two options. If we are animal welfarists like Singer and Regan, we should take (7) as a serious requirement and get to work on what is, surely, a ridiculously onerous if not impossible task. We would be required to intervene in natural processes on a massive scale, basically whenever animals' lives are threatened by other animals or adverse environmental conditions. But if we are "environmentalists," we should take the argument as a *reductio* of the animal welfarist's position. That is, *because* the task is thoroughly daunting in scope and probably immoral on independent grounds, we should reject the notion that non-human animals are bearers of intrinsic value. This means that you are *either* an animal welfarist *or* an environmentalist. Animal welfarists believe that the primary locus of value is the individual animal, while environmentalists think it is the species (or some other "collectivity"). And, crucially for the environmentalist, the interests of the species can trump those of some of the individuals it comprises when the two conflict. This, of course, is not true in the human case, which is why premise (2), above, is false. Without it, however, the animal welfarist's position collapses. Finally, is it worthwhile asking the question: How does Sagoff's reductio of animal welfarism affect the arguments of Donaldson and Kymlicka about animal citizenship? Is the latter theory necessarily individualist and therefore subject to Sagoff's attack?

Ⓔ Conclusion

The debate between the animal welfarist and the environmentalist can be illustrated by returning to the Canadian seal hunt, so let's conclude the chapter with that. There are six species of seal hunted off the Atlantic coast of Canada, but the hunt is overwhelmingly focused on just one of them, the harp seal. In the face of a vigorous international campaign, Canada banned the seal hunt in 1987. The result of the ban was a tripling in the population of seals and, since seals eat fish, a corresponding reduction in the region's fish stocks. Although the federal government has been too quick to blame the virtual collapse of the Atlantic cod fishery on the soaring seal population, there is little doubt that some connection exists between the two (though overfishing by humans is clearly the main culprit). In any case, the hunt is now permitted again. Out of a total population of somewhere between 2 and 5 million individuals, the total allowable catch of harp seals in 2007 was set at 270,000. This is a lucrative industry. In 2006, the hunt generated $33 million.[24]

If the hunt was both wantonly cruel to individual seals and a threat to the long-term survival of the harp seal species, both animal welfarists and environmentalists could stand united in opposition to it. But neither of these was the case. First, despite arresting film footage of cute animals being clubbed to death, there was, according to a 1986 royal commission on the subject, little evidence that the slaughter is cruel:

> The Royal Commission, as well as qualified veterinarians, animal pathologists and biologists who have observed the hunt first-hand attested to the humaneness of the clubbing method when it is carried out properly.... A 1993 report from the Parliamentary Assembly of the Council of Europe also found that clubbing ... is as good as the usual methods of slaughter, causing the animal to die in the course of a few seconds.[25]

Second, the harp seal has never become endangered through the hunt and in fact has thrived as a species.[26] Given these two facts, how can the EU's recent decision to ban Canadian seal products be justified? It appears that the decision cannot be supported by appeal to either environmentalist or animal welfarist principles. The species is not threatened, so the environmentalist has no cause for concern. And the utilitarian animal welfarist cannot justifiably oppose the hunt, because the slaughter is humane. What about Regan? Well, unless it can be shown that the harp seal is the subject-of-a-life in the complex sense he lays out, there are no grounds for opposition here either. Surely we know too little about what goes on in the mind of a harp seal to say with confidence that these animals possess, in Regan's words, "robust beliefs and desires, perception, memory and a sense of the future, including their own future; an emotional life together with feelings of pleasure and pain; preference and welfare interests; the ability to initiate action in pursuit of desires and goals; a psychological identity over time...."[27]

There are two further points to be made here. The first is that there is a good deal of hypocrisy in the position of the EU on this matter. Factory farming of other non-human animal species is alive and well in most parts of Europe. Why has it not been banned? As University of Toronto philosopher Wayne Sumner says, "Put European *foie gras* from force-fed ducks and geese and chorizo from factory farmed pigs on the table, and see who has a clear conscience."[28] Second, the hunt is a crucial feature of the culture of the Canadian Inuit. In fact, the Inuit and the current Canadian government for the most part stand shoulder to shoulder in their efforts to protect the hunt from decisions like that of the EU. Whatever the motives of the government, for the Inuit the reason is simple: they rely on the hunt in order to survive and perpetuate their way of life in a brutally inhospitable natural environment. So there are powerful considerations in support of the hunt stemming from the value of cultural preservation.

This is not to suggest that the hunt raises no further moral questions. For example, seal penises are highly prized in some parts of Asia for what are believed—incorrectly, of course—to be their aphrodisiac properties, and seal furs are worn there as displays of status. Is it permissible to hunt non-human animals—of *any* species—in order to satisfy trivial or irrationally grounded human desires like these, even if doing so does not obviously violate environmentalist or animal welfarist principles?

Sustainabilitarianism and Eating Beef

Source: © iStockphoto.com/Pgiam

Beef cattle near Ghost Lake, AB. Are we morally permitted to eat beef?

Since the first part of the twentieth century, the so-called "livestock revolution" has fundamentally altered the way humans raise and consume their domesticated animals. Prior to
this period, because livestock grazed for the most part on open land, livestock densities were
dictated by farm size. Farms, in turn, tended to be quite small in comparison to today's massive
agri-business outfits. However, since about 1950 demand for meat products has increased
exponentially worldwide. This has led to a change in the way livestock are farmed. In particular, it has led to the rise of industrial feedlots, which now supply nearly half the world's beef
and more than half of its pork and poultry. In Canada these factory farms now dominate the
industry. Some farms in Ontario and Quebec, for example, have more than 10,000 animals.

   The argument of this chapter leaves the morality of eating meat somewhat up in the air.
Whereas animal welfarism provides sound reasons for opposing systems of meat production
that cause unnecessary pain to animals—and this is sufficient to damn much of the livestock
industry—it is difficult to see how it can justify a blanket prohibition on eating meat. But perhaps there is a different way to derive such a prohibition—the problem of the contribution
made by livestock to climate change. In 2006, the UN claimed that livestock accounts for
more greenhouse gases than the entire transportation industry. When ruminants like cows
digest, they produce methane, an extraordinarily potent greenhouse gas that is released into

the atmosphere when the animals belch or fart. If we care at all about reducing atmospheric concentrations of greenhouse gases, then, according to Peter Singer, "there are no excuses left for eating beef."

Tim Flannery, ecologist and Australian of the Year in 2007, disagrees. He says we need to be "sustainabilitarian" about beef-eating, as well as just about everything else we do. As he sees it, the key issue is whether or not the practices we engage in are good for both the body and the planet. And there are farming experiments, like Polyface farm in the US, in which beef is produced in a way that meets the sustainabilitarian requirement. For one thing, Polyface produces beef and other agricultural products in a way that enhances soil health, which in turn boosts the capacity of the land to sequester carbon. Properly managed, cattle can actually contribute to a net decrease in greenhouse gas emissions. Says Flannery:

> We would be best served in such matters if we treat the meat, vegetable, coal and avi-ation industries similarly in our demands that they reduce their emissions. . . . There's no doubt that the livestock industry, like airlines, will need to change profoundly over this century. Neither, however, should be put out of business for ideological rather than sustainabilitarian reasons.

What would this transformed industry look like? Canadians consume nearly 30 kilograms per person of red meat (mainly beef) each year. The advent of factory farming has produced an economy of scale that has made this product readily available to people on all levels of the socio-economic ladder. In the world imagined by Flannery, beef would likely be much more expensive than it is now, since sustainabilitarian principles dictate that a lot less of it be produced. This means that eating it would become the prerogative of the relatively well-off. So we might end up with a two-tier diet, one for the rich and another for the poor. Is this a desirable outcome?

Sources: Dearden and Mitchell 2009, 364–6; Tim Flannery, *Now or Never: Why We Need to Act Now to Achieve a Sustainable Future* (Toronto: HarperCollins, 2009); Michael Pollan, *The Omnivore's Dilemma: A Natural History of Four Meals* (New York: Penguin, 2006).

Study Questions

1. Do you think the contractarian view of moral duties—the idea that our duties flow from "contracts" among us—is unable to provide a solid basis for the ethical treatment of non-human animals? Can the theory be altered to better fit this requirement?

2. Is Singer correct to clam that speciesism is on a moral par with sexism and racism?

3. What do you think of the claim made by Kymlicka and Donaldson that we ought to think more seriously about the political status of non-human animals? Is their three-fold division of these animals correct? How would our own institutions need to evolve to put these ideas into practice?

4. Do you agree with Mark Sagoff's *reductio* of animal welfarism? Why or why not?

Further Reading

Susan Armstrong. 2008. *The Animal Ethics Reader.* London: Routledge.

A. Cochrane. 2010. *An Introduction to Animals and Political Theory.* New York: Palgrave.

Wendy Donner. 1997. "Animal Rights and Native Hunters." In *Canadian Issues in Environmental Ethics.* Ed. Wesley Cragg, Allan Greenbaum, and Alex Wellington. Peterborough, ON: Broadview Press. 153–64.

R. Garner. 2005. *The Political Theory of Animal Rights.* Manchester: Manchester University Press.

Jim Mason. 2007. *The Ethics of What We Eat: Why Our Food Choices Matter.* Great River, NY: Rodale Books.

Michael Pollan. 2006. *The Omnivore's Dilemma: A Natural History of Four Meals.* New York: Penguin.

Tom Regan. 2004. *The Case for Animal Rights.* Berkeley: U of California P.

Peter Singer. 2005. *In Defence of Animals: The Second Wave.* London: Wiley-Blackwell.

CHAPTER 2

Biocentrism

Ecological Intuition Pump

Although there has been life on Earth for some 3.5 billion years, this is, as far as we know, still the only place in the observable universe that has produced living things. There are billions of planets in the observable universe, which suggests that there is something quite special about what has unfolded here. Is the extreme rarity of life in the observable universe a reason to find it beautiful and to respect it? We do often value things for just this reason. Think of how we prize fine jewels or Mount Everest or a newly discovered sketch by Picasso. Such things seem to have a good of their own that we want to protect from the various forces of destruction they face. If you are inclined to think this way about such objects—to believe that their value goes beyond the many ways in which they can simply be *used*—shouldn't the same considerations also apply to all livings things? Life, after all, is much *more* rare than shiny rocks, tall mountains, or pretty drawings. Should it not therefore be protected wherever it is threatened? If you think so, what is the best—that is, the most principled   way to deal with the conflicts that will inevitably arise when the interests of two or more living things collide?

🅐 Introduction

Wildwood Forest covers 137 acres of pristine land on Vancouver Island. Its founder, Merv Wilkinson, manages the forest on the basis of a beautifully clear principle: employ a cut-rate that sustains the integrity of the forest over the long term. Wilkinson harvests the wood in his forest, but he does so in a way that allows it to keep producing commercially valuable wood products in perpetuity. Over the course of 60 to 70 years, he has removed nearly two and a half times the forest's original volume, but there is still 10 per cent more wood there than when he started.[1] Just as important, the forest is home to an amazing diversity of species: brown creeper, pileated woodpecker, pine marten, all kinds of owls, otter, deer, beaver, cougar, and so on. What's most impressive about Wilkinson is the attitude of respect for life he manifests: "So much lives in symbiosis here that the more I study it, the more I'm in wonder at the carefully balanced intricacy of all life in this forest." His friends have tried to capture this state of wonder and awe in the phrase "Attaining Mervana."[2]

The same attitude was expressed famously by the German polymath Albert Schweitzer. In 1915, he identified "reverence for life" as the "fundamental principle of morality":

> A man is really ethical only when he obeys the constraint laid on him to help all life which he is able to succor, and when he goes out of his way to avoid injuring anything living.... To him life as such is sacred.... He.... tears no leaf from its tree, breaks off no flower, and is careful not to crush any insect as he walks. If he works by lamplight on a summer evening, he prefers to keep the window shut and to breathe stifling air, rather than to see insect after insect fall on his table with singed and sinking wings.[3]

The idea and the images that express it are arresting, but Schweitzer never developed the concept of reverence for life in a philosophically rigorous fashion. How might such a development proceed? In Chapter 1, we encountered a specific definition of "environmentalism," the one put forward by Mark Sagoff in his critique of animal welfarism. Sagoff is critical of the notion that individual non-human animals have intrinsic value, but his analysis leaves open the possibility that species do. That is, he does not explicitly reject the idea that some aspects of non-human nature are intrinsically valuable. But what grounds are there for claiming that groups of individual animals have such value?

Let's suppose that we can answer this question by claiming that the group has something like a good of its own, a good we then have a duty to respect. Obviously, a species is not itself sentient or the subject-of-a-life. It follows that a natural entity can have a good of its own in the absence of these characteristics. But if this is right, why should we not widen the circle of moral standing to encompass anything that can plausibly be said to have a good of its own? Setting aside the claim that things like species or ecosystems might qualify for this status (the theme of Chapter 3), it seems plausible to suppose that all living things do qualify. Such, at any rate, is the central claim of **biocentrism**. Biocentrism is the view that living things as such can benefit from and be harmed by human actions and that this fact places moral constraints on how we are permitted to treat them. Indeed, it means that living things have intrinsic value.

Ⓑ Rights for Living Things?

We might take the idea even further. That humans have intrinsic value entails that other humans are not permitted to interfere with them in specific ways, by violating their bodily integrity, for instance. And this moral constraint can be expressed as a legal right: if a person is raped, the rapist can be prosecuted under the law and punished appropriately. In other words, having intrinsic value brings with it the right to specific forms or degrees of legal protection. But if this is so for humans, it should also apply to anything we think is intrinsically valuable. Why, for example, should trees not have rights? For an answer to this intriguing question, we turn to this chapter's first selection, from Christopher D. Stone.

SHOULD TREES HAVE STANDING? TOWARD LEGAL RIGHTS FOR NATURAL OBJECTS

Christopher D. Stone

Throughout legal history, each successive extension of rights to some new entity has been, theretofore, a bit unthinkable. We are inclined to suppose the rightlessness of rightless "things" to be a decree of nature, not a legal convention acting in support of some status quo. . . .

The fact is that each time there is a movement to confer rights onto some new "entity," the proposal is bound to sound odd or frightening or laughable. This is partly because until the rightless thing receives its rights, we cannot see it as anything but a thing for the use of "us"—those who are holding rights at the time. . . . There is something of a seamless web involved: there will be resistance to giving the thing "rights" until it can be seen and valued for itself; yet it is hard to see it and value it for itself until we can bring ourselves to give it "rights"—which is almost inevitably going to sound inconceivable to a large group of people.

The reason for this little discourse on the unthinkable, the reader must know by now, if only from the title of the paper. I am quite seriously proposing that we give legal rights to forests, oceans, rivers, and other so-called "natural objects" in the environment—indeed, to the natural environment as a whole. . . .

I. Toward Rights for the Environment

Now, to say that the natural environment should have rights is not to say anything as silly as that no one should be allowed to cut down a tree. We say human beings have rights, but—at least as of the time of this writing—they can be executed. . . . Thus, to say that the environment should have rights is not to say that it should

have every right we can imagine, or even the same body of rights as human beings have. Nor is it to say that everything in the environment should have the same rights as every other thing in the environment. . . . For a thing to be a holder of legal rights, something more is needed than that some authoritative body will review the actions and processes of those who threaten it. As I shall use the term "holder of legal rights," each of three additional criteria must be satisfied. All three, one will observe, go toward making a thing count jurally—to have a legally recognized worth and dignity in its own right and not merely to serve as a means to benefit "us" (whoever the contemporary group of rights-holders may be). They are, first, that the thing can institute legal actions at its behest; second, that in determining the granting of legal relief, the court must take injury to it into account; and third, that relief must run to the benefit of it. . . .

II. The Rightlessness of Natural Objects at Common Law

Consider, for example, the common law's posture toward the pollution of a stream. True, courts have always been able, in some circumstances, to issue orders that will stop the pollution. . . . But the stream itself is fundamentally rightless, with implications that deserve careful reconsideration.

The first sense in which the stream is not a rights-holder has to do with standing. The stream itself has none. So far as the common law is concerned, there is in general no way to challenge the polluter's actions save at the behest of a lower riparian—another human being—able to show an invasion of his rights. . . .

The second sense in which the common law denies "rights" to natural objects has to do with the way in which the merits are decided in those cases in which someone is competent and willing to establish standing. At its more primitive levels, the system protected the "rights" of the property-owning human with minimal weighting of any values.... Today we have come more and more to make balances—but only such as will adjust the economic best interests of identifiable humans....

The third way in which the common law makes natural objects rightless has to do with who is regarded as the beneficiary of a favourable judgment. Here, too, it makes a considerable difference that it is not the natural object that counts in its own right. To illustrate this point, let me begin by observing that it makes perfectly good sense to speak of, and ascertain, the legal damage to a natural object, if only in the sense of "making it whole" with respect to the most obvious factors. The costs of making a forest whole, for example, would include the costs of reseeding, repairing watersheds, restocking wildlife—the sorts of costs the Forest Service undergoes after a fire. Making a polluted stream whole would include the costs of restocking with fish, water fowl, and other animal and vegetable life, dredging, washing out impurities, establishing natural and/or artificial aerating agents, and so forth. Now, what is important to note is that under our present system, even if a plaintiff riparian wins a water pollution suit for damages, no money goes to the benefit of the stream itself to repair its damages....

None of the natural objects, whether held in common or situated on private land, has any of the three criteria of a rights-holder. They have no standing in their own right; their unique damages do not count in determining outcome; and they are not the beneficiaries of awards. In such a fashion, these objects have traditionally been regarded by the common law, and even by all but the most recent legislation, as objects for man to conquer and master and use—in such a way as the law once looked upon "man's" relationships to African Negroes. Even where special measures have been taken to conserve them, as by seasons on game and limits on timber-cutting, the dominant motive has been to conserve them for us—for the greatest good of the greatest number of human beings. Conservationists, so far as I am aware, are generally reluctant to maintain otherwise. As the name implies, they want to conserve and guarantee our consumption and our enjoyment of these other living things. In their own right, natural objects have counted for little, in law as in popular movements.

As I mentioned at the outset, however, the rightlessness of the natural environment can and should change; it already shows some signs of doing so.

III. Toward Having Standing in Its Own Right

It is not inevitable, nor is it wise, that natural objects should have no rights to seek redress in their own behalf. It is no answer to say that streams and forests cannot have standing because streams and forests cannot speak. Corporations cannot speak either, nor can states, estates, infants, incompetents, municipalities, or universities. Lawyers speak for them, as they customarily do for the ordinary citizen with legal problems. One ought, I think, to handle the legal problems of natural objects as one does the problems of legal incompetents—human beings who have become vegetable. If a human being shows signs of becoming senile and has affairs that he is de jure incompetent to manage, those concerned with his well-being make such a showing to the court, and someone is designated by the court with the authority to manage the incompetent's affairs. . . .

On a parity of reasoning, we should have a system in which, when a friend of a natural object perceives it to be endangered, he can apply to a court for the creation of a guardianship. . . . The potential "friends" that such a statutory scheme would require will hardly be lacking. The Sierra Club, Environmental Defense Fund, Friends of the Earth, Natural Resources Defense Council, and the Izaak Walton League are just some of the many groups that have manifested unflagging dedication to the environment and that are becoming increasingly capable of marshalling the requisite technical experts and lawyers. If, for example, the Environmental Defense Fund should have reason to believe that some company's strip-mining operations might be irreparably destroying the ecological balance of large tracts of land, it could, under this procedure, apply to the court in which the lands were situated to be appointed guardian. As guardian, it might be given rights of inspection (or visitation) to determine and bring to the court's attention a fuller finding on the land's condition. If there were indications that under the substantive law some redress might be available on the land's behalf, then the guardian would be entitled to raise the land's rights in the land's name, that is, without having to make the roundabout and often unavailing demonstration . . . that the "rights" of the club's members were being invaded. . . .

One reason for making the environment itself the beneficiary of a judgment is to prevent it from being "sold out" in a negotiation among private litigants who agree not to enforce rights that have been established among themselves. Protection from this will be advanced by making the natural object a party to an injunctive settlement. Even more importantly, we should make it a beneficiary of money awards.

The idea of assessing damages as best we can and placing them in a trust fund is far more realistic than a hope that a total "freeze" can be put on the environmental status quo. Nature is a continuous theatre in which things and species (eventually man) are destined to enter and exit. In the meantime, coexistence of man and his environment means that each is going to have to compromise for the better of both. Some pollution of streams, for example, will probably be inevitable for some time. Instead of setting an unrealizable goal of enjoining absolutely the discharge of all such pollutants, the trust fund concept would (a) help assure that pollution would occur only in those instances where the social need for the pollutant's product (via his present method of production) was so high as to enable the polluter to cover all homocentric costs, plus some estimated costs to the environment per se and (b) would be a corpus for preserving monies, if necessary, while the technology developed to a point where repairing the damaged portion of the environment was feasible. Such a fund might even finance the requisite research and development. . . .

A radical new conception of man's relationship to the rest of nature would not only be a step toward solving the material planetary problems; there are strong reasons for such a changed consciousness from the point of making us far better humans. If we only stop for a moment and look at the underlying human qualities that our present attitudes toward property and nature draw upon and reinforce, we have to be struck by how stultifying of our own personal growth and satisfaction they can become when they take rein of us. Hegel, in "justifying" private property, unwittingly reflects the tone and quality of some of the needs that are played upon:

A person has as his substantive end the right of putting his will into any and every thing and thereby making it his, because it has no such end in itself and derives its destiny and soul from his will. This is the absolute right of appropriation which man has over all "things."[4]

What is it within us that gives us this need not just to satisfy basic biological wants but to extend our wills over things, to objectify them, to make them ours, to manipulate them, to keep them at a psychic distance? Can it all be explained on "rational" bases? Should we not be suspect of such needs within us, cautious as to why we wish to gratify them? When I first read that passage of Hegel, I immediately thought not only of the emotional contrast with Spinoza but of the passage in Carson McCullers's "A Tree, a Rock, a Cloud" in which an old derelict has collared a 12-year-old boy in a streetcar cafe. The old man asks whether the boy knows "how love should be begun."

> The old man leaned closer and whispered:
> "A tree. A rock. A cloud.
> "The weather was like this in Portland," he said. "At the time my science was begun. I meditated and I started very cautious. I would pick up something from the street and take it home with me. I bought a goldfish and I concentrated on the goldfish and I loved it. I graduated from one thing to another. Day by day I was getting this technique.
> "For six years now I have gone around by myself and built up my science. And now I am a master, Son. I can love anything. No longer do I have to think about it even. I see a street full of people and a beautiful light comes in me. I watch a bird in the sky. Or I meet a traveler on the road. Everything, Son. And anybody. All stranger and all loved! Do you realize what a science like mine can mean?"[5]

To be able to get away from the view that nature is a collection of useful senseless objects is, as McCullers's "madman" suggests, deeply involved in the development of our abilities to love—or, if that is putting it too strongly, to be able to reach a heightened awareness of our own, and others," capacities in their mutual interplay. To do so, we have to give up some psychic investment in our sense of separateness and specialness in the universe. And this, in turn, is hard giving indeed, because it involves us in a fight backwards, into earlier stages of civilization and childhood in which we had to trust (and perhaps fear) our environment, for we had not then the power to master it. Yet in doing so, we—as persons—gradually free ourselves of needs for supportive illusions. Is not this one of the triumphs for "us" of our giving legal rights to (or acknowledging the legal rights of) the blacks and women? ...

The time may be on hand when these sentiments, and the early stirrings of the law, can be coalesced into a radical new theory or myth—felt as well as intellectualized—of man's relationships to the rest of nature. I do not mean "myth" in a demeaning sense of the term but in the sense in which, at different times in history, our social "facts" and relationships have been comprehended and integrated by reference to the "myths" that we are co-signers of a social contract, that the pope is God's agent, and that all men are created equal. Pantheism, Shinto, and Tao all have myths to offer. But they are all, each in its own fashion, quaint, primitive, and archaic. What is needed is a myth that can fit our growing body of knowledge of geophysics, biology, and the cosmos.

In this vein, I do not think it too remote that we may come to regard the Earth, as some have suggested, as one organism, of which mankind is a functional part—the mind, perhaps—different from the rest of nature, but different as a man's brain is from his lungs.

Stone identifies a number of factors that have contributed to the widespread notion that natural objects like trees should have no moral or legal standing. At the heart of this scepticism lie two closely related ideas: (1) natural objects as such cannot genuinely be injured, nor is there a meaningful sense in which they can be beneficiaries of favourable legal judgments; and (2) in our culture the notion that things could be otherwise with natural objects is "unthinkable."

What should we make of these claims? First—addressing (2)—we might point out that at various junctures in our history, the idea that people of colour, women, or human fetuses ought to have moral standing was similarly unthinkable. That a notion challenges conventional ways of drawing the line between those with moral standing and those without is not by itself an argument against that notion. Second—addressing (1)—we might ask what is behind the claim that natural objects cannot be genuinely harmed or benefited by our actions. To see how this sort of idea might be defended, let's look at a clear statement of it by the philosopher Joel Feinberg:

> The sorts of beings that can have rights are precisely those that have interests . . . (1) because a right holder must be capable of being represented and it is impossible to represent a being that has no interests, and (2) because a right holder must be capable of being a beneficiary in his own person and a being without interests is a being that is incapable of being harmed or benefited, having no good or "sake" of its own.[6]

The central idea is that natural objects have no interests, a claim that is synonymous with the idea that they have no good of their own. Is the claim sound?

It is certainly a pervasive feature of what we might call the modern outlook on nature (beginning roughly in the late sixteenth and early seventeenth centuries). According to this way of seeing things, there is a clear separation between the normative and the natural, or between the realm of values and that of facts. There is no meaningful sense in which events in the natural world can be said to go well or badly except relative to the interests of entities that are not merely natural, like humans. If earthquakes, for example, are bad, this is only because they have the potential to cause damage to vital human interests. So far, the claim is relatively innocuous, but everything depends on where exactly we draw the line between the normative and the natural, because once we do this we will be saying that those entities we have defined as merely natural can have no good of their own. And this, of course, is what Feinberg wants to say about most living things. Again, this is not to say that such things have no value, just that they have none apart from the values and interests of those things that clearly do have value.

One way to characterize the modern outlook sketched here is to say that it is relatively parsimonious about which things can have a good of their own. Humans for sure, and perhaps the "higher" animals as well, but not much beyond. But there is a contrary view in the history of philosophy, dating back to the work of Aristotle. Aristotle has what we might call a functional understanding of natural objects. For instance, as he sees it, we can talk meaningfully about the function of a knife: to cut things. This is what it does uniquely or uniquely well. But if this function is the "essence" of the knife, then anything that blunts it—being smashed by a stone, say—is bad for the knife. Sharpening the knife, by contrast, would clearly benefit it. So it makes obvious sense to talk about what is good or bad for the knife. Now, this type of analysis may work well enough for artifacts like knives, but does it apply to natural objects like trees? What is the function of living things? What do they do uniquely or uniquely well? We can begin to answer this question by looking at a standard definition of life:

> The typifying mark of a living system . . . appears to be its persistent state of low entropy, sustained by metabolic processes for accumulating energy, and maintained in equilibrium with its environment by homeostatic feedback processes.[7]

The key to this is the idea that life resists the process of entropy, the second law of thermodynamics, which can be summed up by saying that

> "things wear out." Glasses fall from tables and shatter; cars left untended rust away; houses left without maintenance collapse, but piles of bricks left untended never spontaneously re-arrange themselves into a house. Life seems to defy this process. . . . [A] living organism taking in sustenance from outside sources and growing seems to create order out of disorder. . . . The natural tendency in the Universe at large is for disorder to increase. . . . but life reverses the process, at least temporarily.[8]

In contrast to non-living things, living things actively seek to perpetuate themselves over time, and they do this by incorporating outside energy sources for metabolic processes, resisting harmful environmental influences, and repairing damage to the internal system. This is what makes them relatively self-organized entities. For instance, the human body has a remarkably sophisticated immune system, one that is designed to resist potentially destructive attacks by pathogens, thus preserving the system over time. No non-living thing has anything remotely similar to a capacity like this, while every living thing has something like it. But it is in virtue of precisely such capacities that it makes sense to speak of things going well or badly for a living entity. Severe malnutrition, for example, can reduce the immune system's capacity to ward off pathogens, and because of this it is *bad* for a body to be severely malnourished. If this is right, we should accept the idea that living things have an interest in doing whatever it takes to stay alive. Feinberg and his ilk are therefore operating with an overly restrictive notion of what it takes to have an interest.

As Stone points out, we should also not be seduced by a specific epistemological worry. The worry is that even if it makes some sense to speak of the interests of living things as such, we have no way of knowing exactly what those interests are in any detailed way. We refer all the time to the interests of non-living bodies like states and corporations, but are these interests any easier to determine than those of, say, the grass on my front lawn? For example, at any given time it might be exceedingly difficult to be precise about what "Canada" wants from a certain round of international trade negotiations. By contrast, according to Stone,

> the lawn tells me that it wants water by a certain dryness of the blades and soil— immediately obvious to the touch—the appearance of bald spots, yellowing, and a lack of springiness after being walked on.

Similarly, once we do some ecology, we should find it obvious that a certain stand of trees *wants* biodiversity-destroying clear-cutting stopped, that the Great Barrier Reef off Australia *wants* the process of coral bleaching to stop, that **bioaccumulation**—the build-up of chemical substances in organisms in ever greater quantities as they are passed up the food chain—of toxic substances is contrary to the *interests* of those fish species at the top of aquatic eco-systems' food chains. And so on.

ⓒ Biocentric Egalitarianism

Still, even if we accept the notion that living things as such have interests and therefore ought to have rights, we might wonder just how much has been accomplished here. Rights, after all, are rarely absolute. In the human sphere, we are always trying to strike a balance among conflicting rights claims. So it might be that although we are willing to grant rights to trees, in practice those rights are always overridden by the rights of loggers, or corporations, or citizens when they come into conflict with them. That is, it is theoretically consistent to argue both (a) that all non-human living things ought to have rights and (b) that when these rights come into conflict with the rights of humans (or corporations, etc.), the rights of the latter always override those of the former. To forestall this possibility, we would need to show not just that all non-human living things have moral standing but that they have such standing on an equal basis with humans. This is the thesis of **biocentric egalitarianism**, advanced most forcefully by Paul Taylor. We turn to it now.

THE BIOCENTRIC OUTLOOK ON NATURE

Paul W. Taylor

The biocentric outlook on nature has four main components. (1) Humans are thought of as members of the Earth's community of life, holding that membership on the same terms as apply to all the non-human members. (2) The Earth's natural ecosystems as a totality are seen as a complex web of interconnected elements, with the sound biological functioning of each being dependent on the sound biological functioning of the others.... (3) Each individual organism is conceived of as a teleological centre of life, pursuing its own good in its own way. (4) Whether we are concerned with standards of merit or with the concept of inherent worth, the claim that humans by their very nature are superior to other species is a groundless claim and, in the light of elements (1), (2), and (3) above, must be rejected as nothing more than an irrational bias in our own favour.

The conjunction of these four ideas constitutes the biocentric outlook on nature. In the remainder of this paper, I give a brief account of the first three components, followed by a more detailed analysis of the fourth. I then conclude by indicating how this outlook provides a way of justifying the attitude of respect for nature....

I. Humans as Members of the Earth's Community of Life

When we look at ourselves from the evolutionary point of view, we see that not only are we very recent arrivals on Earth but that our emergence as a new species on the planet was originally an event of no particular importance to the entire scheme of things. The Earth was teeming with life long before we appeared. Putting the point metaphorically, we are relative newcomers, entering a home that has been the residence of others for hundreds of millions of years, a home that must now be shared by all of us together.

The comparative brevity of human life on Earth may be vividly depicted by imagining the geological time scale in spatial terms. Suppose we start with algae, which have been around for at least 600 million years. (The earliest protozoa actually predated this by several billion years.) If the time that algae have been here were represented by the length of a football field (300 feet),

then the period during which sharks have been swimming in the world's oceans and spiders have been spinning their webs would occupy three-quarters of the length of the field; reptiles would show up at about the centre of the field; mammals would cover the last third of the field; hominids (mammals of the family Hominidae) the last two feet; and the species *Homo sapiens* the last six inches.

Whether this newcomer is able to survive as long as other species remains to be seen. But there is surely something presumptuous about the way humans look down on the "lower" animals, especially those that have become extinct. We consider the dinosaurs, for example, to be biological failures, though they existed on our planet for 65 million years. One writer has made the point with beautiful simplicity:

> We sometimes speak of the dinosaurs as failures; there will be time enough for that judgment when we have lasted even for one tenth as long. ... [9]

The possibility of the extinction of the human species, a possibility that starkly confronts us in the contemporary world, makes us aware of another respect in which we should not consider ourselves privileged beings in relation to other species. This is the fact that the well-being of humans is dependent upon the ecological soundness and health of many plant and animal communities, while their soundness and health does not in the least depend upon human well-being. Indeed, from their standpoint the very existence of humans is quite unnecessary. Every last man, woman, and child could disappear from the face of the Earth without any significant detrimental consequence for the good of wild animals and plants. On the contrary, many of them would be greatly benefited. The destruction of their habitats by human "developments" would cease. The poisoning and polluting of their environment would come to an end.

The Earth's land, air, and water would no longer be subject to the degradation they are now undergoing as the result of large-scale technology and uncontrolled population growth. Life communities in natural ecosystems would gradually return to their former healthy state. Tropical rainforests, for example, would again be able to make their full contribution to a life-sustaining atmosphere for the whole planet. The rivers, lakes, and oceans of the world would (perhaps) eventually become clean again. Spilled oil, plastic trash, and even radioactive waste might finally, after many centuries, cease doing their terrible work. Ecosystems would return to their proper balance, suffering only the disruptions of natural events such as volcanic eruptions and glaciation. From these the community of life could recover, as it has so often done in the past. But the ecological disasters now perpetrated on it by humans—disasters from which it might never recover—these it would no longer have to endure.

If, then, the total, final, absolute extermination of our species (by our own hands?) should take place and if we should not carry all the others with us into oblivion, not only would the Earth's community of life continue to exist, but in all probability its well-being would be enhanced. Our presence, in short, is not needed. If we were to take the standpoint of the community and give voice to its true interest, the ending of our six-inch epoch would most likely be greeted with a hearty "Good riddance!"

II. The Natural World as an Organic System

To accept the biocentric outlook and regard ourselves and our place in the world from its perspective is to see the whole natural order of the Earth's biosphere as a complex but unified web of interconnected organisms, objects, and events. The ecological relationships between any community of living things and their environment form an organic whole of functionally interdependent parts. Each ecosystem is a small universe itself in which the interactions of its various species populations comprise an intricately woven network of cause–effect relations. Such dynamic but at the same time relatively stable structures as food chains, predator–prey relations, and plant succession in a forest are self-regulating, energy-recycling mechanisms that preserve the equilibrium of the whole.

As far as the well-being of wild animals and plants is concerned, this ecological equilibrium must not be destroyed. The same holds true of the well-being of humans. When one views the realm of nature from the perspective of the biocentric outlook, one never forgets that in the long run the integrity of the entire biosphere of our planet is essential to the realization of the good of its constituent communities of life, both human and non-human.

Although the importance of this idea cannot be overemphasized, it is by now so familiar and so widely acknowledged that I shall not further elaborate on it here. However, I do wish to point out that this "holistic" view of the Earth's ecological systems does not itself constitute a moral norm. It is a factual aspect of biological reality, to be understood as a set of causal connections in ordinary empirical terms. Its significance for humans is the same as its significance for non-humans, namely, in setting basic conditions for the realization of the good of living things. Its ethical implications for our treatment of the natural environment lie entirely in the fact that our knowledge of these causal connections is an essential means to fulfilling the aims we set for ourselves in adopting the attitude of respect for nature. In addition, its theoretical implications for the ethics of respect for nature lie in the fact that it (along with the other elements of the biocentric outlook) makes the adopting of that attitude a rational and intelligible thing to do.

III. Individual Organisms as Teleological Centres of Life

As our knowledge of living things increases, as we come to a deeper understanding of their life cycles, their interactions with other organisms, and the manifold ways in which they adjust to the environment, we become more fully aware of how each of them is carrying out its biological functions according to the laws of its species-specific nature. But besides this, our increasing knowledge and understanding also develop in us a sharpened awareness of the uniqueness of each individual organism. Scientists who have made careful studies of particular plants and animals, whether in the field or in laboratories, have often acquired a knowledge of their subjects as identifiable individuals. Close observation over extended periods of time has led them to an appreciation of the unique "personalities" of their subjects. Sometimes a scientist may come to take a special interest in a particular animal or plant, all the while remaining strictly objective in the gathering and recording of data.

Non-scientists may likewise experience this development of interest when, as amateur naturalists, they make accurate observations over sustained periods of close acquaintance with an individual organism. As one becomes more and more familiar with the organism and its behaviour, one becomes fully sensitive to the particular way it is living out its life cycle. One may become fascinated by it and even experience some involvement with its good and bad fortunes (that is, with the occurrence of environ-mental conditions favourable or unfavourable to the realization of its good). The organism comes to mean something to one as a unique, irreplaceable individual. The final culmination of this process is the achievement of a genuine understanding of its point of view and, with that understanding, an ability to "take" that point of view. Conceiving of it as a centre of life, one is able to look at the world from its perspective.

This development from objective knowledge to the recognition of individuality, and from the recognition of individuality to full awareness of an organism's standpoint, is a process of heightening our consciousness of what it means to be an individual living thing. We grasp the particularity of the organism as a teleological centrization of the good of living things. Its ethical implications for our treatment of the natural environment lie entirely in the fact that our knowledge of these causal connections is an essential means to fulfilling the aims we set for ourselves in adopting the attitude of respect for nature. In addition, its theoretical implications for the ethics of respect for nature lie in the fact that it (along with the other elements of the biocentric outlook) makes the adopting of that attitude a rational and intelligible thing to do.

IV. The Denial of Human Superiority

The fourth component of the biocentric outlook on nature is the single most important idea in establishing the justifiability of the attitude of respect for nature. Its central role is due to the special relationship it bears to the first three components of the outlook. This relationship will be brought out after the concept of human superiority is examined and analyzed.[10]

In what sense are humans alleged to be superior to other animals? We are different from them in having certain capacities that they lack. But why should these capacities be a mark of superiority? From what point of view are they judged to be signs of superiority, and what sense of superiority is meant? After all, various non-human species have capacities that humans lack. There is the speed of a cheetah, the vision of an eagle, the agility of a monkey. Why should not these be taken as signs of their superiority over humans?

One answer that comes immediately to mind is that these capacities are not as valuable as the human capacities that are claimed to make us superior. Such uniquely human characteristics as rational thought, aesthetic creativity, autonomy and self-determination, and moral freedom, it might be held, have a higher value than the capacities found in other species. Yet we must ask: valuable to whom, and on what grounds?

The human characteristics mentioned are all valuable to humans. They are essential to the preservation and enrichment of our civilization and culture. Clearly, it is from the human standpoint that they are being judged to be desirable and good. It is not difficult here to recognize a begging of the question. Humans are claiming human superiority from a strictly human point of view, that is, from a point of view in which the good of humans is taken as the standard of judgment. All we need to do is look at the capacities of non-human animals (or plants, for that matter) from the standpoint of their good to find a contrary judgment of superiority. The speed of the cheetah, for example, is a sign of its superiority to humans when considered from the standpoint of the good of its species. If it were as slow a runner as a human, it would not be able to survive. And so it would be for all the other abilities of non-humans that further their good but that are lacking in humans. In each case, the claim to human superiority would be rejected from a non-human standpoint.

When superiority assertions are interpreted in this way, they are based on judgments of merit. To judge the merits of a person or an organism one must apply grading or ranking standards to it. (As I show below, this distinguishes judgments of merit from judgments of inherent worth.) Empirical investigation then determines whether it has the "good-making properties" (merits) in virtue of which it fulfills the standards being applied. In the case of humans, merits may be either moral or nonmoral. We can judge one person to be better than (superior to) another from the moral point of view by applying certain standards to their character and conduct. Similarly, we can appeal to non-moral criteria in judging someone to be an excellent piano player, a fair cook, a poor tennis player, and so on.

Different social purposes and roles are implicit in the making of such judgments, providing the frame of reference for the choice of standards by which the non-moral merits of people are determined. Ultimately, such purposes and roles stem from a society's way of life as whole. Now, a society's way of life may be thought of as the cultural form given to the realization of human values. Whether moral or non-moral standards are being applied, then, all judgments of people's merits finally depend on human values. All are made from an exclusively human standpoint.

The question that naturally arises at this juncture is: Why should standards that are based on human values be assumed to be the only valid criteria of merit and hence the only true signs of superiority? This question is especially pressing when humans are being judged superior in merit to non-humans. It is true that a human being may be a better mathematician than a monkey, but the monkey may be a better tree-climber than a human being. If we humans value mathematics more than tree-climbing, that is because our conception of civilized life makes the development of mathematical ability more desirable than the ability to climb trees. But is it not unreasonable to judge non-humans by the values of human civilization, rather than by values connected with what it is for a member of that species to live a good life? If all living things have a good of their own, it at least makes sense to judge the merits of non-humans by standards derived from their good. To use only standards based on human values is already to commit oneself to holding that humans are superior to non-humans, which is the point in question.

A further logical flaw arises in connection with the widely held conviction that humans are morally superior beings because they possess, while others lack, the capacities of a moral agent (free will, accountability, deliberation, judgment, practical reason). This view rests on a conceptual confusion. As far as moral standards are concerned, only beings that have the capacities of a moral agent can properly be judged to be either moral (morally good) or immoral (morally deficient). Moral standards are simply not applicable to beings that lack such capacities. Animals and plants cannot therefore be said to be morally inferior in merit to humans. Since the only beings that can have moral merits or be deficient in such merits are moral agents, it is conceptually incoherent to judge humans as superior to non-humans on the ground that humans have moral capacities while non-humans don't.

Up to this point, I have been interpreting the claim that humans are superior to other living things as a grading or ranking judgment regarding their comparative merits. There is, however, another way of understanding the idea of human superiority. According to this interpretation, humans are superior to non-humans not as regards their merits but as regards their inherent worth. Thus, the claim of human superiority is to be understood as asserting that all humans, simply in virtue of their humanity, have a greater inherent worth than other living things.

The inherent worth of an entity does not depend on its merits.[11] To consider something as possessing inherent worth, we have seen, is to place intrinsic value on the realization of its good. This is done regardless of whatever particular merits it might have or might lack, as judged by a set of grading or ranking standards. In human affairs, we are all familiar with the principle that one's worth as a person does not vary with one's merits or lack of merits. The same can hold true of animals and plants. To regard such entities as possessing inherent worth entails disregarding their merits and deficiencies, whether they are being judged from a human standpoint or from the standpoint of their own species.

The idea of one entity having more merit than another and so being superior to it in merit, makes perfectly good sense. Merit is a grading or ranking concept, and judgments of comparative merit are based on the different degrees to which things satisfy a given standard. But what can it mean to talk about one thing being superior to another in inherent worth? In order to get at what is being asserted in such a claim, it is helpful first to look at the social origin of the concept of degrees of inherent worth.

The idea that humans can possess different degrees of inherent worth originated in societies having rigid class structures. Before the rise of modern democracies with their egalitarian outlook, one's membership in a hereditary class determined one's social status. People in the upper classes were looked up to, while those in the lower classes were looked down upon. In such a society one's social superiors and social inferiors were clearly defined and easily recognized.

Two aspects of these class-structured societies are especially relevant to the idea of degrees of inherent worth. First, those born into the upper classes were deemed more worthy of respect than those born into the lower orders. Second, the superior worth of upper-class people had nothing to do with their merits, nor did the inferior worth of those in the lower classes rest on their lack of merits. One's superiority or inferiority entirely derived from a social position one was born into. The modern concept of a meritocracy simply did not apply. One could not advance into a higher class by any sort of moral or non-moral achievement. Similarly, an aristocrat held his title and all the privileges that went with it just because he was the eldest son of a titled nobleman. Unlike the bestowing of knighthood in contemporary Great Britain, one did not earn membership in the nobility by meritorious conduct.

We who live in modern democracies no longer believe in such hereditary social distinctions. Indeed, we would wholeheartedly condemn them on moral grounds as being fundamentally unjust. We have come to think of class systems as a paradigm of social injustice, it being a central principle of the democratic way of life that among humans there are no superiors and no inferiors. Thus, we have rejected the whole conceptual framework in which people are judged to have different degrees of inherent worth. That idea is incompatible with our notion of human equality based on the doctrine that all humans, simply in virtue of their humanity, have the same inherent worth. (The belief in universal human rights is one form that this egalitarianism takes.)

The vast majority of people in modern democracies, however, do not maintain an egalitarian outlook when it comes to comparing human beings with other living things. Most people consider our own species to be superior to all other species, and this superiority is understood to be a matter of inherent worth, not merit. There may exist thoroughly vicious and depraved humans who lack all merit. Yet because they are human they are thought to belong to a higher class of entities than any plant or animal. That one is born into the species *Homo sapiens* entitles one to have lordship over those who are one's inferiors, namely, those born into other species.

The parallel with hereditary social classes is very close. Implicit in this view is a hierarchical conception of nature according to which an organism has a position of superiority or inferiority in the Earth's community of life simply on the basis of its genetic background. The "lower" orders of life are looked down upon, and it is considered perfectly proper that they serve the interests of those belonging to the highest order, namely humans. The intrinsic value we place on the well-being of our fellow humans reflects our recognition of their rightful position as our equals. No such intrinsic value is to be placed on the good of other animals, unless we choose to do so out of fondness or affection for them. But their well-being imposes no moral requirement on us. In this respect there is an absolute difference in moral status between ourselves and them.

This is the structure of concepts and beliefs that people are committed to insofar as they regard humans to be superior in inherent worth to all other species. I now wish to argue that this structure of concepts and beliefs is completely groundless. If we accept the first three components of the biocentric outlook and from that perspective look at the major philosophical traditions that have supported that structure, we find it to be at bottom nothing more than the expression of an irrational bias in our own favour. The philosophical traditions themselves rest on very questionable assumptions or else simply beg the question. I briefly consider three of the main traditions to substantiate the point. These are classical Greek humanism, Cartesian dualism, and the Judeo-Christian concept of the Great Chain of Being.

The inherent superiority of humans over other species was implicit in the Greek definition of man as a rational animal. Our animal nature was identified with "brute" desires that need the order and restraint of reason to rule them (just as reason is the special virtue of those who rule in the ideal state). Rationality was then seen to be the key to our superiority over animals. It enables us to live on a higher plane and endows us with a nobility and worth that other creatures lack. This familiar way of comparing humans with other species is deeply ingrained in our Western philosophical outlook. The point to consider here is that this view does not actually provide an argument for human superiority but rather makes explicit the framework of thought that is implicitly used by those who think of humans as inherently superior to non-humans.

The Greeks who held that humans, in virtue of their rational capacities, have a kind of worth greater than that of any non-rational being, never looked at rationality as but one capacity of living things among many others. But when we consider rationality from the standpoint of the first three elements of the ecological outlook, we see that its value lies in its importance for human life. Other creatures achieve their species-specific good without the need of rationality, although they often make use of capacities that humans lack. So the humanistic outlook of classical Greek thought does not give us a neutral (non-question-begging) ground on which to construct a scale of degrees of inherent worth possessed by different species of living things.

The second tradition, centring on the Cartesian dualism of soul and body, also fails to justify the claim to human superiority. That superiority is supposed to derive from the fact that we have souls while animals do not. Animals are mere automata and lack the divine element that makes us spiritual beings. I won't go into the now familiar criticisms of this two-substance view. I only add the point that, even if humans are composed of an immaterial, unextended soul and a material, extended body, this in itself is not a reason to deem them of greater worth than entities that are only bodies. Why is a soul substance a thing that adds value to its possessor? Unless some theological reasoning is offered here (which many, including myself, would find unacceptable on epistemological grounds), no logical connection is evident. An immaterial something that thinks is better than a material something that does not think only if thinking itself has value, either intrinsically or instrumentally. Now, it is intrinsically valuable to humans alone, who value it as an end in itself, and it is instrumentally valuable to those who benefit from it, namely humans.

For animals that neither enjoy thinking for its own sake nor need it for living the kind of life for which they are best adapted, it has no value. Even if "thinking" is broadened to include all forms of consciousness, there are still many living things that can do without it and yet live what is for their species a good life. The anthropocentricity underlying the claim to human superiority runs throughout Cartesian dualism. A third major source of the idea of human superiority is the Judeo-Christian concept of the Great Chain of Being. Humans are superior to animals and plants because their Creator has given them a higher place on the chain. It begins with God at the top, and then moves to the angels, who are lower than God but higher than humans, then to humans, positioned between the angels and the beasts (partaking of the nature of both), and then on down to the lower levels occupied by non-human animals, plants, and finally inanimate objects. Humans, being "made in God's image," are inherently superior to animals and plants by virtue of their being closer (in their essential nature) to God.

The metaphysical and epistemological difficulties with this conception of a hierarchy of entities are, in my mind, insuperable. Without entering into this matter here, I only point out that if we are unwilling to accept the metaphysics of traditional Judaism and Christianity, we are again left without good reasons for holding to the claim of inherent human superiority.

The foregoing considerations (and others like them) leave us with but one ground for the assertion that a human being, regardless of merit, is a higher kind of entity than any other living thing. This is the mere fact of the genetic makeup of the species *Homo sapiens*. But this is surely irrational and arbitrary. Why should the arrangement of genes of a certain type be a mark of superior value, especially when this fact about an organism is taken by itself, unrelated to any other aspect of its life? We might just as well refer to any other genetic makeup as a ground of superior value. Clearly, we are confronted here with a wholly arbitrary claim that can only be explained as an irrational bias in our own favour.

That the claim is nothing more than a deep-seated prejudice is brought home to us when we look at our relation to other species in the light of the first three elements of the biocentric outlook. Those elements taken conjointly give us a certain overall view of the natural world and of the place of humans in it. When we take this view, we come to understand other living things, their environmental conditions, and their ecological relationships in such a way as to awake in us a deep sense of our kinship with them as fellow members of the Earth's community of life. Humans and non-humans alike are viewed together as integral parts of one unified whole in which all living things are functionally interrelated. Finally, when our awareness focuses on the individual lives of plants and animals, each is seen to share with us the characteristic of being a teleological centre of life striving to realize its own good in its own unique way.

As this entire belief system becomes part of the conceptual framework through which we understand and perceive the world, we come to see ourselves as bearing a certain moral relation to non-human forms of life. Our ethical role in nature takes on a new significance. We begin to look at other species as we look at ourselves, seeing them as beings that have a good they are striving to realize just as we have a good we are striving to realize. We accordingly develop the disposition to view the world from the standpoint of their good as well as from the standpoint of our own good. Now, if the groundlessness of the claim that humans are inherently superior to other species were brought clearly before our minds, we would not remain intellectually neutral toward that claim but would reject it as being fundamentally at variance with our total world outlook. In the absence of any good reasons for holding it, the assertion of human superiority would then appear simply as the expression of an irrational and self-serving prejudice that favours one particular species over several million others.

Rejecting the notion of human superiority entails its positive counterpart: the doctrine of species impartiality. One who accepts that doctrine regards all living things as possessing inherent worth—the same inherent worth, since no one species has been shown to be either "higher" or "lower" than any other. Now, we saw earlier that, insofar as one thinks of a living thing as possessing inherent worth, one considers it to be the appropriate object of the attitude of respect and believes that attitude to be the only fitting or suitable one for all moral agents to take toward it.

Here, then, is the key to understanding how the attitude of respect is rooted in the biocentric outlook on nature. The basic connection is made through the denial of human superiority. Once we reject the claim that humans are superior either in merit or in worth to other living things, we are ready to adopt the attitude of respect. The denial of human superiority is itself the result of taking the perspective on nature built into the first three elements of the biocentric outlook.

Now, the first three elements of the biocentric outlook, it seems clear, would be found acceptable to any rational and scientifically informed thinker who is fully "open" to the reality of the lives of non-human organisms. Without denying our distinctively human characteristics, such a thinker can acknowledge the fundamental respects in which we are members of the Earth's community of life and in which the biological conditions necessary for the realization of our human values are inextricably linked with the whole system of nature. In addition, the conception of individual living things as teleological centres of life simply articulates how a scientifically informed thinker comes to understand them as the result of increasingly careful and detailed observations. Thus, the biocentric outlook recommends itself as an acceptable system of concepts and beliefs to anyone who is clear-minded, unbiased, and factually enlightened and who has a developed capacity of reality awareness with regard to the lives of individual organisms. This, I submit, is as good a reason for making the moral commitment involved in adopting the attitude of respect for nature as any theory of environmental ethics could possibly have.

I have not asserted anywhere in the foregoing account that animals or plants have moral rights. This omission was deliberate. I do not think that the reference class of the concept, bearer of moral rights, should be extended to include non-human living things. My reasons for taking this position, however, go beyond the scope of this paper. I believe I have been able to accomplish many of the same ends that those who ascribe rights to animals or plants wish to accomplish. There is no reason, moreover, why plants and animals, including whole species populations and life communities, cannot be accorded legal rights under my theory. To grant them legal protection could be interpreted as giving them legal entitlement to be protected, and this, in fact, would be a means by which a society that subscribed to the ethics of respect for nature could give public recognition to their inherent worth.

There remains the problem of competing claims, even when wild plants and animals are not thought of as bearers of moral rights. If we accept the biocentric outlook and accordingly adopt the attitude of respect for nature as our ultimate moral attitude, how do we resolve conflicts that arise from our respect for persons in the domain of human ethics and our respect for nature in the domain of environmental ethics? This is a question that cannot be adequately dealt with here. My main purpose in this paper has been to try to establish a base point from which we can start working toward a solution to the problem. I have shown why we cannot just begin with an initial presumption in favour of the interests of our own species. It is after all within our power as moral beings to place limits on human population and technology with the deliberate intention of sharing the Earth's bounty with other species. That such sharing is an ideal difficult to realize even in an approximate way does not take away its claim to our deepest moral commitment.

Paraphrased, the four theses of biocentric egalitarianism are: (1) humans are members of the Earth's community of living things; (2) the biosphere is a complex web of causally inter-dependent systems; (3) each living thing is a "teleological centre of life"; (4) humans are not superior to other species. Let's look at these claims more closely, focusing most intently on (4), as Taylor himself does. We have already broached the idea contained in (3). All living things have a good of their own—this distinguishes them from non-living things—which they pursue in the manner peculiar to the species of which they are members. "Teleological" means goal-directed, acting for purposes (though not necessarily consciously). At a detailed biochemical level, fish metabolism and plant metabolism are different processes, but both involve taking organic material from outside the organism and using it for the construction and maintenance of cells. This is the *purpose* of metabolic processes, which is why it is not at all obscure to refer to them as goal-directed. Individual fish and plants, in turn, can be thought of as collections of such processes. All such processes are in place in order to preserve the larger organism of which they are a part. Members of both species are therefore "teleological centres of life."

Claims (1) and (2) are best taken together. As such, they make the claim that humans are a mere part of a complex and causally interdependent web of systems that together constitute the biosphere. To help illustrate this idea, let's look at forest ecosystems and the sorts of manage-ment methods that might either harm or benefit them. Forests dominate the physical makeup of Canada, making up more than 43 per cent of the country's total area (arctic tundra and ice fields are second at 28 per cent).[12] This astonishing bounty extends from one end the country to the other: from the balsam fir, white spruce, sugar maple, and eastern hemlock of the Cape Breton Highlands, through the tamarack and jack pine of the interior's Boreal Shield, all the way to the Douglas fir and Sitka spruce that dominate the Pacific Maritime ecozone. In 2007, the total value of Canadian timber exports was $38.2 billion. And non-timber products—such as tree saps, ber-ries, mushrooms, understorey plants, and wild rice—are worth $725 million annually.[13] Beyond the resource potential of the forests, the quantity and diversity of life they support is truly astonishing. More than 90,000 species are directly or indirectly dependent on forest habitats in Canada, and this number includes 65 per cent of the species considered at risk in the country.[14]

The dominant method of harvesting trees in Canada is clear-cutting. This method of timber extraction can have profoundly negative effects on the entire forest ecosystem, beginning with the foundation of that system, the soil. Soil contains all manner of organic materials, bacteria, insects, and fungi. When it is degraded or eroded, none of the other species that depend on the forest ecosystem can thrive. Remember Merv Wilkinson? This is what he says about the effects of clear-cutting in the forests of British Columbia:

> When you clear-cut and burn your forest, it destroys the soil, and that's a crime against nature. . . . Once they [the clear-cutters] are through, there's no more forest, period. The soil is so devastated it won't grow trees. We have 3.7 million hectares (9.2 million acres) of insufficiently reforested land because of malpractice in the logging industry. . . . Given the average growth rate of B.C. forests, that same land. . . . under my system would have 3,400 people employed in the logging end, to say nothing of processing.[15]

This comment contains a number of strands worth drawing out further. First, the language is strongly reminiscent of Stone: to talk of "crimes against nature" and "malpractice" in the logging industry is implicitly to invoke the idea that the legal rights of natural objects have been violated in the process of clear-cutting the BC forests. Nor is this problem fully redressed through replanting. The problem with the replanting that follows clear-cutting is that forest companies plant fast-growing and commercially valuable monocultures. They are breeding grounds for insects, pests, and diseases, which are in turn combatted with insecticides and herbicides.

Second, with the forest's foundation—the soil—degraded, the rest of the species that depend on the forest are also placed at risk. From the standpoint of biodiversity, a replanted timber monoculture is decidedly not a forest. This is an illustration of Taylor's thesis (2): life in the natural world is intimately interconnected such that to degrade or eliminate one element of an ecosystem—depending on the systemic importance of that element—is potentially to place at risk every species in the system. In speaking of crimes against nature, Wilkinson is clearly not interested just in soil and its constituent life forms for their own sake. Rather, he is upset by a practice that threatens that whole symbiotic forest network.

Finally, the reference to the livelihood of loggers is important. Taylor's thesis (1) states that humans are a part of Earth's interconnected systems. It follows that it would be a mistake to think either that human values and interests are the only ones that count or that they don't count at all. Humans are a living part of the whole, and they too need to "make a living." But they must do so in a manner that respects the capacity of the forest—and this does not mean just the trees the forest contains—to regenerate itself.[16] The type of sustainable forestry Merv Wilkinson practises achieves this. It not only allows the whole forest bio-community to regenerate over time, but also provides steady and meaningfully remunerated work to lots of people. At the end of the day, all people who depend on the forest should take a page from the Itza Maya's book of forestry management. The Itza Maya are a Guatemalan forest people who understand the intricate connectedness of the forest's life forms and the extent to which their own survival depends on respecting that intricacy. They say:

> Listen for the sound of the jaguar. When there are no more jaguars there will be no more forest. And then there will be no more Maya.[17]

Let's move on to Taylor's thesis (4), the denial that humans are morally superior to any other living species.

Ⓓ Human Superiority and Inter-species Conflict

Thesis (4) is the denial that humans are morally superior to other species. The superiority claim is that when our interests come into conflict with the interests of anything else, we invariably win. What could have made us believe we are morally superior? We could claim that we are just inherently superior and that's all there is to it. But that approach has little to recommend it. Alternatively, we could claim that we are superior in virtue of possessing some capacity or set of capacities that sets us apart from other species. In Chapter 1, we investigated this sort of claim but rejected it on the grounds that it is arbitrary to say that a capacity possessed by one species but not others makes that species superior to others in the moral sense. This is a central component of Singer's attack on speciesism, and Taylor clearly subscribes to it as well.

But here a possible challenge to Taylor emerges. Is it not arbitrary to draw the circle of moral standing around living things, thus excluding non-living things? In order to do this, Taylor needs to show that what marks life off from non-life *matters morally*. Think about the processes we have looked at in this chapter, those that define the anti-entropic activity of living things. For example, all living things metabolize organic matter whereas non-living things do not. Therefore, the former have moral standing and the latter lack it. Is this an arbitrary—i.e., philosophically ungrounded—distinction? Surely not, because Taylor could say that possession of the criterion he has isolated—life—is simply a precondition of having interests at all and having interests is necessary for moral standing. This would give him grounds for criticizing Singer's account of moral standing as overly restrictive—confined as it is to sentient things—without inviting the charge of arbitrariness.

The more serious challenge to Taylor's position comes from the observation that the interests of humans and those of other species come into conflict all the time, most notably from the fact that we need to eat them in order to survive. How can this sort of practice be justified without abandoning the claim—the essence of Taylor's thesis (4)—that humans are not morally superior to any other living things? After all, to be an egalitarian—biocentric or otherwise—by definition commits one to the view that everything with moral standing has it to an equal degree. Christopher Belshaw thinks that Taylor can be rescued from this line of attack:

> We do not signal a lack of respect merely by killing some other creature or by allowing it to die when we could, without serious inconvenience, give it assistance. . . . Is reverence or respect really possible as long as we continue, more or less, to put ourselves first?[18]

Belshaw suggests that in order to answer this question, Taylor needs generous recourse to the notion of a necessary evil. Thus, although humans must often kill in order to survive, they should, presumably, do so with a heavy heart, recognizing that they are undermining the real interests of another living thing. This is plausible. Suppose, for example, that you were forced to kill another human purely in self-defence. You could claim that doing so is compatible with thinking of that other person as morally equal to you, and this recognition would then cause you to feel genuine regret for the act you performed. And in any case, biocentric egalitarianism surely does place *some* constraints on our behaviour. If the only way to fulfill our luxury wants (for things we do not need to live or even thrive) is to undermine the basic interests of other living things (their interests in having things they do need to live or thrive), then we are not permitted to fulfill such wants. It can be difficult to work out the distinction between luxury wants and basic interests in a very detailed way, but in many cases the distinction is clear enough for practical purposes.

But there is another point to make here. That biocentric egalitarianism does not require us to erase ourselves is surely a strength of the theory. What the example of sustainable forestry management, explored above, shows is that we can both respect the interests of all living things and make a living from those things. And the imperative to be sensitive to the interests of all living things places real constraints on the kind of living we can expect to make or on how many of us can make such a living, so we are not given carte blanche with respect to natural "resources." At the same time, however, we should recognize that we need to kill to survive and that we are probably morally justified in doing so, even if we are good biocentric egalitarians.

(E) Conclusion

There are at least two additional possible worries for biocentric egalitarianism. The first has to do with the claim that we have a duty to respect anything that has a good of its own. Belshaw, for example, thinks that this is an unexplained argumentative move. But is it possible to be sceptical here without subscribing to a general, and more obviously implausible, scepticism on this front? When talking about animals, for instance, Belshaw says this:

> Imagine someone who says, "I agree that if I beat this animal it will feel great pain. But why does this give me any reason not to beat it?" This person sounds like a psychopath or a monster. The question seems perverse.[19]

We can agree that the question seems perverse, but why should we allow that a similar question applied to, say, a majestic Douglas fir would not seem equally perverse? Of course, one could claim that animals have, while trees lack, moral standing and that this is because the former are sentient, the latter not. But Taylor has, presumably, given us good grounds for questioning that claim, and in any case Belshaw admits that it makes sense to speak of the interests of trees. Why then is it not perverse to deny that the fact that the tree has a clear interest in staying alive gives us a reason to refrain from destroying it? That reason might not win the day; it might not be *overriding* as philosophers sometimes put it. But this is true of all moral reasons, including the reasons we have for refraining from beating animals. If the only way to save a child from being torn to pieces by a dog is to beat the dog, then it is permissible to do so. But this is compatible with saying that in both cases—beating animals and destroying trees—if we perform the action *wantonly,* we are morally blameworthy.

The second worry provides a bridge to Chapter 3. Again, it has been well put by Belshaw. What, really, is so special about life?

> Only living things can be killed, but not only living things can be destroyed. Perhaps only living things can be injured, but non-living things can quite easily be damaged. And surely thoughtless harm and wanton destruction are bad wherever they occur, whether to animals, plants, sculptures or greenhouses.[20]

Is the wanton destruction of a statue or a great work of art bad, apart from the value attached to it by humans? Can such things be said to have interests, and if not, what grounds could there be for the claim that such actions are bad? These are intriguing questions, but we are interested in a different, though related, set of questions. Is it plausible to extend the circle of moral standing further than we have done so far so as to include things like ecosystems, species, the biosphere, perhaps the Earth itself? We have seen that Stone, although he is clearly most interested in individual natural objects—like trees—is already gesturing in this direction. On what grounds could we justify this extension? This brings us to ecocentrism and deep ecology.

CASE STUDY

Respect for Nature and the Mount Polley Mine Disaster

Source: THE CANADIAN PRESS/Jonathan Hayward

Mount Polley, British Columbia. If we came to see nature as intrinsically worthy of respect, would we likely see fewer large-scale industrial disasters like this?

Taylor's biocentric egalitarianism is a conscious attempt to apply to all living things a moral concept—respect—that has hitherto been applicable only to persons. The idea of respect gets its most sustained and famous expression in the philosophy of Kant. For Kant, all persons as such are ends in themselves and therefore have a fundamental dignity, which moral agents

Continued

are required to respect. We possess this dignity in virtue of our rational nature, the capacity we have to live and act on the basis of self-given reasons. This capacity, in turn, is the ground of human freedom or autonomy. At the bottom of the moral duty to respect persons is the recognition that all—and only—beings that are autonomous ought to be treated this way. For example, it is impermissible to lie to or coerce rational natures, because in doing so you undermine their ability to formulate and act on the best available self-given reasons. When you lie to someone, you conceal crucial information from them, and when you coerce them, you simply bypass their reasons for acting. So respecting someone involves protecting him or her from certain kinds of treatment.

But if the idea of respect is so intimately tied to concepts like rationality and autonomy, how can we have respect for something like individual plants? If successful, what the arguments of this chapter show is that the connection between these concepts is merely contingent, that the way Kant thought of respect is merely one possibility among many. In our culture, the idea of respect is motivationally powerful. If we are generally decent people, when we are enjoined to treat a thing with respect we will generally be motivated to protect it in appropriate ways. So if we can divorce the concept of respect from those of autonomy and rationality while retaining its motivational force, we will have achieved something philosophically worthwhile. This, broadly, is what Taylor is up to, and he is clear about the extent to which we have failed to adopt a respectful attitude to the rest of living nature. As a measure of this failure, consider what he says about how Earth's plant and animal communities might get along without us:

> The destruction of their habitats by human "developments" would cease. The poisoning and polluting of their environment would come to an end. The Earth's land, air, and water would no longer be subject to the degradation they are now undergoing as the result of large-scale technology and uncontrolled population growth.

This is a scathing commentary on our disrespect for nature. Is it warranted? A recent example may provide some insight. On 4 August 2014, there was a major disaster at the Mount Polley gold and copper mining operation in the Cariboo region of south-central British Columbia. What occurred was a breach in one of the mine's tailings ponds, huge pools containing water and chemicals that are left over from the extraction of the minerals. Nearly 5 billion litres of toxic slurry went crashing into nearby lakes and rivers. Predictably, the president of the company responsible—Bryan Kynoch of Imperial Metals Corporation—downplayed the event, declaring that it had caused no significant contamination of drinking water in the region. This quick response highlights our reluctance to think about the environmental impacts of our activities on species other than our own, or those we depend upon directly (in this case the sockeye salmon, for instance). In contrast to Kynoch's blithe assurances, Megan Thompson, aquatic ecologist and limnologist, argues that we need to look at the spill's effects on the solids below the surface waters in the region's lakes and rivers. As she puts it, "there are many things in the tailings that could impact lakes and rivers, especially if those substances did not naturally occur in the aquatic systems prior to the spill."

The truth is that we have no idea what sorts of long-term impacts the spill will have on the many species in the area, but there is no reason to be optimistic or naive about these impacts.

Even if humans were not at the top of the food chain, this unfolding scenario should give us great cause for worry if we believe that all of Earth's living things are worthy of our respect—and our protection. First Nations groups see this clearly. Neskonlith chief Judy Wilson has evicted Imperial Metals from the proposed Ruddock Creek zinc and lead mine as a result of its apparent negligence in the Mount Polley disaster. Given the damage we are inflicting on the rest of nature through our industrial activities, isn't it high time to start thinking about and acting toward all living things as though they mattered as much as we do?

Sources: "First Nation Will Evict Mining Company after Massive Spill Contaminated Area Water," at http://thinkprogress.org/climate/2014/08/14/3471118/first-nation-evict-mining-company/; "British Columbia Declares a Local State of Emergency after Massive Mine Waste Spill," at http://thinkprogress.org/climate/2014/08/07/3468421/british-columbia-mine-spill-emergency/; Draper and Reed 2009, 294.

Study Questions

1. Think hard about the four theses of Taylor's biocentric egalitarianism. Should this list be shortened or extended? Would you alter any of these theses to make the view more palatable? Which ones and why?

2. What do you think about "A Tree, A Rock, A Cloud," the parable invoked by Stone toward the end of his paper? What is he trying to express here? Do you think the ideas he is trying to get at could have been more clearly stated in less poetic or allegorical language?

3. Is it possible to be as egalitarian in our biocentrism as Taylor seems to imply? Are all species on a moral par?

4. Is it time to begin thinking of living things like trees as having "rights" that can be defended, if necessary, in courts? We already do this for some corporations, so why not trees or ecosystems?

Further Reading

Nicholas Agar. 1997. "Biocentrism and the Concept of Life." *Ethics* 108, no. 1: 147–68.

Robin Attfield. 1981. "The Good of Trees." *Journal of Value Inquiry* 15: 35–54.

Charles S. Cockle. 2005. "The Value of Microorganisms." *Environmental Ethics* 27, no. 4: 375–90.

Karaan Durland. 2008. "The Prospects for a Viable Biocentric Egalitarianism." *Environmental Ethics* 30, no. 4: 401–16.

Kenneth E. Goodpaster. 1978. "On Being Morally Considerable." *Journal of Philosophy* 75 (June): 308–25.

James P. Sterba. 1998. "A Biocentrist Strikes Back." *Environmental Ethics* 20, no. 4: 361–76.

CHAPTER 3

Ecocentrism and Deep Ecology

Ecological Intuition Pump

Living things survive by incorporating energy from outside and using it to maintain, power, and repair themselves. They put energy to *work* for themselves. Plants, for instance, convert solar energy into sugars, building biomass that is then passed along to other members of the community of living things. In this fashion, each living thing is able to resist, for a while at least, being broken down into a state of disorder. Rocks don't do this kind of thing; they are not self-organized wholes. This is why living things have, and non-livings things lack, interests. But is it not then arbitrary to deny that ecosystems are enough like individual living things to have interests? They too incorporate energy from outside and use it to build and maintain sub-systems like specific (and relatively stable) predator–prey relations. An ecosystem is really just a huge network of energy transfer among its members, and just as with individual living things, it is a relatively closed, self-organized whole. Having gone this far, why not try to *identify* with the interests of such a thing? Can you imagine yourself acting so as to protect the integrity of a threatened forest ecosystem? Could you *feel* as strongly about the need to do this as you might in deciding to protect the interests of a threatened (human) loved one?

Ⓐ Introduction

Rex Weyler is one of the co-founders of Greenpeace, an environmental movement that began in Vancouver in 1971. The original targets of Greenpeace's activism were the testing of nuclear weapons and the whaling industry, but their focus has since expanded and altered. They now concentrate on the state of our oceans, the problem of climate change, genetic engineering, and other issues. Weyler himself currently writes a monthly column on Greenpeace's website entitled "Deep Green" in which he reflects on "the roots of activism, environmentalism, and Greenpeace's past, present, and future."[1] As the title of the column indicates, that history and the current self-understanding of the movement are rooted in the philosophy of deep ecology, part of this chapter's focus. **Deep ecology**, at least as practised by members of Greenpeace, emphasizes self-realization through identification with nature, green consumerism, voluntary simplicity, bioregionalism or "living in place," and the protection and restoration of wild spaces. And it urges people to express all of these commitments through direct political action.

But before getting to deep ecology, we need to examine its immediate predecessor in the movement of ideas we have been tracing so far in this book. This idea is **ecocentrism**, the notion that the objects of primary moral concern are ecosystems. Like both animal welfarism and biocentrism, ecocentrism is anti-anthropocentric. That is, it is opposed to the claim that the only entities that have moral standing or intrinsic value are humans and that everything else possesses, at best, a value that is merely derivative of human values. But here the affinities end, for the ecocentrist is equally opposed to both psychocentrism and individualism. **Psychocentrism** is the view that what matters morally is the capacity to have some sort of psychological experience, the ability to feel pain and pleasure for instance. Individualism is the view that the primary objects of moral concern are individual plants or animals. Animal welfarism—at least Singer's version of it and probably Regan's as well—is both psychocentric and individualist, while biocentrism, though it eschews psychocentrism, is individualist. Both, therefore, ought to be rejected from the standpoint of ecocentrism. This makes ecocentrism a genuine alternative to anything we have examined so far.

B The Land Ethic

The best place to start is with Aldo Leopold, originator of the **land ethic**. Leopold spent most of his professional life as a forest manager. Through his work in the field, he became intimately familiar with the workings of ecosystems. Leopold has had an immensely powerful influence on the environmental movement since he first wrote *A Sand County Almanac*—from which the following selection has been taken—more than 60 years ago. Although outside the mainstream of academic philosophy, he is a staunch critic of his—and our—culture's dominant tradition of environmental thought and practice.

THE LAND ETHIC

Aldo Leopold

When god-like Odysseus returned from the wars in Troy, he hanged all on one rope a dozen slave-girls of his household whom he suspected of misbehaviour during his absence. This hanging involved no question of propriety. The girls were property. The disposal of property was then, as now, a matter of expediency, not of right and wrong.

Concepts of right and wrong were not lacking from Odysseus's Greece: witness the fidelity of his wife through the long years before at last his black-prowed galleys clove the wine-dark seas for home. The ethical structure of that day covered wives but had not yet been extended to human chattels. During the 3000 years that have since elapsed, ethical criteria have been extended to many fields of conduct, with corresponding shrinkages in those judged by expediency only.

I. The Ethical Sequence

This extension of ethics, so far studied only by philosophers, is actually a process in ecological evolution. Its sequences may be described in ecological as well as in philosophical terms. An ethic, ecologically, is a limitation on freedom of action in the struggle for existence. An ethic, philosophically, is a differentiation of social from anti-social conduct. These are two definitions of one thing. The thing has its origin in the tendency of interdependent individuals or groups to evolve modes of cooperation. The ecologist calls these symbioses. Politics and economics are advanced symbioses in which the original free-for-all competition has been replaced, in part, by cooperative mechanisms with an ethical content.

The complexity of cooperative mechanisms has increased with population density and with the efficiency of tools. It was simpler, for example, to define the anti-social uses of sticks and stones in the days of the mastodons than of bullets and billboards in the age of motors.

The first ethics dealt with the relation between individuals; the Mosaic Decalogue is an example. Later accretions dealt with the relation between the individual and society. The Golden Rule tries to integrate the individual into society, democracy to integrate social organization to the individual.

There is as yet no ethic dealing with man's relation to land and to the animals and plants that grow upon it. Land, like Odysseus's slave-girls, is still property. The land-relation is still strictly economic, entailing privileges but not obligations.

The extension of ethics to this third element in human environment is, if I read the evidence correctly, an evolutionary possibility and an ecological necessity. It is the third step in a sequence. The first two have already been taken. Individual thinkers since the days of Ezekiel and Isaiah have asserted that the despoliation of land is not only inexpedient but wrong. Society, however, has not yet affirmed their belief. I regard the present conservation movement as the embryo of such an affirmation.

An ethic may be regarded as a mode of guidance for meeting ecological situations so new or intricate, or involving such deferred reactions, that the path of social expediency is not discernible to the average individual. Animal instincts are modes of guidance for the individual in meeting such situations. Ethics are possibly a kind of community instinct in-the-making.

II. The Community Concept

All ethics so far evolved rest upon a single premise: that the individual is a member of a community of interdependent parts. His instincts prompt him to compete for his place in the community, but his ethics prompt him also to cooperate (perhaps in order that there may be a place to compete for).

The land ethic simply enlarges the boundaries of the community to include soils, waters, plants, and animals, or collectively: the land.

This sounds simple: Do we not already sing our love for and obligation to the land of the free and the home of the brave? Yes, but just what and whom do we love? Certainly not the soil, which we are sending helter-skelter downriver. Certainly not the waters, which we assume have no function except to turn turbines, float barges, and carry off sewage. Certainly not the plants, of which we exterminate whole communities without batting an eye. Certainly not the animals, of which we have already extirpated many of the largest and most beautiful species. A land ethic of course cannot prevent the alteration, management, and use of these "resources," but it does affirm their right to continued existence and, at least in spots, their continued existence in a natural state.

In short, a land ethic changes the role of *Homo sapiens* from conqueror of the land-community to plain member and citizen of it. It implies respect for his fellow-members and also respect for the community as such.

In human history, we have learned (I hope) that the conqueror role is eventually self-defeating. Why? Because it is implicit in such a role that the conqueror knows, *ex cathedra*, just what makes the community clock tick, and just what and who is valuable, and what and who is worthless, in community life. It always turns out that he knows neither, and this is why his conquests eventually defeat themselves.

In the biotic community, a parallel situation exists. Abraham knew exactly what the land was for: it was to drip milk and honey into Abraham's mouth. At the present moment, the assurance with which we regard this assumption is inverse to the degree of our education.

The ordinary citizen today assumes that science knows what makes the community clock tick; the scientist is equally sure that he does not. He knows that the biotic mechanism is so complex that its workings may never be fully understood.

That man is, in fact, only a member of a biotic team is shown by an ecological interpretation of history. Many historical events, hitherto explained solely in terms of human enterprise, were actually biotic interactions between people and land. The characteristics of the land determined the facts quite as potently as the characteristics of the men who lived on it. . . .

III. The Ecological Conscience

Conservation is a state of harmony between men and land. Despite nearly a century of propaganda, conservation still proceeds at a snail's pace; progress still consists largely of letterhead pieties and convention oratory. On the back 40, we still slip two steps backward for each forward stride.

The usual answer to this dilemma is "more conservation education." No one will debate this, but is it certain that only the volume of education needs stepping up? Is something lacking in the content as well? It is difficult to give a fair summary of its content in brief form, but as I understand it, the content is substantially this: obey the law, vote right, join some organizations, and practise what conservation is profitable on your own land; the government will do the rest.

Is not this formula too easy to accomplish anything worthwhile? It defines no right or wrong, assigns no obligation, calls for no sacrifice, implies no change in the current philosophy of values. In respect of land use, it urges only enlightened self-interest. Just how far will such education take us? . . .

To sum up: we asked the farmer to do what he conveniently could to save his soil, and he has done just that, and only that. The farmer who clears the woods off a 75 per cent slope, turns his cows into the clearing, and dumps its rainfall, rocks, and soil into the community creek is still (if otherwise decent) a respected member of society. If he puts lime on his fields and plants his crops on contour, he is still entitled to all the privileges and emoluments of his Soil Conservation District. The district is a beautiful piece of social machinery, but it is coughing along on two cylinders because we have been too timid, and too anxious for quick success, to tell the farmer the true magnitude of his obligations. Obligations have no meaning without conscience, and the problem we face is the extension of the social conscience from people to land.

No important change in ethics was ever accomplished without an internal change in our intellectual emphasis, loyalties, affections, and convictions. The proof that conservation has not yet touched these foundations of conduct lies in the fact that philosophy and religion have not yet heard of it. In our attempt to make conservation easy, we have made it trivial.

IV. Substitutes for a Land Ethic

When the logic of history hungers for bread and we hand out a stone, we are at pains to explain how much the stone resembles bread. I now describe some of the stones that serve in lieu of a land ethic. One basic weakness in a conservation system based wholly on economic motives is that most members of the land community have no economic value. Wildflowers and songbirds are examples. Of the 22,000 higher plants and animals native to Wisconsin, it is doubtful whether more than 5 per cent can be sold, fed, eaten, or otherwise put to economic use. Yet these creatures are members of the biotic community, and if (as I believe) its stability depends on its integrity, they are entitled to continuance.

When one of these non-economic categories is threatened, and if we happen to love it, we invent subterfuges to give it economic importance. At the beginning of the century, songbirds were supposed to be disappearing. Ornithologists jumped to the rescue with some distinctly shaky evidence to the effect that insects would eat us up if birds failed to control them. The evidence had to be economic in order to be valid.

It is painful to read these circumlocutions today. We have no land ethic yet, but we have at least drawn nearer the point of admitting that birds should continue as a matter of biotic right, regardless of the presence or absence of economic advantage to us.

A parallel situation exists in respect of predatory mammals, raptorial birds, and fish-eating birds. Time was when biologists somewhat overworked the evidence that these creatures preserve the health of game by killing weaklings, or that they control rodents for the farmer, or that they prey only on "worthless" species. Here again, the evidence had to be economic in order to be valid. It is only in recent years that we hear the more honest argument that predators are members of the community and that no special interest has the right to exterminate them for the sake of a benefit, real or fancied, to itself. Unfortunately, this enlightened view is still in the talk stage. . . .

Some species of trees have been "read out of the party" by economics-minded foresters because they grow too slowly or have too low a sale value to pay as timber crops: white cedar, tamarack, cypress, beech, and hemlock are examples. In Europe, where forestry is ecologically more advanced, the non-commercial tree species are recognized as members of the native forest community, to be preserved as such, within reason. Moreover some (like beech) have been found to have a valuable function in building up soil fertility. The interdependence of the forest and its constituent tree species, ground flora, and fauna is taken for granted.

Lack of economic value is sometimes a character not only of species or groups but of entire biotic communities: marshes, bogs, dunes, and "deserts" are examples. Our formula in such cases is to relegate their conservation to government as refuges, monuments, or parks. The difficulty is that these communities are usually interspersed with more valuable private lands; the government cannot possibly own or control such scattered parcels. The net effect is that we have relegated some of them to ultimate extinction over large areas. If the private owner were ecologically minded, he would be proud to be the custodian of a reasonable proportion of such areas, which add diversity and beauty to his farm and to his community.

In some instances, the assumed lack of profit in these "waste" areas has proved to be wrong, but only after most of them had been done away with. The present scramble to re-flood muskrat marshes is a case in point.

There is a clear tendency in American conservation to relegate to government all necessary jobs that private landowners fail to perform. Government ownership, operation, subsidy, or regulation is now widely prevalent in forestry, range management, soil and watershed management, park and wilderness conservation, fisheries management, and migratory bird management, with more to come. Most of this growth in governmental conservation is proper and logical, some of it is inevitable. That I imply no disapproval of it is implicit in the fact that I have spent most of my life working for it. Nevertheless, the question arises: What is the ultimate magnitude of the enterprise? Will the tax base carry its eventual ramifications? At what point will governmental conservation, like the mastodon, become handicapped by its own dimensions? The answer, if there is any, seems to be in a land ethic, or some other force that assigns more obligation to the private landowner.

Industrial landowners and users, especially lumbermen and stockmen, are inclined to wail long and loudly about the extension of government ownership and regulation to land, but (with notable exceptions) they show little disposition to develop the only visible alternative: the voluntary practice of conservation on their own lands.

When the private landowner is asked to perform some unprofitable act for the good of the community, he today assents only with outstretched palm. If the act costs him cash, this is fair and proper, but when it costs only forethought, open-mindedness, or time, the issue is at least debatable. The overwhelming growth of land-uses subsidies in recent years must be ascribed, in large part, to the government's own agencies for conservation education: the land bureaus, the agricultural colleges, and the extension services. As far as I can detect, no ethical obligation toward land is taught in these institutions.

To sum up: a system of conservation based solely on economic self-interest is hopelessly lopsided. It tends to ignore, and thus eventually to eliminate, many elements in the land community that lack commercial value but that are (as far as we know) essential to its healthy functioning. It assumes, falsely I think, that the economic parts of the biotic clock will function without the uneconomic parts. It tends to relegate to government many functions eventually too large, too complex, or too widely dispersed to be performed by government.

An ethical obligation on the part of the private owner is the only visible remedy for these situations.

V. The Land Pyramid

An ethic to supplement and guide the economic relation to land presupposes the existence of some mental image of land as a biotic mechanism. We can be ethical only in relation to something we can see, feel, understand, love, or otherwise have faith in. The image commonly employed in conservation education is "the balance of nature." For reasons too lengthy to detail here, this figure of speech fails to describe accurately what little we know about the land mechanism. A much truer image is the one employed in ecology: the biotic pyramid. I shall first sketch the pyramid as a symbol of land and later develop some of its implications in terms of land use.

Plants absorb energy from the sun. This energy flows through a circuit called the biota, which may be represented by a pyramid consisting of layers. The bottom layer is the soil. A plant layer rests on the soil, an insect layer on the plants, a bird and rodent layer on the insects, and so on up through various animal groups to the apex layer, which consists of the larger carnivores.

The species of a layer are alike not in where they came from or in what they look like, but rather in what they eat. Each successive layer depends on those below it for food and often for other services, and each in turn furnishes food and services to those above. Proceeding upward, each successive layer decreases in numerical abundance. Thus, for every carnivore there are hundreds of his prey, thousands of their prey, millions of insects, uncountable plants. The pyramidal form of the system reflects this numerical progression from apex to base. Man shares an intermediate layer with the bears, raccoons, and squirrels, which eat both meat and vegetables.

The lines of dependency for food and other services are called food chains. Thus, soil-oak-deer-Indian is a chain that has now been largely converted to soil-corn-cow-farmer. Each species, including ourselves, is a link in many chains. The deer eats a hundred plants other than oak and the cow a hundred plants other than corn. Both, then, are links in a hundred chains. The pyramid is a tangle of chains so complex as to seem disorderly, yet the stability of the system proves it to be a highly organized structure. Its functioning depends on the cooperation and competition of its diverse parts.

In the beginning, the pyramid of life was low and squat, the food chains short and simple. Evolution has added layer after layer, link after link. Man is one of thousands of accretions to the height and complexity of the pyramid. Science has given us many doubts, but it has given us at least one certainty: the trend of evolution is to elaborate and diversify the biota.

Land, then, is not merely soil; it is a fountain of energy flowing through a circuit of soils, plants, and animals. Food chains are the living channels that conduct energy upward; death and decay return it to the soil. The circuit is not closed; some energy is dissipated in decay, some is added by absorption from the air, some is stored in soils, peats, and long-lived forests, but it is a sustained circuit, like a slowly augmented revolving fund of life. There is always a net loss by downhill wash, but this is normally small and offset by the decay of rocks. It is deposited in the ocean and, in the course of geological time, raised to form new lands and new pyramids.

The velocity and character of the upward flow of energy depend on the complex structure of the plant and animal community, much as the upward flow of sap in a tree depends on its complex cellular organization. Without this complexity, normal circulation would presumably not occur. Structure means the characteristic numbers, as well as the characteristic kinds and functions, of the component species. This interdependence between the complex structure of the land and its smooth functioning as an energy unit is one of its basic attributes.

When a change occurs in one part of the circuit, many other parts must adjust themselves to it. Change does not necessarily obstruct or divert the flow of energy; evolution is a long series of self-induced changes, the net result of which has been to elaborate the flow mechanism and to lengthen the circuit. Evolutionary changes, however, are usually slow and local. Man's invention of tools has enabled him to make changes of unprecedented violence, rapidity, and scope.

One change is in the composition of floras and faunas. The larger predators are lopped off the apex of the pyramid; food chains, for the first time in history, become shorter rather than longer. Domesticated species from other lands are substituted for wild ones, and wild ones are moved to new habitats. In this worldwide pooling of faunas and floras, some species get out of bounds as pests and diseases, others are extinguished. Such effects are seldom intended or foreseen; they represent unpredicted and often untraceable readjustments in the structure. Agricultural science is largely a race between the emergence of new pests and the emergence of new techniques for their control.

Another change touches the flow of energy through plants and animals and its return to the soil. Fertility is the ability of soil to receive, store, and release energy. Agriculture, by overdrafts on the soil, or by too radical a substitution of domestic for native species in the superstructure, may derange the channels of flow or deplete storage. Soils depleted of their storage, or of the organic matter that anchors it, wash away faster than they form. This is erosion. Waters, like soil, are part of the energy circuit. Industry, by polluting waters or obstructing them with dams, may exclude the plants and animals necessary to keep energy in circulation.

Transportation brings about another basic change: the plants or animals grown in one region are now consumed and returned to the soil in another. Transportation taps the energy stored in rocks, and in the air, and uses it elsewhere; thus, we fertilize the garden with nitrogen gleaned by the guano birds from the fishes of seas on the other side of the equator. Thus, the formerly localized and self-contained circuits are pooled on a worldwide scale.

The process of altering the pyramid for human occupation releases stored energy, and this often gives rise, during the pioneering period, to a deceptive exuberance of plant and animal life, both wild and tame. These releases of biotic capital tend to becloud or postpone the penalties of violence.

This thumbnail sketch of land as an energy circuit conveys three basic ideas:

1. That land is not merely soil.
2. That the native plants and animals kept the energy circuit open; others may or may not.
3. That man-made changes are of a different order than evolutionary changes and have effects more comprehensive than is intended or foreseen.

These ideas, collectively, raise two basic issues: Can the land adjust itself to the new order? Can the desired alterations be accomplished with less violence?

Biotas seem to differ in their capacity to sustain violent conversion. Western Europe, for example, carries a far different pyramid than Caesar found there. Some large animals are lost; swampy forests have become meadows or plowland; many new plants and animals are introduced, some of which escape as pests; the remaining natives are greatly changed in distribution and abundance. Yet the soil is still there and, with the help of imported nutrients, still fertile, and waters flow normally; the new structure seems to function and to persist. There is no visible stoppage or derangement of the circuit.

Western Europe, then, has a resistant biota. Its inner processes are tough, elastic, resistant to strain. No matter how violent the alterations, the pyramid, so far, has developed some new *modus vivendi*, which preserves its habitability for man and for most of the other natives. Japan seems to present another instance of radical conversion without disorganization. . . . Most other civilized regions, and some as yet barely touched by civilization, display various stages of disorganization, varying from initial symptoms to advanced wastage.

This almost worldwide display of disorganization in the land seems to be similar to disease in an animal, except that it never culminates in complete disorganization or death. The land recovers, but at some reduced level of complexity and with a reduced carrying capacity for people, plants, and animals. Many biotas currently regarded as "lands of opportunity" are in fact already subsisting on exploitative agriculture, that is, they have already exceeded their sustained carrying capacity. Most of South America is overpopulated in this sense.

In arid regions, we attempt to offset the process of wastage by reclamation, but it is only too evident that the prospective longevity of reclamation projects is often short. In our own west, the best of them may not last a century.

The combined evidence of history and ecology seems to support one general deduction: the less violent the man-made changes, the greater the probability of successful readjustment in the pyramid. Violence, in turn, varies with human population density; a dense population requires a more violent conversion. In this respect, North America has a better chance for permanence than Europe, if she can contrive to limit her density.

This deduction runs counter to our current philosophy, which assumes that because a small increase in density enriched human life, an indefinite increase will enrich it indefinitely. Ecology knows of no density relationship that holds for indefinitely wide limits. All gains from density are subject to a law of diminishing returns.

Whatever may be the equation for men and land, it is improbable that we as yet know all its terms. Recent discoveries in mineral and vitamin nutrition reveal unsuspected dependencies in the up-circuit: incredibly minute quantities of certain substances determine the value of soils to plants, of plants to animals. What of the down-circuit? What of the vanishing species, the preservation of which we now regard as an aesthetic luxury? They helped build the soil; in what unsuspected ways may they be essential to its maintenance? . . .

VI. Land Health and the A–B Cleavage

A land ethic, then, reflects the existence of an ecological conscience, and this in turn reflects a conviction of individual responsibility for the health of the land. Health is the capacity of the land for self-renewal. Conservation is our effort to understand and preserve this capacity.

Conservationists are notorious for their dissensions. Superficially, these seem to add up to mere confusion, but a more careful scrutiny reveals a single plane of cleavage common to many specialized fields. In each field, one group (A) regards the land as soil and its function as commodity-production; another group (B) regards the land as a biota and its function as something broader. How much broader is admittedly in a state of doubt and confusion.

In my own field, forestry, Group A is quite content to grow trees like cabbages, with cellulose as the basic forest commodity. It feels no inhibition against violence; its ideology is agronomic. Group B, on the other hand, sees forestry as fundamentally different from agronomy because it employs natural species and manages a natural environment rather than creating an artificial one. Group B prefers natural reproduction on principle. It worries on biotic as well as economic grounds about the loss of species like chestnut and the threatened loss of the white pines. It worries about a whole series of secondary forest functions: wildlife, recreation, watersheds, wilderness areas. To my mind, Group B feels the stirrings of an ecological conscience.

In the wildlife field, a parallel cleavage exists. For Group A, the basic commodities are sport and meat; the yardsticks of production are ciphers of take in pheasants and trout. Artificial propagation is acceptable as a permanent as well as a temporary recourse—if its unit costs permit. Group B, on the other hand, worries about a whole series of biotic side issues. What is the cost in predators of producing a game crop? Should we have further recourse to exotics? How can management restore the shrinking species, like prairie grouse, already hopeless as shootable game? How can management restore the threatened rarities, like trumpeter swan and whooping crane? Can management principles be extended to wildflowers? Here again it is clear to me that we have the same A–B cleavage as in forestry.

In the larger field of agriculture, I am less competent to speak, but there seem to be somewhat parallel cleavages. Scientific agriculture was actively developing before ecology was born, hence a slower penetration of ecological concepts might be expected. Moreover, the farmer, by the very nature of his techniques, must modify the biota more radically than the forester or the wildlife manager. Nevertheless, there are many discontents in agriculture, which seem to add up to a new vision of "biotic farming."

Perhaps the most important of these is the new evidence that poundage or tonnage is no measure of the food value of farm crops; the products of fertile soil may be qualitatively as well as quantitatively superior. We can bolster poundage from depleted soils by pouring on imported fertility, but we are not necessarily bolstering food value. The possible ultimate ramifications of this idea are so immense that I must leave their exposition to abler pens.

The discontent that labels itself "organic farming," while bearing some of the earmarks of a cult, is nevertheless biotic in its direction, particularly in its insistence on the importance of soil flora and fauna.

The ecological fundamentals of agriculture are just as poorly known to the public as in other fields of land use. For example, few educated people realize that the marvellous advances in technique made during recent decades are improvements in the pump rather than the well. Acre for acre, they have barely sufficed to offset the sinking level of fertility.

In all of these cleavages, we see repeated the same basic paradoxes: man the conqueror versus man the biotic citizen; science the sharpener of his sword versus science the searchlight on his universe; land the slave and servant versus land the collective organism. Robinson's injunction to Tristram may well be applied, at this juncture, to *Homo sapiens* as a species in geological time:

> Whether you will or not
> You are a King, Tristram, for you are
>   one
> Of the time-tested few that leave the
>   world,
> When they are gone, not the same place
>   it was.
> Mark what you leave.

It is inconceivable to me that an ethical relation to land can exist without love, respect, and admiration for land and a high regard for its value. By value, I of course mean something far broader than mere economic value; I mean value in the philosophical sense.

Perhaps the most serious obstacle impeding the evolution of a land ethic is the fact that our educational and economic system is headed away from, rather than toward, an intense consciousness of land. Your true modern is separated from the land by many middlemen and by innumerable physical gadgets. He has no vital relation to it; to him it is the space between cities on which crops grow. Turn him loose for a day on the land, and if the spot does not happen to be a golf links or a "scenic" area, he is bored stiff. If crops could be raised by hydroponics instead of farming, it would suit him very well. Synthetic substitutes for wood, leather, wool, and other natural land products suit him better than the originals. In short, land is something he has "outgrown."

Almost equally serious as an obstacle to a land ethic is the attitude of the farmer for whom the land is still an adversary, or a taskmaster that keeps him in slavery. Theoretically, the mechanization of farming ought to cut the farmer's chains, but whether it really does is debatable.

One of the requisites for an ecological comprehension of land is an understanding of ecology, and this is by no means co-extensive with "education"; in fact, much higher education seems deliberately to avoid ecological concepts. An understanding of ecology does not necessarily originate in courses bearing ecological labels; it is quite as likely to be labelled geography, botany, agronomy, history, or economics. This is as it should be, but whatever the label, ecological training is scarce.

The case for a land ethic would appear hopeless but for the minority that is in obvious revolt against these "modern" trends. The "key-log" which must be moved to release the evolutionary process for an ethic is simply this: quit thinking about decent land use as solely an economic problem. Examine each question in terms of what is ethically and aesthetically right, as well as what is economically expedient. A thing is right when it tends to preserve the integrity, stability, and beauty of the biotic community. It is wrong when it tends otherwise. . . .

I have purposely presented the land ethic as a product of social evolution because nothing so important as an ethic is ever "written." Only the most superficial student of history supposes that Moses "wrote" the Decalogue; it evolved in the minds of a thinking community, and Moses wrote a tentative summary of it for a "seminar." I say tentative because evolution never stops.

The evolution of a land ethic is an intellectual as well as emotional process. Conservation is paved with good intentions that prove to be futile, or even dangerous, because they are devoid of critical understanding either of the land or of economic land use. I think it is a truism that as the ethical frontier advances from the individual to the community, its intellectual content increases.

The mechanism of operation is the same for any ethic: social approbation for right actions, social disapproval for wrong actions.

By and large, our present problem is one of attitudes and implements. We are remodelling the Alhambra with a steam shovel, and we are proud of our yardage. We shall hardly relinquish the shovel, which after all has many good points, but we are in need of gentler and more objective criteria for its successful use.

In *A Sand County Almanac*, there is a magnificent essay in which Leopold offers us a recent moral history of our species as he saws through a huge oak tree that had been struck by lightning. The saw bites through ring after concentric ring of the tree, each ring corresponding to a specific event in this history. Nearing the central ring, Leopold comes to the racial conflict that spawned the American Civil War. At this point he raises the core question for environmental ethics:

> Our saw now cuts the 1860's, when thousands died to settle the question: Is the man–man community lightly to be dismembered? They settled it, but they did not see, nor do we yet see, that the same question applies to the man–land community.[2]

There are two closely related questions to be addressed in Leopold's analysis. First, what does he mean by "the land"? Leopold himself says that "the land" should be thought of as the community of soils, waters, plants, and animals. In other words, ecosystems. An **ecosystem** is a geographically specific collection of plants and animals interacting among themselves and with the non-living, or abiotic, things—such as rocks, soil, and climate—of that area. The **biosphere**, then, is the totality of interlocking ecosystems on Earth. It is crucial to note the way in which, as a description of the entity to which we might have duties, this goes beyond the biocentric outlook. Recall from Chapters 1 and 2 that the key question here has to do with moral standing. Which things or kinds of things can we have duties *toward* and not simply duties *regarding*? Or, what things have intrinsic value? Biocentrism draws the moral circle around living or biotic things. Ecocentrism goes a step further to draw it around systems that contain both biotic and abiotic elements. Soil and water, although they contain many living things, are not themselves alive. So the primary objects of potential moral concern on this view are ecosystems.

The second question focuses on the "ethic" part of the land ethic. The most basic way to describe an ethical system is to ask which actions it permits and which it does not. Or, which actions are right and which ones are wrong? Leopold is crystal clear about this:

> A thing is right when it tends to preserve the integrity, stability, and beauty of the biotic community. It is wrong when it tends otherwise.

What does it mean to act so as to preserve the integrity, stability, and beauty of an eco-system? To answer this question, we need to do a bit of ecosystem ontology. Of the three qualities a properly functioning ecosystem might display—integrity, stability, and beauty—integrity would appear to be the most fundamental. First, Leopold actually says that the stability of a biotic community "depends on its integrity." And second, it is at least plausible to suppose that beauty does as well, although Leopold does not explicitly assert this (it would in any case be odd to claim that beauty was foundational). So what is ecosystem integrity?

To describe something as displaying integrity is to say that it is a unified and ordered whole. The originator of the concept of an ecosystem, Sir Arthur Tansley, said in 1935 that the "currency" of the "economy of nature" is *energy*, and this is how Leopold also understands the notion.[3] This is how energy flow is described by J. Baird Callicott in an influential study of the land ethic:

> A description of the ecosystem begins with the sun.... Solar energy flows through a circuit called the biota. It enters the biota through the leaves of green plants and courses through plant-eating animals, and then on to omnivores and carnivores. At last the tiny fraction of solar energy converted to biomass by green plants remaining in the corpse of a predator, animal feces, plant detritus, or other dead organic material is garnered by decomposers—worms, fungi, and bacteria.[4]

The activity of these last humble members of the community then return key chemical nutrients to the soil so that it can support the growth of more plant life. The system gradually loses energy and must therefore be re-energized by the sun. For this reason, the system cannot strictly be said to be closed, but it *is* relatively closed. Over the long course of an ecosystem's history, its various elements will have co-evolved such that the relations between them have achieved a kind of balance. This can be seen most clearly in food chains, where the population of various species is held at a more or less constant rate over time—despite short-term fluctuations up or down—by the activities of members of other species. Balanced predator–prey relations are a key component of any healthy ecosystem, but so too is a steady climate, an adequate supply of soil nutrients, suitable availability of water, and so on.

So, to return to ethics, our duties to ecosystems might involve both (a) not interfering in such balanced systems to a degree that damages the energy flow and (b) rectifying whatever damage to such systems our actions have already produced by restoring them to something resembling their pre-interference state. The ecocentrist is not, however, committed to the view that ecosystems should be prevented from changing. Leopold recognizes that change over time is integral to them. Nevertheless, there is an important distinction to be made between the "slow and local" changes that happen through natural evolutionary means and human-induced changes, which are of "unprecedented violence, rapidity, and scope."

The call to refrain from interfering unduly with wild nature is at the heart of **preservationism**, and it is no accident that many preservationists see their principles rooted in Leopold's land ethic. Preservationism is often contrasted with **conservationism**. Both ideas stress protecting natural areas, but conservationists generally think that achieving such protection is compatible with allowing some human uses of wild spaces. Moreover, they believe it is permissible to derive economic benefits from wild nature if this can be done without damaging it. Preservationists, by contrast, believe that the primary goal of nature protection has to do with keeping ecosystems in as pristine a condition as possible, and this often means prohibiting human activities in them altogether. That is, preservationists think that the maintenance of ecosystem integrity rather than the balancing of human interests and the protection of nature should be the main focus of environmental policy. The distinction is obviously important, but it should come with a warning: in much of the literature in ecology and environmental ethics, the precise meaning of these terms is not fastidiously observed. For example, one often finds references to the "goals of conservation" even though the author clearly has preservationist principles in mind.

One of the most important activities of the preservationist movement has been to advocate strongly for protected wilderness areas like national parks. The phenomenon of national parks illustrates a kind of schizophrenia in us: are we preservationists or conservationists regarding these areas? This is probably not a particularly important confusion to resolve, however, as long as the parks function to protect ecosystem integrity and biodiversity. Ethics is, on one construal, about finding rationally acceptable principles to govern the way we *restrain* ourselves in our actions. And, as Monica E. Mulrennan has said, "parks are an important locus for the exercise of such restraint."[5] So we can get a better idea of how ecocentrism can be put into practice by looking at how we think of our national parks. In Canada, there are 43 national parks, the first of which was Banff and the most recent Ukkusiksalik in Nunavut. The long-term goal is for Canada to have 55 national parks. This sounds impressive, but the existing 43 parks represent just

3 per cent of Canada's land mass.[6] Some ecologists and environmentalists believe that in order to truly protect biodiversity, countries should aim to have 12 per cent of their territory legally protected from large-scale human interference. This figure is based on the Brundtland Report, a 1987 UN–commissioned document.[7] However, the national parks system *has* played a role in protecting endangered species. Wood Buffalo National Park, for instance, has been a refuge for whooping cranes, a species that, prior to the establishment of the park, had been on the verge of extinction.[8] Still, according to the UN, Canada needs to quadruple the land area under the legal protection afforded by the park system if it is to see results like this on a really meaningful scale.

The Canada National Parks Act, the law governing the maintenance of parks, makes explicit reference to the concept of ecological integrity. It defines integrity, somewhat confusingly, as "a condition . . . that includes abiotic components, . . . the composition and abundance of native species and biological communities, [and] rates of change and supporting processes." To promote integrity, Parks Canada has adopted a strategy of **ecosystem-based management** (EBM). Some of the principles of EBM are to maintain ecosystems in their natural conditions through practices such as prescribed burnings of forest areas; to recognize that target ecosystems are connected to contiguous ecosystems and that this will affect how one manages them; to maintain and restore the diversity of genes, species, and communities native to a particular region; and to limit human use of the target system.[9]

Let's focus for a moment on this last principle, the idea of restricting human use of an ecosystem. Presumably, one of the tests for how much human use of the park should be allowed is the degree to which that use upsets the integrity of the park's ecosystems. There are many examples of human activities interfering with the natural functioning of a park's ecosystem. In Banff National Park, for example, the predator–prey relation between wolves and elk has been thrown out of balance by large-scale human activities, in this case the tourism industry. Elk populations have soared in the past decade or so, mainly because wolves are much less tolerant of human presence than elk are. For the elk, we humans are like shields: with us around, the wolves will not come near them. And true to the nature of a whole of interdependent parts, the imbalance in one part of the system has effects elsewhere. Thus, the population of moose has declined dramatically as that of elk has risen. This is largely because elk are carriers of the liver fluke parasite, which is deadly to moose but not to elk. More elk means more of the parasite in the region, and this is bad news for moose.[10]

From the standpoint of the land ethic and the concept of ecological integrity buttressing the Parks Canada mission, this one example of a human-induced ecosystem imbalance is a clear moral failing on our part. Every time something like this happens, a duty to rectify the situation—to restore the ecosystem's previous balance—is placed on us. Again, this is not to moralize: it is only to hold Parks Canada to its word about the philosophical underpinnings of the national park system.

Ⓒ Going Deep

Taking the land ethic seriously entails, according to Leopold, changing "the role of *Homo sapiens* from conqueror of the land-community to plain member and citizen of it." At several points in his analysis, Leopold makes reference to the kind of alteration of affections, imagination, and judgment we need to undergo in order to live this way. As he puts it, it is "inconceivable" that "an ethical relation to land can exist without love, respect, and admiration for land, and a high regard for its value." These attitudes and beliefs are the true "foundations of conduct." We need a radical reorientation of our lives away from the economic and exploitative stance we currently have to the land. The "deep ecologists" have responded directly to this challenge, so let's turn to an analysis of their development of the basic principles of ecocentrism.

IDENTIFICATION AS A SOURCE OF DEEP ECOLOGICAL ATTITUDES

Arne Naess

I. Deep Ecology

In the 1960s, two convergent trends made headway: a deep ecological concern and a concern for saving deep cultural diversity. These may be put under the general heading "deep ecology" if we view human ecology as a genuine part of general ecology. For each species of living beings, there is a corresponding ecology. In what follows, I adopt this terminology, which I introduced in 1973. The term "deep" is supposed to suggest explication of fundamental presuppositions of valuation as well as of facts and hypotheses. Deep ecology, therefore, transcends the limit of any particular science of today, including systems theory and scientific ecology. Deepness of normative and descriptive premises questioned characterize the movement....

The shallow ecological argument carries today much heavier weight in political life than the deep. It is therefore often necessary for tactical reasons to hide our deeper attitudes and argue strictly homocentrically.... As an academic philosopher raised within analytic traditions, it has been natural for me to pose the questions: How can departments of philosophy, our establishment of professionals, be made interested in the matter? What are the philosophical problems explicitly and implicitly raised or answered in the deep eco-

logical movement? Can they be formulated so as to be of academic interest?

My answer is that the movement is rich in philosophical implications. There has, however, been only moderately eager response in philosophical institutions. The deep ecological movement is furthered by people and groups with much in common. Roughly speaking, what they have in common concerns ways of experiencing nature and diversity of cultures. Furthermore, many share priorities of lifestyle, such as those of "voluntary simplicity." They wish to live "lightly" in nature. There are of course differences, but until now the conflicts of philosophically relevant opinion and of recommended policies have, to a surprisingly small degree, disturbed the growth of the movement.

In what follows, I introduce some sections of a philosophy inspired by the deep ecological movement. Some people in the movement feel at home with that philosophy or at least approximately such a philosophy, others feel that they, at one or more points, clearly have different value priorities, attitudes, or opinions. To avoid unfruitful polemics, I call my philosophy "Ecosophy T," using the character T just to emphasize that other people in the movement would, if motivated to formulate their world view and general value priorities, arrive at different ecosophies: Ecosophy "A," "B," ... "Z."

By an "ecosophy" I here mean a philosophy inspired by the deep ecological movement. The ending -sophy stresses that what we modestly try to realize is wisdom rather than science or information. A philosophy, as articulated wisdom, has to be a synthesis of theory and practice. It must not shun concrete policy recommendations but has to base them on fundamental priorities of value and basic views concerning the development of our societies.

Which societies? The movement started in the richest industrial societies, and the words used by its academic supporters inevitably reflect the cultural provinciality of those societies. The way I am going to say things perhaps reflects a bias in favour of analytic philosophy intimately related to social science, including academic psychology. It shows itself in my acceptance in Ecosophy T of the theory of thinking in terms of "gestalts." But this provinciality and narrowness of training does not imply criticism of contributions in terms of trends or traditions of wisdom with which I am not at home, and it does not imply an underestimation of the immense value of what artists in many countries have contributed to the movement.

II. Self-Realization. But Which Self?

The Empirical Self of each of us is all that he is tempted to call by the name of me. But it is clear that between what a man calls me and what he simply calls mine the line is difficult to draw. We feel and act about certain things that are ours very much as we feel and act about ourselves. Our fame, our children, the work of our hands, may be as dear to us as our bodies are and arouse the same feelings and the same acts of reprisal if attacked. . . .

How do we develop a wider self? What kind of process makes it possible? One way of answering these questions: There is a process of ever-widening identification and ever-narrowing alienation that widens the self. The self is as comprehensive as the totality of our identifications. Or, more succinctly: Our Self is that with which we identify. The question then reads: How do we widen identifications?

Identification is a spontaneous, non-rational, but not irrational, process through which the interest or interests of another being are reacted to as our own interest or interests. The emotional tone of gratification or frustration is a consequence carried over from the other to oneself: joy elicits joy, sorrow, sorrow. Intense identification obliterates the experience of a distinction between ego and alter, between me and the sufferer. But only momentarily or intermittently: If my fellow being tries to vomit, I do not, or at least not persistently, try to vomit. I recognize that we are different individuals. . . .

A high level of identification does not eliminate conflicts of interest: Our vital interests, if we are not plants, imply killing at least some other living beings. A culture of hunters, where identification with hunted animals reaches a remarkably high level, does not prohibit killing for food. But a great variety of ceremonies and rituals have the function to express the gravity of the alienating incident and restore the identification. Identification with individuals, species, ecosystems, and landscapes results in difficult problems of priority. What should be the relation of ecosystem ethics to other parts of general ethics?

There are no definite limits to the broadness and intensity of identification. Mammals and birds sometimes show remarkable, often rather touching, intra-species and cross-species identification. Konrad Lorenz tells of how one of his bird friends tried to seduce him, trying to push him into its little home. This presupposes a deep identification between bird and man (but also an alarming mistake of size). In certain forms of mysticism, there is an experience of identification with every life form, using this term in a wide sense. Within the deep ecological movement, poetical and philosophical expressions of such experiences are not uncommon. In the shallow ecological movement, intense and wide identification is described and explained psychologically. In the deep movement, this philosophy is at least taken seriously: reality consists of wholes that we cut down rather than of isolated items that we put together. In other words: there is not, strictly speaking, a primordial causal process of identification but one of largely unconscious alienation, which is overcome in experiences of identity. To some "environmental" philosophers, such thoughts seem to be irrational, even "rubbish." This is, as far as I can judge, due to a too narrow conception of irrationality.

The opposite of identification is alienation, if we use these ambiguous terms in one of their basic meanings. The alienated son does perhaps what is required of a son toward his parents, but as performance of moral duties and as a burden, not spontaneously, out of joy. If one loves and respects oneself, identification will be positive, and, in what follows, the term covers this case. Self-hatred or dislike of certain of one's traits induces hatred and dislike of the beings with which one identifies.

Identification is not limited to beings that can reciprocate: Any animal, plant, mountain, ocean may induce such processes. In poetry this is articulated most impressively, but ordinary language testifies to its power as a universal human trait.

Through identification, higher-level unity is experienced: from identifying with "one's nearest," higher unities are created through circles of friends, local communities, tribes, compatriots, races, humanity, life, and, ultimately, as articulated by religious and philosophic leaders, unity with the supreme whole, the "world" in a broader and deeper sense than the usual.

I prefer a terminology such that the largest units are not said to comprise life and "the not living." One may broaden the sense of "living" so that any natural whole, however large, is a living whole.

This way of thinking and feeling at its maximum corresponds to that of the enlightened, or yogi, who sees "the same," the atman, and who is not alienated from anything. The process of identification is sometimes expressed in terms of loss of self and gain of Self through "self-less" action. Each new sort of identification corresponds to a widening of the self and strengthens the urge to further widening, furthering Self-seeking. This urge is in the system of Spinoza called *conatus in suo esse perseverare*, striving to persevere in oneself or one's being (*in se, in suo esse*). It is not a mere urge to survive but to increase the level of acting out (*ex*) one's own nature or essence and is not different from the urge toward higher levels of "freedom" (*libertas*). Under favourable circumstances, this involves wide identification.

III. Wideness and Depth of Identification as a Consequence of Increased Maturity

Against the belief in fundamental ego–alter conflict, the psychology and philosophy of the (comprehensive) Self insist that the gradual maturing of a person inevitably widens and deepens the self through the process of identification. There is no need for altruism toward those with whom we identify. The pursuit of self-realization conceived as actualization and development of the Self takes care of what altruism is supposed to accomplish. Thus, the distinction egoism–altruism is transcended.

The notion of maturing has to do with getting out what is latent in the nature of a being. Some learning is presupposed, but thinking of present conditions of competition in industrial, economic-growth societies, specialized learning may inhibit the process of maturing. A competitive cult of talents does not favour Self-realization. As a consequence of the imperfect conditions for maturing as persons, there is much pessimism or disbelief in relation to the widening of the Self and more stress on developing altruism and moral pressure.

The conditions under which the self is widened are experienced as positive and are basically joyful. The constant exposure to life in the poorest countries through television and other media contributes to the spread of the voluntary simplicity movement. But people laugh: What does it help the hungry that you renounce the luxuries of your own country? But identification makes the efforts of simplicity joyful, and there is not a feeling of moral compulsion. The widening of the self implies widening perspectives, deepening experiences, and reaching higher levels of activeness (in Spinoza's sense, not as just being busy).

Joy and activeness make the appeal to Self-realization stronger than appeal to altruism. The state of alienation is not joyful and is often connected with feelings of being threatened and narrowed. The "rights" of other living beings are felt to threaten our "own" interests.

The close connection between trends of alienation and putting duty and altruism as a highest value is exemplified in the philosophy of Kant. Acting morally, we should not abstain from maltreating animals because of their sufferings but because of its bad effect on us. Animals were to Kant, essentially, so different from human beings that he felt we should not have any moral obligations toward them. Their unnecessary sufferings are morally indifferent, and norms of altruism do not apply in our relations to them. When we decide ethically to be kind to them, it should be because of the favourable effect of kindness on us—a strange doctrine.

Suffering is perhaps the most potent source of identification. Only special social conditions are able to make people inhibit their normal spontaneous reaction toward suffering. If we alleviate suffering because of a spontaneous urge to do so, Kant would be willing to call the act "beautiful" but not moral. And his greatest admiration was, as we all know, for stars and the moral imperative, not spontaneous goodness. The history of cruelty inflicted in the name of morals has convinced me that increase of identification might achieve what moralizing cannot: beautiful actions. Natural environments offered a much needed rational and economic justification for processes of identification that many people already had more or less completed. Their relative high degree of identification with animals, plants, landscapes were seen to correspond to factual relations between themselves and nature. "Not man apart" was transformed from a romantic norm to a statement of fact. The distinction between man and environment, as applied within the shallow ecological movement, was seen to be illusory. Your Self crosses the boundaries.

When it was made known that the penguins of the Antarctic might die out because of the effects of DDT upon the toughness of their eggs, there was a widespread, spontaneous reaction of indignation and sorrow. People who never see penguins and who would never think of such animals as "useful" in any way insisted that they had a right to live and flourish and that it was our obligation not to interfere. But we must admit that even the mere appearance of penguins makes intense identification easy.

Thus, ecology helped many to know more about themselves. We are living beings. Penguins are too. We are all expressions of life. The fateful dependencies and interrelations that were brought to light, thanks to ecologists, made it easier for people to admit and even to cultivate their deep concern for nature and to express their latent hostility toward the excesses of the economic growth of societies.

IV. Living Beings Have Intrinsic Value and a Right to Live and Flourish

This perhaps rather lengthy philosophical discourse serves as a preliminary for the understanding of two things: first, the powerful indignation of Rachel Carson and others who, with great courage and stubborn determination, challenged authorities in the early 1960s and triggered the international ecological movement. Second, the radical shift . . . toward more positive appreciation of non-industrial cultures and minorities—also in the 1960s and expressing itself in efforts to "save" such cultures and in a new social anthropology.

The second movement reflects identification with threatened cultures. Both reactions were made possible by doubt that the industrial societies are as uniquely progressive as they usually had been supposed to be. Former haughtiness gave way to humility or at least willingness to look for deep changes both socially and in relation to nature. . . . How can these attitudes be talked about? What are the most helpful conceptualizations and slogans?

One important attitude might be thus expressed: "Every living being has a right to live." One way of answering the question is to insist upon the value in themselves, the auto-telic value, of every living being. This opposes the notion that one may be justified in treating any living being as just a means to an end. It also generalizes the rightly famous dictum of Kant, "Never use a person solely as a means." Identification tells me: if I have a right to live, you have the same right. . . .

Deep ecology is an attempt to emphasize or perhaps radicalize themes latent in ecocentrism. One of the most important of these themes has to do with how and why we ought to conceive of ourselves as fully natural beings. Here, as elsewhere, deep ecologists are unapologetically metaphysical. Whereas we have seen Leopold talk about the affective foundations of sound environmental conduct, the deep ecologists want to talk about the very nature of the environmentally aware self. The goal of developing environmental awareness is self-realization, but this is not to be interpreted in a narrowly egoistic way, because self-realization cannot be achieved without also bringing about a profound *identification* with the Other, in this case wild nature:

> Identification is a spontaneous, non-rational, but not irrational, process through which the interest or interests of another being are reacted to as our own interest or interests. . . . Identification is not limited to beings which can reciprocate: Any animal, plant, mountain, ocean may induce such processes.

In the *Almanac*, Leopold has an essay called "Thinking Like a Mountain," a title that nicely captures the spirit of this call to identify with aspects of inanimate nature. In spite of the title's reference to "thinking," however, the attitude Leopold and Naess are recommending here can be understood better as a species of *love*.

What does it mean to love nature? It is difficult to be precise about what is involved in a process that is avowedly non-rational, but we can get our bearings here by looking at what many philosophers think is the essence of love between persons. For example, according to Descartes, identification with the interests of the beloved is at the heart of this emotion. When we love someone or something, we

> consider ourselves henceforth as joined with what we love in such a manner that we imagine a whole, of which we take ourselves to be only one part, and the thing loved to be the other.[11]

This is, in some ways, a fairly standard way of thinking of love, and it has a great deal of intuitive appeal. If we like it, we can think of deep ecology as posing the following challenge. If we are able to accept that this kind of radical identification with the Other is possible in interpersonal love, why should we not also accept the possibility of a richer and wider understanding of the Other such that it encompasses all of wild nature? If we can accept this, then wild nature is a possible object of love. The really hard part, it might be suggested, is learning to love *at all*. Once we have made the first step beyond the narrowly confined ego, the sky, quite literally, is the limit.

For Naess, expanding our vision this way is part of a process of increasing "maturity." If we can achieve it, we will be that much closer to endorsing his key principles. In particular, we will understand that wild nature does not belong to us in the crude economic sense, that it is "worth defending, whatever the fate of humans," and that it has value independently of human interests. And if we accept these principles, we will not only be committed to the idea that ecosystems as such have intrinsic value, we should also be prepared to defend wild nature wherever it is threatened by exploitative human activity. The connection between deep ecology and radical environmentalism—with its often confrontational stance toward policies and practices that are judged to degrade the environment—is not accidental. Imagine, by way of comparison, the actions you would be willing to take to protect a person you loved from harm.

Ⓓ Three Objections and Responses

Ecocentrism and deep ecology have encountered some fairly pointed criticism over the years. We will concentrate on three broad avenues of critical attack and see whether there is a way for defenders of these views to respond.

The first objection focuses on the metaphysical status of ecosystems and the parts that comprise them. The trouble arises with the ecocentrist claim that the whole—the ecosystem itself—is *more real* than its constituent parts. Consider the following claim, which expresses a view explicitly endorsed by many ecocentrists and deep ecologists:

> Viewed from the standpoint of modern [ecology] each living thing is a dissipative structure, that is, it does not endure in and of itself but only as a result of the continual flow of energy in the system. . . . From this point of view the reality of individuals is problematic because they do not exist *per se* but only as local perturbations in this universal energy flow. . . . Consider a vortex in a stream of flowing water. The vortex is a structure made of an ever-changing group of water

molecules. It does not exist as an entity in the classic Western sense; it exists only because of the flow of water through the stream. . . . In the same sense the structures out of which the biological entities are made are transient, unstable entities with constantly changing molecules dependent on a constant flow of energy to maintain form and structure.[12]

It is uncontroversial to claim that some things depend for their existence on other things. But to assert that dependent things are as such *less real* than the things on which they depend is surely false. The problem here is that it looks difficult to explain clearly the notion that reality comes in degrees, such that some things have more of it than others. This idea has its roots in Plato's metaphysics according to which the Forms of things are more real than their instantiations in this world. But there are very few philosophers now who accept this aspect of Plato's philosophy, so it is problematic to label it as the "classic Western" understanding of the concept of reality and then to rely on this understanding to argue that the parts of ecosystems are less real than those systems themselves.

If we reject the theory of Forms, then it would seem that *everything* is dependent to some degree on other things for its existence: individual plants and animals on specific features of their habitats, those habitats themselves on regional climatic patterns, climate on larger geophysical and geochemical processes, these processes on still larger geological and even interplanetary forces, and so on. If this picture—suitably filled out—is right, then, according to the ecocentrist, none of the stuff comprising this totality exists per se. That is surely an unwelcome result, but if we resist it, we should reject the appeal to degrees of reality that generates it. In any case, it is not obvious that ecocentrism and deep ecology require this metaphysics. The claim that ecosystems ought to be of primary moral concern does not need the support of Platonic metaphysics.

The second objection to ecocentrism and deep ecology has to do with the way ecosystems are alleged to work. We have seen that the defining feature of an ecosystem is its integrity, its internal order. Some ecologists and their philosophical followers have challenged this view of ecosystems. Hettinger and Throop say this:

> The more radical proponents of what we call the "ecology of instability" argue that disturbance is the norm for many ecosystems and that natural ecosystems do not tend towards mature, stable, integrated states. On a broad scale, climatic changes show little pattern, and they ensure that over the long term natural systems remain in flux. On a smaller scale, fires, storms, droughts, shifts in the chemical compositions of soils, chance invasions of new species, and a wealth of other factors continually alter the features of natural systems in ways that do not create repeating patterns of return to the same equilibrium states.[13]

The key to the notion of integrity is that an ecosystem can reach a state—a "mature state"—in which its internal components are fully integrated and that, like a biological organism, it will then tend to both resist external pressures to upset this balance and seek to return to this balance after having undergone some fragmentation or disturbance. Proponents of the ecology of instability, by contrast, claim that these notions of integrity and stability are, at bottom, human projections onto a reality that does not have such attributes. For example, it has been shown that most ecosystems are subject to large fluctuations over time in the populations of particular species. Some ecologists have gone so far as to claim that interacting populations of species display the characteristics of a technically *chaotic* system.[14]

Ecosystems are not comprised of perfectly harmonized elements whose mode of mutual interaction is fixed once and for all. Since they are, just like the living things that make them up, powered by an external energy source, the flow of energy and nutrients through them varies over time, and this "encourages the evolution of diversity" in them. The reason is that the constantly changing conditions do not allow one species to wipe out the competition: when a species is on the cusp of doing this, conditions will change again, depriving it of the advantage it had previously enjoyed over its competitors.[15] So we should not apply concepts like "equilibrium" and "harmony" to ecosystems to the extent that such terms imply *stasis*.

Nevertheless, it is incorrect to speak of such systems as strictly chaotic. Chaotic systems are, by definition, easily disturbed and changed into something different (indeed, to speak of them as "systems" at all is somewhat forced). But there are degrees to which ecosystems are able to resist change or perturbation—from fire, invasive species or pests, and climate change, for example—and to recover from such things when they are damaged by them. **Inertia** describes the ability of an ecosystem to withstand change, whereas **resilience** refers to its ability to recover after change. Ecosystems with a high degree of inertia and resilience are *relatively stable* structures. To capture the idea that they are neither fully chaotic nor in full equilibrium, many ecologists use the term **dynamic equilibrium** to describe ecosystems.[16] If we understand ecosystem integrity this way, it would appear that we can preserve the concept.

Indeed, Leopold himself understood this. We have seen that he recognized that slow change is integral to ecosystems. What he opposed, in the name of the land ethic, were human-induced changes that were happening with "unprecedented violence, rapidity, and scope." Anthropogenic climate change, for example, can reduce a forest's ability to control pests that eat away at its trees, thus reducing the forest's resilience to fires. Such changes clearly threaten ecosystem integrity and stability. No finding in ecology forces us to abandon these practically useful notions.

The third and final objection to ecocentrism and deep ecology was originally made by Regan. Regan calls the land ethic a kind of "environmental fascism."[17] The idea is that by identifying groups, wholes, or systems as the primary objects of moral concern, ecocentrism and deep ecology allow us to treat individuals—especially human and non-human animals—in ways that we might think are unfair or even vicious on independent grounds. For example, we sometimes cull individual members of a species—in a national park, say—in order to benefit the species as a whole. But assuming that the human population is too high, how, asks Callicott, "can we consistently exempt ourselves from a similar draconian regime?"[18] Hence the charge of "fascism." The land ethic, at least as presented by Leopold himself, does not appear to provide a satisfying response to this criticism, a failing that would seem to doom the theory. However, Callicott does not think the land ethic should be saddled with the charge. As he sees it, the land ethic asks us to expand our ethical awareness outward so that it encompasses the interests of more and more entities. But as we move outwards, we do not simply abandon the moral commitments made at earlier stages.

Blow up a balloon, and you will notice that as it expands, it leaves no trace of its previous dimensions. All you have is the newly achieved outer limits of the balloon. That, says Callicott, is the *wrong* image to capture the movement of ethical consciousness we are contemplating. It is better to think of it as similar to the outward expansion of a tree as it grows. The tree, unlike the balloon, leaves traces of its outward journey in the form of rings emanating from the centre. If we believe there are good reasons to think of individual animals and plants—to say nothing of humans—as bearers of intrinsic value, then ecocentrism and deep ecology ask us only to *add* another set of entities to this class, not to replace the older members of the class with the newer ones. Having done so, what happens when, say, the interest of an individual animal comes into conflict with that of an ecosystem?

> As a general rule, the duties correlative to the inner social circles to which we belong eclipse those correlative to the rings farther from the heartwood.[19]

The obvious problem with this is that it provides an answer to a very pressing worry—the charge of environmental fascism—only by stripping the theory of its capacity to serve as a *radical* critique of environmental anthropocentrism. If this is indeed the general rule by which we ought to govern ourselves in our interactions with wild nature, then human interests will very often override the interests of ecosystems. This is why some, especially among the deep ecologists, see Callicott's resolution of the problem as betraying the promise of the land ethic. For them, it is better to accept the charge of environmental fascism than to water the theory down in this fashion.

E Conclusion

Chapters 1–3 provide an analysis of moral standing as applied to environmental ethics, an analysis that reaches its end point with ecocentrism and deep ecology. We have moved from the idea that only sentient beings and their interests count intrinsically to the idea that the interests of ecosystems do as well. It might be thought that in the course of this philosophical journey we have stretched the concept of moral standing to the breaking point. This worry can show up in at least two related ways. The first we have just seen with Callicott's response to Regan's claim that ecocentrism is fascistic. The more entities, or kinds of entities, we grant moral standing to, the more intractable is the problem of resolving conflicts among those entities. It is easy to profess a principled egalitarianism here and say that all beings with moral standing have it to an equal degree. But when interests do conflict, there is a tendency to pull back from this result and assert that some entities—think of Callicott's innermost circles—are simply more considerable, morally speaking, than others. This need not leave everything just as it was before, because it is doubtless a genuine advance to say that trees or ecosystems possess more than merely extrinsic value. Anthropocentrism and ecocentric egalitarianism are surely not the only options here.

But positions occupying a middle ground between these two extremes are nevertheless peculiarly unstable. When confronted with the fact of moral conflict, they will *tend* to collapse to one extreme or the other. Employing the distinction between basic (or vital) needs and non-basic (or luxury) needs is one way to meet this challenge and give everything with moral standing its due. So when my non-basic need to wash my hair with shampoo that has been tested on rabbits conflicts with the basic needs of those animals not to be tortured, I lose. Whether or not we can resolve all environmental conflicts this way—those between the interests of individuals and wholes, humans and non-humans, plants and animals, and so on—is an open question. But again, it is a question made more difficult the wider we cast the net of moral considerability.

The second worry about moral standing has to do specifically with ecocentrism and deep ecology. Chapters 1 and 2 make it clear that something has moral standing only if it has interests. There is a meaningful sense in which individual animals and plants have interests. The former, many of them anyway, are sentient and therefore capable of suffering; the latter are, in Taylor's phrase, "teleological centres of life." But can we talk meaningfully of the interests of ecosystems? It seems that in order to establish this claim, we have to push the analogy between ecosystems and organisms very far. James Lovelock thinks we should do this. Lovelock refers to the biosphere by the name of the ancient Greek goddess, Gaia, and calls his view **Gaia Theory**. He thinks that just like a living organism, both individual ecosystems and the collection of interconnected ecosystems—the biosphere—are self-regulating, organized wholes. Both are *literally* alive.

Lovelock would have us think of the biosphere as similar to a cell. In a cell, the membrane is the boundary between the living thing and what lies outside of it. Inside the membrane, energy is circulated and stored in the process of resisting entropic breakdown. Now think of the upper atmosphere as the membrane of the biosphere. Inside this membrane, the energy of the sun is circulated through and stored in the super-organism's various subsystems, a process that helps the biosphere as a whole resist entropic disorder. It does this by regulating Earth's climate and geochemistry in a way that allows for the continuance of life.

If this is right and it makes sense to speak of ecosystems and Gaia this way, then there is a sense in which these systems, just like trees, are teleologically ordered things. However, if we cannot make full sense of this idea, it would seem that we thereby forfeit the ability to say that ecosystems and Gaia have interests. And then, of course, it is difficult to see how we can still talk of such things having moral standing.

Tallgrass Prairie as an "Endangered Space"

Source: Living Prairie Museum

Dropseed at the Live Tallgrass Prairie Museum in Winnipeg, MB. Should endangered spaces like this be expanded?

Although he was American, Aldo Leopold was well aware of the environmental challenges facing Canada. In 1949, he had this to say about us:

> In Canada . . . a representative series of wilderness areas can and should be kept. . . . It will be contended of course that no deliberate planning to this end is necessary; that adequate areas will survive anyhow. All recent history belies so comforting an assumption. . . . To what extent Canadians . . . will be able to see and grasp their opportunities is anybody's guess.

Leopold is claiming that the laissez-faire approach that has so far dominated our thinking about wilderness must be abandoned in favour of much more aggressive preservationist policies. What "recent history" might Leopold be referring to here? The text is unclear on this question, but we can make some plausible suggestions. For instance, we could point to the transformation of the tallgrass prairie into what is now prime farmland in southern Manitoba and Saskatchewan.

In a collection of studies of Canada's wild spaces, *Endangered Spaces*, Monte Hummel urges us to adopt an unflinchingly ecocentric approach to the wild. We should diminish the focus on endangered species and focus primarily, as the book's title indicates, on endangered *spaces*, whole systems that have become, or are in danger of becoming, ecologically degraded. The Canadian tallgrass prairies were once one of the most productive places on Earth. Early European visitors to the area marvelled at the abundance of life there: raspberries, strawberries, currants, long grasses, cranberries, oak trees, pronghorn antelope, prairie wolves, grizzly bears, deer, and, of course, bison. Bison had been hunted for a long time here, but after about 1830, the hunt became a slaughter, reaching up to 300,000 animals a year. In just 40 years, by 1870, the plains bison, which had previously reached a population of nearly 60 million, were extinct in Manitoba. This devastation reverberated throughout the tallgrass prairie ecosystem. Populations of the bison's traditional predators, the wolves and grizzlies, dropped dramatically. With the loss of the bison, elk and deer were then over-hunted, and their numbers crashed too.

In the twentieth century, the ecosystem was further degraded with the development of ranching, agriculture, and mining. These developments led to the virtual elimination of the native tallgrass habitat, which is home to many of the original species. Now less than 1 per cent of these plant species remains. To appease hunters, exotic species like wild turkey, pheasant, and grey partridge were introduced. Wetlands were drained to make room for more agriculture. Fires, which maintain the prairie ecosystem by recycling nutrients and eliminating dead biomass, were suppressed. According to David Gauthier and J. David Henry, human activities have completely transformed prairie ecology:

> The original stable condition of soil and grass—maintained by large grazers, fire and periodic drought—[has] given way to an intensively managed, *inherently unstable system*. . . . [T]he Canadian prairies have become one of the most endangered natural habitats in Canada. . . . Biological diversity has decreased. . . . Organic matter and biomass have reduced by 50 percent since the land was broken, a fact masked by the current heavy use of commercial fertilizers. [emphasis added]

Of course, what has happened on the prairies is that all that natural productivity has been harnessed to commercial ends. In place of the original biological diversity, just four strains of wheat account for 75 per cent of the entire wheat crop, and that crop dominates the landscape. And the region is in danger of losing even more species. Currently, 25 per cent of the bird and mammal species listed as endangered in Canada inhabit the Prairie provinces.

As Leopold reminds us, we cannot afford to simply sit back and allow ecosystems like this to perish. Gauthier and Henry are unequivocal: "The terrestrial and wetland ecosystems of the prairies are indeed endangered spaces." Deep ecologists might say that recognition of this fact demands swift political action. The plea should be explicitly ecocentric: "We must manage the remaining resources and surviving productivity of the prairies as *functional ecological entities*." Among other things, this entails demanding that governments (a) "establish a system of protected native prairie ecosystems"; (b) develop restoration plans for damaged or threatened ecosystems; and (c) conduct exhaustive inventories of remaining native prairie

Continued

lands. The ultimate goal is to maintain what biological diversity remains in the prairies and to restore as much of what has been lost as we can. Doing this could be one way of showing, in Leopold's phrase, that we think of ourselves not as "conqueror of the land community" but as "plain member and citizen of it."

Sources: David A. Gauthier and J. David Henry, "Misunderstanding the Prairies," in Hummel 1989, 183–95; Government of Canada Committee on the Status of Wildlife in Canada, www.cosewic.gc.ca; Leopold 1949.

Study Questions

1. What are the main differences between ecocentrism and deep ecology?
2. Describe and expand on Leopold's notion of the "land pyramid." Do you think this is a helpful metaphor for understanding how nature works?
3. Do we know enough about what it means to preserve the "integrity, stability and beauty" of ecosystems? In particular, should beauty be on this list? What exactly is ecosystemic beauty? What sacrifices would we have to make to put the ideal of the land ethic into practice?
4. Should Canada expand its National Park system? How large should it be and why?

Further Reading

Murray Bookchin. 1988. "Social Ecology versus Deep Ecology." *Socialist Review* 18, no. 3: 11–29.

J. Baird Callicott. 1987. *Companion to a Sand County Almanac.* Madison: U of Wisconsin P.

Bill Devall and George Sessions. 1985. *Deep Ecology: Living as If Nature Mattered.* Salt Lake City: Peregrine Smith Books.

D.W. Lauer. 2002. "Arne Naess on Deep Ecology." *Journal of Value Inquiry* 36: 109–15.

Andrew Light. 2000. "Ecological Restoration and the Cult of Nature." In *Restoring Nature: Perspectives from the Social Sciences and Humanities.* Ed. P. Gobster and B. Hall. Washington: Island Press.

Part 

Challenges and New Directions

Part II, Challenges and New Directions, looks at some ways in which the manner of framing the issues in environmental ethics—as laid out in Part I—has been or is being upset. But the views expressed also represent new ways of thinking about our relationship to the natural world, so they are not aimed solely at the work of previous philosophical ethicists. The chapters present six distinct challenges, each of which picks up on and disputes a core element of the traditional approach. The first (Chapter 4) comes from the standpoint of neo-liberal economics. It is appropriately placed first because it challenges the first and most important move of environmental ethics: its *rejection* of the idea that only human interests count or are even real. For the economist, human interests are the only game in town. The second challenge comes from the environmental pragmatist (Chapter 5), who claims that the theoretical dispute among philosophers about who or what has moral standing is a dangerous distraction. What we need is environmental action, and we need it before we come to a resolution on theoretical matters. The third challenge comes from ecofeminists (Chapter 6), who argue that our ecological crisis is the product of a patriarchal culture that conceives of women and nature as passive objects, existing to be dominated and controlled by men. Only when we overturn this mode of thinking can we reorient ourselves to nature in a sustainable manner. The next challenge comes from environmental aesthetics (Chapter 7). It takes aim at our tendency to neglect the role that judgments of natural beauty play in determining how we ought to treat the environment. The environmental views of First Nations peoples in Canada are examined next (Chapter 8). Here, an attempt is made to discover what ecological wisdom these cultures contain, without romanticizing or idealizing them. In the next chapter (Chapter 9), we look at the challenge to the dominant consequentialist and deontological approaches in traditional environmental ethics posed by the new sub-field of environmental virtue ethics. Finally, we look at the school of thought known as "social ecology" founded by Murray Bookchin (Chapter 10). As we will see, the ideas that fit under this heading draw together many of the concepts discussed elsewhere in this part of the book.

CHAPTER | 4

Economics and Ecology

Ecological Intuition Pump

Most of us value friendship, meaningful work, material prosperity, recreation, our specific place in a tradition of cultural practices, beautiful things, wild nature, and much more. Our tendency is to think of such things as being valuable in distinct ways. But suppose an economist-philosopher, call him Friedtotle, declared one day that he had isolated the very *essence* of valuable things! He decided to call this valuable thing *manna*, and he developed an instrument that could detect the precise quantity of manna in all the things we value. We now realized that it was not friendship or wild nature per se that we valued but only the manna in them. The discovery was revolutionary because it meant that we could rearrange our lives so that they would contain as much manna as possible. If we found, for example, that we could not obtain very much wild nature manna, it did not matter because as it happened there was lots of material prosperity manna to be had. All we needed to do was add more of *this* stuff to our lives. In fact, this stuff seemed relatively abundant, so many people decided to trade away their other manna-containing things for increased quantities of it. How do you think we should view Friedtotle's manna? In spite of its apparent absurdity, why are we so seduced by the idea that all value is ultimately the same or can be expressed in the same way? Are there many valuable things or just one?

Ⓐ Introduction

Canadian environmentalist David Suzuki has a story about a group of environmental activists standing in an ancient forest and confronting some loggers. They were met by the CEO of the company that had been given a licence to clear-cut the site. Pointing to a giant tree, the CEO said to the environmentalists:

> You see that tree over there? It doesn't have any value until it's cut down. Unless, of course, you tree huggers decide you'll pay money to save it so that you can enjoy it. Think your cronies can raise enough money to save the entire forest?[1]

The CEO is expressing, albeit somewhat crudely, a common view about how we ought to value nature. It is the standpoint of **neo-liberal economics**, according to which there should be as few constraints as possible on the operations of the free market. Associated with this view is that idea that the Earth and its natural capital are resources that should be exploited in the service of an ever-growing economy. The reason we need to consider a view like this is that it constitutes the most radical challenge to environmental ethics. Recall the movement of ideas laid out in Chapters 1–3. We began with a consideration of the claim that non-human animals ought to have moral standing and ended with the suggestion that so too should eco-systems and possibly even the biosphere—Gaia—as a whole. That is a very radical suggestion, but the neo-liberal economic approach to environmental ethics rejects the very starting point of this exercise. That starting point is premised on the rejection of anthropocentrism, the view that only human interests have intrinsic value. Neo-liberal economics is *unapologetically anthropocentric*. So unless we understand what its central claims are and perhaps find some way of undermining its strict anthropocentrism, the project of environmental ethics may not get off the ground at all.

At its most basic, economics is the study of the many ways we appropriate scarce resources to satisfy our wants and thus manage our "homes" ("economy" derives from the Greek word for house: *oikos*). Beyond this basic description, we will see that there is more than one way to understand the economic relation between humans and the natural goods and services they appropriate and use. For the moment, however, it will suffice to draw out two key ideas from the CEO's statement. First, natural entities like trees have value *only* as economic resources. This means that such entities are literally valueless except insofar as they can be bought and sold in a market. Their value is whatever price they fetch there. Second, this does not mean that those who want to preserve nature have no effective way of expressing this desire. It's just that the only way to do so is to show that you are willing to pay for preservation. The economic mode of valuation is the only game in town, but, so goes the claim, it provides a level playing field for all would-be "users" of natural resources. This is a crucial claim. Economic anthropocentrism is not intrinsically opposed to the protection of the environment. From this standpoint, it is possible that we enact the most stringent environmental policies imaginable *if this is what people value*. It's just that there is no standpoint outside of *our* interests to which we can appeal in arguing for such protection.

These are commonplace ideas in a culture, like ours, where economic values have a certain pride of place in public discourse. As we will see, they are both open to dispute. But before we look into why this is the case, it is important to notice what value underpins them: that of *efficiency*. It is a brute fact of life that we need to appropriate goods and services from the nat-ural world in order to survive but that the natural world is stingy in supplying such things. So we had better not go about our business in a wasteful manner. And, so goes the tale, the most efficient mechanism ever devised for appropriating and allocating nature's goods and services is the system of pricing that defines a free market. What exactly do we mean by efficiency here? Economists have a somewhat unusual understanding of this term, one based on the work of Italian economist Vilfredo Pareto (1848–1923). Pareto's key innovation in economic theory is to define the efficiency of a particular state of affairs in such a way that everyone whose interests are involved in that state of affairs gets taken into account. More technically, the claim is that a state of affairs is the best we can achieve—i.e., the most efficient—if there is at least one person who prefers it to any (feasible) alternative and nobody who prefers the alternative to it. The people have in this case achieved **Pareto Optimality**. Nor do we need to achieve optimality thus construed in order to compare two possible proposals. Suppose there are two states of affairs, A and B. Neither is Pareto Optimal, but A is closer to it than B. A is therefore Pareto Superior to B, and B is Pareto Inferior to A.

Think of what goes on in a simple trade between two people. Suppose I own a vintage 1972 Paul Henderson Team Canada hockey sweater but I'm desperately short on cash. You, on the other hand, are flush with cash but struggling to find materials for your burgeoning collection of hockey memorabilia. Somehow we find each other, and I sell you the Henderson sweater for $300. For me, the $300 is more valuable than the sweater, while for you the reverse is true. So from the standpoint of Pareto efficiency, we have moved from a state of affairs that was inferior—I with surplus goods relative to my bundle of wants, you with surplus cash relative to yours—to one that is superior. At least one of us prefers the post-trade state of affairs to the pre-trade state of affairs (in fact we both do), and (obviously) neither of us prefers the latter to the former. Assuming there is no further desirable trade to be made here, we have reached a state of Pareto Optimality.[2] As far as the spread of consumption goods across a population is concerned, we have achieved what many consider a kind of distributional nirvana: full *efficiency*.

Ⓑ Optimal Thinking

There is something undeniably elegant at work in this example. As long as I owned the sweater and you obtained your money legally, nobody can complain about what has transpired between us. We have reduced waste in the allocation of our respective goods. But if we are pleased with the result at the level of a simple two-person trade, why should we not generalize the model? That is, if we are concerned that nobody's interests are undermined and nothing is being wasted in our exchanges, should we not allocate *all* goods and services through free trade? Or, since there may be some goods—like military protection—that are best provided by governments rather than the market, that as much of such allocation as possible should be accomplished through market mechanisms? If we answer either of these questions in the affirmative, we are committed to the view that most, and possibly all, goods and services— including environmental ones—have a price. For a vigorous defence of this view, let's turn to our first selection of the chapter.

PEOPLE OR PENGUINS: THE CASE FOR OPTIMAL POLLUTION

William F. Baxter

I start with the modest proposition that in dealing with pollution, or indeed with any problem, it is helpful to know what one is attempting to accomplish. Agreement on how and whether to pursue a particular objective, such as pollution control, is not possible unless some more general objective has been identified and stated with reasonable precision. We talk loosely of having clean air and clean water, of preserving our wilderness areas, and so forth. But none of these is a sufficiently general objective: each is more accurately viewed as a means rather than as an end.

With regard to clean air, for example, one may ask, "How clean?" and "What does clean mean?" It is even reasonable to ask, "Why have clean air?" Each of these questions is an implicit demand that a more general community goal be stated—a goal sufficiently general in its scope and enjoying sufficiently general assent among the community of actors that such "why" questions no longer seem admissible with respect to that goal.

If, for example, one states as a goal the proposition that "every person should be free to do whatever he wishes in contexts where his actions do not interfere with the interests of other human beings," the speaker is unlikely to be met with a response of "why." The goal may be criticized as uncertain in its implications or difficult to implement, but it is so basic a tenet of our civilization—it reflects a cultural value so broadly shared, at least in the abstract—that the question "why" is seen as impertinent or imponderable or both.

I do not mean to suggest that everyone would agree with the "spheres of freedom" objective just stated. Still less do I mean to suggest that a society could subscribe to four or five such general objectives that would be adequate in their coverage to serve as testing criteria by which all other disagreements might be measured. One difficulty in the attempt to construct such a list is that each new goal added will conflict, in certain applications, with each prior goal listed, and thus each goal serves as a limited qualification on prior goals.

Without any expectation of obtaining unanimous consent to them, let me set forth four goals that I generally use as ultimate testing criteria in attempting to frame solutions to problems of human organization. My position regarding pollution stems from these four criteria. If the criteria appeal to you and any part of what appears hereafter does not, our disagreement will have a helpful focus: which of us is correct, analytically, in supposing that his position on pollution would better serve these general goals. If the criteria do not seem acceptable to you, then it is to be expected that our more particular judgments will differ, and the task will then be yours to identify the basic set of criteria upon which your particular judgments rest. My criteria are as follows:

1. The spheres of freedom criterion stated above.
2. Waste is a bad thing. The dominant feature of human existence is scarcity—our available resources, our aggregate labours, and our skill in employing both have always been, and will continue for some time to be, inadequate to yield to every man all the tangible and intangible satisfactions he would like to have. Hence, none of those resources, or labours, or skills, should be wasted—that is, employed so as to yield less than they might yield in human satisfactions.
3. Every human being should be regarded as an end rather than as a means to be used for the betterment of another. Each should be afforded dignity and regarded as having an absolute claim to an even-handed application of such rules as the community may adopt for its governance.
4. Both the incentive and the opportunity to improve his share of satisfactions

should be preserved to every individual. Preservation of incentive is dictated by the "no-waste" criterion and enjoins against the continuous, totally egalitarian redistribution of satisfactions, or wealth, but subject to that constraint, everyone should receive, by continuous redistribution if necessary, some minimal share of aggregate wealth so as to avoid a level of privation from which the opportunity to improve his situation becomes illusory.

The relationship of these highly general goals to the more specific environmental issues at hand may not be readily apparent, and I am not yet ready to demonstrate their pervasive implications. But let me give one indication of their implications. Recently, scientists have informed us that use of DDT in food production is causing damage to the penguin population. For the present purposes, let us accept that assertion as an indisputable scientific fact. The scientific fact is often asserted as if the correct implication—that we must stop agricultural use of DDT—followed from the mere statement of the fact of penguin damage. But plainly it does not follow if my criteria are employed.

My criteria are oriented to people, not penguins. Damage to penguins, or sugar pines, or geological marvels is, without more, simply irrelevant. One must go further, by my criteria, and say: Penguins are important because people enjoy seeing them walk about rocks, and furthermore, the well-being of people would be less impaired by halting use of DDT than by giving up penguins. In short, my observations about environmental problems will be people-oriented, as are my criteria. I have no interest in preserving penguins for their own sake.

It may be said, by way of objection to this position, that it is very selfish of people to act as if

each person represented one unit of importance and nothing else was of any importance. It is undeniably selfish. Nevertheless, I think it is the only tenable starting place for analysis for several reasons.

First, no other position corresponds to the way most people really think and act—that is, corresponds to reality. Second, this attitude does not portend any massive destruction of non-human flora and fauna, for people depend on them in many obvious ways and they will be preserved because and to the degree that humans do depend on them. Third, what is good for humans is, in many respects, good for penguins and pine trees—clean air for example. So that humans are, in these respects, surrogates for plant and animal life.

Fourth, I do not know how we could administer any other system. Our decisions are either private or collective. Insofar as Mr Jones is free to act privately, he may give such preferences as he wishes to other forms of life: he may feed birds in winter and do less with himself, and he may even decline to resist an advancing polar bear on the ground that the bear's appetite is more important than those portions of himself that the bear may choose to eat. In short, my basic premise does not rule out private altruism to competing life forms. It does rule out, however, Mr Jones's inclination to feed Mr Smith to the bear, however hungry the bear, however despicable Mr Smith.

Insofar as we act collectively on the other hand, only humans can be afforded an opportunity to participate in the collective decisions. Penguins cannot vote now and are unlikely subjects for the franchise—pine trees more unlikely still. Again, each individual is free to cast his vote so as to benefit sugar pines if that is his inclination. But many of the more extreme assertions that one hears from some conservationists amount to tacit assertions that they are specially appointed representatives of sugar pines and hence that their preferences should be weighted more heavily than the preferences of other humans who do not enjoy equal rapport with "nature." The simplistic assertion that agricultural use of DDT must stop at once because it is harmful to penguins is of that type.

Fifth, if polar bears or pine trees or penguins, like men, are to be regarded as ends rather than means, if they are to count in our calculus of social organization, someone must tell me how much each one counts, and someone must tell me how these life forms are to be permitted to express their preferences, for I do not know either answer. If the answer is that certain people are to hold their proxies, then I want to know how those proxy-holders are to be selected: self-appointment does not seem workable to me.

Sixth, and by way of summary of all the foregoing, let me point out that the set of environmental issues under discussion—although they raise very complex technical questions of how to achieve any objective—ultimately raise a normative question. What ought we to do? Questions of ought are unique to the human mind and world—they are meaningless as applied to a non-human situation.

I reject the proposition that we ought to respect the "balance of nature" or to "preserve the environment" unless the reason for doing so, express or implied, is the benefit of man.

I reject the idea that there is a "right" or "morally correct" state of nature to which we should return. The word "nature" has no normative connotation. Was it "right" or "wrong" for the Earth's crust to heave in contortion and create mountains and seas? Was it "right" for the first amphibian to crawl up out of the primordial ooze? Was it "wrong" for plants to reproduce themselves and alter the atmospheric composition in favour of oxygen? For animals to alter

the atmosphere in favour of carbon dioxide both by breathing oxygen and eating plants? No answers can be given to these questions, because they are meaningless questions.

All this may seem obvious to the point of being tedious, but much of the present controversy over environment and pollution rests on tacit normative assumptions about just such non-normative phenomena: that it is "wrong" to impair penguins with DDT but not to slaughter cattle for prime rib roasts. That it is wrong to kill stands of sugar pines with industrial fumes but not to cut sugar pines and build housing for the poor. Every man is entitled to his own preferred definition of Walden Pond, but there is no definition that has any moral superiority over another, except by reference to the selfish needs of the human race.

From the fact that there is no normative definition of the natural state, it follows that there is no normative definition of clean air or pure water—hence no definition of polluted air—or of pollution—except by the reference to the needs of man. The "right" composition of the atmosphere is one that has some dust in it and some lead in it and some hydrogen sulphide in it—just those amounts that attend a sensibly organized society thoughtfully and knowledgeably pursuing the greatest possible satisfaction for its human members.

The first and most fundamental step toward solution of our environmental problems is a clear recognition that our objectives are not pure air or water but rather some optimal state of pollution. That step immediately suggests the question: How do we define and attain the level of pollution that will yield the maximum possible amount of human satisfaction?

Low levels of pollution contribute to human satisfaction, but so do food and shelter and education and music. To attain ever lower levels of pollution, we must pay the cost of having less of these other things. I contrast that view of the cost of pollution control with the more popular statement that pollution control will "cost" very large numbers of dollars. The popular statement is true in some senses, false in others; sorting out the true and false sense is of some importance. The first step in that sorting process is to achieve a clear understanding of the difference between dollars and resources. Resources are of vital importance; dollars are comparatively trivial.

Four categories of resources are sufficient for our purposes: at any given time a nation, or a planet if you prefer, has a stock of labour, of technological skill, of capital goods, and of natural resources (such as mineral deposits, timber, water, land, etc.). These resources can be used in various combinations to yield goods and services of all kinds—in some limited quantity. The quantity will be larger if they are combined efficiently, smaller if combined inefficiently. But in either event, the resource stock is limited, the goods and services that they can be made to yield are limited; even the most efficient use of them will yield less than our population, in the aggregate, would like to have.

If one considers building a new dam, it is appropriate to say that it will be costly in the sense that it will require x hours of labour, y tons of steel and concrete, and z amount of capital goods. If these resources are devoted to the dam, then they cannot be used to build hospitals, fishing rods, schools, or electric can openers. That is the meaningful sense in which the dam is costly. Quite apart from the very important question of how wisely we can combine our resources to produce goods and services is the very different question of how they get distributed—who gets how many goods? Dollars constitute the claim checks that are distributed among people and that control their share of national output. Dollars

are nearly valueless pieces of paper except to the extent that they do represent claim checks to some fraction of the output of goods and services. Viewed as claim checks, all the dollars outstanding during any period of time are worth, in the aggregate, the goods and services that are available to be claimed with them during that period—neither more nor less. . . .

The point is this: many people fall into error upon hearing the statement that the decision to build a dam, or to clean up a river, will cost X money. This is a wealthy country, and we have lots of money. But you cannot build a dam or clean a river with X million—unless you also have a match, you can't even make a fire. One builds a dam or cleans a river by diverting labour and steel and trucks and factories from making one kind of good to making another. The cost in dollars is merely a shorthand way of describing the extent of the diversion necessary. If we build a dam for X million, then we must recognize that we will have X million less housing and food and medical care and electric can openers as a result.

Similarly, the costs of controlling pollution are best expressed in terms of the other goods we will have to give up to do the job. This is not to say the job should not be done. Badly as we need more housing, more medical care, more can openers, and more symphony orchestras, we could do with somewhat less of them, in my judgment at least, in exchange for somewhat cleaner air and rivers. But that is the nature of the trade-off, and analysis of the problem is advanced if that unpleasant reality is kept in mind. Once the trade-off relationship is clearly perceived, it is possible to state in a very general way what the optimal level of pollution is. I would state it as follows:

People enjoy watching penguins. They enjoy relatively clean air and smog-free vistas. Their health is improved by relatively clean water and air. Each of these benefits is a type of good or service. As a society, we would be well advised to give up one washing machine if the resources that would have gone into that washing machine can yield greater human satisfaction when diverted into pollution control. We should give up one hospital if the resources thereby freed would yield more human satisfaction when devoted to elimination of noise in our cities. And so on, trade-off by trade-off, we should divert our productive capacities from the production of existing goods and services to the production of a cleaner, quieter, more pastoral nation up to—and no further than—the point at which we value more highly the next wash-ing machine or hospital that we would have to do without than we value the next unit of environmental improvement that the diverted resources would create.

Now, this proposition seems to me unassailable but so general and abstract as to be unhelpful—at least unadministerable in the form stated. It assumes we can measure in some way the incremental units of human satisfaction yielded by very different types of goods. . . . But I insist that the proposition stated describes the result for which we should be striving—and again, that it is always useful to know what your target is even if your weapons are too crude to score a bull's eye.

Baxter's conclusion is striking. For many people, the idea that there can be an "optimal" level of a clearly bad thing—like pollution—is counterintuitive. It appears to turn an important aspect of our normative world on its head. For if we strive to achieve an optimal level of something, are we not thereby assuming that the thing has some sort of positive value for us? But how can we say that we value pollution to any extent? Surely it is an unambiguously bad thing? Such intuitions, although deeply entrenched in our thinking, are, according to the economist, deeply confused.

Let's focus on the second of Baxter's four criteria supporting our judgments about how to organize ourselves. It states a commitment to the value of efficiency. The point is put in terms of avoiding waste, and it is clearly a normative claim, one concerning how we *ought* to organize ourselves. Because of the fact of scarcity, we cannot afford to be wasteful, so we should seek out a state of goods allocation that is Pareto Optimal, or at the very least we should be seeking to allocate goods in a way that is Pareto Superior to the current state of allocation (assuming the latter is not already Pareto Optimal).

But the commitment to efficiency does not by itself get us all the way to the conclusion about optimal levels of pollution. To get there, we need to add two further claims. The first is that anthropocentrism—according to which the only wants that matter, or are real, are human wants—is correct. Humans may want to preserve penguins or rainforests or wetlands from destruction, but if so, it is still *our* wants that determine what ought to be done. Were these wants entirely absent from the total set of human wants, there could be no justifiable basis for undertaking such preservation. We cannot make decisions by trying to construe the "wants" of penguins, to say nothing of rainforests or wetlands.

Second, all goods are substitutable. Economists sometimes describe humans as seeking out "consumption bundles," collections of goods and services that, ideally at least, meet their specific wants. Imagine Sam faced with two bundles of goods, *A* and *B*. These bundles are identical except that *B* contains 15 per cent more of a specific good, say health insurance, than *A*. Clearly, Sam should consider bundle *B* more desirable than bundle *A*. However, *A*'s short-coming *can* be rectified by an equivalent amount of some other desired good. For example, it might contain a salary that is 15 per cent higher than the one in bundle *B*. Whatever the precise numbers, the point is that there is always some way of reconfiguring the bundles so that Sam sees them as equivalent. This is the **substitutability principle**. Another way of stating the principle is to claim that no good is "priceless." That is, there is no good whose loss cannot in principle be compensated for by swapping an appropriate amount of some other good for it.

Now we can draw our conclusion about optimal pollution. Because of the truth of anthropocentrism, we need consider only human wants. Clearly, we want pollution control. But we want lots of other things too, and controlling pollution costs money. If as a society we divert some of our money into pollution control, then precisely that amount of money is unavailable for other projects—highway construction, public schools, opera houses, and so on. So if we decide we want to control pollution, it is crucial to know *how much we want this*. In other words, we need to think about the opportunity costs of controlling pollution to a specific degree. What are we willing to give up in order to control pollution just so much? Note that as soon as we frame the question this way, we are squarely in the territory of the concept of substitutability. We are reasoning on the assumption that decisions concerning pollution control are essentially about trade-offs among goods. All we can do, according to Baxter, is try to allocate our scarce resources in such a way that our various wants are satisfied with the least waste possible. But arriving at this state of affairs—at Pareto Optimality—is, depending on the precise structure of our wants, entirely compatible with allowing some pollution to persist.

Ⓒ The Allure of Cost-Benefit Analysis

The claim about pollution can then be generalized to all environmental goods and services. But how precise can our measurements be here? Baxter himself seems unsure about our ability to "measure in some way the incremental units of human satisfaction yielded by very different types of goods." But other economists are not this cautious. They have developed a tool to help weigh the appropriate trade-offs: **cost-benefit analysis (CBA)**. For any proposed project—for instance, implementation of a regulation requiring companies that emit pollution to put scrubbers in their smokestacks—it is necessary to add up both the benefits of the proposal and its costs and weigh the two against each other. If aggregate benefits exceed aggregate costs, it is rational to go ahead with the project. If not, it is not. Once you have accepted the basic idea that everything can be priced, mathematics does all the rest.

It is relatively easy to see how much it would cost if companies were forced to implement some new piece of pollution-controlling technology. Sticking with Baxter's analysis, we can express this cost as a function of opportunities foregone. Because the technology has a precise monetary cost, the company may not be able to expand its workforce as planned or pay its executives quite so much. Similarly, it is relatively easy to come up with a monetary estimate of the benefits of constructing a dam to provide hydropower. But how exactly do we measure the benefits to society of reduced pollution or the costs to society of the dam project? Pollution control might improve human health, and a dam might threaten some local indigenous cultural practices. Such things do not *appear* to have a market value.

This has not stopped economists from trying to measure the value of environmental protection. In fact, they make a key distinction between the "use value" and the "non-use value" of natural goods and services. Use value is relatively easy to determine. For example, when the *Exxon Valdez* spilled 11 million gallons of oil into Alaska's Prince William Sound in 1989, the local fishing economy was damaged significantly. In 1994, a jury awarded local fishers $300 million in compensatory damages, based on estimates of income losses.[3] Something similar is bound to happen with regard to the 2010 oil leakage in the Gulf of Mexico, which caused even more extensive damage to the local fisheries. But what about less tangible harms in disasters like these, such as the loss of a pristine natural environment and the damage done to non-commercially useful species? What about values that are aesthetic or spiritual or symbolic? Economists insist that the non-use value attached to the existence of unspoiled natural environments can indeed be measured. To this end, they have attempted to determine people's "willingness to pay" for the protection of natural environments.

Willingness to pay can be expressed in several ways. Sometimes economists will try to compute the value of clean air, for instance, by looking at the difference in price between homes in neighbourhoods with clean air and those located where the air is polluted. Another method brings us back to the lumber company CEO encountered in the rainforest by Suzuki and his friends. He challenges them to express their commitment to environmental protection by funding groups that work for it. Or people can be asked, in surveys and questionnaires, how much they are willing to see their taxes go up in order to provide for the implementation of governmental regulations designed to protect the environment. So one way to determine the *precise* value of a Pacific coast rainforest in its pristine state—permanently protected from development or exploitation, at least on the massive and disruptive scale involved in logging—is to aggregate the amount of money people are willing to put into the coffers of environmental NGOs or the government (or both) to fund its protection.

To test the plausibility of these ideas, let's look at an example of how a CBA might proceed. The Great Whale Complex, also known as James Bay II, called for the construction of a hydroelectric dam with a capacity of 3212 megawatts near Hudson Bay, about 1000 kilometres north of Montreal. The idea for the project was put forward by Premier Robert Bourassa in 1989 and was ultimately scuttled by Premier Jacques Parizeau in 1994. Bourassa put the economic case tersely: "Quebec is a vast hydro-electric power plant in the bud, and every day millions of potential kilowatt hours flow downhill and out to sea. What a waste!"[4] The dam would, according to its proponents—the provincial government, Hydro-Québec, the Quebec Federation of Labour, investment bankers, and others—have clear and massive economic benefits: hydroelectricity is a relatively cheap and reliable form of energy, the surplus energy generated could be sold to the northeastern US, the construction project itself would inject up to $2 billion per year into the provincial economy, lower rates could be charged to electricity-intensive industries (like aluminum smelters and magnesium refining), the construction region would be provided with roads and other infrastructural improvements, and so on. All of these benefits can be expressed in dollars and cents.

What about the project's costs? Its opponents—a collection of environmentalists, NGOs, indigenous groups, and energy-industry competitors—pointed out that the project would result in the loss of aquatic and terrestrial habitat. This loss would likely have a negative effect on migratory birds, caribou, beluga whales, and rare freshwater seals, among other species in the James Bay region. Also, because it would result in a drop in the Great Whale River of more than two metres, the dam would cause changes in water flow and sediment load that could affect the entire marine environment. The upshot of all this disruption would likely be a reduction in the biological productivity of estuaries and offshore eelgrass beds, crucial habitat for the Canada goose.[5] Moreover, organic matter in flooded areas converts to toxic methylmercury, which can then spread through the entire marine food web. Local Cree and Inuit groups therefore saw the dam as a threat to both their health and the viability of culturally entrenched subsistence fishing practices.[6] Finally, opponents pointed to the opportunity costs of the project. By going the hydroelectric route, Quebec would lose the opportunity to explore alternative energy sources and energy-saving programs and technologies. These were deemed to be "more efficient sources of permanent employment than the short term construction activity" involved in building the dam.[7] In sum, opponents were arguing on the basis of the values of biodiversity, lost opportunity to develop alternative forms of energy, ecosystem integrity, human health, the integrity of indigenous cultural practices, lost income due to pollution of resources, and more.

Some of this—the opportunity costs of failing to develop alternative energy sources, for instance—can be classified as use value. But most of the values just enumerated—such as ecosystem health and cultural integrity—are clearly of the non-use variety. Can we put a price on them? It appears that we can do so indirectly by looking at concerned parties' willingness to pay for them. As it happens, the project was actively opposed by many environmental NGOs: Greenpeace, Friends of the Earth, Solidarity Foundation, Audubon Society, Sierra Club, Coalition for a Public Debate on Energy, Canadian Arctic Resources Committee, among others. No surveys or questionnaires were sent out to the people of Quebec to ask them how much they would be willing to pay in taxes devoted to the preservation of the James Bay region, but this surely could have been done. If it had been, the aggregate cost of the project—the total amount people would have been willing to give to the NGOs and government for the express purpose of preventing development—could have been expressed solely and precisely in monetary terms. Or so goes the claim.

Ⓓ Problems with Cost-Benefit Analysis

Although there was a massive environmental review of James Bay II involving six different committees and commissions, at the end of the day no strict CBA of the project, along the lines just suggested, was made. But as we have seen, the opposition *was* ultimately successful, and the project was shelved. That implies that the arguments actually made in opposition to the project were deemed more appropriate than a CBA would have been. This, in turn, suggests that there may be deep philosophical problems with CBA and the narrow economic outlook on the environment that it expresses. To see what some of these problems might be, we will turn to this chapter's second contribution. Then we will return to the Great Whale Complex.

ETHICS, PUBLIC POLICY, AND GLOBAL WARMING

Dale Jamieson

There are many uncertainties concerning anthropogenic climate change, yet we cannot wait until all the facts are in before we respond. All the facts may never be in. New knowledge may resolve old uncertainties, but it may bring with it new uncertainties. It is an important dimension of this problem that our insults to the biosphere outrun our ability to understand them. We may suffer the worst effects of global warming before we can prove to everyone's satisfaction that they will occur.[8]

The most important point I wish to make, however, is that the problem we face is not purely scientific. It cannot be solved by the accumulation of scientific information alone. Science has alerted us to a problem, but the problem also concerns our values. It is about how we ought to live, and how humans should relate to one another and to the rest of nature. These are problems of ethics and politics as well as problems of science.

In the first section, I examine what I call the "management" approach to assessing the impacts of climate change and our responses to them. I argue that this approach cannot succeed, for it does not have the resources to answer the most fundamental questions that we face. In the second section, I explain why the problem of anthropogenic global change is to a great extent an ethical problem, and why our conventional value system is not adequate for addressing it. . . .

I. Why Management Approaches Must Fail

From the perspective of conventional policy studies, anthropogenic climate change and its attendant consequences are problems to be "managed." Management techniques mainly are drawn from neoclassical economic theory and are directed toward manipulating behaviour by controlling economic incentives through taxes, regulations, and subsidies.

In recent years, economic vocabularies and ways of reasoning have dominated the discussion of social issues. Participants in the public dialogue have internalized the neoclassical economic perspective to such an extent that its assumptions and biases have become almost invisible. It is only a mild exaggeration to say that in recent years debates over policies have largely become debates over economics.

The Environmental Protection Agency's draft report "Policy Options for Stabilizing Global Climate" is a good example.[9] Despite its title, only one of report's nine chapters is specifically devoted to policy options, and in that chapter only "internalizing the cost of climate change risks" and "regulations and standards" are considered. For many people, questions of regulation are not distinct from questions about internalizing costs. According to one influential view, the role of regulations and standards is precisely to internalize costs, thus (to echo a parody of our forefathers) "creating a more perfect market." For people with this view, political questions about regulation are really disguised economic questions.[10]

It would be both wrong and foolish to deny the importance of economic information. Such information is important when making policy decisions, for some policies or programs that would otherwise appear to be attractive may be economically prohibitive. Or in some cases, there may be alternative policies that would achieve the same ends and also conserve resources.

However, these days it is common for people to make more grandiose claims on behalf of economics. Some economists or their champions believe not only that economics provides important information for making policy decisions but also that it provides the most important information. Some even appear to believe that economics provides the only relevant information. According to this view, when faced with a policy decision, what we need to do is to assess the benefits and costs of various alternatives. The alternative that maximizes the benefits minus the costs is the one we should prefer. This alternative is "efficient" and choosing it is "rational."

Unfortunately, too often we lose sight of the fact that economic efficiency is only one value, and it may not be the most important one. Consider, for example, the idea of imposing a carbon tax or a market in emissions permissions (i.e., "cap and trade") as a policy response to the prospect of global warming. What we think of this proposal may depend to some extent on how it affects other concerns that are important to us. Equity is sometimes mentioned as one other such concern, but most of us have very little idea about what equity means or exactly what role it should play in policy considerations.

One reason for the hegemony of economic analysis and prescriptions is that many people have come to think that neoclassical economics provides the only social theory that accurately represents human motivation. According to the neoclassical paradigm, welfare can be defined in terms of preference satisfaction, and preferences are defined in terms of choice behaviour. From this, many (illicitly) infer that the perception of self-interest is the only motivator for human beings. This view suggests the following "management technique": if you want people to do something give them a carrot, but if you want them to desist, give them a stick.

Many times the claim that people do what they believe is in their self-interest is understood in such a way as to be a circular proposition, and therefore unfalsifiable and trivial. On the one hand, we know that something is perceived as being in a person's self-interest because the person pursues it; and if the person pursues it, then we know that the person must perceive it to be in his or her self-interest. On the other hand, if we take the assertion that people act in their own interests to be an empirical claim, it appears to be false. If we look around the world we see people risking or even sacrificing their own interests in attempts to overthrow oppressive governments or to realize ideals to which they are committed. Each year more people die in wars fighting for some perceived collective good than die in criminal attempts to further their own individual interests. It is implausible to suppose that the behaviour (much less the motivations) of a revolutionary, a radical environmentalist, or a friend or lover can be revealed by a benefit-cost analysis.

It seems plain that people are motivated by a broad range of concerns, including concern for family and friends, and religious, moral, and political ideals. And it seems just as plain that people sometimes sacrifice their own interests for what they regard to be a greater, sometimes impersonal, good.

People often act in ways that are contrary to what we might predict on narrowly economic grounds, and moreover, they sometimes believe that it would be wrong or inappropriate even to take economic considerations into account. Many people would say that choosing spouses, lovers, friends, or religious or political commitments on economic grounds is simply wrong. People who behave in this way are often seen as manipulative, untrustworthy, without character or virtue. One way of understanding some environmentalists is to see them as wanting us to think about nature in the way that many of us think of friends and lovers—to see nature not as a resource to be exploited but as a partner with whom to share our lives.

Some may think that I have exaggerated the dominance of narrow, economistic approaches to policy-making. Neoclassical economics is in retreat, it might be said, in the wake of the Great Recession of 2008. Paul Krugman has recently reported that some economists are talking about a discipline in crisis.[11] Indeed, it might be thought that the hegemony of the neoclassical paradigm has been eroding since the late 1990s, as indicated by Nobel Prizes awarded to Amartya Sen (in 1998), Daniel Kahneman (in 2002), and Elinor Ostrum (in 2009). I am skeptical. The events of 2008 were hardly unprecedented in demonstrating the inability of conventional neoclassical models to predict dramatic changes in the economy. From 1973 to 1974, stocks lost 48 per cent of their value; in 1987 the Dow plunged nearly 23 per cent in a single day for no apparent reason. A series of "bubbles" that simply could not exist according to some influential theories have occurred throughout the world since the 1980s. Intellectually, there has always been dissatisfaction with the neoclassical paradigm, sometimes even very close to the heart of the discipline. Despite this, the President's Council of Economic Advisors is still in business, unchecked by a Council of Philosophical (or Ethical) Advisors. Public decisions of great consequence continue to be made on the basis of shallow and misleading indicators such as gross domestic product (GDP), rather than on the basis of broader considerations such as quality of life or impacts on fundamental planetary systems. The "straw man" of neoclassical economics seems to me to have quite a lot of blood left in him.

What I have been claiming in this section is that it is not always rational to make decisions solely on narrow economic grounds. Although economic efficiency may be a value, there are other values as well, and in many areas of life, values other than economic efficiency should take precedence. I have also suggested that people's motivational patterns are complex and that exploiting people's perceptions of self-interest may not be the only way to move them. This amounts to a general critique of viewing all social issues as management problems to be solved by the application of received economic techniques. There is a further reason why economic considerations should take a back seat in our thinking about global climate change: there is no way to assess accurately all the possible impacts and to assign economic values to alternative courses of action. A greenhouse warming will have impacts that are so broad, diverse, and uncertain that conventional economic analysis is practically useless.

Consider first uncertainties about the potential impacts, some of which I have already noted. Even if the IPCC [Intergovernmental Panel on Climate Change] is correct in supposing that global mean surface temperatures will increase between 1.1 and 6.4°C during this century, there is still great uncertainty about the impact of this warming on regional climates. One thing is certain: the impacts will not be homogeneous. Some areas will become warmer, some will probably become colder, and overall variability is likely to increase. Precipitation patterns will also change, and there is much less confidence in the projections about precipitation than in those about temperature. These uncertainties about regional effects make estimates of the economic consequences of climate change radically uncertain.

There is also another source of uncertainty regarding these estimates. In general, predicting human behaviour is difficult. The difficulties are especially acute in the case that we are considering because climate change will affect a wide range of social, economic, and political activities. Changes in these sectors will affect emissions of greenhouse gases, which will in turn affect climate, and around we go again.[12] Climate change is itself uncertain, and its human effects are even more radically so. It is for reasons such as these that in general, the area of environment and energy has been full of surprises.

A second reason why the benefits and costs of the impacts of global climate change cannot reliably be assessed concerns the breadth of the impacts. Global climate change will affect all regions of the globe. About many of these regions—those in which most of the world's population lives—we know very little. Some of these regions do not even have monetized economies. It is ludicrous to suppose that we could assess the economic impacts of global climate change when we have so little understanding of the global economy in the first place.

Finally, consider the diversity of the potential impacts. Global climate change will affect agriculture, fishing, forestry, and tourism. It will affect "unmanaged" ecosystems and patterns of urbanization, as well as international trade and relations. Some nations and sectors may benefit at the expense of others. Moreover, there will be complex interactions among these effects. For this reason we cannot reliably aggregate the effects by evaluating each impact and combining them by simple addition. We have no idea what the proper mathematical function would be for aggregating them—if the idea of aggregation even makes sense in this context. It is difficult enough to assess the economic benefits and costs of small-scale, local activities. It is almost unimaginable to suppose that we could aggregate the diverse impacts of global climate change in such a way as to dictate policy responses.

In response to skeptical arguments like the one that I have given, it is sometimes admitted that our present ability to provide reliable economic analyses is limited, but then it is asserted that any analysis is better than none. I think that this is incorrect and that one way to see this is by considering an example.

Imagine a century ago a government had done an economic analysis in order to decide whether to build its national transportation system around the private automobile. No one back then could have imagined the secondary effects: the attendant roads, the loss of life, the effects on wildlife and on communities; the impact on air quality, noise, travel time, and quality of life. Given our inability to reliably predict and evaluate the effects of even small-scale technology (e.g., the artificial heart),[13] the idea that we could predict the impact of global climate change reliably enough to permit meaningful economic analysis seems fatuous indeed.

When our ignorance is so extreme, it is a leap of faith to say that some analysis is better than none. A bad analysis can be so wrong that it can lead us to do bad things, outrageous things— things that are much worse than what we would have done had we not tried to assess the costs and benefits at all (this may be the wisdom in the old adage that "a little knowledge can be a dangerous thing").

What I have been arguing is that the idea of managing global climate change is a dangerous conceit. The tools of economic evaluation are not up to the task. However, the most fundamental reason why management approaches are doomed to failure is that the questions they can answer are not the ones that are most important and profound. The problems posed by anthropogenic global climate change are ethical as well as economic and scientific. I will explain this claim in the next section.

II. Ethics and Global Change

Since the end of World War II, humans have attained a kind of power that is unprecedented in history. While in the past entire peoples could be destroyed, now all people are vulnerable. While once particular human societies had the power to upset the natural processes that made their lives and cultures possible, now people have the power to alter the fundamental global conditions that permitted human life to evolve and that continue to sustain it. While our species dances with the devil, the rest of nature is held hostage. Even if we step back from the precipice, it will be too late for many or even perhaps most of the plant and animal life with which we share the planet.[14] Even if global climate can be stabilized, the future may be one without wild nature.[15] Humans will live in a humanized world with a few domestic plants and animals that can survive or thrive on their relationships with humans.

Such possibilities pose fundamental questions of morality. They concern how we ought to live, what kinds of societies we want, and how we should relate to nature and other forms of life. Seen from this perspective, it is not surprising that the discipline of economics cannot tell us everything we want to know about how we should respond to global warming and global change. Economics may be able to tell us how to reach our goals efficiently, but it cannot tell us what our goals should be or even whether we should be concerned about reaching them efficiently.

Values are more objective than mere preferences.[16] A value has force for a range of people who are similarly situated. A preference may have force only for the individual whose preference it is. Whether or not someone should have a particular value depends on reasons and arguments. We can rationally discuss values, while preferences may be rooted simply in desire, without supporting reasons.

A system of values may govern someone's behaviour without these values being fully explicit. They may figure in people's motivations and in their attempts to justify or criticize their own actions or those of others. Yet it may require a theorist or a therapist to make these values explicit.

A system of values is generally a cultural construction rather than an individual one.[17] It makes sense to speak of contemporary American values, or those of eighteenth-century England or tenth-century India. Our individual differences tend to occur around the edges of our value system. The vast areas of agreement often seem invisible because they are presupposed or assumed without argument.

I believe that our dominant value system is inadequate and inappropriate for guiding our thinking about global environmental problems, such as those entailed by climate changes caused by human activity. This value system, as it impinges on the environment, can be thought of as a relatively recent construction, coincident with the rise of capitalism and modern science, and expressed in the writings of such philosophers as Francis Bacon, John Locke, and Bernard Mandeville.[18] It evolved in low-population-density and low-technology societies, with seemingly unlimited access to land and other resources. This value system is reflected in attitudes toward population, consumption, technology, and social justice, as well as toward the environment. . . .

Today, we face the possibility that the global environment may be destroyed; yet no one will be responsible. This is a new problem. It takes a great many people and a high level of consumption and production to change the Earth's climate. It could not have been done in low-density, low-technology societies. Nor could it have been done in societies like ours until recently. London could be polluted by its inhabitants in the eighteenth century, but its reach was limited. Today no part of the planet is safe. Unless we develop new values and conceptions of responsibility, we will have enormous difficulty in motivating people to respond to this problem.

Some who seek quick fixes may find this concern with values frustrating. A moral argument will not change the world overnight. Collective moral change is fundamentally cooperative rather than coercive. No one will fall over, mortally wounded, in the face of an argument. Yet if there is to be meaningful change that makes a difference over the long term, it must be both collective and thoroughgoing. Developing a deeper understanding of who we are, as well as how our best conceptions of ourselves can guide change, is the fundamental issue that we face.

What Jamieson calls "management approaches" to environmental issues generally rely on a disputed normative theory—namely, utilitarianism. Utilitarianism is sometimes called a "maximizing" theory. For example, if the production of human happiness is our ethical goal, then we ought to produce as *much* of this good as we can. This formula works for anything we identify as our ethical goal. So it is entirely possible that some of the proponents of James Bay II believed that development of the hydroelectric dam would have been better for Quebec society as a whole even though key interests of the local Cree and Inuit would be undermined in the process. Building the dam, on this understanding, simply produces *more* happiness for more people than not building it. But is this a morally acceptable way of reasoning? It begins to look problematic if the interests that are undermined are defined as fundamental rights. Should the rights of individuals, or cultural groups, get overlooked or undermined in the effort to maximize overall welfare? Most critics of utilitarianism think not.

The possibility that the effort to maximize welfare might engender such conflicts raises an interesting dilemma for the economist. How can one be both a utilitarian and committed to achieving Pareto Optimality? Recall that a Pareto Optimal state of affairs is one endorsed by *everyone* whose interests are at stake. But surely the local Cree and Inuit would not describe the state of affairs had the dam been constructed this way, since their subsistence fishing practices would thereby have been irreparably damaged. No problem, says the economist, because adversely affected parties can always be compensated in ways that, in principle, will satisfy them. This claim relies on the substitutability principle. An adequate amount of cash and/or, say, infrastructural development will render the Cree and Inuit's post-project bundle of consumption goods equivalent—or even superior—to their pre-project bundle.

Even if this were an acceptable way of treating the concerns of the Cree and Inuit, it is crucial to note that in practice economists have thought it sufficient to show merely that adversely affected parties *could have been* compensated, not that they actually must be compensated. This is known as "hypothetical compensation": as long as it is possible in a "perfect system of transfers" to compensate losers, the project is justified on the grounds that net benefits exceed net costs.[19] In contrast, we might believe that as a matter of *right*, people are entitled to be treated, or not interfered with, in certain ways. Just as people have a right to due legal process, we might say that they have a right to clean air or clean water, or to the integrity of a certain cultural practice, or even to the existence of unspoiled wilderness. If so, these rights cannot be overridden by appeal to net benefits, hypothetical or otherwise. Thus, in theoretical terms, the problem with hypothetical compensation is that it forces us to sacrifice principles of Pareto efficiency on the altar of utilitarian maximizing. This will almost invariably infringe some of the rights and interests of those minority groups.

However, even if compensation were not merely hypothetical but instead was invariably paid out, there would be reason to object to CBA. Here, our criticism goes to the heart of the substitutability principle. There may be some values or goods that are considered priceless. This does not mean that they have an infinite monetary value but that it makes no sense to ascribe monetary value to them at all.[20] Many of the values put forward by opponents of James Bay II can be treated this way. According to Stephen Kelman, the chief problem with attempting to assign market values to non-market goods is that the latter are inevitably cheapened in the process. By way of analogy, Kelman asks us to consider the distinction between sex as the consummation of love and sex that is bought. Our intuitions are that the value of sex is simply misunderstood and degraded when it is thought of as the sort of thing that can be bought and sold. Similarly, goods like life, health, and a clean environment can be bought and sold, but their true value is misunderstood and degraded when they are.[21]

E Conclusion

We can conclude by raising three general problems with the neo-liberal economic model applied to environmental problems.

The first is the difficulty of accurate costing. Remember that Paul Henderson sweater I sold you? We assumed that neither of us, or any third party, had cause to complain about what transpired between us. Another way to put this is that the price of the sweater was exactly right, that it reflected the true cost of the good. But this is not always the case in the larger economic world, and this is why a "perfect system of transfers" is almost never found there. In a perfect system of transfers, price is a true reflection of cost, but there are factors that make it unlikely that this perfect system will appear. These factors are often referred to as **market failures**. Among them are the existence of monopolies, the activities of which drive prices of certain goods higher than they should be, and a lack of informational transparency so that consumers are not, for instance, always aware of the dangers a certain product might pose to them.

But the most important market failure for our purposes is underpricing due to the creation of **negative externalities**. This happens when the production of a good imposes costs that are not reflected in the price of the good. If the production of a certain good causes pollution as a by-product and the pollution causes people to get sick, the cost of their sickness—to themselves and to the health care system—will not be reflected in the price of the good. The result is that the good is cheaper than it should be and the market will therefore overproduce it.[22]

For example, there is good evidence to suggest that the market overproduces cars. In Canada, smog caused largely by carbon dioxide (CO_2) emissions from cars kills some 5000 people per year. Add to this time lost in traffic jams, the adverse environmental effect of urban sprawl, and the effects of greenhouse gas emissions on the climate, and the real cost of all those cars we drive can start to look pretty large. Indeed, according to one estimate, the full cost of driving in Canada should be fixed at $200 billion per year. To put this number in perspective, it is approximately 10 times the amount that would be required to eliminate hunger and malnutrition across the world.[23] Think about this the next time you are inclined to complain about high gas prices. The point to emphasize is that economic analysis, whether or not it takes the form of CBA, must include **full-cost pricing** if it is to provide an honest picture of the practical options before us. Full-cost pricing effectively internalizes what had previously been externalized costs. When this happens, the price of a good really does reflect its cost. However, since in practice this almost never happens, the purely economic approach to dealing with environmental issues should be viewed with deep suspicion.

The second problem has to do with the **value monism** of economics. Value monism is the view that there is one single kind of value that is most fundamental. Any other sort of value needs to be "translated" into this one true value-language in order to make real sense. Another way to put this is to say that on this view all values are commensurable, which the *Oxford English Dictionary* defines as "measurable by the same standard." Of course, the preferred language of economics is dollars and cents. But does the idea that money is the foundational value not distort our thinking about environmental issues? It is interesting to note that the guidelines for the review of James Bay II required Hydro-Québec to incorporate "the knowledge and opinions of all communities that inhabit the region" into its analysis. More specifically, in its analysis Hydro-Québec was required to define the environment, in part, "in accordance with the acquired knowledge of the local Cree and Inuit."[24] In other words, it was prohibited from presenting its case for development as though the only value that mattered was the economic one.

The claim is not that economic values are irrelevant, only that they are not the only relevant values or sources of knowledge. Thus, the review guidelines were implicitly endorsing **value pluralism**—the idea that there are multiple ways of valuing, all with equal legitimacy—and rejecting value monism. Arguably, the main reason the project was scuttled is that when forced to justify it in these diverse terms, its proponents simply failed to convince most stakeholders. This is a crucial point, because it puts rational public debate—rather than mathematical analysis—at the centre of environmental decision-making. Another way of putting this is to say that we are consumers but we are also citizens. And since we do not think or behave the same way in each of these roles, it is unwise to try to reduce one to the other. In particular, it is misguided to reduce our thinking as citizens—where we deliberate in principled ways about the public good—to our thinking as consumers, where we are often just out to get the best deal we can for ourselves.[25]

With these points in mind, let's return to Suzuki's lumber company CEO. Both of his key claims now look to be mistaken. First, it is not the case that trees have value only as commodities. It is also possible to talk sensibly about non-use value, whether this value is aesthetic, spiritual, or symbolic. Even hard-bitten economists should admit this, but they will likely not agree that the goods these values express are fundamentally non-marketable. This brings us to the second claim, that the economic mode of valuation is the only game in town. We have seen that it is not the case that monetary value is fundamental. There are other ways of valuing natural things, and this value does not have to be reduced to, or expressed in, monetary terms. The James Bay II example shows us that this sort of value monism is untenable.

The final problem has to do with the dominant way economists look at the future. This is extremely important for environmental ethics, since so much of this enterprise is concerned with the way we leave the natural world for future people. Think about climate change (the subject of Chapter 11). If we are to address this problem robustly, so that we leave people of the future with a non-degraded climate, we need to spend some money now. That is, because there will be foreseeable costs of climate change down the road, it is reasonable to set aside money now to meet them. But how much should we spend in order to achieve a better state of affairs in the (possibly far) future? Economists respond to challenges like this with the **discounting** theory of value, according to which a future good is worth less to us in the present than the same good received now is worth to us and that in investing in that future good now we should therefore spend less on it than we would if we were actually receiving it now. In the abstract this sounds convoluted, but a simple example should clarify it. Imagine you want to buy a new car in 10 years. Again, you are in this case foreseeing a cost down the road for which you reasonably want to begin preparing yourself now. The question is, how much should you set aside for the car now? We can answer by identifying three deliberative steps. First, determine how much the car is likely to cost in 10 years. Second, calculate how much you can safely expect to get in interest per annum between now and then. This is the *discount rate*. Third, set aside just enough to grow the amount you need given the discount rate. The amount you set aside is the *discount cost* of the car, which will obviously be less than the full cost of a car right now. This way of reasoning looks attractive, but there is a problem with it as applied to issues that are more complex than simple consumer purchases.

To see the problem, come back to climate change. Peter Singer has calculated that at a discount rate of 5 per cent, we should spend only $14.20 to prevent every $100 of losses in 40 years from now (roughly, 2055). So, we should spend no more than $142 million now to prevent every $1 billion of expected losses in 2050.[26] This sounds big, but for a global GDP of (roughly) $70 trillion, is it? It looks as though when it comes to costing climate change it lowballs things dramatically. For example, economist Sir Nicholas Stern says that we need to spend 2 to 4 per cent of global GDP per year on climate change if we are to do anything meaningful about it. Now just *1 per cent* of $70 trillion is $700 billion. So the 5 per cent discount rate gives us a shortfall of over $699 billion, and that is a far less drastic shortfall than the one we would get if we compared our expenditure to the 2 to 4 per cent of global GDP required, rather than the conservative 1 per cent I have chosen here. What this means is that everything depends on which discount rate is chosen. The higher the rate, the less we have to invest in the present (since the rate affects how much our investment will grow).

The more general point is that when it comes to the far future, finding the correct rate is highly speculative. Some economists use 5 per cent, some closer to 0 per cent. The choice will alter the CBA dramatically. However, the choice of a specific discount rate is an *ethical* one, not a purely economic one. This is what makes the problem as applied to environmental issues so much more difficult than it is as applied to simple consumer purchases. With the latter, we simply work with a discount rate that already exists (as set by the bank, for example). But in the case of environmental goods we need to decide what discount rate to adopt and this choice is predicated on how much we *ought* to value certain goods. The value we place on life (for example) might be *infinite*, not because this is what the math tells us but because we think that as a matter of morality we should not put a price on human life. This does not mean we should simply refuse to set aside money to protect future lives, but rather that we should, like Stern himself, use a very low discount rate when we judge that human lives, livelihoods, or other fundamental interests are at stake in our policy choices. A very low rate means that we must spend a great deal right now combatting extreme threats to the future like climate change.

CASE STUDY

Aquaculture and the Economic Growth Imperative

Source: © Bill Brooks/Alamy

Atlantic salmon fish farm, Dark Harbour, NB. Will technology save us as we draw down nature's capital?

There has been a distinct change in the patterns of our fish consumption over the past 50 years or so. We have, according to scientists, gradually been "fishing down the food chain." That is, we have been exhausting the world's supplies of bigger, more commercially valu- able fish and turning instead to smaller, less commercially valuable fish. The result is that nearly three-quarters of the world's commercially valuable fish species—swordfish, marlin, tuna, shark, and cod, for example—are overfished. One result of this is the proliferation of commercially worthless species like jellyfish, whose natural predators are being removed from marine ecosystems. What is driving this phenomenon? At the most general level, the answer is clear: the idea that our economy must inevitably and perpetually *grow*. This is the economic growth imperative, another key assumption of neo-liberal economics.

The imperative is grounded in two factors. First, the human population continues to grow. Second, our individual wants are unlimited. Let's focus on this second factor. Remember Sam and his bundle of goods? It is always possible to imagine this bundle enhanced in some way so as to satisfy Sam's wants better or more fully. This can be a matter of adding new goods, adding more of a good he already possesses, or simply improving the quality of his goods. On this model, Sam's wants are not static. They are never defined once and for all by any bundle of goods he currently enjoys. He is always looking for a quantitative or qualitative upgrade. So the economy must always grow to meet his expanding needs.

Sometimes we respond to the havoc perpetual growth causes with technological fixes that are themselves environmentally problematic. For example, a key response to the depletion of natural fish stocks has been the development of aquaculture, the farming of aquatic organ- isms in "controlled" marine environments. In Canada, this practice began in the 1950s and is currently most heavily concentrated in British Columbia and New Brunswick. Aquaculture has become big business. In 2005, farmed fish production was valued at $715 million, accounting for 25 per cent of the total value of fisheries production in the country. But there are serious concerns about it, among them the following. First, the diseases and parasites that proliferate in farmed fish populations because of their dense stocking can be transmitted to wild popula- tions. Second, in an effort to combat these diseases, fish are given antibiotics, which can then leach out into the surrounding marine ecosystem. Third, strong evidence suggests that pink salmon smolts on the Pacific coast are being weakened by sea lice coming from fish farms. And finally, fish farms in British Columbia produce an amount of organic pollution equivalent to the quantity of sewage released into the ocean by a city of 500,000 people.

The industry is addressing some of these issues. For example, since many of the problems are caused by the unintended interaction between farmed and wild animals or ecosystems, the industry has begun experimenting with closed-tank technologies. It is too early to assess the extent to which the industry will be able to cope with all the problems. But however it turns out, we should be clear about the cultural forces driving developments such as this.

The imperative of growth seems inescapable in large part because we associate cease- less growth with the concept of progress, something that seems undeniably good. However, Canadian writer Ronald Wright has argued that this sort of thinking often gets us into "progress traps," reliance on technological fixes for our environmental problems that initially seem good but can often have extremely problematic long-term consequences. It is foolish, according to Wright, to live by drawing down nature's goods and services through ever-increasing economic

Continued

growth in the blithe assumption that when we run out of natural capital, technology will come to the rescue. Historically speaking, this view is completely unfounded. Instead, a cavalier approach to the consumption of natural capital has often led to civilizational crisis or even collapse.

This points to a big problem with the economic model generally—namely, that it provides no grounds on which to criticize people's wants rationally. Should Sam's wants be accepted uncritically? Sam is just one person, but if we all demand to have all our wants satisfied and this can happen only at the cost of massive social and/or ecological disruption, then surely we have good grounds for refusing to satisfy these wants. And in any case, does perpetual upgrade lead to more happiness? There are good reasons to think that it does not, at least not above a certain level of material security. If this is the case, then perhaps other values—such as poverty alleviation and environmental conservation—should override the growth imperative when they conflict.

The claim that our economy can grow ceaselessly is contrary to the manner in which every other species manages its "economy." This insight has inspired many economists to reject the basic assumptions of the neo-liberal model on which we have been focusing in this chapter. In its place they have developed **ecological economics**, a model that sees humans as mere parts of the larger biophysical reality. Prominent among these new economists is Clive Splash, who claims that the pro-growth stance of neo-liberal economics "fails to address the . . . nature of the widespread environmental change we are undertaking." Only economic policies that deny the growth imperative can address these changes adequately. Given the dominance in our culture of the idea that perpetual economic growth is both necessary and desirable, it does not take too much imagination to see how radical a proposal this is. So what does aquaculture represent: proof that perpetual economic growth is possible or yet another progress trap?

Sources: Freeman III 1998, 294; Dearden and Mitchell 2009, 285–8; Draper and Reed 2009, 327–32; Hackett and Miller 2008, 264–8; Ronald Wright, *A Short History of Progress* (Toronto: Anansi Press, 2004); Clive Splash, "The Economics of Climate Change Impact à la Stern," *Ecological Economics* 63, no. 4: 706–13.

Study Questions

1. Is it correct to say that a tree (for example) has no value until it is cut down and commodified?
2. Jamieson argues that "management" approaches to certain environmental issues are misguided. Why does he say this and do you agree with him?
3. A key assumption of neo-liberal economics is that economies must continually grow. What do you think of the opposing notion that we should construct a "steady-state" or "ecological" economics? If we put this idea into practice, how would some of our other values alter as a result?
4. Do you agree with Baxter that it makes sense to think about "optimal" levels of things like pollution? Why or why not?

Further Reading

Mark Anielski. 2009. *The Economics of Happiness*. Vancouver: New Society Publishers.

Herman Daly. 1997. *Beyond Growth: The Economics of Sustainable Development*. Boston: Beacon Press.

Herman Daly and Joshua Farley. 2009. *Ecological Economics: Principles and Applications*. Washington: Island Press.

Joseph Heath. 2001. *The Efficient Society: Why Canada Is as Close to Utopia as It Gets*. Toronto: Penguin.

Mark Sagoff. 2004. *Price, Principle, and the Environment*. Cambridge: Cambridge UP.

CHAPTER | 5

Environmental Pragmatism

Ecological Intuition Pump

A large and ever-increasing number of us now live in urban, suburban, or exurban areas. In these places, at least as they are usually constructed in North America, nature is carefully controlled. Although we can experience bits of it within the city, it is for the most part set apart from us, something we have to drive to. This can give us the feeling that *we* are separate from wild nature, an experience that can distort our understanding both of ourselves and of wild nature. Environmental pragmatists ask us to imagine new ways of living with, and in, nature. Can you imagine building cities so that they incorporate nature more fully than they currently do? Think of farm towers—skyscrapers devoted to the production of food—nestled in among the bank towers. Or large urban spaces where no cars or overhead flights are allowed and that act as huge bird sanctuaries or free-roaming spaces for *other* animals. Imagine post-carbon cities powered entirely by a "micro-grid" of geothermal, wind, and solar energy. The sights, sounds, and smells of such a place would be entirely different from those that currently assail us in our cities. In effect we would be blurring the boundary between artifacts like cities and the rest of the natural world. Wouldn't life in such a place cause us to think differently about nature? And how could such new experiences fail to provide us with new bases for our theoretical under-standing of the relation between humans and the rest of nature?

Ⓐ Introduction

Look at a satellite photograph of the Gulf of Mexico in the late spring, and you will notice something strange: a thick, swirling band of green hugging much of the coastline of Texas, Mississippi, and Louisiana. It is an algae bloom covering some 10,000 square kilometres. This part of the gulf has been subject to extensive **eutrophication**. The process begins when an excess of phosphates enters the marine body. In the case of the Gulf, this happens mainly as runoff into the Mississippi River (which drains into the Gulf) from Midwest farms using fertilizers loaded with nitrogen and phosphorous. It can also be caused by treated human waste water, which contains abundant quantities of the offending chemicals, being discharged into the water. The chemicals stoke algae growth, and the algae's bacterial decomposers then consume much of the water's available oxygen. Many marine animals—especially bottom-dwelling fish species—are starved of oxygen as this process unfolds, which is why unchecked eutrophication can create massive "dead zones" in lakes and oceans.

Something similar, alas, is happening to Lake Winnipeg, the centre of Canada's second-largest watershed. Lake Winnipeg is enormous. With a surface area of nearly 24,000 square kilometres, it covers close to 4 per cent of Manitoba. Somehow, the federal government decided that the lake was not its responsibility, and it has since been difficult to determine exactly who *is* responsible for scientifically monitoring and stewarding it. For nearly 30 years, the lake was ignored by federal scientists even though eutrophication had been proceeding in it all along. Appalled by this persistent bureaucratic oversight, Mike Stainton, a former employee of the federal Department of Fisheries and Oceans, formed the Lake Winnipeg Research Consortium. Its goal, as reported by Allan Casey in his book *Lakeland*, was to "bring real stewardship back to the lake." To this end, Stainton and his colleagues

> enlisted citizens who lived and worked on Lake Winnipeg: cottagers, sailing club people, commercial fishers, pig farmers, First Nations band councils, citizens of Winnipeg, Gimli, Selkirk, and other municipalities.[1]

The group is raising funds to monitor water quality in the lake and to lobby governments to help repair the damage done to it. These people are well on the way to saving the lake from the fate that has befallen a vast swath of the Gulf of Mexico and other bodies of water. The efforts of the consortium encapsulate this chapter's theme. We might be sceptical of the idea that every member of the consortium is motivated by the same value to steward the lake. In particular, it seems far-fetched to assume they all believe that the various ecosystems of the Lake Winnipeg watershed possess intrinsic value. More likely, their motivating values are diverse: economic, cultural, aesthetic, spiritual, scientific, and so on. But they have all made paramount the *practical* goal of preservation. They are, whether or not they self-identify this way, environmental pragmatists.

Environmental pragmatism is a relatively new movement in environmental ethics. Three themes surface consistently in the work of environmental pragmatists: (1) a rejection of the search for intrinsic value, associated with a denial that our values require the sort of "deep" grounding or justification that appeals to intrinsic value are meant to supply; (2) the espousal of value pluralism and a corresponding rejection of value monism (as we will see, this theme also involves a rejection of the notion that an acceptable environmental ethic must be fully non-anthropocentric and holist); and (3) a de-emphasis on the importance of theoretical debate, coupled with a focus on consensus-building with a goal to influencing public policy on important environmental issues.[2] Above all, the pragmatist wants us to *get things done* and is declaring that previous environmental ethics is simply inadequate to this task. Let's look at these three themes in detail.

Ⓑ Two Problems with Intrinsic Value

There are three key elements in the concept of intrinsic value. The first is that intrinsic value is non-relational; it is a self-owned or inherent property of the thing possessing it. Second, the intrinsically valuable thing is not of merely instrumental value. Third, intrinsic value is not bestowed *by* valuers—humans, for example. These three ideas suggest what we might call the radical *independence* of the intrinsically valuable thing. That which is intrinsically valuable is self-sufficient. However, in environmental ethics the most important way in which the concept of intrinsic value is used is to pick out that class of things that has moral standing. Here, the claim that a thing possesses intrinsic value is meant to act as a constraint on us. There are certain things we may not do to or with such things, or there may be certain actions we are morally required to take on their behalf.

What is the connection between the idea of constraints and that of radical independence? Think of it in the human case. To say that humans are intrinsically valuable is to claim, with Kant, that they are "ends in themselves" and not merely a means to someone else's ends. Their valuing activity is independent of both the natural world and the valuations of other valuers. This fact about them, Kant thinks, places on us a duty to respect their capacity to set their own ends in life, to live according to self-imposed ideals (as long as these ideals are not themselves morally objectionable). We go beyond the human case by dropping the reference to *self-imposed* ends or ideals (since presumably only humans have them) to show that an intrinsically valuable thing simply *has* its own interests or ends. We can go even further and say that, just like us, they *strive after* or pursue these ends, even if they do so non-consciously. Because their ends are set by their own natures and not by us (or anything else), we are morally required to respect such things, to refrain from infringing upon their interests or to actively promote such interests (or both). This is the sense in which making a claim of intrinsic value halts the process of moral justification. If successful, it provides a rationally compelling justification for action.

In her article later in this chapter, Jennifer Welchman offers some pointed criticisms of the concept of intrinsic value, especially of the claim that what is intrinsically valuable is uniquely choiceworthy. Before getting there, however, we will examine two other problems with appeals to intrinsic value. The first focuses on what the appeal is meant to do—namely, provide a *foundation* for our values. Suppose I believe that x is valuable. If I am asked to provide a reason for this claim, I might say that it is valuable because it helps me achieve or acquire something else of value, y. I might again be asked to provide grounds for this believing that y is valuable and might, again, refer beyond y itself to something else, z, from which y receives its value. So x and y are merely instrumentally valuable, and the same, presumably, might be true for z as well. However, so goes the argument, passing-the-value-buck cannot go on indefinitely. At some point we must arrive at a thing that is valuable in itself.

Aristotle, famously, made an argument like this, claiming that "happiness" (the Greek word is *eudaimonia*, which some prefer to translate as "flourishing") is the only intrinsically valuable thing. Whereas pleasure, intelligence, and honour may be worthy things to pursue, they are so only because and insofar as they lead to happiness. Happiness, however, is not like this: we want it for its own sake. Here, the idea of self-sufficiency is paramount. The intrinsically valuable thing, as such, does not depend on any other thing for the value it possesses. The claim is that the value being passed along in the chain of instrumentally valuable things must originate somewhere, otherwise it would not be *in* the instrumentally valuable things at all. Therefore there must be *something* that is intrinsically valuable.

Aquinas made an analogous argument about God. He reasoned that the chain of natural causes could not extend indefinitely backward, because in that case it could never have gotten started on its *forward* trajectory. So there must be a first or originating cause, and this cause is God.

Both arguments—those of Aristotle and Aquinas—are question-begging. That is, both contain in their premises the claim that there *must be* a first cause or an intrinsic value. But this is just what the argument is meant to prove. Why, we might ask, can a chain of natural causes not be traced indefinitely backward? Similarly, we might insist that all values can be justified by appeal to some other value or values. Here is how Anthony Weston puts the point:

> We justify a value by articulating the supporting role it plays with respect to other values, which in turn play a supporting role with respect to it, and by referring to the beliefs which make it natural, which it in turn makes natural by reaffirming those choices and models which link it to the living of our lives.[3]

This sort of view supports a more general claim about the justification of values known as **value coherentism**. The idea is that a value system is *internally* justified: any element in the system is justified to the extent that it fits in with—coheres with—other elements in that system. The view is usually opposed to **value foundationalism**, the position that to be justified, a value system must be anchored by one value—or cluster of values—that is itself, somehow, *self*-justifying. As we have seen, the appeal to intrinsic value performs this anchoring function for values. Weston's claim is that the whole project of "grounding" values in this fashion is pointless. We will never agree about our assignments of intrinsic value, and we don't need them in any case. We can get along just as well merely by "relating" our values to one another as well as to our beliefs.

This brings us to the second problem with intrinsic value. If there were intrinsic value, how could we detect it? The twentieth-century philosopher G.E. Moore argued that such value must be "intuited" rather than picked out by ordinary sensation or logical processes. Much has been made of the following thought experiment, dubbed the "last man" argument, in support of the claim that some things are intrinsically valuable. Suppose there were just two living individuals left on Earth: one man and one very beautiful tree. The last man knows he is just about to die, and he decides that before dying he will destroy the tree as well. The claim is that we intuitively judge the world he brings into being, the one with no life, to be inferior to the one whose existence he discontinues—the one with some life. So we think he is blameworthy or vicious for destroying the tree. But how could we make these judgments unless we also believe that living things do not get their value from us or from anything else? The world with some life is objectively better than the one with no life, *even though there are no humans around to make this judgment.*

Of course, one response is simply to deny that the world without the tree is inferior to the one with it. Another, more subtle approach has been suggested by Dale Jamieson:

> While the thought-experiment stipulates all valuers out of existence there are still some left hanging around. For we who are contemplating the world without valuers are ourselves valuers, and indeed we are contemplating the loss of something we find very valuable. Even if it is stipulated that we will never experience this world in either its preserved or its destroyed state, we are already experiencing these states in our imagination, and it is plausible that this is what governs our response to this thought experiment.[4]

The problem is epistemological. The thought experiment is not a reliable way of picking out intrinsic value, because it does not, and we cannot, remove valuers entirely from consideration. We cannot conceive of a valuable world that contains *no* valuers, for trying to do so presupposes our presence in the picture *precisely as valuers*. So there is an apparently insuperable obstacle to detecting intrinsic value. It may be out there, but we can't know that it is. No other way of describing how we get in touch with intrinsic value in the environmental sphere has fared much better than the last man argument.

In sum, we should reject appeals to intrinsic value because (1) we don't require them as foundations for our value judgments and (2) even if we did, we have no way of discovering them.

Appeals to intrinsic value have another flaw, according to environmental pragmatists. It is that such appeals generally rely on value monism. Recall the discussion of economics in Chapter 4. There we criticized the economic approach to nature, especially in the form of cost-benefit analysis, because it assumes that there is just one way of expressing value: the monetary way. But monism shows up in other approaches to environmental ethics too. It is tempting to think that extreme anthropocentrism—the economic approach—and extreme non-anthropocentrism—ecocentrism, say—are about as distinct as two theories could be. However, we might instead think of them as simply two faces of the same philosophical coin. Each denies that the other understands what things are of ultimate value, but they agree to the deeper claim that there is *just one* kind of thing that has ultimate value. One side says this thing is human wants, the other that it is the interests of ecosystems or the biosphere. (An ecocentrist is not required to say this, since she might think that humans, for instance, are *also* intrinsically valuable. Think of Callicott's image of the tree rings, from Chapter 3. But she could say it, and many do say it.) Rejecting such monism entails embracing value pluralism, the topic of our next section.

Ⓒ Value Pluralism: Reclaiming the Land Ethic

What would the founder of the land ethic, Aldo Leopold, think of this attempt to steer our way clear of appeals to intrinsic value? As it happens, there is something of a battle over the soul of Leopold in the literature. One side—the biocentrists, ecocentrists, and deep ecologists—think that the land ethic is telling us that what Leopold calls "the land" is the bearer of ultimate environmental value. This value is radically independent of humans and their needs. We have seen this interpretation of the land ethic in Chapter 3. On the other side stand the environmental pragmatists. The battleground is the famous claim that "a thing is right when it tends to preserve the integrity, stability, and beauty of the biotic community. It is wrong when it tends otherwise." Here is what the pragmatists think about the remark:

> Leopold was making the ... point that because of the complexity of the interrelationships in nature, and because there are so many different values exemplified in nature, the only way to manage to protect *all* of these diverse and pluralistic values is to protect the integrity of community processes.[5]

Notice the relatively modest claim: there are lots of different values, or valuable things, in the natural world, and none of them are, as it were, morally or metaphysically privileged relative to any of the others. The contrary claim would be that ecosystems as such have intrinsic value and that this is the value that the land ethic seeks to protect. The various things that go on in ecosystems—from the circulation of energy and nutrients to the various symbiotic relations that have grown up among species over the long course of evolution—are merely instrumental to the maintenance of this larger whole. To the extent that we advocate Gaia Theory, we could say instead that individual ecosystems themselves are only instrumentally valuable to the maintenance of Gaia or the biosphere, but at some point the value buck has to stop.

Leopold began his career as a forest *manager*, and he never really abandoned this perspective, even though he did alter his views about how best to manage the land. Indeed, as Anthony Weston argues in this chapter's first article, Leopold is best seen as engaging in a kind of philosophical ground-clearing. Leopold invites us to investigate the many and diverse ways we can value the natural world. He not only provides us with a richer understanding of that world but also compels us to do environmental ethics in an entirely novel way. Let's turn to this provocative analysis.

BEFORE ENVIRONMENTAL ETHICS

Anthony Weston

Since we now look at the evolution of the values of persons mostly from the far side, it is easy to miss the fundamental contingency of those values and their dependence upon practices, institutions, and experiences that were for their time genuinely uncertain and exploratory. Today we are too used to that easy division of labour that leaves ethics only the systematic tasks of "expressing" a set of values that is already established and abandons the originary questions to the social sciences. As a result, ethics is incapacitated when it comes to dealing with values that are now entering the originary stage. Even when it is out of its depth, we continue to imagine that systematic ethics, such as the ethics of the person, is the only kind of ethics there is. We continue to regard the contingency, open-endedness, and uncertainty of "new" values as an objection to them, ruling them out of ethical court entirely, or else as a kind of embarrassment to be quickly papered over with an ethical theory.

This discussion has direct application to environmental ethics. First and fundamentally, if environmental ethics is indeed at an originary stage, we can have only the barest sense of what ethics for a culture truly beyond anthropocentrism would actually look like. The Renaissance and the Reformation did not simply actualize some pre-existing or easily anticipated notion of persons but rather played a part in the larger co-evolution of respect for persons. What would emerge could only be imagined in advance in the dimmest of ways or not imagined at all. Similarly, we are only now embarking on an attempt to move beyond anthropocentrism, and we simply cannot predict in advance where even another century of moral change will take us.

Indeed, when anthropocentrism is finally cut down to size, there is no reason to think that what we will have or need in its place will be something called non-anthropocentrism at all— as if that characterization would even begin to be useful in a culture in which anthropocentrism had actually been transcended. Indeed, it may not even be any kind of "centrism" whatsoever, that is, some form of hierarchically structured ethics. It is already clear that hierarchy is by no means the only option.[6]

Second and correlatively, at this stage exploration and metaphor are crucial to environmental ethics. Only later can we harden originary notions into precise analytic categories. Any attempt to appropriate the moral force of rights language for (much of) the trans-human world, for example, ought to be expected from the start to be imprecise, literally confused. (Consider "animal rights." The very concept of animal seems to preclude one of them being one of "us,"

i.e., persons, i.e., rights-holders.) It need not be meant as a description of prevailing practice; rather, it should be read as an attempt to change the prevailing practice. Christopher Stone's book *Should Trees Have Standing? Toward Legal Rights for Natural Objects*, for example, makes a revisionist proposal about legal arrangements; it does not offer an analysis of the existing concept of rights.[7]

Something similar should be understood when we are invited to conceive not only animals or trees as rights-holders but also the land as a community and the planet as a person. All such arguments should be understood to be rhetorical, in a non-pejorative, pragmatic sense: they are suggestive and open-ended sorts of challenges, even proposals for Deweyan kinds of social reconstruction, rather than attempts to demonstrate particular conclusions on the basis of premises that are supposed to already be accepted.[8] The force of these arguments lies in the way they open up the possibility of new connections, not in the way they settle or "close" any questions. Their work is more creative than summative, more prospective than retrospective. Their chief function is to provoke, to loosen up the language, and correspondingly our thinking, to fire the imagination: to open questions, not to settle them.

The founders of environmental ethics were explorers along these lines. Here I want, in particular, to reclaim Aldo Leopold from the theorists. Bryan Norton reminds us, for example, that Leopold's widely cited appeal to the "integrity, stability, and beauty of the biotic community" occurs in the midst of a discussion of purely economic constructions of the land. It is best read, Norton says, as a kind of counterbalance and challenge to the excesses of pure commercialism, rather than as a criterion for moral action all by itself. Similarly, John Rodman has argued that Leopold's work should be read as an environmental ethic in process, complicating the anthropocentric picture more or less from within, rather than as a kind of proto-system, simplifying and unifying an entirely new picture, that can be progressively refined in the way that utilitarian and deontological theories have been refined over the last century.[9] Leopold insists, after all, that

> the land ethic [is] a product of social evolution.... Only the most superficial student of history supposes that Moses "wrote" the Decalogue; it evolved in the mind [and surely also in the practices!] of the thinking community, and Moses wrote a tentative summary of it.... I say "tentative" because evolution never stops.[10]

It might be better to regard Leopold not as purveying a general ethical theory at all but rather as simply opening some questions, unsettling some assumptions, and prying the window open just far enough to lead, in time, to much wilder and certainly more diverse suggestions or "criteria."

Third and more generally, as I put it above, the process of evolving values and practices at originary stages is seldom a smooth process of progressively filling in and instantiating earlier outlines. At the originary stage, we should instead expect a variety of fairly incompatible outlines coupled with a wide range of proto-practices, even social experiments of various sorts, all contributing to a kind of cultural working-through of a new set of possibilities. In environmental ethics, we arrive at exactly the opposite view from that of J. Baird Callicott, for example, who insists that we attempt to formulate, right now, a complete, unified, even "closed" (his term) theory of environmental ethics. Callicott even argues that contemporary environmental ethics should not tolerate more

than one basic type of value, insisting on a "univocal" environmental ethic.[11] In fact, however, as I argued above, originary stages are the worst possible times at which to demand that we all speak with one voice. Once a set of values is culturally consolidated, it may well be possible, perhaps even necessary, to reduce them to some kind of consistency. But environmental values are unlikely to be in such a position for a very long time. The necessary period of ferment, cultural experimentation, and thus multi-vocality is only beginning. Although Callicott is right, we might say, about the demands of systematic ethical theory at later cultural stages, he is wrong—indeed wildly wrong—about what stage environmental values have actually reached....

Space for some analogues to the familiar theories does remain in the alternative environmental ethics envisioned here. I have argued that although they are unreliable guides to the ethical future, they might well be viewed as another kind of ethical experiment or proposal rather like, for example, the work of the Utopian socialists. However unrealistic, they may, nonetheless, play a historical and transitional role, highlighting new possibilities, inspiring reconstructive experiments, even perhaps eventually provoking environmental ethics" equivalent of a Marx.

It should be clear, though, that the kind of constructive activity suggested by the argument offered here goes far beyond the familiar theories as well. Rather than systematizing environmental values, the overall project at this stage should be to begin co-evolving those values with practices and institutions that make them even unsystematically possible. It is this point that I now want to develop by offering one specific example of such a co-evolutionary practice. It is by no means the only example. Indeed, the best thing that could be hoped, in my view, is the emergence of many others. But it is one example, and it may be a good example to help clarify how such approaches might look and thus to clear the way for more.

A central part of the challenge is to create the social, psychological, and phenomenological preconditions—the conceptual, experiential, or even quite literal "space"—for new or stronger environmental values to evolve. Because such creation will "enable" these values, I call such a practical project "enabling environmental practice."

Consider the attempt to create actual, physical spaces for the emergence of transhuman experience, places within which some return to the experience of and immersion in natural settings is possible. Suppose that certain places are set aside as quiet zones, places where automobile engines, lawnmowers, and low-flying aeroplanes are not allowed and yet places where people will live. On one level, the aim is modest: simply to make it possible to hear the birds, the winds, and the silence once again. If bright outside lights were also banned, one could see the stars at night and feel the slow pulsations of the light over the seasons. A little creative zoning, in short, could make space for increasingly divergent styles of living on the land—for example, experiments in recycling and energy self-sufficiency, Midgleyan mixed communities of humans and other species, serious "re-inhabitation" (though perhaps with more emphasis on place and community than upon the individual re-inhabiters), the "ecosteries" that have been proposed on the model of monasteries, and other possibilities not yet even imagined.[12]

Such a project is not Utopian. If we unplugged a few outdoor lights and rerouted some roads, we could easily have a first approximation in some parts of the country right now. In gardening, for example, we already experience some semblance of mixed communities. Such practices as beekeeping, moreover, already provide a model for a symbiotic relation with the "biotic community." It is not hard to work out policies to protect and extend such practices.

Enabling environmental practice is, of course, a practice. Being a practice, however, does not mean that it is not also philosophical. Theory and practice interpenetrate here. In the abstract, for example, the concept of "natural settings," just invoked, has been acrimoniously debated, and the best-known positions are unfortunately more or less the extremes. Social ecologists insist that no environment is ever purely natural, that human beings have already remade the entire world, and that the challenge is really to get the process under socially progressive and politically inclusive control. Some deep ecologists, by contrast, argue that only wilderness is the "real world."[13] Both views have something to offer. Nevertheless, it may be that only from within the context of a new practice, even so simple a practice as the attempt to create "quiet places," will we finally achieve the necessary distance to take what we can from the purely philosophical debate and also to go beyond it toward a better set of questions and answers.

Both views, for example, unjustly discount "encounter." On the one hand, non-anthropocentrism should not become anti-anthropocentrism: the aim should not be to push humans out of the picture entirely but rather to open up the possibility of reciprocity between humans and the rest of nature. Nevertheless, reciprocity does require a space that is not wholly permeated by humans either. What we need to explore are possible realms of interaction. Neither the wilderness nor the city (as we know it) is "the real world," if we must talk in such terms. We might take as the most "real" places the places where humans and other creatures, honoured in their wildness and potential reciprocity, can come together, perhaps warily but at least openly.

The work of Wendell Berry is paradigmatic of this kind of philosophical engagement. Berry writes, for example, of "the phenomenon of edge or margin, that we know to be one of the powerful attractions of a diversified landscape, both to wildlife and to humans." These margins are places where domesticity and wildness meet. Mowing his small hayfield with a team of horses, Berry encounters a hawk who lands close to him, watching carefully but without fear. The hawk comes, he writes,

> because of the conjunction of the small pasture and its wooded borders, of open hunting ground and the security of trees. . . . The human eye itself seems drawn to such margins, hungering for the difference made in the country-side by a hedgy fencerow, a stream, or a grove of trees. These margins are bio-logically rich, the meeting of two kinds of habitat.[14]

The hawk would not have come, he says, if the field had been larger, or if there had been no trees, or if he had been ploughing with a tractor. Interaction is a fragile thing, and we need to pay careful attention to its preconditions. As Berry shows, attending to interaction is a deeply philo-sophical and phenomenological project as well as a practical one—but nonetheless, it always revolves around and refers back to practice. Without actually maintaining a farm, he would know very little of what he knows, and the hawk would not—could not—have come to him.

Margins are, of course, only one example. They can't be the whole story. Many creatures avoid them. It is for this reason that the spotted owl's survival depends on large tracts of old-growth forest. Nonetheless, they are still part of the story—a part given particularly short shrift, it seems, by all sides in the current debate.

It is not possible in a short article to develop the kind of philosophy of "practice" that would be necessary to work out these points fully. However, I can at least note two opposite pit-falls in speaking of practice. First, it is not as if we come to this practice already knowing what values we will find or exemplify there. Too often the notion of practice in contemporary philosophy has degenerated into "application," that is, of prior principles or theories. At best, it might provide an opportunity for feedback from practice to principle or theory. I mean something more radical here. Practice is the opening of the "space" for interaction, for the re-emergence of a larger world. It is a kind of exploration. We do not know in advance what we will find. Berry had to learn, for example, about margins. Gary Snyder and others propose Buddhist terms to describe the necessary atti-tude, a kind of mindfulness, attentiveness. Tom Birch calls it the "primary sense" of the notion of "consideration."[15]

On the other hand, this sort of open-ended practice does not mean reducing our own activ-ity to zero, as in some form of quietism. I do not mean that we simply "open, and it will come." There is not likely to be any single and simple set of values that somehow emerges once we merely get out of the way. Berry's view is that a more open-ended and respectful relation to nature requires constant and creative activ-ity—in his case, constant presence in nature, constant interaction with his own animals, maintenance of a place that maximizes mar-gins. Others will, of course, choose other ways. The crucial thing is that humans must neither monopolize the picture entirely nor absent ourselves from it completely but rather try to live in interaction, to create a space for genuine encounter as part of our ongoing reconstruction of our own lives and practices. What will come of such encounters, what will emerge from such sustained interactions, we cannot yet say.

No doubt it will be argued that Berry is necessarily an exception, that small un-mechanized farms are utterly anachronistic, and that any real maintenance of margins or spaces for encounters is unrealistic in mass society. Perhaps. But these automatically accepted common-places are also open to argumentation and experiment. Christopher Alexander and his colleagues, in *A Pattern Language* and elsewhere, for example, make clear how profoundly even the simplest architectural features of houses, streets, and cities structure our experience of nature—and how they can be consciously redesigned to change those experiences. Windows on two sides of a room make it possible for natural light to suffice for daytime illumination. If buildings are built on those parts of the land that are in the worst condition, not the best, we thereby leave the most healthy and beautiful parts alone, while improving the worst parts. On a variety of grounds, Alexander and his colleagues argue for the presence of both still and moving water throughout the city, for extensive common land—"accessible green," sacred sites, and burial grounds within the city—and so on. If we build mindfully, they argue, maintaining and even expanding margins is not only possible but easy, even with high human population densities. . . .[16]

I [have] offered only the barest sketch of enabling environmental practice: a few examples, not even a general typology. To attempt a more systematic typology of its possible forms at this point seems to me premature, partly because ethics has hitherto paid so little attention to the cultural constitution of values that we have no such typology and partly because the originary stage of environmental values is barely underway.

Moreover, enabling environmental practice is itself only one example of the broader range of philosophical activities invited by what I call the co-evolutionary view of values. I have not denied that even theories of rights, for instance, have a place in environmental ethics. However, it is not the only "place" there is, and rights themselves, at least when invoked beyond the sphere of persons, must be understood (so I argue) in a much more metaphorical and exploratory sense than usual. This point has also been made by many others, of course, but usually with the intention of ruling rights talk out of environmental ethics altogether. A pluralistic project is far more tolerant and inclusive. Indeed, it is surely an advantage of the sort of umbrella conception of environmental ethics I am suggesting here that nearly all of the current approaches may find a place in it.

Because enabling environmental practice is closest to my own heart, I have to struggle with my own temptation to make it the whole story. It is not. Given the prevailing attitudes, however, we need to continue to insist that it is part of the story. Of course, we might still have to argue at length about whether and to what degree enabling environmental practice is "philosophical" or "ethical." My own view, along pragmatic lines, is that it is both, deeply and essentially. Indeed, for Dewey the sustained practice of social reconstruction—experimental, improvisatory, and pluralistic—is the most central ethical practice of all. But that is an argument for another time. It is, nevertheless, one of the most central tasks that now calls to us.

Weston thinks that the main task of environmental philosophers now is to "enable environmental practice." We are at a very early stage in the process of understanding nature—the science of ecology is still quite immature, for example—and so we should not expect a single notion of what is ultimately worthy of protection in the natural world to be available to us yet. We are, in a sense, at a stage where the most important thing to do is simply to *listen* to the various ways humans have over the ages given expression to value in nature. For instance, some pragmatists insist that we should look more carefully at what poets of nature have to tell us. Let's follow up that suggestion and consider "Dark Pines under Water" by Canadian poet Gwendolyn MacEwen:

> This land like a mirror turns you inward
> And you become a forest in a furtive lake;
> The dark pines of your mind reach downward,
> You dream in the green of your time,
> Your memory is a row of sinking pines.
> Explorer, you tell yourself this is not what you came for
> Although it is good here, and green.
> You had meant to move with a kind of largeness,
> You had planned a heavy grace, an anguished dream.
> But the dark pines of your mind dip deeper
> And you are sinking, sinking, sleeper
> In an elementary world;
> There is something down there and you want it told.[17]

MacEwen presents an arresting vision of self-discovery through immersion in and contemplation of nature. To the extent that we took nature to be merely a resource—we had "meant to move with a kind of largeness" on the land—we are being invited to consider the possibility that this sort of exercise might reveal an alternative way of living. MacEwen does not tell us what exactly we might discover in the process, though she lures us with the promise of something "elementary." Perhaps we should get out into the wild and see for ourselves, and then—just maybe—we will be inclined to think of wild nature as something worth preserving rather than simply using. This insight, in turn, could make us see the value in aiding environmentalists as they work to change public policy ("There is something down there and you want it told"). The important thing is that we may not have come to this point without having read the poem, and we can do all of this while being entirely unconcerned about whether or not the dark pines are intrinsically valuable.

In a separate article, Anthony Weston has argued that we don't need to "ground" our various experiences of and attempts to express nature's value. Poetry, biography, fiction, and, of course, philosophy instead offer us multi-faceted "portraits" of ourselves and nature.[18] Any attempt to reduce this pluralism to a single kind of value will inevitably oversimplify the many ways in which we apprehend nature. In an important sense, this will falsify the picture of nature we obtain, thereby distorting our view about how best to relate to nature. There may come a time when we understand enough about ecology and ourselves to allow a more monistic environmental ethic to occupy centre stage, but we are not there yet. Our understanding of the many values of nature is still at an "originary stage," and we therefore must not attempt to force this understanding into a single mould. This, argue the pragmatists, is the real lesson of the land ethic.

Ⓓ The Importance of Building Consensus

Of course, environmental pragmatists do not espouse some purely abstract fidelity to the rich diversity of natural values, as though respecting diversity itself were the cardinal value. The claim, rather, is that we will never get the sort of consensus we require if we insist on forcing everyone to speak the same value language. People have plural ways of encountering nature and of valuing it, but many of these ways nevertheless express an underlying respect for wild nature. If we cannot accept these ways of valuing unless and until they prove their non-anthropocentric and holist bona fides, we will not be able to muster the consensus required for meaningful political action. And, again, the goal is to get things done on behalf of nature.

Some pragmatists argue that the move to provide deep theoretical justifications for our moral views only results in hardening those views to the point where it is more difficult to achieve agreement with people holding opposing views. Imagine Jim, Gerry, and Jill, each of whom holds a different view about whether a company ought to be allowed to develop a ski resort on a piece of privately owned but environmentally sensitive land. Jim owns the land and wants to lease it to the company, Jill is an environmental activist, and Gerry is a municipal official in the town where the proposed development is to take place. Jim wants the development to go forward on a large scale. Gerry is cautiously supportive of development but wants it to proceed on a more limited scale than what Jim is proposing. Jill thinks that valuable and sensitive ecosystems will be damaged with any development and so opposes the project on any scale. Each of the three can appeal to a moral principle in support of his or her views: Jim to the inviolability of libertarian property rights, Jill to the intrinsic value of ecosystems, and Gerry to utilitarianism (he thinks the greater good will be served by limited development).[19]

Discussing a similar case, Paul Thompson claims that the participants to such disputes are more likely to reach a workable solution precisely to the extent that they refrain from providing the theoretical arguments just sketched:

> [T]his analysis shows why the policy standoff can become protracted. The availability of incompatible moral justifications for each position can form the basis for a brand of self-righteousness on the part of each interest that bodes ill for a consensus solution of the problem.[20]

We have all seen something like this happen. When people believe their position is deeply grounded, they can become absolutists. Indeed, part of the point of providing philosophical foundations for beliefs and values is to make the latter more difficult to dismantle. This process, however, can lead to self-righteousness and a tendency to vilify those on the other side of the dispute. It is the sort of confrontation that is nearly always a **zero-sum game**: someone seeks to win, the result of which is that someone else will lose. By contrast, a real commitment to pluralism is a **non-zero-sum game**: with sufficient good will, all interested parties can win. But building consensus is dangerous. If we eschew appeals to deep theory in justification of our values, won't we become morally rudderless in the face of alternative views? Jennifer Welchman's response to this challenge, in the chapter's next article, is that we can avoid such an outcome and achieve consensus on important environmental issues if we think of ourselves as "satisficers."

ENVIRONMENTAL PRAGMATISM

Jennifer Welchman

Environmental pragmatism is a relatively new philosophical approach to the ethical evaluation of human interactions with natural entities and systems, loosely based upon the pragmatic moral philosophy of John Dewey, whose architects include philosophers such as Bryan Norton, Anthony Weston, Andrew Light, Ben Minteer, and Paul Thompson. While the way each builds upon Dewey's pragmatic moral philosophy differs in important respects, their common commitment to a pragmatic conception of values grounds their shared conceptions of environmental ethics.

Environmental pragmatists do not accept the view, common to biocentrists and ecocentrists, that Western ethical traditions lack the resources to ground moral arguments to protect and conserve non-human natural entities and systems. Consequently, environmental pragmatists do not call for the creation of radical new principles of value at odds with those central to our traditional normative systems. Nor do they interpret "environmental ethics" as the project of constructing and defending new non-anthropocentric moral theories as rivals to the systems we now possess. Instead, environmental pragmatists tend to view environmental ethics as a branch of applied ethics, analogous to bioethics and to business and professional

ethics. Consequently, they see the role of moral philosophy in environmental matters to be one of helping to clarify issues and values in such a way as to allow those holding divergent views to reach a workable consensus that preserves and promotes, as equitably as possible, the values of all involved.

Environmental pragmatists have often argued that an important benefit of their approach is that it sidesteps potentially divisive issues arising from philosophical arguments about the theoretical foundations of the values we assign nature. Critics have charged that any advantages conferred by the approach are exaggerated. For example, J. Baird Callicott rejects the suggestion that biocentrists and ecocentrists have been guilty of a "preoccupation with theory" to the exclusion of practical problem-solving. He argues it is a mistake to suppose that "it makes no difference to environmental practice and policy whether we think of nature as having intrinsic value or only instrumental value."[21] Callicott is dubious about Bryan Norton's "convergence hypothesis" (that individuals engaged in open, equitable, and reflective discussion about environmental controversies will tend to converge on common solutions whatever the values they assign nature), which he views as little more than an "article of faith."[22] And further, Callicott thinks it by no means clear that environmental pragmatism's practical approach constitutes an alternative, rather than a supplement, to theoretical inquiry. Properly speaking, he argues, theory and practice are "complementary, not competitive." And we should never forget that "theory does make a difference to practice."[23]

Callicott's objections are not without merit. If environmental pragmatism is to be a genuine philosophical alternative to biocentric, ecocentric, and/or environmental virtue ethics, as opposed to a mere grab-bag of convenient strategies for getting stakeholders with differing views to talk constructively together, environmental pragmatism must develop and defend a theoretical platform that unifies and explains its peculiar approach to environmental ethics. Though I cannot speak for environmental pragmatists generally about how this is to be done, Dewey's moral philosophy provides a common ground upon which a reasonably representative account can be constructed.

I. Pragmatism on Values

Following Dewey, environmental pragmatists put little reliance upon the intrinsic/instrumental value distinction that has been central to so many discussions of the value or values of nature.[24] This is not because they consider theoretical inquiry into the intrinsic value of natural entities and systems uninteresting or unimportant but rather because they object to certain presuppositions about intrinsic value that have framed these inquiries. One particularly important example is the common presupposition that intrinsic value is best understood in opposition to instrumental value.[25] Most monographs and textbooks on environmental ethics begin discussions of the value of nature with a distinction between "intrinsic" value, the value a thing is said to have in virtue of its own

inherent properties, independent of relations to other things, and "instrumental" value, the value a thing is said to have as a means or instrument for the achievement of other intrinsically valuable ends. If one pauses to reflect upon the conflations of value claims involved, one can readily understand why pragmatists (and not only pragmatists) have raised objections.[26]

A claim that a thing is intrinsically valuable is a metaphysical claim about the nature of the properties upon which the value supervenes (or attaches). Specifically, it is a claim that these are properties internal to the thing itself and independent of any extrinsic properties it has in virtue of its relations to things external to it. By contrast, a claim that a thing is valuable instrumentally is not a metaphysical claim. It says nothing about the status of the properties to which that value attaches. It is instead a claim about the end value or choiceworthiness of the thing in question—whether the thing is choiceworthy for its own sake or rather as a means to some further end or goal. Neither sort of value claim entails the other.

Consider a few examples: an instance of pleasure, a work of art, and an endangered species such as the woodland caribou living in the Little Smoky forest outside Jasper National Park in western Alberta.[27] Each might be valued in virtue of either their intrinsic or their extrinsic properties. Does anything follow about the kind of value that we must assign them in each case?

Take an instance of pleasure. Assume its value supervenes upon its intrinsic properties. What does this tell one about whether or how someone might find it choiceworthy? A person might value that instance of pleasure for its own sake or, alternately, might value it solely as a means of quieting a crying child. The properties that make it an end in itself in one context are precisely the same properties that make it instrumentally valuable in the other.

Or consider a work of art. Imagine one has been given an oval mask made of unpainted, greying wood with simply carved facial features unremarkable either for their design or craftsmanship. Such a mask could possess a certain aesthetic merit in virtue of its intrinsic formal qualities. But since its formal qualities are not remarkable, neither would its merit in virtue of those qualities alone be remarkable for most recipients of such a gift. Now let us assign it some extrinsic properties. Imagine one learns that the object in question is a nineteenth-century Inuit ceremonial mask, created and used for celebratory feasts. Extrinsic properties of this sort would transform most recipients' estimation of the gift. But how? Some recipients will value it instrumentally for the high price it would fetch at an auction. Others will scorn such considerations—some of them cherishing the mask in virtue of its special historical associations, others as a memento of their friendship with the giver. That the value assigned in each case supervenes primarily upon the object's extrinsic properties does not determine which sort of value the recipients will assign it.

Finally, consider the caribou. We might value them in virtue of their intrinsic properties as either sentient or living organisms or, alternately, in virtue of their extrinsic properties: their evolutionary history or their roles in Canadian immigrant and/or Aboriginal culture. Does the metaphysical status of the salient properties determine the sort of values supervening upon them? Again, the answer is no. We may cherish and try to preserve them for their own sakes, value them instrumentally as means to our ends, or both. Since there is no necessary relationship between the values of things as ends or means and the nature of properties on which those values supervene, the status of properties a thing possesses does not by itself determine whether that thing can or

should be cherished as an end or instrumentally valued as a means. What determines this is the perspective of the evaluator.

This leads us to a second presupposition about intrinsic values with which pragmatists take issue, the common presumption that unlike instrumental values, the end-values that things possess in virtue of their properties (internal or external) is absolute—that is, wholly independent of context. From a pragmatic perspective, to value a thing as a "means" is to value it from a perspective within which functional considerations take priority. To value a thing as an "end" is to value it from a perspective within which, for the moment at least, functionality is not a priority. Our situation is such that we can allow ourselves to focus on the qualities that might make the thing choiceworthy to us for its own sake, as opposed to being choiceworthy for its functional efficacy in helping us to resolve some practical problem. But though end-value assignments are independent of functional considerations, they are not independent of the likes, tastes, concerns, or beliefs of the agent who makes that assignment.

This might seem to invite the worrying conclusion that our value judgments could all change from moment to moment, with each passing change in our moods, whims, situations. But happily this is not the case. Neither our physical natures nor the settled dispositions of our characters change from moment to moment. Our lives are organized around long-term projects and commitments that lend further stability to the perspectives we take when evaluating things in our environments. As a result, for each of us, some things, activities, or persons will be stable, unchanging, relatively speaking, "absolute" ends that will not change with changes in our particular circumstances. And because we are all members of the same species, these ends will overlap. There will be certain ends that are, relatively speaking, absolute ends for us all. And if we are members of the same communities or cultural traditions, the overlap will be greater still.

However, since no two people (or two groups) are ever quite the same, however closely tied by common nature and cultural nurture, no two individuals' (or groups') perspectives are ever quite the same. Different individuals and groups will differ in their value assignments. Some will cherish old-growth forests and wild species as things to be promoted or preserved (i.e., cherish them as ends), while others will not. Some who cherish these wild entities will cherish them in virtue of their intrinsic properties (e.g., being subjects of a life or constituents of intrinsically valuable wholes), others in virtue of their extrinsic properties (e.g., scientific importance or cultural significance). These differences, though often reconcilable, are not eradicable. There is no neutral public perspective from within which perspectives can be impartially compared. Differences about what we should assign end-values are as intractable as differences about the genesis of the universe and its meaning or purpose. For those who value equity and justice, this entails that our response to disagreements about the values of nature should be the same as that we adopt toward disagreements between religious faiths. When disputes arise, we should aim at mutual tolerance and equitable accommodation of all the diverging interests and values in play, as far as is practically possible, whatever our own views may be.

II. Pragmatism and Practical Reasoning

As pragmatists understand values, human values are plural, irreducible, and non-intersubstitutable. That is, people assign end-values to many different kinds of things. The end-values assigned are not all of one type (thus the plurality is not reducible to some one single end-value). And those end-values cannot be substituted one for another (e.g., supplying a person with opportunities for aesthetically valuable experiences does not replace, or necessarily make good, lost opportunities for autonomous self-determination). Human values originate with human beings and thus are "anthropogenic." But this does not entail the view that human values must necessarily be "anthropocentric," with end-value assignments restricted to human beings alone. Most people assign nature at least some non-instrumental values in virtue of a range of extrinsic properties (e.g., spiritual, aesthetic, religious, scientific, and/or historic). A smaller, yet substantial, number cherish nature as valuable for its intrinsic properties alone. This pluralist conception of values underpins the model of practical reasoning that environmental pragmatists tend to employ when decisions have to be made about human interactions with natural entities and systems in particular cases. Specifically, pragmatists tend to eschew "maximizing" and "optimizing" models in favour of "satisficing."

Broadly speaking, there are three goals toward which practical reasoning may be directed: maximizing, optimizing, or satisficing. If one is a value monist who supposes that the object of practical reasoning is always to promote just one kind of valuable quality or state of affairs, then one will naturally suppose that rational choice is choice that maximizes enjoyment of that quality or state of affairs. Pluralists, who deny that there is any quality or state of affairs that is valuable as an end, will just as naturally arrive at a different conclusion: that the practically rational choice is the choice that brings about the optimal set of what is valued.

Pluralists cannot adopt a maximizing conception of the goal of practical reasoning, because maximization is only practical in situations where just one value is at stake. When playing chess, for example, one's overall objective is to avoid losing to one's opponent. So one can choose moves with a view to maximizing this outcome. But in any situation where one has a plurality of non-interchangeable objectives, practicality dictates a different strategy. If the values in play are few, the most rational strategy may be optimization. Say, for example, you have only six hours in which to obtain a birthday gift for a friend with the $50 you have allotted for the purpose. Your friend is an avid and indiscriminate collector of Edmonton Oilers memorabilia. Here, it is possible to opti mize for the three values in play: time, money, and memorabilia. You find the best-equipped shop you can reach within the six hours and buy her as many items as you can afford.

Whether one is a value monist or a pluralist, a third strategy—satisficing—will sometimes be a rational strategy for resolving particular problems. Satisficing is a simplified form of practical reasoning, first developed by an economic theorist, Herbert Simon, as a model for rational decision-making in real-life situations where decision-makers do not know what all their options are or how much time they would need to expend to identify and assess all their options (supposing they actually had time enough to accomplish this).[28] Because our options are not readily apparent to us, we must investigate without any assurance that the benefits of an extended search will outweigh the costs involved. In these sorts of circumstances, Simon argued that it is rational to simplify our decision-making by (1) setting minimum "aspiration levels" for the benefits we wish to obtain, prior to commencing our search for options by which to achieve them, and (2) stopping our search upon discovery of the first option that meets one's aspiration levels without unacceptable costs.[29]

It is generally agreed that satisficing can be rational for the speedy realization of "local" objectives in the service of maximizing or optimizing the individual's or organization's longer-term or "global" projects. That is, it is rational to settle for a suboptimal or merely "good enough" solution of a particular problem when doing so allows us to pursue our global objectives more efficiently. But it is not generally thought rational to settle for suboptimal outcomes at the level of our more global objectives (whatever they might be). So many argue that satisficing is a rational strategy for practical reasoning, not as a general rule, but only in special cases where attempting to maximize or optimize local objectives would be counterproductive (or suboptimal). Some pragmatists would accept this in principle. But all will argue that the variety of values at stake in significant real-life decisions makes it impossible to achieve the kind of certainty about what our options are, and what the consequences of their adoption would be, that is essential if optimizing our more global objectives is not to prove a counterproductive strategy.

When we are trying to decide what careers to pursue, where to live or with whom, whether to have children or how to educate them, which community objectives to support or oppose, optimizing is rarely a practical objective. Consequently, we satisfice. We set minimum thresholds for the non-intersubstitutable values we aim to realize. Then we use them as a basis for assessing our options as we begin to investigate what these options may be. We continue the search for options as long as time allows, reasoning correctly that in circumstances that preclude our determining which option (if any) actually is optimal, we can view any option that is "good enough" as a rational choice.

Since satisficing decision-making relies upon predetermined minimum thresholds for certain valued ends or objectives, great care must be taken in deciding what these thresholds are to be. We must be mindful of two important facts about human psychology: (1) our short-term interests often blind us to our longer-term interests, and (2) our own perspectives often blind us to values central to other persons' perspectives. Setting our minimum thresholds too quickly in our own cases or without adequate consultation with others in cases of collective decision-making is apt to cause us to set our thresholds for some values too low and to fail to set any at all for others. As a result, we are liable to find that we have failed to ensure that we will have adequate opportunities to realize the former and that we have irrevocably lost opportunities ever to realize the latter. For this reason, rational satisficers, individually and collectively, will want to avoid setting their aspiration levels in ways that irrevocably close off options to realize values that may not seem immediately pertinent to a decision context, whether individual or collective. It is for the same reason that environmental pragmatists believe the most important contribution philosophers can make to environmental ethics and policy is to help ensure that such deliberations do not end prematurely and with irrevocable consequences. They often favour, as Callicott correctly notes, a "bottom-up, rather than a top-down approach" in which philosophers concentrate on helping encourage breadth of vision of the values at stake in particular controversies rather than on theoretical questions about the justification of different schemes for ranking values abstractly considered.[30] Thus, environmental pragmatists' suggestions about the practical contributions philosophers can make to the resolution of such controversies should not be taken as an indication of their rejection of value theory per se. It should instead be understood as dictated by the theories of value and of practical reasoning they find most persuasive.

III. Valuing Nature: A Case Study

To illustrate the practical implications of adopting the environmental pragmatist outlook, let us consider a current controversy: whether and/or how to try to protect Alberta's Little Smoky herd of woodland caribou, a species endangered throughout Canada. Logging has played an important role in their decline, since logging has decreased the habitat available to them while simultaneously increasing predation by wolves. Logging operations create gaps in and roadways through the forest. These attract elk and deer to regions deeper within the forest than any of these animals would otherwise travel. The wolves follow the elk and deer into the forest, where they also have access to the slower-moving caribou. Predation has escalated, further diminishing the herd.

Readily identifiable options by which to respond to this situation include: increased restrictions on logging, culling of the local wolf packs annually, some combination of these first two options, and allowing the herd to die out. But before we try to assess these options, we must decide what our minimum aspirations should be for this and any similar situations. In cases like this, a multiplicity of irreducibly plural personal and social values supervene upon the natural entities and systems involved. If, as seems likely, the woodland caribou and their old-growth forest home are important constituents of a healthy environment, their instrumental values are such that no rational individual could be wholly indifferent to them. Both are also non-instrumentally valuable to many people: to some in virtue of their cultural and spiritual associations, to others in virtue of their aesthetic qualities, and to still others in virtue of their being teleological centres of life and/or components of larger ecological wholes. Since whatever strategy we ultimately adopt will affect a wide range of persons (as well as the caribou, wolves, and forest), it is essential that dialogue about where to set our minimum aspiration levels should not exclude anyone whose contribution can alert us to values that might be at risk. Discussion should not, for example, be limited to those with only one kind of stake in the issue, whether economic, cultural, recreational, spiritual, or moral.

A few decades ago, the loss of the Little Smoky caribou as a by-product of commercial resource extraction might have been judged to be as insignificant an event as was the accidental extirpation of the swift fox from Canada's prairies through agricultural development. Ecological science has taught us that such by-products of anthropogenic disturbances are never insignificant events, even when assessed for their impact on anthropocentric values alone. Non-anthropocentric-value theorists, such as Callicott, Holmes Rolston, Paul Taylor, and Tom Regan, among others, have made important contributions to fuller, more reflective value assessments of

environmental policies by providing those who value non-human nature for its own sake with clear and coherent means of articulating their views. Nor have they been alone. Aboriginal activists' contributions in helping the dominant, non-Aboriginal culture to recognize the values nature has from the perspectives of Aboriginal people have also been particularly important for improving the dialogue among Canadians on environmental issues.

All of these contributions have helped to enrich our collective understanding of the plurality of values at stake as Albertans struggle with the problem of the threats their activities pose to the survival of the Little Smoky and its caribou. But for the environmental pragmatist, no one of these values can or should necessarily trump any of the others. As long as equity is one of our common cherished values, our aspirations must be set in such a way as to achieve equitable accommodation, insofar as is practical, of all the diverging interests and values at stake. This is not, as another critic of environmental pragmatism has charged, either a counsel of despair or an unholy compromise with human apathy, arrogance, or greed.[31] It is instead an appreciative response to the richness and complexity of the many values of nature.

A consistent theme in the work of environmental pragmatists is the need to find solutions to environmental problems on the ground, at the level of particular communities:

> Pragmatic necessity implies that any analysis of [environmental] problems that does not facilitate the formation of broader community and action to address problems is philosophically flawed.... [M]ultiple ethical viewpoints must be integrated into a community of interest and hope.[32]

Welchman makes the point too: environmental ethics is chiefly concerned with allowing those "holding divergent views to reach a workable consensus that preserves and promotes, as equitably as possible, the values of all involved." On this understanding, there are two core principles of applied environmental ethics: first, get the job done; second, do so in a way that incorporates the interests of all stakeholders as much as possible. The real novelty in Welchman's analysis, however, is the role she gives to the concept of satisficing. Satisficing is an important concept because it strikes a mean between high-minded idealism and anything-goes relativism.

That is, satisficing allows us to avoid two potential dangers. The first—that of idealism or absolutism—will likely prevent the resolution of our environmental problems because the contending parties will have hardened views that admit of little compromise. The second danger—relativism or moral rudderlessness—insists too strongly on the value of compromise and may therefore allow us to act in ways that violate relevant values. Either we get nothing done or we get the wrong things done. The satisficer does not require ideal solutions to environmental problems, but she will also fix her "minimum aspiration levels" in such a way that the right problems are solved in the right way. And, again, the key to ensuring that this happens is that as many divergent viewpoints as possible are taken into account in framing and attempting to solve environmental problems.

The effort to protect Alberta's Little Smoky herd of woodland caribou is an elegant example of this process at work. Here is another. In the Temagami region of northeastern Ontario, there are some marvellous stands of old-growth red and white pine trees. At one time, the trees covered nearly 6 million hectares of land in southeastern Canada and the eastern United States. Now the area has been reduced to less than 1 per cent of this total, which means the ecosystem is officially endangered. In spite of this, in 1988 the provincial government made a decision to allow extensive logging in the Temagami region. The problem is that at the time the area was the subject of a land claims action on the part of the local First Nations group, the Teme-Augama Anishinabai (TAA). The plans to clear-cut the area were made without any consultation with this group. There were extensive blockades by the Natives and their environmentalist allies, with many arrests. Eventually, the original logging plans were dropped, and the area has since seen some creative attempts at community-based management of the forest resource.[33]

For example, in 1991 the TAA and the province of Ontario set up the Wendaban Stewardship Authority (WSA). In 1994, the WSA set out a 20-year plan (extending to 2014) for community-based forest management. The executive of the WSA is comprised of 12 members, 6 from the TAA and 6 representing the province (although none of the latter are public servants). The WSA explicitly endorses a holistic approach to land management. As Monica Mulrennan reports, this approach is based on four principles: "sustained life, sustainable development, co-existence between Aboriginal and non-Aboriginal peoples, and public involvement in the activities of the authority." When it was first established, many were sceptical that it would work, but—although there are still many problems to be worked out in the management of the forest resource—progress has been made in the development of practices that cohere with the four principles. In particular, effective regimes have been established to control access to the forest via tourism, recreation, timber extraction, and so on.[34]

In its 1996 report, the Royal Commission on Aboriginal Peoples addressed the activities of the WSA. This is what it had to say:

> Whatever its future, there have been several positive lessons from this experiment in shared jurisdiction. Not only has the WSA generated support and collaboration among a multitude of conflicting interests, at both the regional and local levels, it has also proven that Aboriginal and non-Aboriginal people can work together on issues of land and resource management. That in itself is a major accomplishment.[35]

A second example of community-based environmental management, again from the Temagami region, concerns the Elk Lake Community Forest Project (ELCF). Elk Lake is a tiny town (population 550) just northwest of Temagami. The mandate of the ELCF is to harvest the local forest resource in a sustainable way. Its representatives draw on the entire gamut of local stakeholders: forestry, mining, tourism, environmentalist, and labour groups. Listening to the group's manager, Paul Tufford, we can hear very clearly the themes of environmental pragmatism:

> To my way of thinking this is the way it should go—to have local people getting involved in decision-making. A lot of decisions have been driven by people far outside the area. . . . When someone else was making the decisions, local people felt that their input didn't matter. . . . Most of the people want to live here and have no interests in seeing the place mowed down, or hunted and fished out.[36]

There are critics of initiatives such as these. For example, it has been pointed out that the control such bodies actually have over their resources is compromised by the fact that the provincial government still holds the relevant purse strings. Because operating budgets are entirely in governmental hands, the idea that there is real local control of the resource can seem somewhat artificial. Still, there is no doubt that the bodies have done some important work, and they surely have significant symbolic importance as well (as the royal commission pointed out). For our purposes, what is most important is that these two examples illustrate the close connection between two key pragmatist values: the value of getting things accomplished and that of doing so while respecting and incorporating the interests of all stakeholders.

E Conclusion

Questions remain about environmental pragmatism—and here are three. First, we might balk at abandoning altogether the concept of intrinsic value as the pragmatist insists we should.

Although Welchman gives us good reasons to question the idea that intrinsically valuable things are uniquely choiceworthy, surely there is something about how we react to the "last man" thought experiment that challenges this claim. Perhaps a biocentrist could argue that the world with some life is better than the world with none because life as such is of value. And he or she could then go on to explain that such value is simply a function of the living thing's efforts to subsist over time, to resist the entropic forces reducing everything else to disorder. If we are inclined to this view, it is difficult to see how it can be spelled out fully without admitting in the process that some things—here, living things—are foundationally valuable and are so entirely independently of human valuers.

Here is a second worry, related to the first. Is it really necessary to be as resolutely anti-theoretical as the environmental pragmatist recommends? Indeed, is it necessary to reject value monism? The main goal is to get things done, but is it obvious that a commitment either to doing environmental theory or, more radically, to being a value monist is an impediment to this goal? We have seen in Chapter 3 that one of the main planks of deep ecology is a commitment to political activism on behalf of wild nature. Deep ecologists themselves certainly do not believe that having one's environmental values rooted in deep theory impedes useful political action—on the contrary. And both deep ecology and the ecocentrist view from which it springs are monistic theories, or at least that is how most pragmatists understand them. Perhaps environmental pragmatists have offered us *sufficient conditions* for meaningful political action: a commitment to value pluralism, decentralization of decision-making, consensus-building, and so forth. But it does not follow that these are *necessary conditions* for such action, and many value monists would deny that they are.

Finally, we might question whether we really ought to incorporate *all* relevant values and interests in our environmental decision-making. Consider, by way of analogy, the claim that in deciding how we ought to treat members of a certain visible minority, there are some views we should ignore on moral grounds. There might be a group of racists among us who argue vociferously that such people ought to be deported forthwith. Since the racists are part of the larger society, there is a meaningful sense in which they are stakeholders in the controversy. But surely we want to say that their views simply ought not to count at all in the formulation of relevant public policy.

Similarly, there may be some groups in society that have an interest in exploiting a natural resource in an environmentally harmful, wasteful, and unsustainable manner. If their views also reflect the sentiments of a large portion of the public, should we really give them a corresponding weight in our decision-making? Some residents in the Temagami region, for instance, have argued that the various stewardship bodies that have been formed over the years still give too much consideration to purely extractive uses of the forest resource. Of course, the pragmatist could just decide to ignore problematic views, but the problem is how to justify this manoeuvre. If proper environmental policy is rooted in local voices and concerns, then excluding local views that happen to be extreme seems arbitrary. Can Welchman's invocation of "minimum aspiration levels" defuse this problem? How?

CASE STUDY |

Protecting Canada's Freshwater Resource

Freshwater Lake and South Ingonish Bay, Cape Breton Highlands, NS. Is Canada's fresh water for sale?

One way that environmental pragmatists have sought to achieve meaningful consensus on environmental policy is to insist on the importance of due political process. The environmental pragmatist Emery Castle, for example, writes:

> The essence of democracy pertains to the process by which values and objectives are selected and implemented. Natural resource management requires public participation.

One revealing object of analysis in this regard is the process by which agendas are set for the management of a particular resource. This is crucial, because it indicates which issues make it onto the table for real political consideration and how they get there. Melody Hessing, Michael Howlett, and Tracy Summerville have argued persuasively that the dominant style of agenda-setting in Canadian natural resource and environmental policy is "inside initiation." On this model, agenda-setting is dominated by "influential groups with special access to decision makers." In particular, there is a tendency in Canada for agendas to be set within the tight circle of state officials and large-scale economic producers. This fact, in turn, means that the interests and concerns of the wider public are neglected. These authors cite examples ranging

Continued

from pollution regulations to waste disposal siting and the licensing of hunters and guides. But the management of our freshwater resources, which they don't discuss, is an even more dramatic example of inside initiation.

Canada contains 20 per cent of the Earth's fresh water. Nearly 22 per cent of our land surface is covered by lakes, rivers, and wetlands. In a world in which freshwater resources are dwindling while demand for them is rising, people around the globe have their eyes on this abundance. Nowhere is this more true than in the United States, where climate change is bound to cause profound water shortages in the southwest. The question is, should we sell our water to the United States? Many people think that water is like any other commodity. If we can make a buck selling it, why shouldn't we? Others counter that since water is necessary for life, it should be treated differently from other commodities. One danger of starting to sell bulk water to the United States is that once we begin doing so, we will not, under the terms of the North American Free Trade Agreement (NAFTA) Axbe allowed to discontinue, or even slow, the flow. According to NAFTA, Canada is legally bound to maintain a proportionate supply of a given resource once we start selling it. The amount we sell can go up, but it can never go down, even if we ourselves begin experiencing shortages of the resource.

But an even more potent threat to sovereignty over our resources emerged after NAFTA was originally drafted. Officials from Canada, the United States, and Mexico created the Security and Prosperity Partnership of North America (SPP), the purpose of which was to create the conditions for a single North American economic bloc. If successful, this would have brought into being an economic entity capable of out-competing China and Europe. Part of the plan was to consider natural resources as belonging to the continent as a whole. So, for example, Canada's water would have become North America's water. In leaked minutes from a meeting of the Independent Task Force on the Future of North America, a body devoted to promoting the ideals of SPP, it was asserted that "no item—not Canadian water, not American anti-dumping, not Mexican oil—is off the table; rather, contentious or intractable issues will simply take time to ripen politically."

Now, whatever we think of sharing our freshwater resources with the United States, there is cause to be concerned about the *process* by which all of this is unfolding. As Maude Barlow, national chairperson of the Council of Canadians, has documented, the scheme was the brainchild of the big-business community in Canada, which wanted to appease American security concerns post-9/11. Yet, as Barlow writes, "the SPP has never been brought to the legislatures of the three countries for debate or political oversight." This is a prime example of inside initiation. Yet the issue of selling Canada's bulk fresh water is surely important enough to warrant a much wider debate than it has so far received. Environmental pragmatism, with its laudable insistence on consensus-based political processes, provides a much needed corrective to the sort of short-sighted, economically driven environmental policies that elite groups may want to impose on us.

Sources: Maude Barlow, *Blue Covenant: The Global Water Crisis and the Coming Battle for the Right to Water* (Toronto: McClelland and Stewart, 2007); Emery N. Castle, "A Pluralistic, Pragmatic and Evolutionary Approach to Natural Resource Management," in Light and Katz 1996, 231–50 (241); Melody Hessing, Michael Howlett, and Tracy Summerville, *Canadian Natural Resource and Environmental Policy* (Vancouver: UBC Press, 2005).

Study Questions

1. Think hard about the notion of "satisficing" in the context of some of our more pressing environmental issues (the advisability of building oil pipelines like Northern Gateway, for example). Do you think adopting this principle can help us work through these issues more effectively? Why or why not?

2. Is the environmental pragmatist correct to say that we should set aside theoretical debates and simply try to solve environmental problems? Is it possible to do the latter without engaging in the former?

3. Do you agree with the multi-faceted approach to understanding nature that Weston advocates? Can we learn as much, or even more, about how to structure our relationship with the natural world from poetry as from science?

4. What are the philosophical and practical benefits and drawbacks of thinking in terms of intrinsic value? Can we really do without this concept, as some have suggested?

Further Reading

Kevin Elliot. 2007. "Norton's Conception of Sustainability: Political, Not Metaphysical." *Environmental Ethics* 29, no. 1: 3–22.

Wendy Lynne Lee. 2008. "Environmental Pragmatism Reconsidered: Human-Centeredness, Language and the Future of Aesthetic Experience." *Environmental Philosophy* 5, no. 1: 9–22.

Andrew Light and Eric Katz, eds. 1996. *Environmental Pragmatism*. London: Routledge.

Peter Lucas. 2002. "Environmental Ethics: Between Inconsequentialist Philosophy and Unphilosophical Consequentialism." *Environmental Ethics* 24, no. 4: 353–69.

Jason Scott Robert. 2000. "Wild Ontology: Elaborating Environmental Pragmatism." *Ethics and the Environment* 5, no. 2: 191–209.

CHAPTER 6

Ecofeminism

Ecological Intuition Pump

For much of Western history, maleness was seen as superior to femaleness. Being male was associated with being closer to other the "higher" things: the products of culture, reason, the public sphere. Being female, by contrast, was associated with the "lower" things: nature, the emotions, the private sphere. Many exponents of this view believed we could derive a plan for social organization from this basic opposition, with men occupying most positions of power and prestige. Feminists have shown us that the oppositions are ideological constructs used by men simply to dominate women. Focus for a moment on the association of women with nature. Some have argued that we can use it to show that women in fact have a privileged role to play in helping us solve the ecological crisis. All we need to do is flip the traditional assignment of value and claim that the closeness of women to nature makes them, in a way, superior to men. What should we think of this sort of move? Is it flawed because it retains the old nature/culture, female/male oppositions? Does it imply that women have some sort of essence that they do not in fact have? Most important, if we reject the idea that women are closer to nature than men, does this mean we should also reject the idea that women—because of their unique *experiences*—might have something special to offer to environmental ethics? Is this not to throw the baby out with the bathwater?

Ⓐ Introduction

In 1973, a spontaneous protest arose in a series of villages in northern India aimed at protecting a vital ecological resource: the forests on which many village people relied for their day-to-day needs. The villages, scattered across a huge portion of the state of Uttar Pradesh, which skirts the Nepalese border to the north, were able to subsist for centuries because of a delicate balance established between the people and the land. The forests provided food and fuel and also stabilized the soil, while the villagers harvested the area's resources sustainably. But in the early 1970s, the Indian government began awarding logging contracts in the area to commercial firms that were not interested in going about their business in a way that sustained the forests over time. As often happens, this move was justified in the name of bringing the benefits of economic development to the region. Events came to a head in April 1973 in the Reni region. The government lured the village men away from the villages by inviting them to talks, then allowed loggers to enter the forests and begin cutting.

The village women responded to this provocation by entering the forests and interposing their bodies between the trees and the loggers, in this way preventing the loggers from cutting the trees. Thus began a powerful social movement with women activists at its centre. The movement was dubbed the "Chipko Movement." In Hindi, *chipko* refers to embracing or hugging, so these women were the original "tree-huggers." The technique is based on Gandhian principles of non-violent confrontation, and in this case it worked. Although it took some seven years, the Indian government eventually responded by imposing a 15-year moratorium on logging in the area. The Chipko Movement has since spread all over India, to the states of Bihar, Himachal Pradesh, Rajasthan, Karnataka, and beyond. The Chipko slogan is, "Ecology is permanent economy."[1]

What the Chipko Movement highlights is that in many parts of the world, women are on the front lines of the struggle to preserve the integrity of subsistence cultures and the essential connection to the land on which such cultures often depend. They are the ones who must work to meet the basic and vital needs of their groups, and they are therefore the ones who must confront most directly the potent threat of the commercialization of their resources. Of course, such commercialization is usually described and advertised as "development" for the region in question. But, as Vandana Shiva argues, in most cases a better word for it is "maldevelopment." Shiva is talking about Africa, but the point applies equally to India:

> In actual fact there is less water, less fertile soil, less genetic wealth as a result of the development process. Since these natural resources are the basis of nature's economy and women's survival economy their scarcity is impoverishing women and marginalized peoples in an unprecedented manner. Their new impoverishment lies in the fact that resources which supported their survival were absorbed into the market economy while they themselves were excluded and displaced by it.[2]

Something similar may be happening in the Canadian North. Along with other circumpolar countries, Canada is rushing to establish sovereignty in a region that because of accelerating climate change is likely to be ice-free in the summer within 20 years and that contains 20 per cent of the world's oil. But there are fears that the knowledge of the land and its sensitive ecosystems possessed by the Dene, Northern Cree, and Inuit—including, of course, the women members of these groups—will be ignored in the process of extracting the resources. Canadian philosopher John Ralston Saul summarizes the danger this way: "If you look coolly at today's arguments over sovereignty, they are still about cutting through the Arctic, not living in it."[3]

Two themes come together in these examples. First, there is the fact of the threat to subsistence or marginalized cultures and ecologies by the globalized process of economic development. Second, there is the claim that in many cases women are in a privileged position both to perceive this threat for what it is and to respond to it. How exactly are these two ideas or themes connected? This is a complex question, one that this chapter will attempt to answer by way of presenting the relatively new branch of environmental ethics known as "ecofeminism." Ecofeminism is itself a complex cluster of positions within environmental ethics, but what all such positions have in common is the claim that there is some kind of connection between sexism—the domination and subordination of women by men—and **naturism**, the domination and subordination of nature by humans.

Ⓑ Sexism and Naturism: Making the Connection

The term "ecofeminism" was coined by Françoise d'Eaubonne in 1973, but the most influential recent articulation of the position was provided by Karen Warren in our chapter's first article.

THE POWER AND THE PROMISE OF ECOLOGICAL FEMINISM
Karen J. Warren

As I use the term in this paper, ecological feminism is the position that there are important connections—historical, experiential, symbolic, theoretical—between the domination of women and the domination of nature, an understanding of which is crucial to both feminism and environmental ethics. I argue that the promise and power of ecological feminism is that *it provides a distinctive framework both for reconceiving feminism and for developing an environmental ethic that takes seriously connections between the domination of women and the domination of nature.* I do so by discussing the nature of a feminist ethic and the ways in which ecofeminism provides a feminist and environmental ethic. I conclude that any feminist theory *and* any environmental ethic that fails to take seriously the twin and interconnected dominations of women and nature is at best incomplete and at worst simply inadequate.

I. Feminism, Ecological Feminism, and Conceptual Frameworks

Whatever else it is, feminism is at least the movement to end sexist oppression. It involves the elimination of any and all factors that contribute to the continued and systematic domination or subordination of women. While feminists disagree about the nature of and solutions to the subordination of women, all feminists agree that sexist oppression exists, is wrong, and must be abolished.

A "feminist issue" is any issue that contributes in some way to understanding the oppression of women. Equal rights, comparable pay for comparable work, and food production are feminist issues wherever and whenever an understanding of them contributes to an understanding of the continued exploitation or subjugation of women. Carrying water and searching for firewood are feminist issues wherever and whenever women's primary responsibility for these tasks contributes to their lack of full participation in decision-making, income-producing, or high-status positions engaged in by men. What counts as a feminist issue, then, depends largely on context, particularly the historical and material conditions of women's lives.

Environmental degradation and exploitation are feminist issues because an understanding of them contributes to an understanding of the oppression of women. In India, for example, both deforestation and reforestation through the introduction of a monoculture-species tree (e.g., eucalyptus) intended for commercial production are feminist issues because the loss of indigenous forests and multiple species of trees has drastically affected rural Indian women's ability to maintain a subsistence household. Indigenous forests provide a variety of trees for food, fuel, fodder, household utensils, dyes, medicines, and income-generating uses, while monoculture-species forests do not. Although I do not argue for this claim here, a look at the global impact of environmental degradation on women's lives suggests important respects in which environmental degradation is a feminist issue.

Feminist philosophers claim that some of the most important feminist issues are *conceptual* ones: these issues concern how one conceptualizes such mainstay philosophical notions as reason and rationality, ethics, and what it is to be human. Ecofeminists extend this feminist philosophical concern to nature. They argue that, ultimately, some of the most important connections between the domination of women and the domination of nature are conceptual. To see this, consider the nature of conceptual frameworks.

A *conceptual framework* is a set of *basic* beliefs, values, attitudes, and assumptions that shape and reflect how one views oneself and one's world. It is a socially constructed lens through which we perceive ourselves and others. It is affected by such factors as gender, race, class, age, affectional orientation, nationality, and religious background. Some conceptual frameworks are oppressive. An *oppressive conceptual framework* is one that explains, justifies, and maintains relationships of domination and subordination. When an oppressive conceptual framework is *patriarchal*, it explains,

justifies, and maintains the subordination of women by men.

I have argued elsewhere that there are three significant features of oppressive conceptual frameworks: (1) value-hierarchical thinking, i.e., "up-down" thinking, which places higher value, status, or prestige on what is "up" rather than on what is "down"; (2) value dualisms, i.e., disjunctive pairs in which the disjuncts are seen as oppositional (rather than as complementary) and exclusive (rather than as inclusive) and which place higher value (status, prestige) on one disjunct rather than on the other (e.g., dualisms that give higher value or status to that which has historically been identified as "mind," "reason," and "male" than to that which has historically been identified as "body," "emotion," and "female"); and (3) logic of domination, i.e., a structure of argumentation that leads to a justification of subordination.

The third feature of oppressive conceptual frameworks is the most significant. A logic of domination is not *just* a logical structure. It also involves a substantive value system, since an ethical premise is needed to permit or sanction the "just" subordination of that which is subordinate. This justification typically is given on grounds of some alleged characteristic (e.g., rationality), which the dominant (e.g., men) have and the subordinate (e.g., women) lack.

Contrary to what many feminists and ecofeminists have said or suggested, there may be nothing *inherently* problematic about "hierarchical thinking" or even "value-hierarchical thinking" in contexts other than contexts of oppression. Hierarchical thinking is important in daily living for classifying data, comparing information, and organizing material. Taxonomies (e.g., plant taxonomies) and biological nomenclature seem to require some form of "hierarchical thinking." Even "value-hierarchical thinking" may be quite acceptable in certain contexts. (The same may be said of "value dualisms" in non-oppressive contexts.) For example, suppose it is true that what is unique about humans is our conscious capacity to radically reshape our social environments (or "societies"), as Murray Bookchin suggests.[4] Then one could truthfully say that humans are better equipped to radically reshape their environments than are rocks or plants—a "value-hierarchical" way of speaking.

The problem is not simply *that* value-hierarchical thinking and value dualisms are used but *the way* in which each has been used *in oppressive conceptual frameworks* to establish inferiority and to justify subordination.[5] It is the logic of domination, *coupled with* value-hierarchical thinking and value dualisms, that "justifies" subordination. What is explanatorily basic, then, about the nature of oppressive conceptual frameworks is the logic of domination.

For ecofeminism, that a logic of domination is explanatorily basic is important for at least three reasons.

First, without a logic of domination, a description of similarities and differences would be just that—a description of similarities and differences. Consider the claim "Humans are different from plants and rocks in that humans can (and plants and rocks cannot) consciously and radically reshape the communities in which they live; humans are similar to plants and rocks in that they are both members of an ecological community." Even if humans are "better" than plants and rocks with respect to the conscious ability of humans to radically transform communities, one does not *thereby* get any *morally* relevant distinction between humans and non-humans or an argument for the domination of plants and rocks by humans. To get *those* conclusions, one needs to add at least two powerful assumptions, viz., (A2) and (A4) in argument A below:

(A1) Humans do, and plants and rocks do not, have the capacity to consciously and radically change the community in which they live.

(A2) Whatever has the capacity to consciously and radically change the community in which it lives is morally superior to whatever lacks this capacity.

(A3) Thus, humans are morally superior to plants and rocks.

(A4) For any X and Y, if X is morally superior to Y, then X is morally justified in subordinating Y.

(A5) Thus, humans are morally justified in subordinating plants and rocks.

Without the two assumptions that *humans are morally superior* to (at least some) non-humans, (A2), and that *superiority justifies subordination*, (A4), all one has is some difference between humans and some non-humans. This is true *even if* that difference is given in terms of superiority. Thus, it is the logic of domination, (A4), that is the bottom line in ecofeminist discussions of oppression.

Second, ecofeminists argue that, at least in Western societies, the oppressive conceptual framework that sanctions the twin dominations of women and nature is a patriarchal one characterized by all three features of an oppressive conceptual framework. Many ecofeminists claim that historically, within at least the dominant Western culture, a patriarchal conceptual framework has sanctioned the following argument B:

(B1) Women are identified with nature and the realm of the physical; men are identified with the "human" and the realm of the mental.

(B2) Whatever is identified with nature and the realm of the physical is inferior to ("below") whatever is identified with the "human" and the realm of the mental; or, conversely, the latter is superior to ("above") the former.

(B3)   Thus, women are inferior to ("below") men; or, conversely, men are superior to ("above") women.

(B4)   For any X and Y, if X is superior to Y, then X is justified in subordinating Y.

(B5)   Thus, men are justified in subordinating women.

If sound, argument B establishes *patriarchy*, i.e., the conclusion given at (B5) that the systematic domination of women by men is justified. But according to ecofeminists, (B5) is justified by just those three features of an oppressive conceptual framework identified earlier: value-hierarchical thinking, the assumption at (B2); value dualisms, the assumed dualism of the mental and the physical at (B1) and the assumed inferiority of the physical vis-à-vis the mental at (B2); and a logic of domination, the assumption at (B4), the same as the previous premise (A4). Hence, according to ecofeminists, insofar as an oppressive patriarchal conceptual framework has functioned historically (within at least dominant Western culture) to sanction the twin dominations of women and nature (argument B), both argument B and the patriarchal conceptual framework, from whence it comes, ought to be rejected.

Of course, the preceding does not identify which premises of B are false. What is the status of premises (B1) and (B2)? Most, if not all, feminists claim that (B1)—and many ecofeminists claim that (B2)—has been assumed or asserted within the dominant Western philosophical and intellectual tradition. As such, these feminists assert, as a matter of historical fact, that the dominant Western philosophical tradition has assumed the truth of (B1) and (B2).

Ecofeminists, however, either deny (B2) or do not affirm (B2). Furthermore, because some ecofeminists are anxious to deny any ahistorical identification of women with nature, some ecofeminists deny (B1) when (B1) is used to support anything other than a strictly historical claim about what has been asserted or assumed to be true within patriarchal culture—e.g., when (B1) is used to assert that women properly are identified with the realm of nature and the physical. Thus, from an ecofeminist perspective, (B1) and (B2) are properly viewed as problematic though historically sanctioned claims: they are problematic precisely because of the way they have functioned historically in a patriarchal conceptual framework and culture to sanction the dominations of women and nature.

What *all* ecofeminists agree about, then, is the way in which the *logic of domination* has functioned historically within patriarchy to sustain and justify the twin dominations of women and nature. Since *all* feminists (and not just ecofeminists) oppose patriarchy, the conclusion given at (B5), all feminists (including ecofeminists) must oppose at least the logic of domination, premise (B4), on which argument B rests—whatever the truth-value status of (B1) and (B2) *outside of* a patriarchal context.

That *all* feminists must oppose the logic of domination shows the breadth and depth of the ecofeminist critique of B: it is a critique not only of the three assumptions on which this argument for the domination of women and nature rests, viz., the assumptions at (B1), (B2), and (B4); it is also a critique of patriarchal conceptual frameworks generally, i.e., of those oppressive conceptual frameworks that put men "up" and women "down," allege some way in which women are morally inferior to men, and use that alleged difference to justify the subordination of women by men. Therefore, ecofeminism is necessary to *any* feminist critique of patriarchy and, hence, necessary to feminism (a point I discuss again later).

Third, ecofeminism clarifies why the logic of domination, and any conceptual framework that gives rise to it, must be abolished in order both to make possible a meaningful notion of difference that does not breed domination and to prevent feminism from becoming a "support" movement based primarily on shared experiences. In contemporary society, there is no one "woman's voice," no *woman* (or *human*) is a woman (or human) of some race, class, age, affectional orientation, marital status, regional or national background, and so forth. Because there are no "monolithic experiences" that all women share, feminism must be a "solidarity movement" based on shared beliefs and interests rather than a "unity in sameness" movement based on shared experiences and shared victimization.[6] In the words of Maria Lugones, "Unity—not to be confused with solidarity—is understood as conceptually tied to domination."[7] Ecofeminists insist that the sort of logic of domination used to justify the domination of humans by gender, racial or ethnic, or class status is also used to justify the domination of nature. Because eliminating a logic of domination is part of a feminist critique—whether a critique of patriarchy, white supremacist culture, or imperialism—ecofeminists insist that *naturism* is properly viewed as an integral part of any feminist solidarity movement to end sexist oppression and the logic of domination that conceptually grounds it....

II. Ecofeminism as a Feminist and Environmental Ethic

All the props are now in place for seeing how ecofeminism provides the framework for a distinctively feminist and environmental ethic. It is a feminism that critiques male bias wherever it occurs in ethics (including environmental ethics) and aims at providing an ethic (including an environmental ethic) that is not male-biased—and it does so in a way that satisfies the preliminary boundary conditions of a feminist ethic.

First, ecofeminism is quintessentially anti-naturist. Its anti-naturism consists in the rejection of any way of thinking about or acting toward non-human nature that reflects a logic, values, or attitude of domination. Its anti-naturist, anti-sexist, anti-racist, anti-classist (and so forth, for all other "isms" of social domination) stance forms the outer boundary of the quilt: nothing gets on the quilt that is naturist, sexist, racist, classist, and so forth.

Second, ecofeminism is a contextualist ethic. It involves a shift *from* a conception of ethics as primarily a matter of rights, rules, or principles predetermined and applied in specific cases to entities viewed as competitors in the contest of moral standing *to* a conception of ethics as growing out of what Jim Cheney calls "defining relationships," i.e., relationships conceived in some sense as defining who one is. As a contextualist ethic, it is not that rights, or rules, or principles are *not* relevant or important. Clearly, they are in certain contexts and for certain purposes. It is just that what *makes* them relevant or important is that those to whom they apply are entities in *relationship with* others.

Ecofeminism also involves an ethical shift *from* granting moral consideration to non-humans *exclusively* on the grounds of some similarity they share with humans (e.g., rationality, interests, moral agency, sentiency, right-holder status) *to* "a highly contextual account to see clearly what a human being is and what the non-human world might be, morally speaking, *for* human beings."[8] For an ecofeminist, *how* a moral agent is in relationship to another becomes of central significance, not simply *that* a moral agent is a moral agent or is bound by rights, duties, virtue, or utility to act in a certain way.

Third, ecofeminism is structurally pluralistic in that it presupposes and maintains difference—difference among humans as well as between humans and at least some elements of non-human nature. Thus, while ecofeminism denies the "nature/culture" split, it affirms that humans are both members of an ecological community (in some respects) and different from it (in other respects). Ecofeminism's attention to relationships and community is not, therefore, an erasure of difference but a respectful acknowledgement of it.

Fourth, ecofeminism reconceives theory as theory in process. It focuses on patterns of meaning that emerge, for instance, from the storytelling and first-person narratives of women (and others) who deplore the twin dominations of women and nature. The use of narrative is one way to ensure that the content of the ethic—the pattern of the quilt—may/will change over time, as the historical and material realities of women's lives change and as more is learned about women–nature connections and the destruction of the non-human world.

Fifth, ecofeminism is inclusivist. It emerges from the voices of women who experience the harmful domination of nature and the way that domination is tied to their domination as women. It emerges from listening to the voices of indigenous peoples . . . who have been dislocated from their land and have witnessed the attendant undermining of such values as appropriate reciprocity, sharing, and kinship that characterize traditional Indian culture. It emerges from listening to voices of those who, like Nathan Hare, critique traditional approaches to environmental ethics as white and bourgeois and as failing to address issues of "Black ecology" and the "ecology" of the inner

city and urban spaces.[9] It also emerges out of the voices of Chipko women who see the destruction of "earth, soil, and water" as intimately connected with their own inability to survive economically.[10] With its emphasis on inclusivity and difference, ecofeminism provides a framework for recognizing that what counts as ecology and what counts as appropriate conduct toward both human and non-human environments are largely matters of context.

Sixth, as a feminism, ecofeminism makes no attempt to provide an "objective" point of view. It is a social ecology. It recognizes the twin dominations of women and nature as social problems rooted both in very concrete, historical, socio-economic circumstances and in oppressive patriarchal conceptual frameworks that maintain and sanction these circumstances.

Seventh, ecofeminism makes a central place for values of care, love, friendship, trust, and appropriate reciprocity—values that presuppose that our relationships to others are central to our understanding of who we are.[11] It thereby gives voice to the sensitivity that in climbing a mountain, one is doing something in relationship with an "other," an "other" whom one can come to care about and treat respectfully.

Lastly, an ecofeminist ethic involves a reconception of what it means to be human and in what human ethical behaviour consists. Ecofeminism denies abstract individualism. Humans are who we are in large part by virtue of the historical and social contexts and the relationships we are in, including our relationships with non-human nature. Relationships are not something extrinsic to who we are, not an "add on" feature of human nature; they play an essential role in shaping what it is to be human. Relationships of humans to the non-human environment are, in part, constitutive of what it is to be a human.

By making visible the interconnections among the dominations of women and nature, ecofeminism shows that both are feminist issues and that explicit acknowledgement of both is vital to any responsible environmental ethic. Feminism *must* embrace ecological feminism if it is to end the domination of women, because the domination of women is tied conceptually and historically to the domination of nature.

A responsible environmental ethic also *must* embrace feminism. Otherwise, even the seemingly most revolutionary, liberational, and holistic ecological ethic will fail to take seriously the interconnected dominations of nature and women that are so much a part of the historical legacy and conceptual framework that sanction the exploitation of non-human nature. Failure to make visible these interconnected, twin dominations results in an inaccurate account of how it is that nature has been and continues to be dominated and exploited and produces an environmental ethic that lacks the depth necessary to be truly *inclusive* of the realities of persons who at least in dominant Western culture have been intimately tied with that exploitation, viz., women. Whatever else can be said in favour of such holistic ethics, a failure to make visible ecofeminist insights into the common denominators of the twin oppressions of women and nature is to perpetuate, rather than overcome, the source of that oppression.

This last point deserves further attention. It may be objected that as long as the end result is "the same"—the development of an environmental ethic that does not emerge out of or reinforce an oppressive conceptual framework—it does not matter whether that ethic (or the ethic endorsed in getting there) is feminist or not. Hence, it simply is *not* the case that any adequate environmental ethic must be feminist. My argument, in contrast, has been that it *does* matter, and for three important reasons.

First, there is the scholarly issue of accurately representing historical reality, and that, ecofeminists claim, requires acknowledging the historical feminization of nature and naturalization of women as part of the exploitation of nature.

Second, I have shown that the conceptual connections between the domination of women and the domination of nature are located in an oppressive and, at least in Western societies, patriarchal conceptual framework characterized by a logic of domination. Thus, I have shown that failure to notice the nature of this connection leaves at best an incomplete, inaccurate, and partial account of what is required of a conceptually adequate environmental ethic. An ethic that *does not* acknowledge this is simply *not* the same as one that does, whatever else the similarities between them.

Third, the claim that, in contemporary culture, one can have an adequate environmental ethic that is *not* feminist assumes that, in contemporary culture, the label *feminist* does not add anything crucial to the nature or description of environmental ethics. I have shown that at least in contemporary culture, this is false, for the word *feminist* currently helps to clarify just *how* the domination of nature is conceptually linked to patriarchy and, hence, how the liberation of nature is conceptually linked to the termination of patriarchy. Thus, because it has critical bite in contemporary culture, it serves as an important reminder that in contemporary sex-gendered, raced, classed, and naturist culture, an unlabelled position functions as a privileged and "unmarked" position. That is, without the addition of the word *feminist*, one presents environmental ethics as if it has no bias, including male-gender bias, which is just what ecofeminists deny: failure to notice the connections between the twin oppressions of women and nature *is* male-gender bias.

One of the goals of feminism is the eradication of all oppressive sex-gender (and related race, class, age, affectional preference) categories and the creation of a world in which *difference does not breed domination*—say, the world of 4001. If in 4001 an "adequate environmental ethic" is a "feminist environmental ethic," the word *feminist* may then be redundant and unnecessary. However, this is *not* 4001, and in terms of the current historical and conceptual reality, the dominations of nature and of women are intimately connected. Failure to notice or make visible that connection [in the present] perpetuates the mistaken (and privileged) view that "environmental ethics" is *not* a feminist issue and that *feminist* adds nothing to environmental ethics.

I have argued in this paper that ecofeminism provides a framework for a distinctively feminist and environmental ethic. Ecofeminism grows out of the felt and theorized-about connections between the domination of women and the domination of nature. As a contextualist ethic, ecofeminism refocuses environmental ethics on what nature might mean, morally speaking, *for* humans and on how the relational attitudes of humans to others—humans as well as non-humans—sculpt both what it is to be human and the nature and ground of human responsibilities to the non-human environment. Part of what this refocusing does is to take seriously the voices of women and other oppressed persons in the construction of that ethic. . . .

A re-conceiving and *re-visioning* of both feminism and environmental ethics are, I think, the power and promise of ecofeminism.

Warren's key claim is that there is *some* connection between naturism and sexism. But what is the nature of this connection? Is it conceptual? If so, sexism and naturism would not coexist in a merely contingent or accidental manner. Rather, sexist beliefs or practices are necessarily also naturist and vice versa. It follows that if one is opposed to either one of these, then to be rationally consistent one should also be opposed to the other. Warren certainly seems to endorse this way of looking at the issue. As she says, ecofeminists "argue that, ultimately, some of the most important connections between the domination of women and the domination of nature are conceptual." One way to parse this is to say that there is a "single pattern of thought used to justify the domination of women and the domination of nature."[12]

If a single pattern of thought is indeed operating in the two cases, it would follow that if one were criticizing one of them, one would also be criticizing the other, whether one were fully aware of this fact or not. So why not make one's critical commitments explicit and criticize both sets of beliefs and practices openly? Besides, one way to strengthen the argument against an objectionable mode of thought and practice is to show that it infects more of our thinking and behaviour than might appear at first sight. For the conceptual claim to go through, it would have to be shown that men dominate nature and women in the same way, for the same reasons, and that the domination is equally wrong in both cases.[13] But some philosophers claim that Warren is simply assuming this is the case. We can, these philosophers insist, allow that men have historically dominated both women and nature. Moreover, it is surely the case that many of those men themselves see a connection between these two forms of domination. Even so, it would not follow that opposing men's sexism requires us—on pain of violating a standard of rational consistency—to oppose their naturism or vice versa.

To see what seems to have gone wrong in Warren's argument, consider the following example, by Levin and Levin:

> The Nazis spoke of Jews as vermin and a disease infecting Germany. In other words, the Nazis created a conceptual link between anti-Semitism and the germ theory of disease. It hardly follows that anyone who opposed Nazism had to oppose the germ theory of disease. . . . In effect . . . Warren is assuming that if a person or group of people A thinks B is similar to C, and for this reason treats B and C similarly, opposition to A's treatment of B *requires* opposition to A's treatment of C.[14]

But Levin and Levin claim that no such *requirement* exists in the case just described, so this—allegedly—stands as a counter-example to the conceptual claim. Presumably, what we would say to the Nazis in this example is that they were simply wrong in having described Jews the way they did. More particularly, we would say that there are morally important differences between people—in this case Jews—and germs such that it is impermissible to do to one what it is permissible to do to the other. Similarly, we might say that there are morally important differences between people—in this case women—and the rest of nature such that it is impermissible to do to one what it is permissible to do to the other.

In other words, although we can agree that it is impermissible for men to dominate women, it must be demonstrated *independently* that it is impermissible for humans to dominate nature (if this is indeed true). Whatever else we might think of him, the person who works assiduously and sincerely to better the lot of women but who also believes that non-sentient nature is a mere resource is not being obviously inconsistent. If asked to justify the two beliefs or practices, he might reply that only sentient beings have moral standing, a fact that justifies fundamentally different attitudes to women and non-sentient nature. By contrast, the claim that people and non-sentient nature have equal moral standing is a substantive moral doctrine that, as we have seen in Chapter 3, is highly controversial. It follows that a sexist is not *necessarily* a naturist or vice versa.

Ⓒ Overcoming Dualisms

Still, there can be little doubt that *as a matter of fact*, thinkers in the Western tradition have seen both women and nature as relatively passive objects, there to be controlled in some fashion by men. This fact provides the ecofeminist with a possible response to the criticism we examined in the previous section. Ecofeminists can allow that the connection between naturism and sexism is not conceptual if by "conceptual" we mean that the connection obtains in every possible world. We can all presumably agree that the connection is not *logically* necessary. But, so goes the response, it is nevertheless a very deep and persistent feature of our culture. Indeed, Val Plumwood thinks that "rationalism"—the view that reason is our most important attribute—is at the heart of the problem and that the glorification of reason extends right back to the ancient roots of the Western philosophical tradition.

The key problem lies in what ecofeminists have identified as a tendency to create dualisms or oppositions and then to claim that one half of these dualisms or oppositions is superior to the other. So on one side of the moral ledger we have men, reason, the public sphere, and culture, while on the other we have women, the emotions, the private sphere, and nature. Since the first set of terms is deemed to be superior to its counterpart in the second set, we are justified in treating the latter as *subordinate*. So it is permissible for women to be dominated by men, for reason to dominate the emotions, for the public sphere to dominate the private sphere, and for culture to dominate nature. Nor are these acts of domination unconnected: women, it is alleged, may be dominated precisely *because* they are—in contrast to men—incorrigibly emotional and natural, mere guardians of domestic space. This is the "logic of domination" examined by Warren. There are reasons to be sceptical about the extent to which the concept of domination can do this much analytic work, but Warren has surely identified a significant problem.

How do we overcome this way of thinking? Plumwood suggests that the best way to situate the question is to consider it in the context of feminism's successive "waves." The first wave of feminism was directed at abolishing more blatant forms of discrimination and inequality between the genders. So, for example, the evident injustice of unequal pay for work of equal value was targeted. But the focus on equality was seen as insufficient. Here is how Joseph Desjardins puts the point:

> The problem with this view is that in a culture in which masculine traits and characteristics dominate, equality for women amounts to little more than requiring women to adopt these dominant male traits. In effect women can be equal to men only if they become masculine, and to the degree that strong cultural forces work against this, women always fall just a little short of full equality.[15]

From the standpoint of environmental ethics, what this first-wave focus on mere economic and social equality can produce is a tendency to disregard the potentially negative consequences of our actions on the environment. For example, Canadian environmental geographer Maureen G. Reed has done very thorough and insightful work on the gender of labour practices in British Columbia's forestry industry. What her research has revealed is that women are systemically discriminated against in this industry in two distinct though related ways. First, they are for the most part simply prevented from getting the jobs that pay the most—as scalers, heavy equipment operators, and so on. This exclusion is based on the biased claim that this sort of work is too physically demanding for women. Second, and largely because of the first factor, women are confined to low-paying jobs—many of them minimum-wage—or to tending the domestic hearth.[16]

Strictly from the perspective of the first wave, the injustice here is glaring. There is an ideology of gender in operation, the practical result of which is that women are treated unfairly relative to men. The solution would be more equality in the local workforce. However, the ecofeminist critique of this view would point to the fact that in order to achieve this otherwise laudable result, women would in effect have to adopt "masculine" behaviours and attitudes.

The reason this would not be welcome is that those behaviours and attitudes are, in part, what allow for an uncritical and ecologically destructive practice. Many of the small communities Reed studies are engaged in clear-cut logging, something most environmentalists consider an intrinsically unsustainable mode of timber extraction. So the first-wave approach to gender equality is flawed in that it is *insufficiently critical* of dominant beliefs and attitudes about—as well as practices toward—the natural world.

Hence the need for a second wave of ecofeminism. Here we encounter an inversion of terms in the dualisms and oppositions we have been examining. So we retain the basic oppositions between men/women, reason/emotion, public/private, and culture/nature, but we invert the value assessments involved. That is, we now consider the second term in each pair to be superior to the first.[17] Women, nature, the emotions, and the private sphere are now viewed as distinctively valuable. This is important because of the link between (a) the claim that a certain mode of apprehending the world is inferior to its counterpart in the dualistic structure and (b) the idea that the latter is therefore justified in dominating or subordinating the former.

After the revaluation or inversion, it *looks* as though such domination can no longer be justified (though we will see that things are not so simple). According to Vandana Shiva, the revaluation in question amounts to recovery of the "feminine principle." This refers to women's *essential* connection to the Earth and its productive capacities. Here, a distinctively feminine "way of knowing and being" is unambiguously raised above that of the masculine, which Shiva describes as a "culture of death."[18] The previous dualism has not been abolished, merely turned on its head. However, Plumwood—and others—have attacked this move, labelling it "the feminism of uncritical reversal" and claiming that it actually reinforces, and perhaps even deepens, the oppression of women and nature.[19] Why is this?

ⓓ Ecofeminism Now

According to Warren and Plumwood, the best way forward is through the development of a third wave of ecofeminism. In feminism generally, the third wave rejects the notion that there is some single feminine ideal and embraces instead the culturally wide-ranging ways in which women define themselves. The problem with waves one and two is that each is insufficiently critical of some important facet of patriarchal thinking. But ecofeminism has come a long way since Warren's article first appeared. For a clear articulation of where the movement stands now, we turn to our chapter's second article.

ECOFEMINISTS WITHOUT BORDERS: THE POWER OF METHOD

Trish Glazebrook

I. Introduction

In 1990, Karen Warren published "The Power and Promise of Ecological Feminism" in *Environmental Ethics*. Its power and promise is that "it provides a distinctive framework both for reconceiving feminism and for developing an environmental ethic which takes seriously connections between the domination of women and the domination of nature."[20] To what extent has ecofeminism realized this power and lived up to this promise?

Warren's paper articulated the elements of conceptual frameworks that function as logics of domination: dualism, ordered into a hierarchy, that serves to justify subordination. For example, in phallic logic, the dualisms man/woman and man/nature privilege man over woman and nature, and accordingly justify what d'Eaubonne, who invented the word "l'eco-féminisme" in 1974, identified as a double exploitation of nature in an excess of production, and of women in an excess of reproduction. Patriarchy, understood as a phallic logic of domination, is thus uncovered beyond its function as a system of social organization and as a way of thinking. Modern philosophy, unsurprisingly since its founding father, Descartes, was a scientist, models itself on the sciences—that is, takes objectivity as its epistemological and ontological standard: knowledge is quintessentially the understanding of material objects from a universal perspective. Hence some ecofeminists find the root cause of the contemporary global logic of domination in modern science.[21]

I have shown elsewhere that these aetiologies are complementary, since modern science is the evolved appropriation of one interpretation of Aristotle's distinction between *physis* (nature) and *technê* (production),[22] and that modern science and technology are inseparable,[23] even in thought, in the contemporary global logic of domination that is a logic of capital in contrast to a logic of care.[24] The current age is one of massive environmental destruction enabled by technoscientific engineering. Since the ideology of objectivity precludes contextualization, including cultural and historical location, the contemporary paradigm of knowledge devalues the very historical analysis that might

uncover the limitations of that paradigm to expose the truth that capitalist patriarchy could be intentionally displaced by a different, less destructive logic. Moreover, since objectivity is definitively disinterested impartiality, there is no place in its logic for care—that is, for the emotional intelligence that has been traditionally contrasted with "reason" and denigrated as a feminine trait in the philosophical tradition.[25] Care has the potential to structure public and global institutions on ethical principles rather than neo-liberal principles of market, competition, individualism, and profit that take the fulfillment of human life to be individual accumulation of private wealth. The promise of ecofeminism is that it enables just such critique. Its power to do so lies in its methods. These methods indicate that in order for environmentalists and feminists not to be working with a deficient conceptual framework, feminists must be environmentalists, and environmentalists must be feminists.

II. Feminists as Environmentalists

Using the method of narrative voice that Warren articulated, I can assess, on the basis of my lived experience as a successful academic and self-supporting single mother by choice, that feminism has indeed made great gains.[26] Yet I am not typical. I am part of an elite group of white, middle-class, university-educated women in the Global North. I do not say "developed countries." That would imply progress, but technological advances in, for example, medicine are not *better* so much as *better for some*. In a small village in northern Ghana, a group of women came to my house to ask for ice to apply to the chest of a mother whose heart they feared would explode with sorrow at the funeral of her two-year-old daughter who died of malaria. It was all I could do not to throw away the malaria pills I had brought for myself and my four-year-old son in the shame of my privilege. Massive technological accomplishments have been accompanied by so little advance in ethics, politics, justice, and social responsibility. Now my son is 11. We recently visited a hospital Emergency Department in Texas as I feared his severe abdominal pain indicated life-threatening appendicitis. Despite full insurance coverage, the visit cost roughly $1600 USD. Next time I fear for my son's life, should I gamble on the seriousness of his illness? If I have no coverage? Lack of access to health care is an issue in one of the richest, most-developed countries, and extreme poverty can be experienced in any country that fails to provide a social safety net.

What then is poverty? As Shiva says, "development perceives poverty only in terms of an absence of Western consumption patterns, or in terms of cash incomes and therefore is unable to grapple with self-provisioning economies."[27] The assumption that all cultures want or need to follow the development path of the Global North suggests that its reality is something toward which every culture should strive. Many cultures have epistemologies, ontologies, and economies—from the Greek, rules of the house—that have sustained their communities across generations. Yet *subsistence economies* is a disparaging term. Valuing such economies does not idealize the peasant experience, but allows us to understand that capitalist cultures have not freed themselves from sexism, racism, and other "isms" of domination. Nor have capitalist cultures eliminated non-capital exchange systems; families, for example, draw heavily on unpaid labour. Capital economies could not in fact function without the support of this unpaid, largely female labour. Waring has shown that women's labour more generally is invisible in mainstream economic reckoning because women's food production, for example, eludes traditional economic indices by feeding

the family rather than being traded in the market.[28] Globally, in 2010, about 2.4 billion people lived on less than USD$2 a day.[29] Pollution, climate change, land grabbing, conflict, and other impacts of global capital threaten and destroy their fishery and farming livelihoods even as accrued profits count as economic growth in a system of little, if any, accountability.

The feminism of the Global North in the 1970s failed to address these issues from the Global South, or the feminization of poverty and exploitation of women's labour. Johnson-Odim has argued that feminism in the Global North "narrowly confines itself to [an almost singularly anti-sexist struggle] against gender discrimination."[30] Women in the Global South share issues like racism and economic exploitation with men and want solidarity across genders and a feminism that addresses postcolonial economics, politics, and social issues. Even in the North, feminism failed to address diversity issues and marginalized diversity perspectives. Lugones and Spelman's dialogue on the inclusion of diverse perspectives was followed by bell hooks's argument that white feminist perspectives are one-dimensional and exclusionary.[31] Given that white privilege only functions through exploitation—and Northern privilege through exploitation of the South—white feminist complacency in the Global North concerning anything outside their own experience looks like and functions as complicity.

Ecofeminism ruptures this complacency. In a world of biodiversity loss, ecosystem destruction, and climate change, ecophilosophers—and thus ecofeminists—cannot deny the need for global perspectives. Shiva brought this reality to white feminists in 1988 when she identified women's survival crises in India. Warren has argued for "taking empirical data seriously."[32] Travelling beyond the borders of the West, I am immediately confronted by the overwhelming reality that women's lives throughout the world are under threat from environmental degradation. I am sure they are in the Global North also; it may just be harder to see through the buffered systems of consumer culture. As a feminist, then, I have to be an environmentalist. The empirical data cry out for it; as Warren suggests, I cannot but take them seriously.

Ecofeminism reconceives feminism as necessarily international in perspective because the agenda is larger and more pressing than originally envisioned by the "second wave" of feminism. Johnson-Odim shows that it is not enough to include international feminists in feminist discourses of the Global North.[33] Rather, women in the Global South must set the agenda. This imperative is not directed at these women, but at women of the Global North, that they reject conceptual and ethical onanism.

III. Environmentalists as Ecofeminists

The data also show that environmentalism is necessarily feminism, for at least six reasons. First, women's health suffers with environmental degradation more than men's. Environmental toxins and pollutants have particularly detrimental impacts on women's bodies, especially reproductive functions.[34] Increased rates in miscarriages, birth defects, and cancers associated with reproductive organs were instigating factors in Erin Brokovitch's work, as well as in Lois Marie Gibbs's activism at Love Canal that led to the establishing of the Superfund, the largest program globally for environmental clean-up.[35] Ignoring gender in environmentalism can delay detection and remediation, thus risking everyone's health; women's disproportionate health burden also breaches principles of distributive justice.

Second, most of the world's farmers are women. Women in many cultures work closely with their natural environment in increasingly shouldering primary responsibility for meeting family daily living needs. According to the UN's Food and Agriculture Organization, women produce between 60 per cent and 80 per cent of the food in developing countries.[36] In Ghana, women were growing 70 per cent of food crops in 2003 (GPRS I), but 87 per cent by 2010.[37] As noted above, women's agricultural labour is largely invisible in the systems used in, for example, development and structural adjustment programs. Recognition of women's agricultural contribution is crucial for effective response to growing challenges in food security. Women face compound challenges of climate change, land tenure, and land grabbing that displaces subsistence agriculture with cash crops. Programs of large-scale monoculture production of cash crops are capital-intensive and technology-driven. Women with no bank account or access to credit do not have the resources necessary to participate. If they lose access to the land they farm and become wage labourers in corporate, usually male-owned farms, they often lack job security, child-care support, and maternity benefits, and are vulnerable to sexual abuse. They rarely have access to the food they produce for markets elsewhere, which they could not afford anyway. The environmental issues surrounding adjustment policies with respect to land use are accordingly deeply entangled with gender justice issues. Failure to require, for example, gender impact assessments and budgets precludes identifying and addressing these issues, and thus exacerbates global hunger.

Third, environmental degradation increases women's lived experience of poverty.[38] Deforestation and desertification, for example, oblige women to labour harder for longer hours to meet family daily living needs: they walk further to collect wood for heating and cooking and water for daily use. Since these tasks typically fall to daughters, their education can be compromised. In contexts where women's weekly hour structure includes cooking, caring for children, making, maintaining, and washing clothes, cleaning, working the fields, and walking up to 40 hours to gather water and fuel, environmental degradation makes their lives untenable. One factor prompting the Chipko Movement in India was the increased suicide rate among women who could no longer bear the increasing labour necessitated by their lives. Thus, principles of distributive justice are doubly compromised, as the costs of environmental degradation are borne disproportionately by women, while profits accrue outside the community to logging companies and extractive industries.

Fourth, women are marginalized but nonetheless significant players in environmental reform and have much wisdom to contribute to sustainable policy and practice. Globally, 80 per cent of environmental activists are women.[39] While contemporary policy processes are inadequate to meet demands of sustainability,[40] women's environmental expertise has supported sustainable practice over long historical periods.[41] Women bring unique perspectives to environmental issues,[42] and the remedial value of their conceptions of nature has been well demonstrated.[43] For example, the slogan of the Chipko Movement is "The forest is our home." The women who began this movement in the 1970s were marginally literate, peasant women whose action was nevertheless powerful enough to change India's national forestry policy. They offer an alternative to the prevailing ideology that values nature instrumentally. Instead, they conceive of nature as a home that warrants respect and care. Likewise, the women of the Green Belt Movement in Kenya moved agricultural policy toward sustainable practice. Feminist analysis of women's collectives

in the Deccan Plateau, detailed below, uncovers an ethic of care that these women practice with respect to their community (they provide women with safe, reliable employment), their environment (they regenerate land made unusable through poor stewardship and protect valuable resources like seed banks), and their livestock (they consider the animals' overall well-being rather than just their economic value).[44] Yet these women's perspectives are overlooked in arguments for nature's intrinsic value, and they are absent from the discourse of mainstream environmental ethics.

In developing nations, women's expertise and practices have been dismissed as folklore, superstition, and "old wives" tales."[45] Yet much traditional women's agriculture is sustainable, so not learning from their wisdom impedes movement toward sustainability. To continue not to value the expertise of women throughout the world whose environmental stewardship is based on knowledge systems developed over generations is to ignore a unique and precious resource and exacerbates the "gender gap" in which women's capacities are eroded while their situations and contributions are excluded from environmental policy-making contexts.[46] Women remain under-represented in decision-making positions and resource-management careers throughout the globe.[47] Environmental philosophers who fail to recognize the value of women's environmental expertise and to acknowledge and support their actions are unwittingly reproducing the logic of domination currently underwriting global practices of environmental devastation and injustice rather than promoting sustainability.

The fifth reason environmental thinkers must be ecofeminist is that women's groups have achieved successes in places where men's groups have failed. The Deccan Development Society (DDS), for example, was formed in the

Deccan Plateau of Southern India in 1983 in recognition that livelihoods of the rural poor can improve only when their environment is regenerated. Initially, the DDS supported men's groups, but the programs failed and the groups fell apart over leadership issues. The DDS began instead to support self-selected groups of women. The women's groups rapidly became self-reliant, and the group in the village of Edakulapalli, for example, now represents 1200 women who manage USD$81,000.[48] These women are not successful because they are inherently nice, caring, or accommodating; rather, the benefits of cooperative effort can be more important than political status. The women were already so overburdened with work that they were prepared to share the labour and the status of leadership. They are proof of what Franciska Issaka, district assemblywoman in Ghana and CEO of the Center for Sustainable Development Initiatives in Bolgatanga, said to me based on her 30 years of experience in development: "If you empower women in a community, they will return the benefits to the whole community." Profits are not as important as meeting the basic needs of the family and the community.

The sixth reason environmentalists must have a feminist agenda is population. Earth First! did a great disservice to environmentalism by publishing Miss Ann Thropy's pseudonymous argument that AIDS benefits population control.[49] Hardin's "lifeboat ethics" are equally cringe-worthy.[50] Ecophilosophers need to work hard to refute charges of ecofascism in light of these publications. As I have shown elsewhere, an ecofeminist approach precludes ecofascism.[51] Sen argues that mainstream Northern environmentalists need to focus on gender relations and women's needs in framing their strategies on population issues.[52] Historically, population programs have seen

women simply as targets for family-planning initiatives. Even where they do not violate women's rights through coercive practice, they reproduce class, race, and gender bias. To be effective, family planning must be framed in terms of women's health, education, and livelihood. Making women's basic needs central to population policy increases human welfare, transforms oppressive gender relations, and reduces population growth.

Environmental movements that fail to respond to the feminist call are working against their own agenda. Deep ecologists, for example, failed to see the value of ecofeminist contributions to environmental philosophy. In 1987, Cheney charged deep ecology with androcentrism, Biehl objected to the deep ecologists' critique of anthropocentrism because it implicated women in the male project of domination over nature, and Zimmerman argued that women's experience of relatedness "gives rise to a morality of caring for the concrete needs of those with whom one is related."[53] Salleh charged that deep ecologists fail to grasp both the ecofeminist epistemological challenge and the practicalities around how much labour would be needed to generate social change.[54] In 1993, she argued further that deep ecologists' defensiveness in response to ecofeminism is suspect and that deep ecology is incapable of social critique because its political attitudes are meaningful only to "white-male, middle-class professionals whose thought is not grounded in the labor of daily maintenance and survival."[55] Slicer argued that there would be no genuine debate between ecofeminists and deep ecologists until the latter actually studied ecofeminist analyses, but ecofeminists continued to fail to get on the conference program during the academic heyday of deep ecology.[56] Deep ecology worked against itself by reinscribing the very logic of oppression that ecofeminism argued must be overcome if nature is ever to be thought about and treated beyond its reduction to mere instrumental value.

IV. Power and Promise

De Beauvoir argued in her *Second Sex* that woman and nature both appear as other to man in the logic of patriarchy.[57] Yet she later criticized the merging of feminism with ecology out of suspicion that the appeal to "traditional feminine values, such as woman and her rapport with nature . . . [is a] renewed attempt to pin women down to their traditional role."[58] As an ecofeminist, I have argued precisely that women's labour is not actualization of a biological essence or destiny, but that care arises relationally in their work.[59] In this neo-Marxist, materialist perspective, women are not inherently or inevitably caring. Yet their capacity to care provides new logics contrary to the destructive logic of capitalist patriarchy.

The most promising aspect in twenty-first-century feminism is precisely the turn to materialism that negotiates nature/nurture by rejecting both that nature reduces women to her bodily functions and that her social construction has no grounding in the material conditions of her lived reality. Salleh's "embodied materialism" accepts neither that woman is closer to nature nor that gender is a purely cultural phenomenon.[60] In the next generation of ecofeminists, Mallory argues that ecofeminist activisms are political sites where meaning is made in a way that opens space for subaltern others, including non-human voices. Later, she argues that some of these voices have sufficient agency and subjectivity to be given ethico-political consideration.[61] Twenty-first-century ecofeminism has real-world impacts and takes environmental philosophy well beyond traditional debates in deep ecology, anthropocentrism, and the land

ethic into political ecology and international and policy debates.

Warren has argued for narrative voice over and against traditional philosophical methods.[62] She subsequently reprises the methodological value of narrative because it gives voice to felt sensitivity and overlooked perspectives, allows meaning to emerge rather than be imposed, and suggests what might count as a conclusion to an ethical situation.[63] She revisits her strategy of "taking empirical data seriously" by presenting data on pressing global issues in women's lives. She revalues emotional intelligence to admit caring,[64] and she shows how all "-isms of domination" (sexism, racism, classism, naturism, etc.) are interconnected and can be woven together into the ecofeminist quilt that functions as the book's driving metaphor.[65] Ecofeminist methods have long been capable of challenging the contemporary logic of domination that is institutionalized in epistemological and ontological paradigms of objectivity and implemented in the technologies of destruction that drive global patriarchal capital. By the twenty-first century, ecofeminists have developed and articulated an alternative way of thinking. They may not have the power to change the world, but they have provided conceptual tools and an alternative discourse and logic of care rather than mastery to those who do.

It is not by accident that the gender decision of the United Nations Framework Convention on Climate Change was taken in 2012 under the executive leadership of Christiana Figueres. At the Conference of Parties in Paris 2015, upon which great hopes for movement in the stalled process of global climate policy are pinned, there is hope that the increasing intervention of women's voices is precisely what will burst the deadlock. Buckingham already argues that there have been substantial shifts in European environmental and equalities policy through groups informed by ecofeminism, and impacts of Chipko women on national policy are noted above.[66] Zimmerman wants to "take hope from the global awakening of the quest for the feminine voice that can temper the one-sidedness of the masculine voice."[67]

Ecofeminists speak without borders and from a logic that empowers care, inclusion, partiality, and justice. Wake up! That voice has been around since the 1970s. In the twenty-first century, it speaks throughout a world where time is running out.

Glazebrook provides us with a clear and brisk update of contemporary ecofeminism at the same time as she shows us how, exactly, the movement might continue to extend its power and live up to its promise. The key is to resist the essentialisms of both first- and second-wave feminism. Glazebrook's essay is rife with examples of how this can be done. Here is one more, from the Canadian context.

The example involves the environmental impacts of large-scale industrial agriculture and the smaller-scale alternatives that have arisen as a challenge to it. Canadian sociologist Martha McMahon has studied small-scale organic farmers on southern Vancouver Island and on the Gulf Islands in BC. Out of her research she has constructed a partly fictionalized picture of such farmers as repositories of ecologically important local knowledge and therefore as a potent force for political change. In the area studied by McMahon, many of the farmers are women. So she has us imagine a struggle between these women and the agribusiness forces pitted against them. In the imagined scenario, the farmers are faced with an ordinance from the city that would prevent them from keeping pigs, a significant source of their livelihood. The city's committees and planners decree that pig farms must meet a minimum acreage standard, and the size they call for is larger than the farms run by the women. So their farms would become illegal.

The responses of McMahon's farmers to the proposed bylaws can be summarized in three points. First, the proposed bylaws are gender-biased. Second, they defy mounting evidence about the ecological benefits of small-scale organic farming. Third, they devalue the specific and local knowledge involved in this style of farming. The women hold up their work as a rooted "cultural activity," in contrast to the a-cultural productive activity of multinational agribusiness.[68] These women therefore find themselves on the front line of a significant ecological struggle, a point that brings us back to feminism's third wave. No reference is made here—either by McMahon herself or by any of the farmers she has studied—to women's natural or essential connection to the Earth. Indeed, some of the forces that have resulted in women being small-scale farmers in the first place are themselves the product of gendered power structures.

Here is how McMahon puts the point:

> The small-scale women farmers whose story is told here were using their socially constructed identities as women and their orientation to community food provisioning to disrupt dominant, classed, and masculinist taken-for-granted ideas

about farming. . . . Too often . . . academics have misread women's ecological activism in terms of claims about women's closeness to a romanticized and essentialized nature in order to dismiss the potential of ecological feminism.[69]

McMahon explicitly disavows these essentialist understandings of ecofeminism. What
she offers instead is a picture of women using their socially constructed identities to fight
for changes in environmental policy. In short, the threefold task is to (1) insist on a non-
essentialized, socially constructed understanding of women's various identities and social
roles; (2) pick out those identities and social roles that have significance for environmental
issues; and (3) employ these identities and social roles *strategically* in the public or political
sphere. That is the power and promise of ecofeminism's third wave.

ⓔ Conclusion

An important question remains about the relation between ecofeminism and environmental
ethics. Some have argued that ecofeminism goes well beyond the major positions in environmental ethics and that these positions are fatally flawed. Let's look briefly at what ecofeminists
have to say about two such positions: biocentrism and deep ecology.

As we have seen in Chapter 2, the leading exponent of biocentrism is Paul Taylor, and
his approach to environmental ethics is explicitly Kantian. One aspect of this view is that
moral principles are the product of reason, not inclination, desire, or emotion. For Kant,
we are obligated to treat other rational beings with respect. This means that we may not
coerce or deceive them, for example. But the obligation is a deliverance of reason. Although
Kant's argument is complex, it can be summarized as follows. Each of us inevitably treats
himself or herself as a source of value. We construct plans, principles, and purposes that
give our lives a measure of organizational unity, coherence, and meaning. To have plans
and principles is to have values that allow us to assess and in some cases resist the welter
of temptations and inclinations that constantly assail us. But because we see ourselves this
way, we should, to be consistent, see other rational beings the same way. We should respect
them as sources of value. The practical fruit of this argument—that we should treat others
respectfully—is not based on a feeling like compassion or kindness. It is simply an appeal
to rational consistency.

Taylor, of course, wants to extend the circle of beings to whom we have a duty of respect to
all living things, since each of them is, if not a source of value, at least a teleological centre of
life. But the justification is the same. We each treat ourselves this way, so to be consistent we
should treat other living things the same way. This is just where the argument goes astray,
according to Plumwood. In her view, it relies on a gendered view of reason as *opposed* to and
superior to feeling and desire. What we need instead is more emphasis on care, love, and
kindness as affective foundations of biocentric ethics.

One way to respond to Plumwood on this point is to note that deliverances of reason have the
advantage of being more *stable* than deliverances of emotion or desire. If we have a reason-based

duty to act a certain way, then the duty holds *whether or not* we feel like acting in accordance with it. To say that the duty is derived from desire and emotion, by contrast, seems to amount to the claim that we should perform the appropriate actions *only if* the desire or emotion moves us. But what if it doesn't? Do we really suppose a duty ceases to be binding as the emotion, desire, or inclination that produced it fades or disappears altogether? Can this way of thinking provide a reliable foundation for our duties? It should be noted, in addition, that neither Kant nor Taylor thinks that emotions, desires, and inclinations are entirely inappropriate. As long as the action is motivated by the rational principle, it can be supported by a wide range of other, non-rational, psychological states.

This brings us to the ecofeminist critique of deep ecology. For ecofeminists, one of the key flaws of most Western ethics is what Plumwood calls the problem of "discontinuity." That is, we humans tend to create dualisms, like the ones examined above, and then suppose that there is utter discontinuity between the pairs thus created. So, for example, nature and culture are put at odds in a radical way such that culture is seen as entirely non-natural and vice versa. The same would go for the other key pairs. Of course, deep ecologists have a similar worry about our tendency to separate ourselves from the natural world, which is why they insist so strongly on the idea that we should seek to *identify* ourselves with natural objects and processes. But so far it appears that ecofeminists and deep ecologists might be philosophical allies. Where is the difference?

According to Plumwood, deep ecologists go astray because of the vagueness of their notion of "identification." They shuttle back and forth between two extreme views, neither of which is acceptable. Sometimes they understand identification of self and nature in quasi-mystical terms. But this is a too complete identification in which the self becomes *absorbed* into nature and is therefore threatened with the loss of its own unique identity. At other times, deep ecologists talk about identification more as *expansion* of the self into nature. For instance, deep ecologists will claim that environmental defence is, for the environmentalist, a form of self-defence. Thus, in spite of its attempts to get beyond selfish human interests, the problem with deep ecology, for Plumwood, is that it is just dressed-up egoism.

The alternative is to see the self as a distinct individual but one that is contained in a web of other things—both natural and cultural—that are related to one another in more or less complex ways. This is the ecofeminist concept of "self-in-relationship":

> It is unnecessary to adopt any of the stratagems of deep ecology . . . in order to provide an alternative to anthropocentrism or human self-interest. This can be better done through the relational account of self, which clearly recognizes the distinctness of nature but also our relationship and continuity with it. On this relational account respect for the other results neither from the containment of self nor from a transcendence of self, but is an *expression* of self in relationship.[70]

The best way for such expression to occur is through the connection *particular* people have—via the bonds of compassion, friendship, love, respect, and so on—to *particular* bits of the natural world: this forest, that animal, this wetland, and so on. This is the point of Warren's repeated references to the highly *contextual* character of ecofeminist ethics. In a process like this, we can both retain our sense of ourselves as distinct individuals and consider ourselves as necessarily bound up with other particular things. This will then have implications for how we ought to treat those other things.

But when the concept of self-in-relationship is parsed this way, it is difficult to see how the position is significantly distinct from that of the deep ecologists *with regard to our duties to nature*. After all, deep ecologists also stress the importance of our affective identification with particular parts of the natural world. Furthermore, they insist that these acts of identification should be translated into meaningful political action when those parts of the natural world are threatened with degradation. Does it matter that there are remnants of "rationalism" in the deep ecological view as long as its proponents have correctly picked out our duties to nature? Plumwood evidently thinks it does, maintaining that deep ecologists have failed to appreciate the ways in which the "inferiorization of both women and nature is grounded in rationalism."[71] But why should they have noticed this, especially since it is, as we have just seen, entirely possible to act responsibly on behalf of nature on strictly deep ecological principles? Is the claim that every environmental philosophy worthy of the name ought to be explicitly ecofeminist in orientation? That is surely a form of theoretical imperialism.

CASE STUDY

Dioxin in Breast Milk and Women as Front-Line Environmentalists

Important environmental issues rarely affect members of just one gender. But what the third wave of ecofeminism highlights is the fact that women may be, as it were, on the front lines of this or that environmental threat and that their presence there can be strategically useful in bringing the issue to wider public attention. That was the case of the women organic farmers (section D, above) as well as the Chipko women (section A, above). Another example has to do with women's role as providers of basic material sustenance to infants through breastfeeding. We *do not* need to say that women are naturally or essentially mothering/nurturing, but many of them do perform this service, so there is a sense in which they—and of course the children they feed—would be directly affected were it the case that the by-products of industrial processes had caused the widespread contamination of breast milk. In the same way that our women organic farmers were in a unique position to mobilize for political change on a particular environmental issue, so too would be women-as-mothers on this scenario. Unfortunately, such contamination is just what has happened to breast milk in Canada, although there is a curious story to tell about why the issue has *not* received serious public attention.

Source: © iStockphoto.com/wethervain

Pulp operation, Corner Brook, NL. Industrial production is directly tied to toxins in our food supply. How much should we tolerate? Whose interests count?

Dioxins are toxic chemicals emitted into the atmosphere as by-products of various industrial processes, such as the incineration of medical waste and the chlorination of wood products in pulp mills. The toxins are highly fat-soluble but very low in water solubility, the result of which is that they accumulate very readily in fat tissue. This, in turn, means that dioxins bioac cumulate—that is, they increase in concentration as they are passed up the food chain into members of species with increasing amounts of fat tissue. We get them mostly from eating eggs, fish, poultry, and dairy products. We might think that consuming these things puts adult humans at the top of the food chain, but infants ingesting breast milk are in a sense one step higher. This is because in the process of lactation, a concentrated dose of dioxins is transferred from a woman's body fat to her breast milk. In sufficiently high concentrations, dioxins can cause certain forms of nervous system disorder, impairment of the immune, endocrine, or reproductive systems, and certain cancers. A report released by the Canadian government in 1990 indicated that infants receive *16.5 times* the "tolerable daily intake" of dioxin. But because they receive this intake only for a short period in their lives, the government insists that there is no significant danger to them.

Is it wise to assume that exposing babies to 16.5 times the tolerable daily intake of these toxins—albeit for perhaps just a few months—is really safe? At the very least, the matter surely deserves closer attention at the level of public policy. And yet it has not received such attention. Why not? It is worth pointing out that governments, including Canada's, have devoted much

Continued

more effort to combatting dioxins in fish, dioxins seeping from bleached cardboard cartons into cow's milk, and dioxins from contaminated animal feed. However, the lifetime cancer risk from dioxins in breast milk is equal to or greater than that from any of these sources. The difference in attention can likely be explained by the fact that current levels of dioxin in breast milk are not considered high enough to outweigh the significant advantages of breastfeeding to infants. Canadian political scientist Kathryn Harrison claims that

> the relevant comparison thus may be net risk (or benefit) rather than environmental risk alone. Indeed, despite the presence of contaminants, physicians, government spokes-persons, breast-feeding advocates and even environmentalists invariably stress that the benefits of breast feeding outweigh the risks.

Remember Baxter's arguments about pollution from Chapter 4? He claimed that it is a mistake to think that any of us really desire an environment entirely free of pollution. Rather, we want all sorts of things, but we live in a world of scarce resources, so we have to make hard choices. We have to prioritize and perhaps settle for a little less of a desired thing than we would get in a perfect world. In short, we should seek *optimal* levels of pollution. In the example we are considering here, we are effectively being asked to apply this logic to breast milk, but there is a peculiar danger in this case. What we are saying is that we will "tolerate" carcinogens in breast milk in order also to have the consumer goods—for example, fresh sheets of white paper—the production of which creates them. That's the cash value of the claim about achieving an optimal balance among our competing desires. However, because we also invest breastfeeding with all sorts of cultural significance, we might be willing to see the levels of dioxins in breast milk go extremely high before we do anything about it.

One indication that this situation poses a real danger is the fact that even environmental groups like Greenpeace have been nervous about fighting this particular battle. Harrison argues that they are reluctant to do so because they sense that they would be defeated if they raised the issue, they don't want to cause alarm among the populace, and they consider the issue too emotionally charged. But consider the effect of this reluctance to engage such a serious environmental problem. We have government officials, breastfeeding advocates, the medical establishment, and environmental groups all unwilling to challenge the status quo because they believe that the alternative will be a call to ban breastfeeding. But that is a false alternative. The political campaign can instead be focused on the *source* of the pollut-ants, on getting them out of the industrial system or at least lowering their concentrations in the environment. However, for this to happen we will need to be forthright about the fact that—because of bioaccumulation—the *primary threat* of dioxins is to infants through con-taminated breast milk.

The powerful array of political forces just mentioned seems paralyzed by fear that in engaging the problem in this direct a manner they will be perceived as anti-breastfeeding. Will their collective inaction allow us to become dangerously complacent about what con-stitutes tolerable levels of dioxin in breast milk? This is an issue for all of us but perhaps most prominently for women, because they are on the front line of the battle to keep dioxin levels in breast milk to a level that is *genuinely* tolerable.

Sources: Kathryn Harrison, "Too Close to Home: Dioxin Contamination of Breast Milk and the Political Agenda," in *This Elusive Land: Women and the Canadian Environment*, ed. Melody Hessing, Rebecca Raglon, and Catriona Sandilands (Vancouver: UBC Press, 2005), 213–42; Government of Canada, "It's Your Health," www.hc-sc.gc.ca/hl-vs/iyh-vsv/environ/dioxin-eng.php; World Health Organization, "Dioxins and Their Effects on Human Health," www.who.int/mediacentre/factsheets/fs225/en/index.html.

Study Questions

1. Assess critically the idea that it is a mistake to essentialize women in the ecological struggle, as second-wave feminism allegedly does. Is essentialism necessarily misguided? Is the third-wave focus on using women's social and political positions strategically a better approach? Why or why not?

2. What do you think of Warren's claim that sexism and naturism are conceptually connected?

3. How does Glazebrook think that ecofeminism should be developed and extended? Can ecofeminism remain an important political and theoretical option in the future of philosophical environmentalism?

4. What do you make of the relationship between ecofeminism and deep ecology?

Further Reading

Heather Eaton and Lois Ann Lorentzen. 2005. *Ecofeminism and Globalization*. Lanham, MD: Rowman and Littlefield.

Patricia Glazebrook. 2002. "Karen Warren's Ecofeminism." *Ethics and the Environment* 7, no. 2: 12–26.

Ariel Kay Salleh. 1984. "Deeper Than Deep Ecology: The Ecofeminist Connection." *Environmental Ethics* 6: 339–45.

Catriona Sandilands. 1999. *Good-Natured: Ecofeminism and the Quest for Democracy*. Minneapolis: U of Minnesota P.

Karen Warren. 1997. *Ecofeminism: Women, Culture, Nature*. Bloomington: Indiana UP.

Michael Zimmerman. 1987. "Feminism, Deep Ecology, and Environmental Ethics." *Environmental Ethics* 9: 21–44.

CHAPTER 7

Environmental Aesthetics

Ecological Intuition Pump

When you marvel at a robin building its nest and call the work—both the sustained, goal-directed effort and the final product—beautiful, what sort of judgment are you making? Are aesthetic judgments like this more like claims of taste or claims about the way the world is? The distinction matters, because in the first case you may not be inclined to work very strenuously to get others to agree with you—saying, in effect, *de gustibus non est disputandum* (you can't dispute taste)—while in the second case you might. And perhaps such agreement is a prerequisite for knowing how to act so as to protect threatened nature. To emphasize the importance of the distinction, look at an aerial photograph of the tar sands in Alberta's Athabasca region. You might be struck by the extent to which our industrial activities have made so much of the land there so *ugly*. Now, what is the ethical force of *that* judgment? Are you willing to say that the large-scale uglification of the landscape is a *reason* for not extracting the resource? If you are, then you must be doing something more than making a mere judgment of taste. But if this is correct, how much weight do you think should be given to your judgment? After all, there are other considerations here, many of them economic. Do we tolerate some assaults on the beauty of the natural world as long as the economic payoff is high enough?

Ⓐ Introduction

Roughing It in the Bush is a classic memoir of early settler life in Canada, first published in 1852. Its author, Susanna Moodie, gives readers a vivid picture of what life in this country must have been like for many early homesteaders. Late in the voyage to Canada, travelling by ship from Grosse-Île to Quebec City, Moodie has this to say about the sight laid out before her:

> Nature has lavished all her grandest elements to form this astonishing panorama. There frowns the cloud-capped mountain, and below, the cataract foams and thunders; wood, and rock, and river combine to lend their aid in making the

picture perfect. . . . The precipitous bank upon which the city lies piled, reflected in the still, deep waters at its base greatly enhances the romantic beauty of the situation. The mellow and serene glow of the autumnal day harmonised so perfectly with the solemn grandeur of the scene around me, and sank so silently and deeply into my soul, that my spirit fell prostrate before it, and I melted involuntarily into tears.[1]

This is a splendid example of the **picturesque** view of natural beauty, still dominant in Moodie's day. The picturesque insists that natural scenes are beautiful to the extent that they are like pictures, especially landscape paintings. As historian John Conron puts it, the picturesque emphasizes those aspects of nature that are "complex and eccentric, varied and irregular, rich and forceful, vibrant with energy."[2] It also emphasizes the pleasure that is derived from contemplating the beautiful. Writing in *the Spectator* in 1712, this is how Joseph Addison expresses the connection between the picturesque and pleasure: "We find the works of Nature more pleasant, the more they resemble those of Art."

To illustrate, let's focus on three key elements of the passage. First, Moodie is observing the scene from a *distance*, specifically from the deck of a ship. This allows her to *frame* what she sees in a manner that nicely fits what she thinks she ought to be seeing—namely, the sort of picture-postcard view of nature dominant in England at the time. It is, essentially, a tourist's view of natural beauty. Second, the scene does indeed seem to contain the combination of elements Conron lists: a certain formal complexity missing in small, beautiful artifacts as well as an intensity that startles without terrifying. In all, a formally interesting and safely uplifting spectacle. Finally, Moodie clearly finds the experience pleasurable *because* it has been made to fit the model of a beautiful picture. That sort of reaction is not an accidental or unwelcome add-on to the experience of the picturesque but an essential component of it.

Now, the memoir is in many ways the story of how Moodie comes to shed or at least qualify this naive view of nature. Because she actually has to get into the rough Upper Canadian bush and build a life, she becomes *immersed* in nature in a way that makes the picture-postcard view of it increasingly untenable for her. For our purposes, however, the key aspect of the picturesque is that it fails to hook up to an adequate environmental ethic. It does not seem to provide, or even point toward, a morally appropriate stance toward nature, so it gives us no guidance on many of the practical environmental problems we face. The swooning Moodie is a mere spectator of pleasantly arranged formal properties, miles away from the thought that she may have to *act* to protect or preserve the beauty she beholds.

Ⓑ Beauty and Duty: Mapping the Terrain

But how could any theory about natural beauty do this? How could the appreciation of natural beauty—the act of making determinate judgments of the form "natural object or scene *x* is beautiful"—lead anyone to do right by nature? To get a handle on the many different ways philosophers have tried to explain this, we turn to this chapter's first article.

AESTHETIC APPRECIATION OF NATURE AND ENVIRONMENTALISM
Allen Carlson

I. Introduction

The aesthetic appreciation of nature has been and continues to be vitally important for environmentalism. Environmental philosopher Eugene Hargrove points out that aesthetic value was extremely influential in the preservation of some of North America's most magnificent environments.[3] Other environmental philosophers agree. J. Baird Callicott claims: "What kinds of country we consider to be exceptionally beautiful makes a huge difference when we come to decide which places to save, which to restore or enhance, and which to allocate to other uses," concluding that "a sound natural aesthetics is crucial to sound conservation policy and land management."[4] The importance of a sound natural aesthetics is echoed by other environmental thinkers. Ned Hettinger sums up his investigation of the significance of aesthetics for the "protection of the environment" by affirming that "environmental ethics would benefit from taking environmental aesthetics more seriously."[5]

II. Traditional Aesthetics of Nature: The Picturesque and Formalism

To understand the relationship between aesthetic appreciation of nature and environmentalism, it is useful to review two significant historical developments that shaped traditional aesthetics of nature, the idea of the picturesque and the formalist theory of art.

The picturesque has its roots in the eighteenth century with the acceptance of nature as an ideal object of aesthetic experience and the separation of its appreciation into three distinct modes: the beautiful, the sublime, and the picturesque. Historian John Conron summarizes the differences: the beautiful tends to be small and smooth but subtly varied, delicate, and fair in colour, while the sublime, by contrast, is powerful, vast, intense, and terrifying. The picturesque is in the middle ground between the sublime and the beautiful, being "complex and eccentric, varied and irregular, rich and forceful, vibrant with energy."[6] Of these three, the picturesque achieved pre-eminence as a model for nature appreciation, in part because it covers the extensive middle ground of the complex, eccentric, varied, irregular, rich, forceful, and vibrant, all of which seem well-suited to nature. Moreover, the idea had grounding in the theories of some earlier aestheticians, who thought that the "works of nature" were more appealing when they resembled works of art.

Indeed, the term "picturesque" literally means "picture-like," and thus the picturesque gave rise to a mode of aesthetic appreciation in which nature is experienced as if divided into scenes—into blocks of scenery. Such scenes aim in subject matter and composition at ideals dictated by the arts, especially landscape painting. Picturesque-influenced appreciation was popularized by William Gilpin, Uvedale Price, and Richard Payne Knight.[7] Under their guidance, the picturesque provided the reigning aesthetic ideal for English tourists, who pursued picturesque scenery in the Lake District and the Scottish Highlands. The picturesque continued throughout the nineteenth century to have a great impact on nature appreciation. In North America, it inspired nature writing and was exemplified in landscape painting. And in the twentieth century, it remained the mode of aesthetic appreciation commonly associated with tourism—that which appreciates the natural world in light of the scenic images of travel brochures and picture postcards.

Even as aesthetic appreciation of nature influenced by the picturesque continued to be extremely popular in the early part of the twentieth century, a related but somewhat distinct approach to nature appreciation was spawned by that period's most influential theory of art: the formalist theory. As developed by British art critic Clive Bell, formalism is basically a theory about the nature of art that holds that what makes an object a work of art is an aesthetically moving combination of lines, shapes, and colours. Bell called this "significant form" and argued that aesthetic appreciation of art is restricted to it, notoriously stating that to "appreciate a work of art we need bring with us nothing but a sense of form and colour."[8]

However, even Bell, whose aesthetic interest was almost exclusively devoted to art, could find aesthetic value in nature when it is experienced, in his words, "with the eye of an artist" by which an appreciator, "instead of seeing it as fields and cottages . . . has contrived to see it as a pure formal combination of lines and colours."[9] Like the tradition of the picturesque, Bell had in mind seeing nature as it might look in landscape paintings, but not exactly the same kind of paintings as those favoured by the picturesque. Understandably, Bell's view was more closely allied with the work of artists of his own time, such as Paul Cézanne. For example, Cézanne's landscape paintings are classics of one kind of formal treatment of the landscape in which nature is represented as patterns of line, shapes, and colours. Throughout the first part of the twentieth century, various artists and schools of painters developed this kind of formal approach to landscape appreciation, and thus it came to dictate a popular way of aesthetically experiencing nature.

Although formalism and the picturesque have somewhat different emphases and take different artistic traditions as their models, they are similar enough to constitute what can be called traditional aesthetics of nature. The approach combines features favoured in picturesque appreciation, such as being, to repeat Conron's words, "varied and irregular," "rich and forceful," and "vibrant with energy," with the prominence of bold lines, shapes, and colours privileged by formalists. It focuses on striking and dramatic landscapes with scenic prospects, such as found in the Rocky Mountains of North America, where rugged mountains and clear water come together to contrast and complement one another.[10] The role of traditional aesthetics of nature in the development of popular appreciation of nature, as well as in the growth of environmental thought and action, is difficult to overestimate.

North America's rich heritage of parks and preserves is in large part the result of the fact that these areas were seen as aesthetically appealing in light of traditional aesthetics of nature. The same is true of many other parts of the world. Concerning America in particular, J. Baird Callicott claims that historically, "aesthetic evaluation . . . has made a terrific difference to American conservation policy and management," pointing out that one of "the main reasons that we have set aside certain natural areas as national, state, and county parks is because they are considered beautiful" and concluding that many "more of our conservation and management decisions have been motivated by aesthetic rather than ethical values."[11]

III. The Failings of Traditional Aesthetics of Nature and the Requirements of Environmentalism

Callicott's claim is no doubt true. However, more recently the relationship between aesthetic appreciation of nature and environmentalism has become less congenial. For example, although Aldo Leopold recognized the historical importance of traditional aesthetics of nature and his viewpoint continues to shape contemporary understanding of the relevance of aesthetic appreciation to environmentalism, he nonetheless expressed some worries about the "taste for country" of the vast majority, who he noted "are willing to be herded in droves through 'scenic' places" and "find mountains grand if they be proper mountains with waterfalls, cliffs, and lakes" but yet find "the Kansas plains ... tedious."[12] Leopold realized that the popular "taste for country," largely the result of traditional aesthetics of nature, had certain limitations and that for aesthetic appreciation of nature to support environmentalist thought and action, aesthetic values must be more in line with environmental values.

Some recent environmental thinkers, following in Leopold's footsteps, have also expressed concern about the aesthetic values embodied in traditional aesthetics of nature.[13] They charge that traditional aesthetics of nature has five major failings in that it endorses appreciation of nature that is: (1) anthropocentric; (2) scenery-obsessed; (3) superficial and trivial; (4) subjective; and (5) morally vacuous.[14] In light of these failings, we can indicate Requirements of Environmentalism for appropriate aesthetic appreciation of nature by contrasting them with solutions or, perhaps better, antidotes. Thus, environmentalism requires appreciation that is: (1) acentric; (2) environment-focused; (3) serious; (4) objective; and (5) morally engaged.[15]

1. *Acentric rather than anthropocentric appreciation.* The charge that traditional aesthetics of nature is anthropocentric or human-centred is directed mainly at the picturesque. There is a sense, of course, in which all aesthetic appreciation is, and must be, from the point of view of a particular human appreciator, but the criticism concerns the specific conception of nature and our relationship to it that seems implicit in traditional aesthetics of nature. Part of this conception involves the anthropocentric thought that nature exists exclusively for us and for our pleasure. Landscape geographer Ronald Rees contends that "the picturesque ... simply confirmed our anthropocentrism by suggesting that nature exists to please as well as to serve us."[16]

Likewise, Canadian environmental philosopher Stan Godlovitch argues that to "justify protecting nature as it is and not merely as it is for us ... a natural aesthetic must forswear the anthropocentric limits that ... define and dominate our aesthetic response."[17] Godlovitch's antidote for anthropocentrism and thus a requirement of environmentalism is to attempt to achieve what he calls an acentric approach to appreciating the natural world. The idea is that an appreciator must strive for an experience that is not from any particular point of view, human or otherwise—what is sometimes called a "view from nowhere." It is far from clear exactly how a human appreciator can adopt such a fully non-anthropocentric viewpoint. Nonetheless, Godlovitch proposes that since "only acentric environmentalism takes into account nature as a whole ... we require a corresponding acentric natural aesthetic to ground it."[18]

2. *Environment-focused rather than scenery-obsessed appreciation.* The second criticism is that traditional aesthetics of nature goes beyond a mere focus on scenery to the point of obsession. And although it may be granted that there is much aesthetic value in scenery, when the point of view becomes an obsession, other less conventionally scenic environments are excluded from appreciation. This problem is especially acute concerning environments that may be ecologically valuable but do not fit the traditional conception of scenic landscapes,

such as prairies, badlands, and wetlands.[19] In "The Aesthetics of Unscenic Nature," environmental aesthetician Yuriko Saito argues that the "picturesque . . . has . . . encouraged us to look for and appreciate primarily the scenically interesting and beautiful parts of our environment" with the result that "those environments devoid of effective pictorial composition, excitement, or amusement (that is, those not worthy of being represented in a picture) are considered lacking in aesthetic values."[20]

One antidote for scenery-obsession, according to environmental philosopher Holmes Rolston, is acknowledging a rather obvious feature of the experience of natural environments. The requirement of environmentalism in this case is satisfied by the recognition that appreciation of nature "requires embodied participation, immersion, and struggle" and that it is a mistake to think of forests, for example, "as scenery to be looked upon," for a "forest is entered, not viewed" and you "do not really engage a forest until you are well within it" and once you are within the "forest itself, there is no scenery."[21]

3. *Serious rather than superficial and trivial appreciation.* The third criticism, that the appreciation endorsed by traditional aesthetics of nature is superficial and trivial, is perhaps the most grave of the five charges. After observing that we continue to admire and preserve primarily "landscapes," "scenery," and "views" according to essentially eighteenth-century standards of taste inherited from Gilpin, Price, and their contemporaries, Callicott claims that our "tastes in natural beauty . . . remain fixed on visual and formal properties" and is "derivative from art." The upshot is that the "prevailing natural aesthetic, therefore, is not autonomous: it does not flow naturally from nature itself; it is not directly oriented to nature on nature's own terms . . . It is superficial and . . . trivial."[22]

As Callicott makes clear, the heart of this criticism lies in the fact that traditional aesthetics of nature is dependent on artistic models and does not treat nature as nature. Thus, the requirement that appreciation be serious rather than superficial and trivial is satisfied when it is "true to nature" in the sense of being directed fully and deeply toward what nature is and the qualities it has. In his groundbreaking essay "Contemporary Aesthetics and the Neglect of Natural Beauty," Ronald Hepburn suggests this requirement of environmentalism by contrasting appreciating a cloud as resembling a basket of washing with appreciating it by realizing "the inner turbulence of the cloud, the winds sweeping up within and around it, determining its structure and visible form." He suggests that the latter experience is "less superficial . . . truer to nature, and for that reason more worth having," since rather than "easy beauty," it is "serious beauty."[23]

4. *Objective rather than subjective appreciation.* The criticism that appreciation grounded in traditional aesthetics of nature is subjective is perhaps more a failing of the picturesque than of formalism, for the former focuses more heavily on the pleasurable experiences of appreciators.[24] The problem is that if traditional aesthetics of nature yields only subjective judgments about nature's aesthetic value, then individuals making environmental decisions may be reluctant to acknowledge its importance, regarding it simply as based on personal whims or relativistic, transient, soft-headed artistic ideals. As Ned Hettinger remarks, "If judgments of environmental beauty lack objective grounding, they seemingly provide a poor basis for justifying environmental protection."[25] Environmental philosopher Janna Thompson concurs: "A judgment of value that is merely personal and subjective gives us no way of arguing that everyone ought to learn to appreciate

something, or at least to regard it as worthy of preservation."[26]

Thus, the objectivity requirement is a particularly important requirement of environmentalism. As Thompson further observes, the "link . . . between aesthetic judgment and ethical obligation fails unless there are objective grounds—grounds that rational, sensitive people can accept—for thinking that something has value."[27] The importance of the requirement is put in even stronger terms by aesthetician Noël Carroll, who contends that "any . . . picture of nature appreciation, if it is to be taken seriously, must have . . . means . . . for solving the problem of . . . objectivity."[28]

5. *Morally engaged rather than morally vacuous appreciation.* The last charge against traditional aesthetics of nature is again especially important regarding environmental thought and action, for environmentalists wish to bring aesthetic appreciation in line with ethical obligations to preserve and maintain ecologically healthy environments. But if traditional aesthetics of nature is morally vacuous, then ultimately there is no significant way of linking, as some environmental philosophers put it, beauty and duty. Ronald Rees contends that in traditional aesthetics of nature, there is "an unfortunate lapse" in that our "ethics . . . have lagged behind our aesthetics," allowing "us to abuse our local environments and venerate the Alps and the Rockies."[29] Landscape historian Malcolm Andrews confirms this, arguing that "the trouble is that the picturesque enterprise in its later stage, with its almost exclusive emphasis on visual appreciation, entailed a suppression of the spectator's moral response."[30]

The problem is that scenery and lines, shapes, and colours seem to support either no moral judgments or else only the emptiest ones, such as the prescription to preserve that which pleases the eye. Thus, the key to satisfying the last requirement of environmentalism, that aesthetic appreciation of nature be morally engaged, lies at least partly in the differences between art-based appreciation and nature appreciation. Philosopher Patricia Matthews points out that in the latter case our "aesthetic assessments take into consideration not only formal elements such as color and design, but also the role that an object plays within a system." This, she concludes, "allows for a complex, deep, and meaningful aesthetic appreciation of nature" such that "facts about . . . environmental impact . . . can affect our aesthetic appreciation" and thus "our aesthetic and ethical assessments of what ought to be preserved in nature may be more harmonious."[31]

IV. Contemporary Aesthetics of Nature and the Requirements of Environmentalism

Different contemporary approaches to aesthetics of nature are frequently classified as either non-cognitive or cognitive. Positions of the first type stress emotional and feeling-related states and responses, such as arousal, affection, reverence, engagement, and mystery. By contrast, positions of the second type contend that knowledge about objects of appreciation is a necessary component of their appropriate aesthetic appreciation.

The most fully developed non-cognitive approach is the aesthetics of engagement.[32] This position rejects much of traditional aesthetics of nature, such as the external, distanced appreciator favoured by the picturesque and formalism, arguing that these approaches involve a mistaken conception of the aesthetic and that this is most evident in aesthetic experience of nature. According to the engagement approach, this conception's distancing and isolating gaze is out of place in appreciation of nature, for it wrongly abstracts both natural objects and appreciators from the environments in which they properly belong and in which appropriate appreciation is achieved.

Rather, the approach recommends that traditional dichotomies, such as between the object and the subject of appreciation, be abandoned, contending that aesthetic experience involves a participatory engagement of the appreciator within the object of appreciation. Thus, it stresses the contextual dimensions of nature and our multi-sensory experience of it, taking aesthetic experience to involve a total "sensory immersion" of the appreciator within the natural world.[33] The foremost proponent of this position, Arnold Berleant, claims that "we cannot distance the natural world from ourselves" and that we perceive nature from within, "looking not at it but being in it," in which case it "is transformed into a realm in which we live as participants, not observers." He concludes that the "aesthetic mark of all such times is . . . total engagement, a sensory immersion in the natural world."[34]

In contrast to non-cognitive approaches are positions classified as cognitive, which are united by the idea that information about the object of appreciation is central to appropriate aesthetic appreciation. They hold that nature, in the words of Yuriko Saito, must be "appreciated on its own terms."[35] The best-known cognitive approach is scientific cognitivism. Like most cognitive positions, which in general reject the idea that aesthetic experience of art provides satisfactory models for appreciation of nature, this view stresses that nature must be appreciated as nature and not as art. Nonetheless, it holds that aesthetic appreciation of nature is analogous to that of art in its character and structure and, therefore, that art appreciation can show some of what is required in adequate appreciation of nature. In appropriate aesthetic appreciation of art, it is essential that works be

experienced as what they are and in light of knowledge about them.

For instance, appropriate appreciation of a work such as Jackson Pollock's *One: Number 31, 1950* requires experiencing it as a painting and moreover as an action painting within the school of mid-twentieth-century American abstract expressionism. Therefore, it must be appreciated in light of knowledge of these artistic traditions. In short, in the case of art, serious, appropriate aesthetic appreciation is informed by art history and art criticism. However, since nature must be appreciated as nature and not as art, scientific cognitivism contends that although the knowledge given by art history and art criticism is relevant to art appreciation, in nature appreciation the relevant knowledge is that provided by natural history, by the natural sciences, especially geology, biology, and ecology.[36] In short, to appreciate nature "on its own terms" is to appreciate it as it is characterized by science.[37]

How then do contemporary approaches to aesthetics of nature fare on the requirements of environmentalism? Both the aesthetics of engagement and scientific cognitivism are clearly superior to traditional aesthetics of nature in a number of ways. An appreciator who is totally sensory-immersed in a natural environment and/ or well informed by scientific knowledge about it contrasts dramatically with a distanced appreciator who focuses only on formalist, picturesque scenery.[38] Concerning the acentric requirement, the aesthetics of engagement's stress on sensory immersion seems to facilitate as acentric a point of view as is humanly possible, since it explicitly calls for abandoning traditional dichotomies, such as between the object of appreciation and the appreciator, and thus it would seem that the appreciator's own particular point of view must also be abandoned. Similarly, scientific cognitivism's reliance on scientific knowledge promotes an acentric point of view similar to that of science, which is an acentric way of knowing.[39] Concerning the environment-focus requirement, the aesthetics of engagement's stress on an appreciator's engaged participation takes into consideration whole environments and explicitly not scenery or formal composition. Likewise, scientific cognitivism's emphasis on environmental sciences focuses appreciation on environments rather than on scenery or formal features. There is no ecological science of scenery or of lines, shapes, and colours. Nor can one be immersed within scenery or within a combination of lines, shapes, and colours.

However, the success of contemporary approaches in meeting the remaining three requirements of environmentalism is more mixed. For example, although total sensory immersion may result in a high level of intensity, it does not seem to require seriousness in the sense of being "true to nature." It only allows for serious appreciation to whatever extent such appreciation is consistent with immersion. In addition, the aesthetics of engagement's dependence on immersion seems to weaken the position concerning objectivity, for abandoning the dichotomy between the object of appreciation and the subject of appreciation will seemingly make it difficult for an appreciator to be objective. Moreover, although the aesthetics of engagement would seem to support moral engagement concerning environmental issues, the position's subjectivity undercuts the possibility of a compelling moral stance, for without objectivity, ethical assessments, even if fuelled by intense engagement, can be dismissed as only expressions of personal feelings.

By contrast, scientific cognitivism's reliance on scientific knowledge promotes appreciation that is serious in the sense of being "true to nature" by means of attending fully to what nature is and the properties it has. Moreover, this promotes an objective viewpoint, since science is a paradigm of objectivity and, although aesthetic judgments based on scientific knowledge are not necessarily as objective as that knowledge itself, they nonetheless have an objective foundation.[40] Concerning the last requirement, scientific cognitivism is less clearly successful, for although its objectivity makes possible a compelling moral stance on environmental issues, it does not require it. Yet it can be argued that the factual character of scientific knowledge yields an environmentally informed response to nature and thus provides a firm basis for moral judgments.[41]

V. Conclusion

In conclusion, five points: First, if we must choose between non-cognitive approaches like the aesthetics of engagement and cognitive approaches such as scientific cognitivism, then on balance, the latter scores somewhat better than the former on the requirements of environmentalism.

Second, however, we do not have to choose between them, since although the two positions have different emphases, there need be no theoretical conflict between them.[42] This is because each position can be understood as defending only necessary, not sufficient, conditions for appropriate aesthetic appreciation.[43] There is perhaps some practical tension between the two approaches owing to the appreciative difficulty of being totally engaged within a natural environment and at the same time taking into account knowledge relevant to its appropriate appreciation. However, this kind of bringing together and balancing feeling and knowing, emotion and cognition, is the very heart of aesthetic experience. It is what we expect in aesthetic appreciation of art; there is no reason why we should expect less in aesthetic appreciation of nature.

The third conclusion, therefore, is that, concerning the requirements of environmentalism, since the aesthetics of engagement is especially strong regarding acentrism and environment-focus and scientific cognitivism is stronger regarding seriousness, objectivity, and perhaps moral engagement, the best alternative is to unite the two positions. This conclusion is endorsed by Holmes Rolston when he asks, "Can aesthetics be an adequate foundation for an environmental ethic?" replying, "Yes, increasingly, where aesthetics itself comes to find and to be founded on natural history, with humans emplacing themselves appropriately on such landscapes."[44]

Given such a unified position, the fourth conclusion is that, concerning the requirements of environmentalism, contemporary aesthetics of nature constitutes a substantial advance over traditional picturesque and formalist aesthetics of nature.

Hence, fifth, unlike traditional aesthetics of nature, contemporary approaches help to bring aesthetic values and environmental values in line with one another. They encourage aesthetic appreciation not simply of scenic landscapes but also of less conventionally scenic but nonetheless aesthetically magnificent and ecologically valuable environments, like deserts, swamps, savannahs, and prairies—indeed every kind of natural environment.[45]

According to Carlson, the five requirements of an ethically responsible environmental aesthetics—that it be acentric, environment-focused, serious, objective, and morally engaged—can be achieved if we combine the two broad theoretical approaches he canvasses. Among its many virtues, Carlson's analysis gives us an important new insight into the shortcomings of Susanna Moodie's apprehension of natural beauty. From the cognitivist position endorsed by Carlson himself, she fails to appreciate nature *as* nature. And from the standpoint of the acentric ideal expressed by the aesthetics of engagement, Moodie also falls short: her experience is entirely too self-absorbed to approach the standards of that ideal. If Carlson is right, then with regard to providing an explanation of how our judgments of natural beauty can be trained on the requirements of environmental ethics, we have clear theoretical winners and losers.

Ⓒ The Dispositive Character of Natural Beauty Judgments

Still, we might wonder how this process—aligning our ethical and natural-aesthetic judgments—really works at the more fine-grained level of individual judgment. So far, it is not obvious precisely why the agent who finds nature beautiful should be *motivated* to act on its behalf. For more insight on this problem, we turn to our next article.

THE MORAL DIMENSIONS OF NATURAL BEAUTY

Ronald Moore

I. Kant's Bold Claim

Midway through *The Critique of Judgement*, Immanuel Kant makes a remarkable claim about a connection between moral and aesthetic value that occurs in the appreciation of natural beauty:

> I willingly admit that the interest in the beautiful of art . . . gives no evidence at all of a habit of mind attached to the morally good, or even inclined in that way. But, on the other hand, I do maintain that to take an immediate interest in the beauty of nature . . . is always a mark of a good soul; and that, where this interest is habitual, it is at least indicative of a temper of mind favourable to the moral feeling that it should readily associate itself with the contemplation of nature.[46]

Can this be true? Are people who appreciate natural beauty more likely, other things being equal, to be morally good than those who don't? If I know absolutely nothing about someone other than that she loves to hike in the wilderness and takes great pleasure in the colours and forms of the trees and flowers she finds there, do I have any reason to believe that she is a morally good person, or even that she is apt to make morally correct judgments with respect to the natural objects she aesthetically admires?

I suspect that many people, perhaps even most, will regard these as easy questions. They will be inclined to agree with Kant's claim on the basis of a widely shared intuition that those who take pleasure in natural beauty are, generally speaking, people who have good values, and good values in the aesthetic sphere are very likely to join up in one way or another with good values in the moral sphere. But on reflection, the alleged connection is anything but obvious. It is, in fact, a point of long-standing and intense controversy among philosophical theorists of value in general. "Separatists" famously differ from "non-separatists"[47] over the alleged independence of aesthetic and moral sensibilities.

It is fair to assume that a minimal requirement of moral goodness is the ability to make sound moral judgments and that, similarly, the minimal requirement of aesthetic astuteness is the ability to make sound beauty judgments. Separatists (Stuart Hampshire, Philippa Foot, et al.) hold that from the fact that something is highly moral (or immoral), nothing at all follows regarding its aesthetic excellence (or lack of it) and, conversely, from the fact that something is beautiful (or ugly), nothing whatsoever follows regarding its moral status.

Non-separatists (Marcia Eaton, Noël Carroll, et al.) hold that aesthetic and moral values are not, after all, conceptually distinct, incompatible, or incommensurable. The case for separatism appears to rest on both a logical claim and empirical evidence. The logical claim is that aesthetic and moral regard are discrete activities that take stock of different features of things, involve different attitudes or dispositions, employ different criteria, and so on. The empirical evidence is that examples abound of moral paragons who are aesthetic dunces and vice versa. What is it about the aesthetic appreciation of natural beauty in particular that can cut against the strength of these claims and vindicate the popular perception that, in the natural domain at least, non-separatism is sound? In this article, I will try to answer that question.

II. Dispositiveness and Holism

It is, first of all, important to recognize that many, if not all, aesthetic judgments of natural beauty are, like moral judgments, not only verdicative, they are dispositive. That is, they not only announce the presence of a value (beneficence, say, or brilliant fluorescence), they tacitly express a disposition toward action (promotion, say, or preservation) on the part of the person making the judgment. The familiar Humean dictum that moral claims don't leave us cold (i.e., that they are conceptually linked with a presumptive inclination to act in a certain way) is equally valid for these aesthetic judgments. To declare, for instance, that a particular river cove is beautiful is not simply to record one more natural feature about it, as one might record the river's depth or speed of flow in a stream-gauging chart. It is, instead, to declare oneself in favour of a certain choice with respect to the cove, a choice that, in the absence of countervailing considerations, one would recommend to others. Like moral judgments, aesthetic judgments of natural beauty are arrayed on a broad scale of intensity of commitment. To say that a certain play of moonlight on water is beautiful may entail no more than that one is positively disposed to enjoin others to appreciate its sensible qualities. To say that a snow leopard is beautiful may entail a more active willingness to preserve, protect, celebrate, or otherwise act on behalf of its continued existence and welfare. But at every point on the scale, the disposition to act is not a moral imperative justified by some overarching principle (such as that in the absence of opposing duties, one should always act to favour those natural objects one finds beautiful). Instead, to find a natural object beautiful is, among other things, to be disposed to adjust the pattern of one's life one way rather than another toward it.

Second, any fair account of value judgments in general will acknowledge that aesthetic and moral concerns are not altogether segregated in our beliefs and practices. We often are disposed to act in ways that reflect a choice as to what kind of life we deem to be most worth living, without any determination of what kinds of value (moral, aesthetic, religious, political, and so on) inform that choice. This is most apparent in relation to what Marcia Eaton has called "meaning of life" issues, issues involved when we are trying to decide what, in a given cultural setting, rationality demands of us in a general way—marshalling our full range of approvals, disapprovals, delights, and abhorrences.[48]

In this arena of reflection, we do well to be sensitive and attentive, to seek appropriateness, to do what is fitting—all of which modes of thought commingle aesthetic and ethical considerations.[49] Through our legislatures, we may decide to designate a large patch of wilderness a national park or provide funding to clean up an oil spill, and the drivers of these decisions are clearly as much aesthetic as they are moral. Of course, it remains open to the stubborn moralist to insist that however many extra-moral values are in play in any judgment, the ultimate outcome of the intersection of values is bound to be a moral choice, simply because morality is the name we give to the all-things-considered directives we reach in life. But this reductive gambit trivializes the separatist/non-separatist debate, turning a genuine issue about ways in which some sorts of values lead to others into a matter of mere semantic fiat.

III. Environmental Cognitivism

To make the case for non-separatism in a way that vindicates Kant's remarkable claim about the special relation of the appreciation of natural beauty to morality, it will be necessary to show that the making of judgments of natural beauty and the making of judgments of artistic beauty involve dispositive vectors that diverge as they interact with other values in the guidance of choice. One way of doing this—the way that has been pursued by a substantial contingent of environmental ethicists—has been to show that there are important differences between natural environments and the art world that demand differences in appropriate modes of aesthetic regard to them, implying moral duties toward the former and not (or not to the same degree) toward the latter. First, there is the matter of range.

The judgment that, say, Mark Rothko's painting *Orange/Red/Orange* is beautiful can be taken to confine itself to a disposition on the part of the person making the judgment to visit and view it, to encourage others to do so as well, to urge its protection, to endorse the use of reproductions of the work in art appreciation classes, and so on. By contrast, the judgment that, say, Mount Robson is beautiful implies a wider disposition—a disposition to do more and on a larger scale—and not only because Mount Robson is a substantially larger physical object than *Orange/Red/Orange*. The natural beauty judgment takes stock of the fact that the object being judged is an element of an environmental whole larger than itself and integrally related to that whole in its aesthetic and physical well-being. The expansiveness of this judgment is not morally neutral. It is, instead, a recognition of the implications of human normative involvement in the larger whole.

The notion of "in-ness" is fundamental to this differentiation. Although there is undeniably a sense in which the reader of a great novel can find herself "in" the novel, this is an obviously different sense from that in which the explorer, plunderer, or simple admirer of environmental settings is in them. This situational discrepancy has evident moral consequences. The recognition that the judge of natural beauty is, qua human animal, part of nature and thus already implicated in the issues of its health and preservation is widely understood to be a fundamental link between aesthetic judgment and moral obligation.[50]

Equally important is the recognition in natural beauty judgments that many of the objects being judged (or parts of the complex being judged) are living things. Aesthetic regard for living things is importantly different from that for inanimate objects, however distinctive or precious they may be. Severe damage to a painting or an architectural landmark can be repaired; damage to an orca or to an ecosystem may be lethal. A living thing is aesthetically appreciated not only as a present sensible object but as a being with a future life and an animating impulse to continue that life. Recognizing this fact is as much a matter of aesthetic appreciation as it is of moral judgment. For it can be, and often is, part of the beauty of a natural being that it is recognized as temporal, ephemeral, fragile, and subject to depredation.

In this sense, aesthetic appreciation of a natural being is more than a delight in the look (or other sensible features) of that being. It is a delight in what that being is, an evolved form of life in a living environment, and not something we perceive as a scene or curious object, something apart from its own, inherent nature.[51] The connection between aesthetic and ethical concerns here is most apparent when we consider what it means to treat the natural world appropriately. Arguably, appropriate treatment entails both the right way of perceiving natural beauty and the right way of responding to what we perceive. It is in this joint entailment that Allen Carlson, among others, has argued that environmental aesthetics firmly aligns itself with ethics (as well as a host of other disciplines).[52]

And, of course, consideration of what it means to treat the natural world appropriately leads inevitably to the recognition that many of the natural objects we find beautiful might, or might not, be available to future generations. There are many reasons beyond the aesthetic ones for wanting to preserve various ingredients of the natural environment for our successors. But the aesthetic reasons count. In regarding a natural object, say a rugged shoreline, as beautiful, we are typically not looking at it simply as a compositionally attractive scene. We are, instead, regarding it as having aesthetic qualities (connected no doubt to the aesthetic qualities of its near neighbours) that we understand to have come into being by natural forces in the past and whose continued being we are disposed to encourage in the future.

That this is an aesthetic, rather than a purely moral, imperative is made clear by Kant himself. When we regard something as beautiful, he says, we are at the same time imputing that affirmation of pleasurable response to everyone.[53] And that means not just you and me but our children and our children's children.

A constant leitmotif in cognitivist discussions of environmental values is the notion of respect for nature. Respect, in this context, means an admiring regard that leads from aesthetic attention to an attitude of moral responsiveness and responsibility. A long line of environmentalists—from David Thoreau to John Muir, to Aldo Leopold, to Ned Hettinger—have insisted that there is no way of aesthetically appreciating the natural environment that does not imply moral consequences. The moral consequences are said to follow because the appropriate apprehension of beauty in environmental settings engenders—or even demands—a respect that recognizes that nature and natural environments have their own modes of well-being, modes that are, however, not entirely independent of our own.[54] As Paul Taylor puts it, once we abandon our anthropocentric outlook and admire the natural world for what it is—a living whole in which all living things are functionally related—

[w]e begin to look at other species as we look at ourselves, seeing them as beings which have a good they are striving to realize just as we have a good we are striving to realize. We accordingly develop the disposition to view the world from the standpoint of their good as well as from the standpoint of our own good.[55]

Now, it may be that the aesthetic admiration of artworks leads to a certain moral disposition as well. But it is clearly not this disposition. We may, upon contemplating a Renaissance masterpiece painting, feel morally disposed to foster its protection, to encourage its viewing by others, perhaps even to support its purchase by a favoured museum. But this cannot be because the beauty we find in it calls up an awareness that it is striving, as we are, to realize a rich and full life in a limited life span.[56]

IV. Non-cognitive Factors

Of course, not all aesthetic respect for natural beauty translates into moral respect. Although I am enormously fond of the appearance of a dandelion in my lawn, that fondness doesn't amount to a moral imperative to preserve it. To generalize the point, we cannot infer from the mere fact that some natural object is deemed beautiful that it, for that reason alone, makes a command on our moral sensibilities. Moreover, some natural objects are beautiful in ways that are entirely detached from living environments—ways that may draw more on formal than on functional excellence. The many gorgeous galactic bodies the Hubble telescope has revealed are causally disconnected from our earthly environment. The grain on my table is (at least in some sense) natural, yet its beauty may be enjoyed in abstraction from its forest home. The play of moonlight on the ripples of a lake's surface may be judged beautiful more because of the imaginative associations it calls up rather than because of any knowledge we may have regarding the moon or lake water as elements of the natural order. When I find such things to be beautiful, my judgment can be largely, or entirely, detached from moral respect for the natural environment. Even so, this doesn't mean that they have no bearing on the development of moral consciousness.

To begin with, patterns, contrasts, and harmonies experienced in nature may suggest modes of organized choice in the construction of a good life. Plato thought that lessons for good living in an individual could be extracted from the perceived patterns of balance and harmony, or their opposites, in political communities (and vice versa). It should not be surprising that similar moral lessons can be built upon a familiarity with harmonious patterns that make for beauty in nature. Marcia Eaton has argued that the perception of patterns and the extrapolation of these patterns, wherever we find them, are fundamental elements in the organization of moral life. Where does one learn about harmony? From music, dance, and painting, no doubt—at least for those who have an ear and an eye for such things. But equally, one may learn about it from patterned arrangements and balanced tensions in nature. Sometimes these will be harmonies apparent in ecosystems or in the lives of their component members.

And sometimes they will be apparent in the mere formal arrays nature presents—the delicate geometrical harmony of mountain laurel blossom, the splendid architecture of the spider web, the grand spiral of the Crab Nebula.

As Eaton points out, the moral lesson one may extract from these perceived patterns is directly related to moral living.

> Harmony—elements fitting together in appropriate and pleasing patterns—is achieved by reconciling oneself to one's role and striving for control not of others or the world, but of oneself as one seeks integrity and meaningful relations with family and friends.... This insight [built on the] ability to experience the world aesthetically, [leads to] sensuous and cognitive satisfaction.[57]

Moreover, aesthetic regard for a natural object may connect it to another natural object, or to an artifact, in a variety of ways—juxtaposition, parity, contrast, and so on. Beauty judgment, in such cases, doesn't depend entirely on what is perceived. Instead, it depends on what reflection and imagination construct out of what is perceived. And these constructive processes themselves can be intimately connected to moral sensibility. Reflection is the business of taking time to consider a thing from various interpretive angles. "It's a glorious sunset, so resplendent in its palette of colours. But then again, it's probably so brilliant only because of the pollution from that nearby coal plant. But even so, it cannot help but conjure up the idea of the ends of things—days, lives, civilizations." And so on. In a way, reflection is bound by its objects, and in a way it is free of them. It turns nature over and over in the mind, never leaving it, but not constrained in its consideration by it alone.

Imagination is simply a more extensive elaboration of this mental process. Emily Brady, who makes imagination a pivotal part of her account of the aesthetic appreciation of natural beauty, offers this example:

[I]n contemplating the bark of a locust tree, visually, I see the deep clefts between the thick ridges of the bark. Images of mountains and valleys come to mind, and I think of the age of the tree given the thickness of the ridges and how they are spaced apart. I walk around the tree, feeling the wide circumference of the bark. The image of a seasoned old man comes to mind, with deep wrinkles from age.[58]

Here, imagination connects one interpretation of a natural object with another, but it does so in a way that departs from the thing itself (the tree) to something else (the old man) that is outside the original orbit of appreciation but importantly analogous to the original percept. As Brady points out, imagining can be done well or ill. When it is done well, however, it can recognize patterns of association between elements of our experience that have significant bearings not only on our aesthetic preferences but on our moral deliberations. Shelley's observation that "a man to be greatly good, must imagine intensely and comprehensively"[59] has been seconded by successive generations of philosophers (most particularly Dewey and Beardsley). The idea here is that apprehension of the way one thing's features call up another's features is a relevant part of aesthetic perception that has a direct bearing on moral judgment, where it is important to be able to see things first one way, then another, and yet another again.

V. Growing up Morally; Growing up Aesthetically

There are countless ways in which cognitive and non-cognitive elements intertwine in the developmental process by which aesthetic observers grow from infant innocence to fully mature appreciation of natural beauty. In some of them, the cognitive elements dominate, in others, the non-cognitive, and in most, both kinds of elements play complementary roles in creating the perspective from which the mature observer makes environmental value assessments. The process of growing up aesthetically is in many ways similar to that of growing up morally, a process whose stages and standard trajectory have become familiar through the work of developmental moral theorists from Rousseau to Kohlberg and Erickson.

The leitmotif in most theories of moral development is that there comes a point when individuals have to select among, or cast off, a long series of prior counsels in crossing the threshold where they have to commit themselves to principles by which they are prepared to live as autonomous ethical persons. There is, I believe, a similar course in the development of aesthetic sensibilities. Our awareness of natural beauty is something we grow into, and we become aesthetic grownups only when we settle on a set of ideas about certain natural things that melds delectation with a disposition to protect and promote the continued existence of those things. And once this normative threshold is crossed, we are committed (or should take ourselves to be committed) with regard to our actions affecting nature, just as we are committed by moral principles with regard to our actions affecting persons.[60]

As I see it, the processes of growing up aesthetically and morally are largely complementary; the acquisition of knowledge and awareness in the one value domain fosters the acquisition of knowledge and awareness in the other. This is not a question of principles of natural aesthetics being derived from principles of ethics or vice versa. Rather, aesthetics in general serves as a propaedeutic to ethics in general, and natural aesthetics serves as a propaedeutic to environmental ethics. The relation between them is something like the relation between ice-skating and ballroom dancing. No one would confuse the two, nor would anyone think that a master of the one would inevitably be a master of the other. But surely in learning the first, one is already part of the way toward learning the second.

By requiring close attention to the sensible features of the natural world, the development of aesthetic sensibility cultivates an appreciation of facts and details that is essential to reckoning value in life generally. By demanding some measure of disengagement from immediate, practical concerns, the contemplation of natural beauty cultivates an ability to appreciate things without having a personal stake in them. By inspiring an appreciation of rarity, natural aesthetics cultivates an affinity for the uncommon features of life—a disposition to protect and foster what is irreplaceable. And by stimulating a reflective awareness of one's own thought processes in the activity of appreciating natural beauty, natural aesthetics cultivates self-reflection and self-appraisal, capacities through which we come to recognize that in making choices, we are always at the same time choosing the persons we shall be.

VI. Conclusion

I have argued that both cognitive and non-cognitive factors can play into the development of mature aesthetic appreciation of natural beauty. And I have argued that all of these factors bear upon the development of moral maturity. Together, these arguments lend plausibility to Kant's claim that people who appreciate natural beauty are thereby more likely than not to be people with good moral values.

But what about the other part of his claim, viz., that people who appreciate artistic beauty are not thereby more likely than others to be morally good? Most of the non-cognitive capacities I have mentioned—formal feature appreciation, reflection, and the elaborative force of imagination—are no less available to art appreciation than they are to nature appreciation. To the extent that these capacities assume primacy in the development of an aesthetically mature perspective on natural beauty, what Kant says about the difference in moral weight between natural and artistic aesthetics will seem implausible. To the extent that the cognitive elements I have spoken of as leading to a sense of respect for natural environments assume primacy in the development of that perspective, it will seem more plausible.

Moral maturity comes about through reflection on one's place in a world in which one is both part of nature and apart from nature. If, as I believe, appreciation of natural beauty and appreciation of artistic beauty are mutually reinforcing,[61] both are bound to contribute to its attainment.

The key insight in Moore's rich analysis is that *by their very nature* beauty judgments, like moral judgments, do not leave us cold. If this is correct, it solves the motivational problem nicely. The philosopher Raimond Gaita has an arresting description of the moral psychology of the mountaineer that fits Moore's claims well. The true mountaineer, according to Gaita, is a lover of nature, a love that finds its highest expression in the love of the mountain. More interesting is what grounds this love and what follows from it. It is grounded in an appreciation of the *beauty* of the mountain, and it issues in *respect* for the mountain. For Gaita, one can ultimately derive a "mountaineering ethics" from this, with substantive principles about how one may and may not climb.[62] We see here an elegant illustration of Moore's insistence on the dispositive character of beauty judgments: the latter provide motivation to pattern one's practical life in one way rather than another. Moreover, if one gets the beauty judgment right, one should also be motivated to do whatever is genuinely in the "interest" of the relevant natural object(s).

Ⓓ Aesthetic Weight and the Preservationist's Dilemma

A crucial question remains unresolved, however. We have seen throughout this book that the moral life is full of tensions and dilemmas. Tough choices often have to be made between competing interests and values. So even if we have (a) the correct theoretical account of natural beauty appreciation and (b) the right account of what motivates us to act in the interests of that which we find beautiful in nature, we still don't know how much weight to give such judgments insofar as they conflict with other judgments. Our final article explores this problem.

AESTHETIC PRESERVATION

Glenn Parsons

I. Reason or Rhetoric?

In the minds of many, when a cherished natural area is threatened with degradation or destruction, the issue that springs most readily to mind is the loss of its aesthetic value. As Holmes Rolston puts it, "Ask people, 'Why save the Grand Canyon or the Grand Tetons?' and the ready answer will be, 'Because they are beautiful. So grand!'"[63] Let's call this idea—preserving nature in its undeveloped state for the sake of its aesthetic value—aesthetic preservation.

Aesthetic values can have a powerful impact in preservation debates. Before-and-after photographs of a development project's alteration of a landscape can have an immediate emotional impact that isn't generated by charts and tables detailing long-term ecological impacts. But is maintaining the aesthetic value that we find in nature really a rationally compelling basis for preserving it from a proposed development that promises practical human benefits, such as employment, profit, and greater convenience? Or is it simply an effective rhetorical device, a way of "pushing people's buttons" and thereby mobilizing them to support a preservationist agenda?[64]

In this essay, I take up this question by exploring three issues facing aesthetic preservation. In my discussion, I will be assuming that at least in some instances, certain aesthetic assessments of nature can be said to be more correct, or appropriate, than others. If this is not the case, aesthetic arguments for preservation will be useless: those who oppose preservation can simply deny that nature has aesthetic value, and there is nothing further to be said. In other words, I assume here that aesthetic value of nature is, at least to some degree, objective.[65]

II. Strong and Weak Aesthetic Preservation

One important issue about aesthetic preservation is what exactly it entails. I have described it as the idea that we should save natural things or areas from degradation or destruction because of their aesthetic value. But this description fails to distinguish between two forms of aesthetic preservation. One form, let us call it strong aesthetic preservation, tells us simply to save natural things from destruction or degradation. A second form, which we can call weak aesthetic preservation, requires us to save natural things only when they are threatened by human actions.

Initially, strong aesthetic preservation may seem like a silly idea. The idea of trying to shield a forest from a volcanic eruption, for instance, seems ludicrous. But much of the destruction that nature brings upon itself is far smaller in scale than a volcanic eruption, and in these cases the idea of "saving nature from itself" is feasible. In fact, there are actual examples of this sort of preservation. Consider the preservation of Yew Tree Tarn, a small lake in the English Lake District.[66] At one point, the lake began draining because of the opening of an underground fault. The National Trust, a charitable organization that protects important natural areas in the United Kingdom, intervened in order to stop this drainage. The trust described its actions as aiming to preserve the beauty of the lake and the surrounding area; in their words, "the area has been landscaped to ensure its beauty is permanent." In intervening to preserve the tarn in its current state, the National Trust was not protecting its aesthetic value from human intrusion but from nature itself.[67]

In fact, strong aesthetic preservation might seem the more logical of our two positions, if we reflect on the fundamental idea behind aesthetic preservation. That fundamental idea is that we should preserve the aesthetic treasures of nature so that others might enjoy them. But if that is the aim, then why should it matter from whence threats to these aesthetic treasures arise? Consider an analogy: imagine that an aesthetically valuable artifact, a great work of art, for instance, or an expensive sports car, is entrusted to your care. If a vandal wanted to spray paint on it or take a sledgehammer to it, you would surely see it as your duty to try to protect it, given that it has been entrusted to your care. If the threat arose from a natural source—a hailstorm or an earthquake, say—why wouldn't you do the same? A threat is a threat, after all, and if the fundamental idea is to preserve what is aesthetically valuable, then it seems one ought to attempt to stop any threat that one can.

However, there is an important objection to strong aesthetic preservation: it is self-defeating.[68] The idea here is that in preserving a natural thing or area in this way, we make certain of its aspects artifactual. In our previously mentioned example of preserving a great artwork or a sports car, it seems reasonable to try to protect it against damage from things such as hailstorms and earthquakes. But this seems reasonable because natural forces are alien to the thing we are trying to protect, which is an artifact. But acting in the same way with respect to things such as lakes is not reasonable in the same way, because hailstorms and geological processes are not alien to natural things such as lakes.

On the contrary, lakes are created, and destroyed, by various geological processes. In the case of Yew Tree Tarn, blocking the natural geological processes that are destroying the lake turns it into something that has come about through the voluntary and intentional agency of human beings—in short, an artifact. Consequently, strong aesthetic preservation cannot deliver what the environmentalist ultimately wants: the preservation not just of things that currently happen to be natural but the preservation of those things as natural. The upshot of this objection, then, is that only the weak version of aesthetic preservation is coherent as a form of nature preservation.

This is not the end of the story, however. The proponent of the strong version of aesthetic preservation can concede the foregoing objection but reply that only her versions of the position can be polemically useful in arguing for preservation. The thinking here is this. The preservationist's basic situation is that she wishes to preserve natural areas that others see as worthless. Before they will support keeping them, these people first want to know what is so valuable about these areas. In reply, the preservationist points to the aesthetic value of these areas. If things go well, this convinces the doubters, who endorse preserving the areas in question on this basis. But when these areas are threatened by destruction from natural forces, however, the preservationist objects: "We ought not to save them now," she insists. When asked why, she answers: "Because then they will no longer be natural areas." But this will do nothing to sway the people that the preservationist is addressing: their starting point was an inability to see any value in nature per se. They do not care how much or how little nature there is: this was the entire reason that the preservationist needed to appeal to aesthetic value in the first place. If this line of thought is correct, aesthetic preservation may not always line up tidily with the aims of environmentalism.

III. Weighing Aesthetic Value

Another set of problems that besets aesthetic preservation involves the significance of the beauty of nature relative to the practical considerations that are invariably arrayed against it in debates over preservation. Discussion of this issue generally shapes up as follows. On the one hand, there are the practical advantages that exploiting a natural area or species will bring: more abundant resources for industry, jobs, more convenient travel, cheaper accommodations, cheaper food or goods for consumers, profit for producers, and so on. On the other hand, there is the aesthetic value that is lost through development. In order for aesthetic preservation to be a generally successful strategy, aesthetic value has to outweigh these practical benefits, at least in some reasonable percentage of cases. But is this the case?

Here we confront the vexing task of ranking different kinds of value. One extreme view is that aesthetic value is always, or almost always, outranked by other forms of value, including the practical benefits typically associated with environmental development. One justification for this view is that aesthetic merit is a superficial kind of value. In developing this sort of case, J. Robert Loftis draws an analogy between our treatment of nature and our treatment of persons.[69] Loftis imagines a doctor with only one donor heart but two patients who need a heart transplant. In such a case, he points out, the doctor should appeal only to medical facts in making her decision.

If she were to give the heart to one patient because that person was more physically attractive than the other, she would be making a serious error. Further, the ground of that error would be clear: treating what is essentially a superficial and unimportant form of value—aesthetic value—as outweighing other considerations. On the basis of this sort of analogy, Loftis concludes that "aesthetic considerations involving nature are weak and cannot motivate the kind of substantial measures environmentalists routinely recommend." "How," he asks, "can environmentalists ask thousands of loggers to give up their jobs and way of life on the basis of aesthetics?"[70]

The heart transplant analogy, however, is a misleading one. It is surely correct to say that the aesthetic value of potential heart transplant recipients does not, and ought not, outweigh medical considerations. But this is an extreme case in which the implications of favouring aesthetic value are as strong as they could possibly be: they are literally a matter of life and death. But typical cases of environmental preservation are not like that: choosing aesthetic value over other considerations does not typically place lives at risk. Choosing aesthetic value always has consequences, of course, and some of them are important: it may reduce income for some or result in others having to move to find new employment, for example. We might accept that considerations of basic human viability always come first and trump aesthetic value but still argue that in some cases preserving aesthetic value is worth the cost.[71] The question for such a view is: In these cases, can we say that the aesthetic value of a natural area justifies the cost of preserving it?

One reason for thinking that it might is that in other areas of life, we do make practical sacrifices for aesthetic value. We spend both private and public money on artworks, for example. These measures have costs: they reduce the income that citizens would otherwise keep for themselves or spend on other things. This seems to show that aesthetic value is not so superficial that any other consideration trumps it: sometimes we do sacrifice other valuable things to attain it. In his discussion, Loftis acknowledges this point but objects that the costs of preserving natural beauty far outweigh the costs of funding art. He estimates the economic costs of forgoing oil development in Alaska's Arctic National Wildlife Refuge, for example, at $800 million. This figure, he points out, is much greater than the annual budget of the National Endowment for the Arts, which is around $100 million. Thus, Loftis would have

us conclude that aesthetic preservation of the environment is simply too expensive. But the financial comparison drawn here is apt to be misleading, in two ways.

First, although the economic benefit of exploiting the Arctic National Wildlife Refuge may be substantial, at least according to Loftis's estimate, this may not be the case in other instances. In some instances, developing a mountain into a ski resort, for example, exploiting nature may produce smaller benefits. Second, it is not clear that even when the costs of environmental preservation are relatively substantial, as in the case of the Arctic National Wildlife Refuge, they are too high. Loftis chooses the annual budget of the National Endowment for the Arts as an indication of the upper limit of sacrifice for aesthetic value, but this number seriously underrepresents the sacrifice made for aesthetic value. For one thing, the National Endowment's budget represents only a part of the total amount spent annually on art, which also includes other public and private funds. We also need to consider here other spending on aesthetic value: consider, for instance, the beautification programs run by virtually every sizable town and city, consisting of the maintenance of local gardens and architectural repair and restoration. And then there is spending on various other aesthetically valuable items: furniture, clothes, houses, cars, and so forth.

On balance, it seems that the view that aesthetic value is so weak that it is never capable of outweighing the pragmatic benefits of development is too extreme. How often it actually succeeds in doing so, however, would seem to be something that we can assess only by examining particular cases.

IV. The Preservationist's Dilemma

In the previous section, we saw that aesthetic preservation requires us to consider the practical benefits that would accrue from that development. But we also need to factor in another variable: the aesthetic value of the developed site itself. Aesthetic preservation requires that sparing nature produce a gain in aesthetic value. In other words, it assumes that nature left untouched is more aesthetically valuable than the development that would replace it.

This assumption is open to question. One might argue that strip mines, urban sprawl, factory farms, and massive dams are actually as aesthetically good as, or better than, the nature they replace. Proponents of one major dam project, for example, defended the development by saying that it would create "one of the world's great scenic wonders."[72] At first glance, this may seem to be a cynical ploy by developers to advance their interests rather than an honest aesthetic judgment. However, there might be a sound basis for the claim that strip mines, urban sprawl, and so on have aesthetic merit.

To a first glance, a strip mine might appear aesthetically poor in virtue of being a huge gash in the earth, emitting a terrible din, and spewing black smoke into the sky. But if we consider it not simply as a gash in the earth but as a large industrial mechanism, it might look quite different. The billowing smoke, loud noise, and roaring fires, rather than obtrusive and marring, may seem indicative of its power and vitality. After all, many industrial machines look aesthetically appealing. Further, even if one finds the smoke, noise, and fire obtrusive and jarring in relation to the landscape, one might even find aesthetic value in this very incongruity. As Yuriko Saito points out, a similar incongruity is often aesthetically appreciated in art, particularly in contemporary works that employ dissonance between elements of the artwork as an artistic technique.[73] And finally, such developments can often be viewed as expressive of certain positive values: hard

work, determination, and vision, for instance. If we see them in these ways, strip mines and their ilk might not look so bad after all.[74]

The proponent of aesthetic preservation will want to respond that even if these considerations show that human developments like strip mines, urban sprawl, and industrial farms have some aesthetic value, this value will, generally speaking, be much less than the aesthetic value possessed by the natural areas that they replace. But this supposed difference in value must rest on purely aesthetic grounds. If it turns out that the preservationist's aesthetic preference for nature, as opposed to strip mines, is ultimately based not on any difference in their aesthetic value per se but on the assumption that nature is more valuable than development, then the preservationist is no further ahead for appealing to aesthetic considerations. The aesthetic assessments of the preservationist will then be "morally charged," as Ned Hettinger puts it.[75]

The aesthetic preservationist faces a dilemma here.[76] On the one hand, she can assess aesthetic value purely on aesthetic considerations. In principle, these considerations are capable of persuading sceptics that nature has greater value than the development that would replace it, but in practice they turn out not to favour nature as unequivocally as one would like, for the reasons discussed above. On the other hand, she can allow ethical judgments to play a role in shaping her aesthetic judgments. For example, she might hold that a natural area is aesthetically superior to a strip mine in part because it is more natural and what is natural is more valuable than what is artifactual. In that case, she can assert that nature generally has more aesthetic value than development, but her aesthetic assessments will cease to persuade the sceptic. For since the sceptic does not agree that nature has greater value, he will not generally find it more aesthetically appealing on this ethical basis. The dilemma means, in short, that for the preservationist, the appeal to aesthetic value either fails to favour nature or else becomes a purely rhetorical device rather than a reason for accepting the preservationist position.

Let us consider the two horns of the dilemma in turn. The first horn is grasped by Janna Thompson.[77] She identifies four features of nature that, she argues, make it by and large aesthetically superior to human development. They are: (1) magnificence and richness in detail; (2) the capacity to change or enhance our way of seeing the world; (3) cultural significance for those who experience it; and (4) the capacity to put things in perspective. Thompson draws her list from reflection on the aesthetic value of artworks, and she explains the presence of its features in certain natural areas by way of analogies to artworks. Thus, just as a great work of architecture is magnificent and rich in detail—an inexhaustible feast for the senses, the intellect, and the imagination—so too are natural areas such as the Grand Canyon. Just as some striking and unusual artworks—the paintings of Van Gogh, for instance—prompt us to see the world around us in a new way, so too do striking and unusual natural areas, such as the eucalyptus forests of southeastern Australia.

Some works of art, such as the masterworks of the Renaissance, provide a way of coming to appreciate a cultural tradition and our place in it, but so can some natural areas: Thompson gives the example of the Merri Creek grassland area, the ancestral home of the Koori Aboriginal people in Australia. Like the works of Leonardo and Raphael, this grassland serves to connect certain people to their own cultural history. Finally, Thompson notes that great works of art force us to re-evaluate our basic values and "pose a challenge to our way of thinking";

analogously, nature can force us to put our daily activities into broader perspective. Nature, as Thompson puts it, "exists as a refuge, or at least as a counterweight to the human-made world."[78] Thompson's list is not meant to be exhaustive, but she does claim that the respects that she cites are sufficient to show that "the preference . . . for undomesticated nature can in many cases be justified."[79]

Thompson's proposal is an attractive one: it sets out a list of criteria for aesthetic excellence that are not "morally loaded" in favour of nature. But once these criteria have been set out, it is not so clear that they actually favour nature.[80] On her account, not all natural areas will turn out to be aesthetically superior to human development. An unremarkable river bluff on the Mississippi, for instance, would probably not possess any of the outstanding features on her list (perhaps only the fourth feature, if that). As such, an aesthetic argument for preserving it would be weak: in that case, perhaps a decent oil refinery or some urban sprawl would not really be much of an aesthetic loss, if any loss at all. This means that the utility of aesthetic preservation will be, in the final analysis, rather limited: it becomes a compelling argument for saving the so-called masterpieces of nature—the Grand Canyon or rare eucalyptus forests. Even in these cases, the aesthetic superiority of nature would be in question if the development in question were impressive enough. This casts some doubt on whether Thompson's criteria really show that, as she says, "the preference . . . for undomesticated nature can in many cases be justified."

What about the second horn of the preservationist's dilemma? If she chooses this option, the preservationist can easily assert that nature has greater aesthetic value than development, but only because ethical assumptions have infected her judgments of aesthetic value. This seems to leave the aesthetic superiority of nature worthless as a means of convincing people that nature has value and so ought to be preserved.

Some philosophers, however, argue that this is overly pessimistic. They maintain that aesthetic value can still make a useful contribution even if it turns out that aesthetic judgments are "morally charged." The basic idea behind this view is the idea that, as Hettinger puts it, "the language of aesthetics is more descriptive in comparison with the more prescriptive language of ethics."[81] An ethical argument for the preservation of an endangered species, for example, will typically rely on ethical concepts such as rights, theoretical principles such as the equal worth of all species, and ecological facts. But these are somewhat abstract. Aesthetic descriptions of such creatures can provide us with greater specific insight into the species we are discussing, allowing us, as Saito puts it, to "supplement our purely conceptual understanding of nature, and . . . strengthen our appreciation of nature's workings."[82]

Hettinger gives the example of the campaign to save the habitat of the Delhi Sands flower loving fly, the only endangered fly species in the United States. The fly's defenders relied on weighty moral arguments that were couched in the abstract language of rights and interests. They argued that the fly possessed value, despite its lack of utility for humans, and hence had the right to exist. But more effective than such arguments, Hettinger suggests, were descriptions of the "spectacular" aesthetic qualities of the fly that pointed out its unusual morphology and its odd, hummingbird-like manner of flying. Ultimately, these aesthetic qualities, in themselves, may not give us any grounds for saving the fly: they may be no more valuable, as aesthetic qualities, than the ones that a decent industrial plant would possess.

However, through reflection on these aesthetic qualities, the strength of the ethical basis for choosing the fly—its inherent value and consequent right to exist—becomes clearer to us. As Hettinger puts it, such aesthetic experience makes possible "a more informed assessment of the moral issues involved" and allows "considerations to be entertained that disputants may not have noticed or appreciated."[83] We might sum up this line of thought as the view that the aesthetic value of nature, even if it does not itself figure in the reasoning behind aesthetic preservation, can still play a useful heuristic role in getting people to consider, and see the force of, that reasoning.

A sceptic might object here, though, that the preservationist is "having her cake and eating it too": she appeals to aesthetic considerations, even though she cannot show that they give us any valid reasons for preferring nature to human development. This, it might be said, is nothing different from the practice, discussed at the start of this essay, of using pictures of aesthetically ruined natural areas to "push people's buttons." There can be no doubt that in virtue of their perceptual immediacy, aesthetic considerations can be more effective in swaying people to action than dry facts and abstract moral concepts. But so using them threatens to shift discussions of preservation away from rational debate and toward a mere rhetorical contest. The preservationist can always simply accept the sceptic's objection with equanimity, of course, embracing the turn to rhetoric over reason. Philosophers should remember that, like it or not, environmental debates are often a matter of both.[84]

In the contest between development and preservation, if we are going to insist that the only considerations that count are aesthetic ones—the first horn of Parsons's dilemma—then we must consider the aesthetic value not just of strip mines and the like but also of all the things we make from the stuff we dig out of those mines. We might consider removing the tops of mountains to get at the coal inside them an aesthetic barbarity. But do we feel the same about the opera house's stunning display of lighting, the electricity for which is provided by that very same coal? Considerations like this might force us to leap to the other horn of the dilemma, the one Parsons himself seems most comfortable with. Here, we insist that ethical considerations come first and that judgments of beauty, though they may support such considerations, cannot stand on their own. This, says Parsons, renders natural beauty "judgments" at best non-rational or rhetorical boosters for properly rational—that is, moral—judgments.

Perhaps neither fate is so bad. If we accept the first horn, it may be that plenty of nature is ultimately preserved rather than developed. That is an empirical question, but surely the more comprehensively we absorb the lessons of ecology, the more we will be apt to decide that larger and larger portions of nature are too beautiful to despoil. As for the other horn, Parsons asks just the right question: What exactly is wrong with rhetorical appeals?

Ⓔ Conclusion

Let's conclude with a brief return to Susanna Moodie. As we have seen, a key component of Moodie's aesthetic judgment is her measured emotional experience of the vista before her. Few philosophers want to discount altogether the role experience plays in proper aesthetic judgment, although most want to place constraints on such experience. The flashpoint for this analysis is the role of imaginative association in the apprehension of natural beauty.

When we contemplate a beautiful natural object or array of objects, the experience is heightened if we attend to the object(s) carefully over time rather than simply looking and turning away. But during the time we are engaged with the object, not just any association of ideas counts as a genuinely mature aesthetic response to it. As Ronald Moore has argued, there are, broadly speaking, two possible paths down which the imagination can take us once it becomes fixed on an object. Recall the poem "Dark Pines under Water," which we looked at in Chapter 5. The poem asks us to imagine what it is like to look at pines reflected in water. Transfixed by this sight, the imagination may wander, from the thought of the reflected pines to that of trees generally and finally to the thought that one's stack of firewood is low. This might lead to the thought that one needs more firewood, which leads to the decision to drive to the local gas station (it stocks firewood) and, ultimately, to worrying about the price of gas. All while lying on the bank of the lake looking at the reflected pines.

On the other hand, one might direct the imagination where the poem's author presumably wants it to go: to contemplation on the unity of all life, on the ease with which this thought can lead one to reassess one's purposes in life, and finally, perhaps, to the idea that one somehow owes the land a measure of respect. Whether or not this last thought leads to more substantive ethical principles will presumably depend on the extent to which one believes the lake's ecosystem to be threatened in its integrity. The key difference between the two paths, then, is that the first strays from the object while the second cleaves to it.[85] This is the sense in which a mature ethical stance toward the natural world can be enhanced and deepened through the correct exercise of the aesthetic imagination.

CASE STUDY

Nunavut and the Reciprocity Thesis

One theme not yet explored in this chapter concerns the relation between artifactual beauty judgments and natural beauty judgments. Can the appreciation of beauty in one sphere help us to appreciate beauty in the other? As you might expect, philosophers are divided on this question. On one side there are those, like Carlson, who think that the objects of the two sorts of judgment are so unlike that any attempt to assimilate one into the other is bound to distort our apprehension of both. The key difference between the two sets of objects is that one of them—the artifactual—is *designed* while the other is not. When we make a considered judgment of an artwork, we are, in large measure, assessing the quality of the object's *craftsmanship* in light of the relevant set of cultural standards. But nature is not designed. So what we need to know to make sound judgments in the two spheres differs radically. In the art world, we need to know about "artistic traditions and styles within those traditions." In the "natureworld," by contrast, we need to know about "the different environments of nature and . . . the systems and elements within those environments." One judgmental sphere is ruled by the art historian and the critic, the other by the ecologist and the naturalist.

Ronald Moore disagrees. He thinks that judgments in the two spheres have a lot to learn from each other. "It is pointless," he says, "to think of natural beauty and artistic beauty as antagonists or isolated from each other if they can be mutually nourishing." This is the **reciprocity thesis**. The key to it, at least as developed by Moore, is that the train of ideas set in motion by attention to a natural object can legitimately be extended to analogous objects in the art world (and vice versa). Let's explore how this might work by thinking of a painting by a founding member of Canada's Group of Seven, Lawren Harris. The painting is called *Mount Thule, Bylot Island*. It depicts a scene from Bylot Island, just off the northeast corner of Baffin Island in Nunavut. Most Canadians will never get to see this part of their country, but it is an aesthetic wonder that can at least be viewed in photos. Harris's painting of it is arresting, but the question for the reciprocity thesis is how appreciation of the two objects—the painting and the reality it depicts—can enhance one another.

Continued

Lawren Harris, *Mount Thule, Bylot Island*, 1930, oil on canvas, 82.0 x 102.3 cm, Collection of the Vancouver Art Gallery, Gift of the Vancouver Art Gallery Women's Auxiliary, VAG 49.6

Mount Thule, Bylot Island, *by Lawren Harris. Can an appreciation of natural beauty better equip us to appreciate artistic beauty and vice versa?*

Here is how one critic describes the painting:

> This . . . is a stylization of mountain shapes, mirrored in water. Those snowcapped peaks, that extraordinary quality of infinite space, that menacing green darkness in the lower lying country: all these devices point up aspects of the Harris approach to Canadian nature. Canada is vividly coloured: it is a powerful young giant. To grasp its essential nature and potential power is not easy. If you were to discover it, line and colour had to be bold and simple, and form had to be expressed in symbolic terms.

One might be impressed by the notion of Canada as a "bold, young giant," and one might also agree with the judgment that the only way to get this point across adequately is in the highly stylized manner perfected by Harris and the other members of the Group of Seven. Next, you might take this idea of power, size, intensity of colour, and so on and find that your experience of the place—whether through photographs or, better, a trip up there—is enhanced by reference to it. While in the North, you might come to learn that the place has been home to one of the world's last remaining hunting peoples—the Inuit—for thousands of years. And you might notice that the presence of the Inuit on the land for all that time has itself shaped, and been shaped by, the area's ecosystems in subtle ways.

This recognition, finally, might make you return to Harris's painting (as well as to other similar artworks) with a more critical eye. What does the absence of the original inhabitants of this place from the painting say about the way descendants of Europeans view the land? Just how should we harness all that pent-up power that *is* Canada in a way that looks to the knowledge and experiences of these peoples? With an increasing focus on "Arctic sovereignty" these days, what can these new possibilities of national self-understanding tell us about the preservationist challenges we are sure to face as we seek to "develop" the Arctic? Because we can be moved to reflect critically on these and related ethical questions through the mutually reinforcing aesthetic appreciation of art *and* nature, there is surely something of value in the reciprocity thesis.

Sources: Brody 2000; Moore 2008; Carlson 2000; Vancouver Art Gallery, *75 Years of Collecting: Lawren Harris*, http://projects.vanartgallery.bc.ca/publications/75years/pdf/Harris_Lawren_18.pdf.

Study Questions

1. What does Moore mean by the "dispositive" character of natural beauty judgments? Do you agree with his way of understanding this issue?
2. Explain Parsons's distinction between strong and weak aesthetic preservation. Which of these two options makes more sense and why?
3. Moore places significant weight on the notion of "maturity," both moral and aesthetic. What do you think of this idea? What implications does it have for moral and aesthetic education?
4. What, if anything, is wrong with the "picturesque" view of natural beauty? Do you think it is true that this view is incapable of attending fully to the moral dimensions of our experiences of nature?

Further Reading

Allen Carlson. 2000. *Aesthetics and the Environment: The Appreciation of Nature, Art and Architecture*. London: Routledge.

Allen Carlson and Glenn Parsons. 2004. "New Formalism and the Aesthetic Appreciation of Nature." *Journal of Aesthetics and Art Criticism* 62: 363–76.

Stan Godlovitch. 1999. "Creativity in Nature." *Journal of Aesthetic Education* 33: 17–26.

Ronald W. Hepburn. 1968. "Aesthetic Appreciation of Nature." In *Aesthetics in the Modern World*. Ed. Harold Osborne. New York: Weybright and Talley.

Ronald Moore. 2008. *Natural Beauty: A Theory of Aesthetics beyond the Arts*. Peterborough, ON: Broadview Press.

CHAPTER 8

First Nations' Perspectives

Ecological Intuition Pump

Co-evolution is the process by which one population of organisms develops traits or practices in response to the traits or practices of another population of organisms, the first group then responding to the second group's response, and so on. Over the long course of their mutual evolution, members of the two groups can become very closely connected in a myriad of subtle ways. For example, ecologists have studied the various ways in which certain herbivorous butterfly species and the plants they feed on have co-evolved so that they now form a kind of ecological "community." The butterflies have become *attuned* to the plants they feed on—and vice versa—in a particularly deep and intimate way. If we find this notion arresting, perhaps we can say something similar about the way in which First Nations peoples have co-evolved with the various ecosystems of the Americas. Except in this case, one part of the relation of attunement involves deep *knowledge* of local ecosystems, all the intricacies of their inner workings built up over the course of thousands of years. These people are parts of a community containing themselves and the land (in Leopold's sense). Perhaps, at a time when we are struggling to develop a healthier relationship with the natural world, we can find resources of knowledge and wisdom in the ecological beliefs and practices of these peoples. But is it possible to do this without romanticizing them?

Ⓐ Introduction

The summer of 1990 was a tumultuous one for relations between Canada's Native and non-Native populations. This was the summer of the Oka crisis, when Mohawk Warriors at Kanesatake blocked highway and bridge access to a small piece of land just west of the island of Montreal, where the Ottawa River joins the St Lawrence. The disputed land has been formally claimed by the local Iroquois since 1717 when Louis XV granted it to the Seminary of St Sulpice as a seigneury. Since then, title to it has been transferred from one governmental or non-governmental entity to another, but the Iroquois have never recognized the legitimacy of any of these titles, including that of the municipality of Oka, Quebec.

The 1990 crisis was precipitated by the decision of the municipality of Oka to expand a golf course, originally built in the mid-twentieth century, into the area. The armed standoff lasted for 78 days, during which time a police officer, Corporal Marcel Lemay, was fatally shot (it is still unclear from which side the bullet came, although a police inquest has declared that it came from a Mohawk weapon). The police tried to choke off supplies of food and medicine to the lands, but the Mohawks managed to smuggle these goods in by canoe under cover of darkness. Eventually, the Canadian army was deployed to deal with the Warriors. The latter decided to stand down after the Canadian government promised to "seriously consider" their grievances.

Was the government true to its word? Historian Olive Patricia Dickason provides this dire assessment of Oka's legacy:

> More than a decade later there was still no agreement on what had been learned from Oka. The old arguments that First Nations have no rights to the land they have lived on for thousands of years, unless they have been asserted by special arrangements, are still alive. In some cases they are more entrenched than ever. Unsolved problems, unlike old soldiers, have a habit of staying and growing instead of fading away.[1]

The Oka crisis is important for environmental ethics because it reminds us that land claims are irreducibly political: they involve conflicts between groups over chunks of the environment and its resources. If we are to respect other people on the basis of a fundamental equality, such conflicts must be resolved in a philosophically principled fashion rather than by appeal to ideology or group prejudice. To this end, it might be suggested that since the residents of the municipality of Oka vastly outnumbered the Mohawks and were overwhelmingly in favour of expanding the golf course, their interests should clearly have prevailed. Utilitarian arguments invoking the principle that we should seek to maximize human welfare might have been brought to bear in support of this claim. When applying the utilitarian calculus, there are always going to be winners and losers, but we can rest content in the knowledge that *overall* we will be maximizing welfare among all affected parties.

Critics of utilitarianism will typically respond to arguments like this by invoking considerations of *justice*. They will insist that there are some kinds of basic interests or needs that cannot be trumped by appeal to considerations of total (group) welfare. Basic human rights are good examples: it is a matter of justice that we not interfere with individuals in certain ways—by torturing them, for instance—even if doing so might increase general welfare. But considerations of justice can also be relevant to some of the interests of *minority groups*. That is, we may want to say that some of the vital or fundamental interests of such groups may not be infringed by appeal to the general welfare. If we were to put the point in favour of the Mohawks' claims this way, how might we proceed?

The most important thing to point out is that the conflict pitted the basic (or vital) environmental interests of one group (the Mohawks) against the non-basic (or luxury) interests of another (the residents of Oka). The disputed land contained an area known as "The Pines," a stand of tens of thousands of trees planted by the Natives over the past hundred years or so to stabilize the area's sandy soil. The Pines had become known to the Natives as "the commons" and constituted the ecological centre of their cultural life. It would have been destroyed by the expanded golf course. So one group bases its claim to the disputed land on the need to protect something that is culturally vital and ecologically sensitive, while the other rests its claim on the need to provide more of a certain luxury good to its members. Moreover, as Laura Westra has argued, maintaining a golf course involves ecologically destructive practices, since those pristine green expanses require huge amounts of fungicides, pesticides, and fresh water.[2]

If we accept the principle that basic interests defeat non-basic interests when the two conflict (a difficult principle to defeat), and in the absence of other compelling considerations, the conflict should clearly have been resolved in favour of the Natives' claims. So why was the non-Native population so intransigent? When we consider the fact that the Mohawks were never even seriously consulted on any of the government's plans to build the golf course, it is tempting to draw the conclusion, as Westra does, that the Oka crisis presents a clear case of **environmental racism** on the part of Canada's non-Natives toward its Natives. Environmental racism is defined as "racism practiced in and through the environment."[3] A prominent example of it is the frequency with which toxic waste sites are placed in neighbourhoods comprised largely of racial minorities. Canada's First Nations have persistently been discriminated against in this fashion, whether through the degradation of land and water resources from waste-dumping or through governments allowing hydroelectric, mining, or forestry companies to exploit their land without adequate consultation.[4]

How do we explain the municipality's proposal to build a golf course on land considered sacred by the Mohawks? It does not make sense unless members of that group are deemed *inferior* in virtue of their membership in the group. This is the very definition of racism. Members of racist majority groups typically resist labels like this, but until a more plausible explanation of the psychology of the relevant non-Native players in the sorry events of the summer of 1990 is put forward, the charge of environmental racism is difficult to avoid.

ⓑ The Circle

The Oka crisis is of signal importance to environmental ethics because it expresses a persistent refusal on the part of non-Natives to listen to and learn from the ecological wisdom of Canada's Native population. Here is how Thomas Berry puts the general point:

> This communion with the natural world, understood with a certain instinctive awareness by tribal peoples, is something that we, with all our science and technology, seem unable to appreciate, even when our very existence is imperiled. As Europeans on this continent, we have had a sense of ourselves as above all other living forms, as the lordly rulers of the continent.... The continent ... would sustain any amount of damage as an inexhaustible store of nourishment and energy for carrying out our divine mission. With supreme shock we discover that our historic mission is not what we thought it was. Beyond that we discover that this continent

is a delicate balance of life systems, that the fuels for our machines are limited.... The Indian now offers to the Euroamerican a mystical sense of the place of the human and other living beings. This is a difficult teaching for us since we long ago lost our capacity for being present to the earth and its living forms in a mutually enhancing manner.[5]

Berry's comments reveal both an opportunity and a danger. The opportunity is to learn and appreciate the ecological wisdom First Nations peoples can impart to a culture—that of Euroamericans—in the midst of a deep ecological crisis. Listening to this message requires the virtues of humility and openness on the part of that culture's members. The danger is that we might, in the course of doing this, offer up one more idealized—and ultimately colonial—version of Native peoples, yet another shallow and demeaning invocation of the "noble savage." In what follows in this chapter, we will attempt to absorb the lesson without romanticizing the culture it comes from. Let's begin with the crucial concept of the sacred circle as articulated by Huron philosopher Georges Sioui.

THE SACRED CIRCLE OF LIFE

Georges Sioui

According to the Sioux holy man Hehaka Sapa, everything done by an Indian is done in a circular fashion, because the power of the universe always acts according to circles and all things tend to be round:

> In the old days, when we were a strong and happy people, all our power came from the sacred circle of the nation and as long as the circle remained whole, the people flourished. The blossoming tree was the living centre of the circle and the circle of the four quarters nourished it. The east gave peace and light, the south gave warmth, from the west came rain, and the north, with its cold and powerful wind, gave strength and endurance. This knowledge came to us from the external world (the transcending world, the universe) and with it, our religion. Everything done by the power of the universe is made in the form of a circle. The sky is circular and I have heard that the Earth is round as a ball and the stars too are round. The wind whirls, at the height of its power. The birds build their nests in a circular way, for they have the same religion as us.... Our teepees (tents) were circular like the nests of the birds, and were always laid in a circle—the circle of the nation, a nest made of many nests, where the Great Spirit willed us to brood our children.[6]

The reality of the sacred circle of life, wherein all beings, material and immaterial, are equal and interdependent, permeates the entire Amerindian vision of life and the universe.

I. The Sacred Circle of Life

Every expression of life, material or immaterial, demands of the Amerindian respect and the spontaneous recognition of an order that, while incomprehensible to the human mind, is infinitely perfect. This order is called the Great Mystery. To the traditional Amerindian, life finds its meaning in the implicit and admiring recognition of the existence, role, and power of all the forms of life that compose the circle. Amerindians, by nature, strive to respect the sacred character of the relations that exist among all forms of life.

Where their human kin are concerned, the Amerindians' attitude is the same: all human beings are sacred because they are an expression of the will of the Great Mystery. Thus, we all possess within ourselves a sacred vision, that is, a unique power that we must discover in the course of our lives in order to actualize the Great Spirit's vision, of which we are an expression. Each man and woman, therefore, finds his or her personal meaning through that unique relationship with the Great Power of the universe. There is no room for a system of organized thought to which the individual is subordinate, in the way that religions or political ideologies are at the service of human and material interests. "The duty of a man," states an Ojibwa medicine man with simplicity and all the depth of his respect for his tradition, "is to work for the Great Spirit."[7]

Human beings have an obligation to discover their own vision, their meaning, their religion; woman, with her special powers of self-purification, recognizes her vision much more easily than man does. With their awareness of the sacred relations that they, as humans, must help maintain between all beings, New World men and women dictate a philosophy for themselves in which the existence and survival of other beings, especially animals and plants, must not be endangered. They recognize and observe the laws and do not reduce the freedom of other creatures. In this way, they ensure the protection of their most precious possession, their own freedom. Long ago, the independence of Amerindians was directly related to the incomparable abundance of food to which they had access. . . .

The sacred circle of life, in which the place of humans is equal to that of the other creatures, albeit marked by a special responsibility, is divided into four quarters. Four is the sacred number in America: there are four sacred directions, four sacred colours, four races of humans, each with its own sacred vision, as well as four ages of human life (childhood, adulthood, old age, then childhood again), four seasons, and four times of day, which are also sacred. Thus the circle operates in cycles of four movements each. When Amerindian officiants perform a sacred ritual whose primary function is purificatory, they mention and address themselves to those times or movements, and when they have covered the entire circle, they speak the words "all my relations," thus acknowledging the relationship between all beings in the universe and their common vision of peace.

II. The Originality of Amerindian History

In relation to the other continents, which are contiguous (with the exception of Australia), the American continent stands apart in the configuration of our world. The peoples of the Old World have always evolved in relative symbiosis, influencing one another in almost every way: economically, culturally, biologically, and so on. By contrast, until its definitive contact with Europe, the New World has been almost completely isolated and so able to devise and preserve ways of being and living that are specifically its own. It has developed according to ideological concepts diametrically opposed to those that animated and motivated the Europeans and the other peoples who followed them here. In 1868, the ethnologist Daniel Garrison Brinton wrote:

Cut off time out of mind from the rest of the world, he [the American Native] never underwent those crossings of blood and culture which so modified and on the whole promoted the growth of the old world nationalities. In his own way he worked out his own destiny and what he won was his with a more than ordinary right of ownership.[8]

The same author also pointed out that, during the thousands of years when North America was free of outside influence, the continent became a surprisingly homogeneous linguistic "region":

From the Frozen [Arctic] Ocean to the Land of Fire, without a single exception, the native dialects, though varying infinitely in words, are marked by a peculiarity in construction which is found nowhere else on the globe, and which is so foreign to the genius of our [English] tongue that it is no easy matter to explain it. It is called by philologists the polysynthetic construction. . . . It seeks to unite in the most intimate manner all relations and modifications with the leading idea, to merge one in the other by altering the forms of the words themselves and welding them together, to express the whole in one word, and to banish any conception except as it arises in relation to others.[9]

According to Université Laval linguist Pierre Martin, this striking particularity is still recognized by modern linguistics. It seems to indicate the existence of a conception of the world common to all Native American cultures; further, it helps to account for the fundamental unity of American philosophy and to explain the absence from America of religious or economic wars. For the Amerindian, life *is* circular, and the circle generates the energy of beings. Life is merely a great chain of relationships among beings. Humans acquire power only to the degree that they can channel and circulate energy (material and spiritual possessions). Pierre Clastres, in his book *La société contre l'état*, reports what Francis Huxley observed among the Amerindians of South America:

It is the Chief's role to be generous and to give away all that is asked of him: in some Indian tribes, the Chief can always be recognized in that he owns less than the others and wears the most shabby ornaments. The remainder is gone in gifts. . . . The situation is exactly the same among the Nambik-wara described by Claude Levi-Strauss. . . . There is no need to cite more examples, for this relationship between Indians and their Chief is consistent right across the continent. Avarice and power are not compatible; to be Chief, one has to be generous.[10]

Amerindians have a fundamental respect for life and for the complementary nature of beings, who are all its forms of expression. They have no desire to affirm their supremacy over any other creature. They do not even domesticate animals, for animals, like humans, possess a spirit and liberty. The Amerindian does not exploit. Daniel Garrison Brinton further observed in America "the entire absence of the herdsman's life with its softening associations. Throughout the continent, there is not a single authentic instance of a pastoral tribe, not one of an animal raised for its milk, and very few for their flesh."[11]

III. Ecological Constraints Peculiar to America

Amerindians deserve no credit, of course, for the experience their continent has undergone. Civilizations are but the products of a chain of circumstances in the destiny of Earth, our common habitat; they are shaped by the constraints of climate and geography (for example, proximity to others or states of isolation). Too often, when historians and social scientists try to explain the origins of human social evolution, they describe communal living as the primordial form of society. Nevertheless, superficial descriptions of this stage of evolution show that modern people have long since lost all notion of the intellectual and spiritual tools with which they could have preserved a dignified image of their past and thus of their own human nature. Moreover, they have been so blinded by a falsified, learned image of their material evolution that they have been unable to realize that the spiritual heritage they encountered on their "discovery" of America was one they too must have possessed at a certain period in the Old World—before they lost it because of the constraints that led to the development of their present type of civilization.

As for the Amerindians, when they found the whites lost on the shores of their continent—the Great Island on the Turtle's back—they were in full possession of the spiritual gifts referred to earlier. Possessions and wealth circulated freely, according to the law of the great circle of relations. This was not so much because of greater morality among Amerindians than among Europeans as because of the physical context—the geomorphological constraints peculiar to America.

Accordingly, all first-hand accounts state that the starving, frightened, intolerant people who began landing here in 1492 were received with respect and humanity. They found such harmony, liberty, and tolerance that many began to leave their homelands for the new world of America.

Although seriously threatened by the wars and diseases their pale visitors brought with them, the Indians, believing in the existence of a plan known only to the Master of Life, never rebelled against their fate. They defended themselves only when it was necessary, because Indians did not know "the art of war": they understood infinitely better the art of peace. Resolutely and generously, they undertook the task of "Americizing the White Man,"[12] that is, doing everything possible to help this new child discover the essential, primordial wisdom of the new Earth-Mother he had just found.

IV. The Amerindian Idea of Creation

All Amerindians worshipped the Great Spirit, the Great Mystery, the Great Power, the Sky, the Master of Life, whom they called either Father or Grandfather, not so as to masculinize the Creator but to represent the ultimate creative and protective force, source of all life and all power.

More concretely, all Amerindians refer to the Earth as their mother, composed like them of body, mind, and spirit. The spirit that governs the Earth and materially produces life is feminine. To the Wendat, the Earth was created by a woman named Aa-ṭaentsic, who came from a celestial world. The Great Turtle took her onto his back and ordered the animals to spread there a small amount of earth brought up from the bottom of the sea. The woman, together with the two sons to whom she soon gave birth, founded and arranged Earth for the human race. The two sons vied with one another to impose their personal notion of what human life should be: one, who was too good, wanted it to be easy, while the other strewed it with obstacles and dangers. Their mother caused balance to prevail, and so the human world became what it is: a place of beauty and order but one where the ordeals that are part of the human condition encourage compassion, a fundamental moral dimension of life.

To an Amerindian, woman represents reason, the being who educates man, orients his future, and anticipates society's needs. Man acknowledges in woman the primordial powers of life and a capacity to understand its laws. As regards the organization and direction of society, the role assigned to woman is in a sense superior to that of man. This is most evident in "matriarchal" Amerindian societies, but it is equally true for those called "patriarchal," including most nomadic societies. These must follow a patrilinear order when founding their institutions—because of man's preponderant role in the material and spiritual quest for the vital necessities—but they are not in fact patriarchal. There is only an outward appearance of masculine power; the sense of closeness to the Earth is reinforced by an awareness of direct dependence on its vital products (which, like humans, are born of the Earth). It may be advanced that the vast majority of the nomadic peoples of America are matricentric in their ideological and spiritual conception of the world.

Unlike Wendat-Iroquois culture, which possesses agriculture and the trading power that it brings, members of nomadic societies are obliged to remain in a purer and more intimate harmony with the spirit of the Earth and its children: animal, plant, and others. Nomads are more closely tied to their mother, the Earth, than are the semi-sedentary. Their culture is centred on lightness, in every sense of the term: that is, it requires an optimal harmony. Their highly complex ideology is more alert to the forces of the universe. They are wise and peaceful, like all those who are close to nature, but in a particularly exemplary way.

V. The Matriarchy

Johann Jakob Bachofen has described the matriarchy, or "gynecocracy," as he termed it, as a stage in human evolution, with any deviation from it representing an imbalance:

> We have stated [previously] that the gynecocracy was the poetry of history; we may add that it represents the period of profane intuition and of religious premonition. It is the time of piety, of superstition, of wise moderation, of equity. All these qualities, engendered by the same principle, are attributed by the Ancients, with surprising unanimity, to the gynecocratic peoples without distinction.[13]

Amerindian societies, so steeped in the great circle of relations, are only very rarely patricentric. To support the thesis of widespread matriarchy among prehistoric Amerindian populations, some authors have tried to establish a close relationship between the process of acculturation following epidemiological depopulation and the erosion of matrilinearity. As historian Shepard Krech III argues:

> Comparative data from outside the Northern Athapascan region suggest fairly precise connections between the intensity of acculturation pressures and the erosion of matrilineal descent principles within two generations, and the substitution of neolocal for matrilocal postmarital residence in the same time span. . . . The more acculturated members of a society may abandon matrilocal practices and matrilineal ideologies and terminology in favor of bilateral and generational systems prior to less acculturated members; or the more acculturated

members may simply not be interested in traditional practices.[14]

According to Amerindian gynocentrist thought, the patriarchal theory of evolution, no matter how refined and intellectualized, is nothing but an apology for racism, sexism, and what we term "androcentrism," defined as an erroneous conception of nature that places man at the centre of creation and denies non-human (and indeed, non-masculine) beings their particular spirituality and their equality in relation to life's balance.

Modern woman longs for the day when "woman's persevering influence, rejecting the false and the conventional which are in a sense man's trademark, will rediscover the true paths of nature."[15] This woman finds her most powerful inspiration in the Amerindian woman. It was a Jesuit, Father Lafitau, who first revealed to the world the essence of democratic American thought when he observed the matricentrist mechanism that is the real basis of Iroquois society, since it represents the key to the remarkable social equilibrium of Amerindian societies. Speaking of Iroquois women, he said in 1724:

> Nothing is more real however than the women's superiority. It is they who really maintain the tribe, the nobility of blood, the genealogical tree, the order of generations and conservation of the families. In them resides all the real authority: the lands, fields and all their harvests belong to them; they are the souls of the councils, the arbiters of peace and war; they hold the taxes and the public treasure; it is to them that the slaves are entrusted; they arrange the marriages; the children are under their authority; and the order of succession is founded on their blood. The men, on the contrary, are entirely

isolated and limited to themselves. Their children are strangers to them.[16]

On the different conceptions of feminine law held by Europeans and Amerindians, this missionary to the Huron and Iroquois noted that in those nations the man must offer a considerable gift to the house of the woman he wants for his wife:

> The present made by the husband to his wife's lodge is a true coemption by which he buys, in some sort, this household's alliance. There is this difference [from the Roman custom of coemption] that the husband gives the present whereas, with the Romans, the wife gave it, giving three marked cents, as the symbol of this coemption. The reason for this difference is that, with our Indians, the wives are mistresses and do not go away from their homes, while, with the Romans, they went into the house and under their husband's jurisdiction, so that they had to buy the right to be mothers of a family.[17]

This deference toward the woman reflects the recognition, in matricentric societies, of a human brotherhood vested in the Earth-Mother, source of respect for personal vision in those societies. Non-Native writers still frequently do not perceive this fundamental cultural difference. Instead, they tend to insist on depicting Amerindian societies, especially hunting societies, as being governed by the naturally more imposing men, whereas the reality is quite different. Bachofen, again, offers his thoughts on gynocentric societies, characterized by order and gentleness, as compared with androcentric societies, characterized by harshness and moral confusion:

The cult of reproductive maternity gave way to that of sterile pleasure. For man, sex is for his pleasure, while for woman it is the duty of procreation sanctified by suffering. Under her reign, perpetuating the species had been the dominant preoccupation, imposed by an exclusively moral force with chastity as a necessary and inevitable condition. Man made selfish debauchery the supreme good, establishing his excess as natural law, using, to satisfy them, the brutal constraint imposed by his more vigorous muscles. This dual, contradictory tendency of the sexes is indisputable; it is still apparent today in everything they say and do; male domination has instilled in our civilization and our morality a profound corruption that erodes and dries up the sources of life, lowering happiness to the basest forms of pleasure. The masculine motto continues to be: War and Lust, the feminine: Continence and Peace.[18]

The "high status" of Amerindian women is not, as some authors have declared, "the result of their control over the tribe's economic organization."[19] The matricentric thought in these societies springs from the Amerindian's acute awareness of the genius proper to woman, which is to instill into man, whom she educates, the social and human virtues he must know to help maintain the relations that are the essence of existence and life. Women do not control anything through some "force" they possess, as Judith K. Brown would have it; they act through the natural intuition that Creation communicates to those who are open to its laws. Man, as Bachofen observes, does not possess this genius for educating: "It is by caring for her child that woman, more than man, learns how to exceed the narrow limits of selfishness, to extend her

solicitude to other beings, to strive to preserve and embellish the existence of others."[20]

When he is in command—which itself is a departure from nature—man can only opt for material wealth, for he lacks woman's intimate understanding of the price and value of life. He bases his power on brute force. Woman, however, naturally opts for peace and stability; her power is founded on the education of the inner strength (the quest for vision). Woman's thoughts are long, out of concern for humanity; man's are short, a result of personal and national pride.

Natural man, that is, one who belongs to a matricentrist society, entrusts his seed to woman, who conceives, nourishes, and educates it. Thus filled with goodness, humanity, and gratitude, man cheerfully carries out his role as protector. Throughout time, a large part of the spiritual burden of gynocentric societies is that Earth, a woman, is under the domination of patriarchal man.

Georgina Tobac, a sage of the Dene nation, crystallizes in one sentence the anguish experienced by the Native when the Earth is assailed by modern man: "Every time the white people come to the North or come to our land and start tearing up the land, I feel as if they are cutting my own flesh; because that is the way we feel about our land. It is our flesh."[21]

When fifteenth- and sixteenth-century Europeans, products of societies in which oppression was the norm if not the rule, came to America, they found territories, villages, towns, and cities organized in such a way as to respect the circular chain of all orders of life, notably the four elements that correspond to the four cosmic directions. While these Native nations were not exempt from the dissent, conflicts of interest, or armed confrontations that are peculiar to human nature, none of them tried to impose the philosophical or religious principle whereby man *and not woman* can and must exploit the non-human beings that are necessary for their survival until they have been exhausted.

The multi-millennial Amerindian experience demonstrated the great circle's civilizing potential, as well as the personal strength of the individual who finds his or her place within it.

The image of the circle clearly dominates Sioui's thinking. Among First Nations thinkers and peoples he is not alone in this. Consider, to begin, the metaphor used by First Nations to describe the continent on which they live. Here is Mary Ellen Turpel:

> First Nations peoples use the expression Turtle Island to refer to North America, which is thought of as a shell of a turtle surrounded by oceans. The images of the protective shell jutting out and the living creature within make a powerful metaphor for the connection with and respect for the land that all First Nations cultures share.[22]

The image of Turtle Island is a powerful one. Fundamentally, it bids us to think of the land as essentially alive and therefore as vulnerable. The people, in turn, live on the back of the turtle, so their prosperity is tied inextricably to the health of the living land. And an island, of course, is *encircled* by water. The water and the weather it brings provide the medium in which the living land moves and flourishes. Everything is *connected* in the metaphor: the turtle itself—its hard outer shell and its soft insides—as well as the people on its back and the medium in which everything is set. This provides a clue to the importance of the circle, emphasizing the connectedness of all things, animate (the turtle, the people, and the spirit world) and inanimate (the water, the shell of the turtle). Another notion brought to mind by the circle is that of *equality*. All points on the circumference of a circle are equidistant from the centre and so are equal with respect to that point.

The image also involves the indispensability of continuous *circulation*. This shows up in the radical social equality that governs the relations of most First Nations societies. This equality manifests itself, as Sioui reports, in the circulation of goods and possessions among members of a tribe. The chief is not distinguished from others in virtue of the material wealth he or she is able to accumulate and display. On the contrary, the chief is distinguished precisely by being the most willing to relinquish these things to others. He or she can thus be thought of as the one who keeps the movement of goods and possessions forever in continuity. Or think of the ubiquity of "talking circles" in First Nations cultures. To resolve an issue, everyone sits in a circle and exchanges points of view in an effort to achieve consensus. No single perspective is taken to be *intrinsically* privileged, unlike with more hierarchically organized social structures. Rather, the emphasis is on the circulation of points of view among the group.

The idea of continual circulation should also remind us of a theme explored in Chapter 3: the way that energy moves through ecosystems. Recall the quote from J. Baird Callicott:

> A description of the ecosystem begins with the sun.... Solar energy flows through a circuit called the biota. It enters the biota through the leaves of green plants and

courses through plant-eating animals and then on to omnivores and carnivores. At last the tiny fraction of solar energy converted to biomass by green plants remaining in the corpse of a predator, animal feces, plant detritus, or other dead organic material is garnered by decomposers—worms, fungi, and bacteria.[23]

We will have much more to say later in the chapter about the possibilities for convergence and reciprocity between First Nations and Western approaches to nature. Here, it is important to notice how well Callicott's comments fit the image of the circle. There is no single thing or type of thing that has a privileged place in the circulation of energy that gives life to the biosphere. There is instead radical *equality* coupled with a profound *interdependence*.

Ⓒ Confronting Myths of the Ecological Amerindian

What are we to make of Sioui's invocation of the Earth-Mother, the "source of respect for personal vision" in matriarchal societies? Reference to a figure like this draws attention to the striking resemblance between the principles of Amerindian ecology and Gaia Theory, the view that the biosphere is a self-regulating super-organism (see Chapter 3, section E, for more on Gaia Theory). At least as articulated by Sioui, it seems that Amerindians *might* be willing to accept this picture, suitably filled out. But let's leave this question to one side (Bruce Morito returns to it briefly, below). The more pressing problem concerns the seemingly essentialist and romanticized view of Amerindians being presented by Sioui. Sioui's preference for Earth-Mother imagery is clearly related to the positive role he assigns to women in First Nations' culture. Here again is what he says about this:

> The matricentric thought in these societies springs from the Amerindian's acute awareness of the genius proper to woman, which is to instill into man, whom she educates, the social and human virtues he must know to help maintain the relations that are the essence of existence and life. Women do not control anything through some "force" they possess, as Judith K. Brown would have it; they act through the natural intuition which Creation communicates to those who are open to its laws.

In spite of his dismissal of Judith K. Brown's views, an uncareful reading of Sioui might suggest that he still supports the idea that the Amerindian woman is essentially or naturally ecologically attuned in a way or to a degree that non-Amerindians and Amerindian men are not. Amerindian women are, after all, said to be more "open" to the laws of Creation than others. This way of thinking might then invite a dangerous—because distorting—mythologizing of Amerindians. Although, as Bruce Morito argues, Sioui himself *cannot* be saddled with this sort of essentialist claim, it has been and remains a tempting one for many people writing about Amerindian cultures. To help sort out the complexities here, let's turn to Morito's analysis.

THE "ECOLOGICAL INDIAN" AND ENVIRONMENTALISM
Bruce Morito

The idea of the ecological Indian has been and continues to be employed for both environmental and political purposes. Environmentalists used the idea in the "Keep America Beautiful" campaign of 1971, capturing it in the image of the "Crying Indian" (Iron Eyes Cody) to advance a critique of the dominant sector's environmental record.[24] At the time of this writing, the movie *Avatar* has been released. It advances the same theme (and is enjoying record attendance). The theme is a critique of Western European–based destructive attitudes toward the environment and toward indigenous cultures. The image of the indigenous person as an environmentally benign presence has become an icon of a counterculture movement intent on criticizing Western values and practices by identifying them as the causes of environmental and social degradation.

This movement, represented by the likes of Montaigne, Lahontan, Thoreau, Muir, Ernest Thompson Seton, Archie Belaney (Grey Owl), has used "the ecological Indian" to advance cultural critique to the point where the concept has become a widespread countercultural icon, representing ecologically better times, a veritable prime of human/environment relations. In the field of environmental ethics and philosophy, this icon represents an alternative to Western European, anthropocentric, and capitalist perspectives, the root causes of our environmental crisis.[25] Baird Callicott was one of the earlier environmental ethicists to cite Aboriginal thought and practice as an alternative to Western forms.[26] Donald Hughes takes Aboriginal world views as a model of an ecologically benign lifestyle, one that recognizes and protects the interconnectedness and inter-dependency of all things, such that people's behaviour leaves the environment unspoiled.[27] He describes Aboriginal people as living in "perfect harmony" with the land. When describing elders' knowledge of plants, he says that they were able to "tell the use of every plant in the ecosystem," indicating an intimate connection with Mother Earth.[28]

Does this "ecological Indian" and Mother Earth imagery in fact represent Aboriginal world views?

I. History and "the Ecological Indian"

Sam Gill argues that prior to the 1800s, it is difficult to find evidence of conceptions of Mother Earth imagery, a key image associated with the idea of the ecological Indian in indigenous peoples' cosmologies. Early "anthropological" accounts or the stories of Aboriginal people were devoid of this conception.[29] Gill identifies 1855 as the date for the "first indication of the earth as a major religious conception" in Aboriginal cultures.[30] In 1877, it had become a popular notion used by Wanapum chief Smoholla and was fostered in American scholarship by the likes of J.W. MacMurray and James Mooney in 1890 and 1896, respectively. According to Gill, the best explanation for the emergence of the conception of the Earth as mother in Aboriginal cultures is that it served a political end; it served as a basis to criticize American expansion by establishing an unassailable moral authority for the Native position. It grounded Aboriginal peoples' right to land by reference to divine authority, thereby undermining US justifications for westward expansion. It became a means of distinguishing Aboriginal forms of grounding land rights from Western European forms. An alleged statement by Tecumseh to General W.H. Harrison of the US army in 1810 during a treaty negotiation is often used emphasize this distinction. Tecumseh is said to have refused to accept Harrison's invitation to sit on a chair, stating, "The sun is my father and the earth is my mother; she gives me nourishment and I repose upon her bosom."[31]

Western European culture has long been fascinated with Aboriginal people, conceiving them as "noble savages,"[32] or "truly natural philosophers,"[33] whose manners and customs are engendered by the pure light of nature. This fascination has also long been exploited (since at least the 1600s), as indicated by the use of the *Jesuit Relations*.[34] These were descriptions of Aboriginal peoples' lives and cultures written by priests who lived with Aboriginal people. They were edited and published during the seventeenth and eighteenth centuries to raise financial support for the Jesuit movement. Thomas Hobbes, John Locke, and others also used various conceptions of the Aboriginal personality to model what they considered a "state of nature," which described the human condition before it was shaped by social, political, and legal orders. When British interests in North America turned from mining and trade to colonization and plantations, the conception of Aboriginal people shifted to that of "the ignoble savage," especially as conflicts with Aboriginal people grew.[35]

Characterizations of Aboriginal people that have shaped much of the dominant society's views, then, were typically formulated to advance some social, political, or religious agenda. European colonizers justified their actions partly on the grounds that because they were ignorant savages, Indians had to be converted to Christianity and adopt European ways. As ignorant but noble savages, they were seen as earlier versions of Europeans who had long since been Christianized and made civil.

One effect of this polarized characterization is illustrated in the attitudes of Duncan Campbell Scott during the early to mid-1900s. Scott was deputy superintendent of Indian Affairs and became the chief instrument of Indian assimilation policies for Canada's Indian Department. He held the historical Indian (the noble savage) in highest esteem, even as he exercised some of the most culturally and psychologically destructive assimilation policies, designed to eradicate "Indian culture," on contemporary Aboriginal people. His conception of the true Indian was heavily romanticized in his poetry (e.g., "Powassan's Drum"), which lamented the loss of the historical wild Indian, and this seemed to be his justification for implementing eradication and assimilation policies on living Aboriginal people. Actual living Indians were not "real" Indians, in his view, so assimilation and eradication of cultural remnants were the only policies that made sense.

A number of questions can be raised. Does "the ecological Indian" accurately represent Aboriginal culture? What impacts on Aboriginal people has the use of the concept had? A number of historical, ethnological, anthropological, and political thinkers (Olive Patricia Dickason; William Cronon; Shepard Krech III; Daniel Francis) have attempted to show that it is a serious distortion. But before describing how the conception is a distortion, it is worth dwelling on its historical development. Bruce Trigger and Wilcomb Washburn explain that a nineteenth-century ideological belief in evolutionary progress contributed to the entrenchment of the idea that Aboriginal people lived in a primitive state. Accordingly, Aboriginal societies were seen as backward predecessors of Europeans, who had long since evolved (progressed) from that state.[36]

Casting the Aboriginal personality as uncorrupted but primitive has long been associated with seeing them as politically and legally naive and lacking genuine reasoning capacity. As European and North American social critics have attempted to find means to effect social and environmental change, they have compared their societies' values, beliefs, and practices against those of an idealized Aboriginal society. During the 1700s, Rousseau proposed the idea of a Golden Age of humanity, a time before civilization when there was a balance "between the indolence of the primitive state and the petulant activity of our vanity."[37] The savage was, for Rousseau, evidence that the purpose of human life is to remain in this state, the "veritable prime of the world." His idea was that Aboriginal people living in a primitive state were "free, good and happy . . . according to their nature" without knowledge of good or evil.[38] They had no use for law or morality and in fact had no conception of either. They lived in a kind of blissful ignorance of evil while enjoying the embrace of Mother Earth.

As "veritable primes of the world," Aboriginal people were used as models of what is good in human nature. One of the more influential contemporary papers advancing a similar critique of Western culture is Annie Booth and Harvey Jacobs's "Ties That Bind: Native American Beliefs as a Foundation for Environmental Consciousness."[39] It is much more nuanced than Rousseau's version, but it has also been used to praise Aboriginal culture as a paradigm of good environmental stewardship in much the same way that Rousseau romanticizes it for purposes of social critique.

What do Aboriginal people themselves say about their relationship to the Earth? Georges Sioui, in his *For an Amerindian Autohistory*, states:

Every expression of life, material or immaterial, demands of the Amerindian respect and the spontaneous recognition of an order that, while incomprehensible to the human mind, is infinitely perfect. This order is called the Great Mystery. To the traditional Amerindian, life finds its meaning in the implicit and admiring recognition of the existence, role, and power of the forms of life that compose the circle. Amerindians, by nature, strive to respect the sacred character of the relations that exist among all forms of life.[40]

This passage suggests that Aboriginal people see themselves as resembling Rousseau's description. A careful reading of Sioui, however, indicates that caution is called for when describing Aboriginal environmental perspectives. First, his account, while focusing on sacredness, does not imply that Aboriginal people are politically and morally naive or innocent. Furthermore, Sioui does not argue that the Aboriginal world view produces perfect harmony with nature. Rather, it implies a responsibility to strive to respect all beings. As will soon become evident, the reason for having to strive to act respectfully, as opposed to simply being respectful by nature, is a complex matter.

Similarly, Gill argues that finer distinctions are needed to describe Native environmental perspectives. For instance, when criticizing ascriptions of Mother Earth cosmogonies to Zuni and Luiseño cultures, he points out that the grammar of Mother Earth language does indeed attribute female characteristics to the Earth but not divine characteristics. When Mother Earth is recognized as creator, she plays a relatively minor role compared to that of a sun father or Earth doctor (male figure).[41] These details and subtleties of Aboriginal thought, Gill suggests, have been occluded by the use of "Mother Earth" imagery.

Even in the colonial records of the European powers, evidence of a complex Aboriginal world view is plentiful. During a council meeting between the Six Nations and the government of Pennsylvania, for instance, Canassatego speaks for the Six Nations. He says it makes no sense for the British to assume that they possess the lands in Maryland, because it is the Six Nations who come out of the ground, while the British come from beyond the sea. For this reason, the British ought to accept the Six Nations as elder brethren (having principal authority). The ideas of authority and creation associated with "coming out of the ground" (coming from Mother Earth) are also closely connected to land rights. They are associated with jurisdictional authority whereby being born of a land entitles one to assert authority over foreigners.

The use of totems, or dodaem, further illustrates the complexity of the Aboriginal/Earth relationship. The *Jesuit Relations* indicate how people identified as beaver, otter, and so on. Treaties, up to the late mid-1800s, were signed with dodaem, and the Commission of Indian Affairs in Albany even identified some Aboriginal visitors as members of the wolf, turtle, and other clans, although usually they would use other designators (e.g., Five Nations, River Indians, Mississauga). Warriors tattooed dodaem on their bodies to enable their opponents to identify who was killing whom. Dodaems could represent the clans to which people belonged (in both Iroquoian and Anishnaabeg societies). The clan system determined the order of social relations (e.g., who could marry whom; how responsibilities in families, communities, nations, and confeder-

acies were assigned). As such, they were deeply connected to systems of social order and functioning. In this way, dodaems, however much they signified a close connection to the Earth, were also highly important to systems of governance and legal process. Dodaemic systems did not represent straightforward intimate and harmonious relationships with the Earth.

Indeed, many representations of dodaem had turtles, wolves, bears, and so on carrying weapons of war (e.g., war clubs, muskets, knives). War among tribes and nations was constant, a situation into which Europeans entered as trading partners; they did not create warfare among Aboriginal peoples. Conflict among themselves, therefore, was not unusual. Moreover, conflict could also occur between human nations and other member nations of the Earth community (as the story about the rose and the rabbits in the next section will indicate). If, therefore, the relationship between the Earth and Aboriginal people is a mother–human relationship, it is one that sometimes involves deep conflict. Thus, the political/legal significance of dodaem and Mother Earth cosmologies, as Georges Sioui states, is that people are responsible for striving to respect the sacredness of others; it is not that Aboriginal people are by their very nature in harmony with Mother Earth.

II. Anishnaabeg and Haudenosaunee Descriptions of the Aboriginal Personality

The picture of the Aboriginal culture we gain from oral tradition takes us further into the complexities of Aboriginal culture. John Borrows, an Anishnaabe legal scholar, helps to describe traditional environmental knowledge and responsibility in this light:

> The once numerous and beautiful roses had suffered a massive decline to

the point where none could be found. A council of animals (including the Anishnaabeg) convened to discuss the matter and determined to search the world for any remaining roses. Meanwhile, the rabbits had been growing fatter and more numerous. After a long search, hummingbird found one remaining frail rose, which he carried back to the council. The rose explained that the rabbits had eaten all the roses, at which point all the animals began grabbing the rabbits by the ears (which is why they have long ears today) and buffeting them. But the rose cried out for them to stop, because the rabbits were not entirely to blame. The others (including the Anishnaabeg) had not been mindful of the rabbits over-eating and so had to accept part of the responsibility.[42]

Borrows's analysis of this story describes a duty to be environmentally responsible in the sense of exercising stewardship responsibilities. Ecological balances can go seriously wrong if stewardship responsibilities are ignored, as it appears was often the case, judging by the number of stories that have to do with stewardship. In "The Rose and the Rabbits," the Anishnaabeg are responsible for controlling the rabbits' behaviour and possibly numbers. When they fail, serious imbalances arise, and everyone suffers. Such stories, then, are about a failure of the Anishnaabeg in their responsibility to act as the Creator expected them to act.

At a more fundamental level, creation stories articulate a relationship between good and evil. The Anishnaabeg creation story places the creation of human beings amidst disaster and ruin (a flood).[43] Kitche Manitou decided to give Sky Woman a mate, through whom she conceived two beings—one pure spirit and the

other a purely physical being. These two beings warred with one another. After Sky Woman again conceived, the Anishnaabeg were created along with Nanabush, the great supernatural teacher and hero. Nanabush came to possess the pipe and the law of peace after a battle with his father, Epingishmook, who had earlier killed Nanabush's mother. Nanabush fought to revenge his mother but was unable to defeat his father. The pipe of peace was smoked in recognition of this stalemate. Strife and war were not defeated to be replaced by a harmonious peace. We also find the four directions (guardian beings who have special powers) doing battle, at times, with the Anishnaabeg. There are battles between Zeegwun (summer) and Bebon (winter) over a woman.[44] Hence, the relationships within nature are constituted as much by strife as by harmony; indeed, the two must be thought of together in Anishnaabeg cosmology, because one does not triumph over the other. In fact, the emergence of human characteristics is explained by reference to the twins, one being evil and the other good. Responsibility for establishing social and political order amidst the chaos created by conflict and strife for Anishnaabeg, then, becomes central to their world view.

Evil deeds motivated the Creator to establish a system of punishments—for example, game animals would abandon the Anishnaabeg when they failed to exercise stewardship. Stories about evil impulses causing greedy and destructive behaviour (e.g., stories of Nanabush and the trickster) tell of the arising of councils that establish social and legal norms to control such impulses. In the struggle between good and evil, unity and disintegration, harmony and discord, the Anishnaabeg find themselves having to institute normative systems of law and morality to establish balances between these opposing forces.

Taiaiake Alfred, a Mohawk (Kanien'kehaka), describes the fundamental Haudenosaunee (Iroquoian) personality in a similar manner. It is a balance of opposites; balance, consequently, is the fundamental principle of good governance.[45] The Haudenosaunee tradition explains (in the Deganawidah epic) how the Great League of Peace, itself based on the Great Law of Peace, was given by the Creator through the Peacemaker and Hiawatha.[46] Sending the Great Law was the Creator's response to the constant wars between members of the Iroquoian nations. The Great Law of Peace and its associated League of Peace (constituted of 50 sachems representing each of the Five Nations) was formed in the wake of the cruel and vicious rule of an Onondaga chief, Tadoda:ho, who was eventually transformed by the Great Law to become one of the founders of the confederacy. The story is about controlling and transforming evil by moral and political principles of good governance. Insofar as this law is designed to overcome the arbitrary and capricious rule by terror or might, it resonates with the rule of law in Western traditions. Indeed, it became the model on which the intercultural relationship between First Nations east of the Mississippi and the British was organized. This system was known as the Covenant Chain.

Even a cursory glance at the records of Indian Affairs relevant to the Chain shows that commissioners and governors were quite aware of Aboriginal law, territorial rights, and the importance of using proper protocol (e.g., wampum, the pipe) during council meetings and at what were called "meetings at the wood's edge." The British utilized existing "covenant" systems of intertribal relations to help form what was known as the "Silver Covenant Chain." Francis Jennings has written a three-volume work on this relationship, aimed at correcting misrepresentations of Aboriginal people's political and military contributions to the history of North America.[47] The Covenant Chain has principally to do with the Haudenosaunee but also with Algonquian nations (e.g., Delaware, Mahikan, Ottawa, Shawnee).

Wampum protocol, a complex system of communication, agreement recording, legitimation, and condolence, was adopted by both British and French. It bound signatories economically, legally, and politically. Territory, usufructory rights, legal suits, complaints over land transactions, and council protocols were governed according to the agreements made under the Covenant Chain. The protocol was, for the most part, defined in accordance with Aboriginal traditions and accepted by colonial officers and governors as well as settlers and traders. It was not imposed by Europeans. Indeed, the colonial government (and in many respects the Board of Trade and government in Britain) respected this protocol and tried to ensure that agents (e.g., Edmond Andros, Peter Schuyler, William Johnson) who understood these protocols and could argue in accordance with them were assigned to negotiate treaties, alliances, and trade agreements. British and French cognizance of Aboriginal protocol, then, was in effect a recognition of the rule of law. For this reason, Europeans in North America had to interact with Aboriginal people as full moral, political, and legal agents; they could not treat them as primitive savages.

III. Implications for Environmental Thought and Ethics

Much of the environmental movement's use of "the ecological Indian" in advancing the cause of environmental protection has been based on a distortion. Now, one response might be, "What is the harm in using such a distortion, when there is so much at stake?" To begin answering this question, it should be noted that this use of a misconception belongs to a pattern. The use of "Gaia" in James Lovelock's Gaia hypothesis is part of this pattern.[48] The notion of Gaia captured the imagination of environmentalists in the 1990s. Gaia is a conceptual device that represents the cybernetic capacity of the Earth, which characterizes the Earth as a feedback mechanism. Lovelock used the cybernetic model to argue that life on Earth cannot be explained by analyzing its component parts alone; rather, the configuration and balance of component parts (chemical distributions) can be explained only by presupposing that the Earth is alive.

Granted, in his epilogue Lovelock waxes poetical, drawing on the association between Gaia (as Mother Earth) and familial ties, but this association is not contained in his analysis. Environmentalists have done the same in an attempt to develop a Gaian ethic that connects an ethic of care and nurturing with ecology. This is not to criticize ethics of care as such; it is to criticize the exploitation of an idea that has nothing to do with caring and with forcing an association to be made. Gaia, as a self-balancing cybernetic system, does not care for people and does not care whether people care. The way in which the Gaia hypothesis is exploited, then, is of a piece with the way "the ecological Indian" has been exploited to serve environmentalist ends.

To expand on the previous point, examining Steve Sapontzis's argument can serve to show how using distorted concepts can be self-defeating.[49] Sapontzis would have us suppress predator activity by supplying predators with needed protein so that they need not hunt. The general principle is to suppress destructive behaviour in an attempt to make the world a harmonious and peaceful place. Sapontzis, to his credit, is consistent with his principle of living in harmony by taking it to its logical conclusion. However, if ecosystems are constituted of both constructive and destructive forces,[50] as Aboriginal cosmologies recognize, then Sapontzis's principle actually commits us to undermining natural relations (constructive–destructive, predator–prey, parasitic–symbiotic, harmonious–discordant). If an environmental ethic is to aim at protecting ecological processes, it cannot aim at creating harmony between all beings whereby all beings can flourish. Indeed, it is only because of the tension created by the play of opposing forces that there is an ecosystem in the first place. To try to eliminate the destructive forces, then, is to try to eliminate the very thing we want to protect.

IV. Consequences for Aboriginal People

Just as applying the principle of harmony can undermine ecosystems, so too can applying the idea of the ecological Indian undermine Aboriginal people's culture and way of life. Sandy Grande explains how the very intent of protecting the ecological Indian harms her people.[51] Using conceptualizations of ecologically noble Aboriginal personalities, she argues, has been a factor in making Aboriginal people invisible. This invisibility is effected through objectifying Aboriginal people as representations of what it means to live in a simpler, bygone era, free of the encumbrances of modern technology and civilization.[52]

This conceptualization contributes to the further domination of Aboriginal people through acts of controlling "language, metaphors and epistemic frames."[53] By so controlling language, the dominant society controls the voice of Aboriginal people and the purposes to which giving voice is directed. Such control determines how Aboriginal people's identities are allowed expression and how they will be recognized in the dominant society. As such, Aboriginal people cannot present who they are (how they self-identify) to others, what their values are, or what their actual environmental commitments are but are expected to use representations that accord with the ends set by those in a dominant position. Rather than protecting and alleviating the harmful effects of colonization, the controlling of language actually contributes to further colonization.

Like ecological relations, the human/nature relationship for Aboriginal people is constituted of opposing forces that need to be balanced rather than harmonized (where there is no destruction). The one-sided conceptualizations used by environmentalists stem from a failure to rigorously exercise epistemic responsibility to examine evidence, knowledge claims, and the implications of those claims. The result of this lack of rigour is to harm and, in the case of Aboriginal people, to perpetuate historical harms.

Were we to recognize the responsibility to balance opposites, the aim of ethics would not be to produce harmonious dwelling, if such dwelling requires suppressing deep-seated conflict and destruction. Neither would the aim of environmental ethics be to achieve a particular state of the Earth, as if we could establish a relationship with the Earth at its veritable prime, which would last forever. This would be to promote a static conception of what the Earth should be. If balancing oppositional forces is core to environmental responsibility, with balancing a dynamic act that takes place continually in an evolutionary context, then environmental responsibility needs to be conceived more as a continuing fluid process rather than as achieving an end-state.

There may still be much to learn from Aboriginal approaches to the Earth–human relationship, but to do so we must abandon the idea that we are responsible for returning the Earth to some pristine state. Returning to a former relationship may, in the end, be implied by following Aboriginal ways, but returning to a particular state cannot be the goal of environmental ethics. If we take Georges Sioui's idea of striving to respect the sacredness of other beings and the tension of opposites, we would frame our moral responsibility more in terms of maintaining relationships in various states. And if relationship, giving and taking, and acknowledging positive and negative forces are accepted as fundamental to ethics, then reciprocity, rather than responsibility to return to a particular state, may be the core principle on which we base environmental ethics. However, the principle of reciprocity may not seem to amount to much because there is no defined end to which we should strive. Since this essay can only provide a sketch of what a more adequate ethic would look like, it is not possible to fully articulate an ethic of reciprocity here.

More problems with the idea of the ecological Indian have been articulated than solutions to environmental problems. What has been presented is a mere sketch of a direction that ethics needs to take if Aboriginal and ecological approaches are to be acknowledged more adequately. When the need for action becomes more pressing, this largely negative conclusion may seem inappropriate. But in the final analysis, if the understanding upon which we act is faulty and self-defeating, then we are ensuring that the ethic that guides our actions will, in the end, cause as much as or more harm than it prevents.

In this chapter's case study, we will return to Morito's crucial notion of how colonizers seek to control language to their benefit. For now, it is important to underline and expand on two of his paper's most important contributions to our understanding of the issues we are investigating: (1) that we can talk about First Nations environmental ethics without romanticizing or essentializing First Nations; and (2) that First Nations cultures *do* have a sound and sustainable environmental ethic, painstakingly worked out over the course of thousands of years occupying this land.

Let's begin with (1). Morito argues that First Nations history should not be characterized as though there was ever some state of blissful harmony among tribes or between First Nations generally and nature. At the core of First Nations identity is a normative claim: that we *ought* to recognize the fact of conflict and work so as to achieve a balance of forces in the face of it. For example, Morito asserts that

> War among tribes and nations was constant, a situation into which Europeans entered as trading partners; they did not create warfare among Aboriginal peoples. . . . Moreover, conflict could also occur between human nations and other member nations of the Earth community. . . . If, therefore, the relationship between the Earth and Aboriginal people is a mother–human relationship, it is one that sometimes involves deep conflict.

This thought nicely defuses the idealized and highly sentimental version of the human–mother relationship according to which the two parties are bonded in a harmonious, nurturing, and caring whole. And this insight also takes care of the problem of essentializing that we seemed to find in Sioui. Morito rightly notes that Sioui is *not* making a descriptive claim about First Nations' relationship to the natural world (and to each other) but a normative one: he is laying out the structure of moral responsibilities for First Nations cultures, not describing how First Nations timelessly are. In fact, it appears as though the urge to foist on First Nations cultures the notion that there was—and/or will be—some historical time in which perfect harmony is achieved is itself a colonial imposition. It is to take a central tenet of Euroamerican mythology—that of the fall from the divine embrace coupled with the hoped-for return to it—and project it into a culturally alien context.

One way to avoid such sterile forms of thought is to remember what should be obvious: that we are all—Euroamericans and members of Aboriginal cultures everywhere—of the same species. A trivial point no doubt, but as E.O. Wilson has argued, one of the key features of this species is that it has historically been a killer of biodiversity on a huge scale wherever it has gone. To take just one example, when the Polynesians first arrived in New Zealand in the late thirteenth century, they encountered what Wilson calls a "vast biological wonderland." Most impressive were the moas, large flightless birds that looked a bit like ostriches and were, as it happens, quite delicious. The Maoris wiped out this species within a few decades and probably also the giant New Zealand eagle that preyed on it. The newly arrived humans also brought rats with them, which wiped out many species of birds, reptiles, and amphibians. In total, more than 20 bird species were eliminated during the first 100 years or so of human habitation on the islands. And in case we were inclined to dismiss this as an isolated case, Wilson goes on:

> The New Zealand event was only the final chapter in mass extinctions that began on islands to the north. What we celebrate in the colonization of Polynesia as a grand historical epic for humanity was for the rest of life a rolling wave of destruction. The vast triangle of archipelagoes that embrace the Pacific are a natural laboratory for the study of extinction.[54]

Nor was North America spared the depredations of the "serial killer of the biosphere."[55] The Americas were colonized in successive waves beginning roughly 14,000 years ago by people in north Asia who crossed what was then a land bridge linking Siberia and Alaska. At the time, North America was populated by a truly fantastic array of megafauna: beavers the size of bears, giant armadillos, massive camels, woolly mammoths, and more. Within a thousand years of the arrival of humans—equipped as they were with bow and arrow and fire—these species were nearly all gone, eventually to be replaced by the smaller and generally more fleet-of-foot species we encounter now.

Today's First Nations are the distant descendants of these people, a fact that should make us ponder the significance of the myth of the rose and the rabbits, analyzed by Morito. However, from all of this it would be incorrect to draw the conclusion that First Nations cultures are as ecologically unsound as Euroamerican culture is. This brings us to point (2), explored at length in the next section.

Ⓓ Traditional Ecological Knowledge

Ecologists have been focusing increasingly on the nature and potential of **Traditional Ecological Knowledge (TEK)**. TEK contains much insight on how to monitor and manage ecosystem processes, harvest sustainably, and acquire and transfer ecological knowledge within and across generations. For example, in British Columbia the Shuswap Interior Salish and the Kwakwaka'waka and Nuu-Chah-Nulth peoples of the northwest coast have evolved complex practices for sustainably harvesting many plant and fungus species. Edible mushrooms are harvested only when mature, and they are cut at the base and the soil is carefully replaced so as to protect the remaining individuals. Many species of root vegetables are harvested, but this is done selectively by size, while smaller roots and propagules are replanted. The soil is carefully tilled, and selective burning is also used. Edible berries, fruits, and nuts are picked only from branches shooting off the bush's main branch, and sometimes the bushes are burned or pruned to encourage future regeneration. Fibrous stems are used for mats, cordage, or baskets, but the stems and leaves are cut only from perennial plants at the end of the growing season. Medicinal plants like mountain valerian are selectively harvested and are often regenerated from fragments left in the ground.[56]

The list of such practices could go on indefinitely, but what needs to be stressed is, first, that the practices work: they provide a fully sustainable source of goods and resources to the people employing them; second, that the practices are the product of a deeply localized knowledge of ecosystem processes, one that has been transmitted down the generations. For instance, the avalanche lily has edible bulbs that are obtained by digging up a whole section of turf, taking out the largest bulbs, and replanting the smaller ones as well as the seed-bearing propagule attached to the lower part of the bulb. Further, an area that has been dug extensively is then left to regenerate for three or four years. This optimizes the productivity *over time* of the bulbs. Finally, the large bulbs selected are left on the ground for a time because this enhances their sweetness. Where did the people learn this lesson? From the grizzly bears, of course. Here is a St'at'imc elder from Mount Currie, BC:

> You've got to go pretty well up to the top of the mountain for it [the bulb], the summit. In a certain time of the year they pick it. . . . The old-timers used to pick it and dry it for winter use. I know the grizzly bears they dig it out too. They use their big claws like that [raking motion], and they just leave it like that in the sun, you know. I guess they must taste good when they're dry. They don't eat it right away. I've watched them. A long time I've watched the grizzly bear, digging it out.[57]

Resource rotation is a particularly widespread practice. James Bay Cree use a four-year rotation schedule for their hunting grounds. This allows populations of beaver to recover. The same technique is used for fishing grounds. In the latter case, they use what ecologists call an "optimal foraging model" that uses information about declining catch per unit to inform decisions about harvesting allotments.[58] The Cree employ a 4 to 7–year rotation management of the beaver populations, a 5 to 10–year scale for lake fish and an 80 to 100–year scale for caribou. Added to this is a management schedule for forestry resources such as fibre and soil. This is an extraordinarily complex achievement. And just as with the British Columbia tribes, the key to sustaining it over time is to maintain close observation of the land itself:

> Among the James Bay Cree successful transmission of bush skills and knowledge depends on the amount of time families spend on the land because of apprentice-based knowledge transmission and the amount of time required for hands-on learning.[59]

Let's bring this back to Morito's analysis. He would have us eschew a "static conception of what the Earth should be." Moreover, he thinks we should emphasize ecosystemic "balance" rather than "harmony." A balanced system contains and manages strife and conflict, while a harmonious system eliminates it. This is exactly the lesson that ecologists are drawing from the study of TEK. As we have seen in Chapter 3, ecologists have generally moved away from an equilibrium-based model of ecosystems—where the system moves toward a stable, harmonious climax community of species—and toward a dynamic equilibrium model. The latter emphasizes the ability of an ecosystem to resist or repel major environmental disturbances.

As far as ecosystem management is concerned—and this is the place where an ecologically informed environmental ethics enters the picture—we should seek to increase resilience rather than achieve a probably mythical equilibrium.

Crucially, ecologists are finding that TEK can show us how to do this. Here are five lessons ecologists have taken from TEK:

1. Management is best carried out by locally derived and enforced rules.
2. Resource use is flexible, employing methods like rotation.
3. Users have developed a system of knowledge that responds sensitively and in a timely fashion to environmental feedbacks such as reductions in catch.
4. A diversity of resources is used in case one or more species collapses or is drastically diminished.
5. "Quantitative yield targets" are eschewed in favour of empirically derived and qualitative methods of ecosystem monitoring and management.[60]

Ⓔ Conclusion

The similarities between Western scientific ecology on the one hand and sustainable First Nations ecological practices on the other are powerful examples of the possibilities for reconciliation and convergence between the two cultures. Working together, we might actually figure out a way to respond to our joint ecological crisis in a manner that provides a unique example for the rest of the world. But there is at least one very potent obstacle to our doing this on a significant level. That obstacle is the psychology underlying the victim/victor duality that has marked, and corroded, our mutual dealings so profoundly.

In her classic study of Canadian literature, *Survival*, Margaret Atwood provides an interesting perspective on the prevalence of this distinction in our literature. One chapter in her book deals with our literary treatment of First Peoples. Literature is an illuminating window onto the general culture: in many ways, it says what we are all thinking, even if we are not quite aware of it (yet). Atwood shows that First Nations people have traditionally taken two roles in literature: victim *and* victor. As victor, they have been cast in the role of Nature the Monster, "torturing and killing white victims." But it is their role as victims themselves that has dominated. Here is a representative statement from a character in *The Fire Dwellers*, a novel by Margaret Laurence. The character is talking about a village on the West Coast:

> ... Indian village, a bunch of rundown huts and everything dusty, even the kids and dogs covered with dust like they were all hundreds of years old which maybe they are and dying which they almost certainly are. And they look at you ... with a sort of inchoate hatred and who could be surprised at it? ... If I were one of them ... I'd sure as hell hate people like me, coming in from the outside.... You don't ask anybody anything. You haven't suffered enough. You don't know what they know. You don't have the right to pry. So you look and then you go away.[61]

This way of looking at a whole people is a powerful trap, both for the victims and the victors, because it is often based on a recognition of genuine grievance. The problem is that if we were to cease seeing members of a specific group as victims, it would mean that we were forgetting or cancelling the grievance. This is, or can be, unpalatable to both parties. But the victim/victor duality is the antithesis of the circle. The circle, as we have seen, is predicated on the twin concepts of respect and equality. Although victors can commiserate with their victims and even hate themselves for being victors, they cannot think of their victims as equals. Another way to put the point is to say that pity and respect are incompatible attitudes or stances to adopt toward someone. To pity someone is to see that person as, in a sense, trapped in an inferior or deplorable state. What you are assuming is that the person can be indelibly marked by the victor's actions. To pity is therefore to deny the other's autonomy. To respect him, on the other hand, is to deny that the person is, in the relevant way, *inevitably* lowered. The guilt and shame felt by the victor in the quote just above do not cancel the inequality between the two parties. In fact, they reinforce it.

Regrettably, this stance still dominates the mutual understandings between First Nations and the non-Native majority (though there are of course exceptions to it). Atwood, however, sees a glimmer of hope in our literary tradition. She looks at a number of poems in which the suffering of First Nations "is not regarded as unavoidable."[62] For instance, in John Newlove's poem "The Pride," whites are actually transformed into Indians, in the process discovering who they, the whites, really are:

> At last we become them
> and they
> become our true forebears, moulded
> by the same wind and rain
> and in this land we
> are their people, come
> back to life again

Unfortunately, all the Natives are dead when the poem's protagonist has this thought. But in his recent book, *A Fair Country*, writer John Ralston Saul develops a strikingly similar idea. Canada, he asserts, is a Métis nation, a single culture that contains both Euroamerican and First Nations elements. If we would only embrace this identity, we would have a unique opportunity to develop a fully sophisticated modern state that is at the same time deeply imbued with First Nations' values and practices. Such a transformation would aid us immeasurably in responding to the many challenges that confront us, not the least of which is the ecological crisis. But so far, rather than incorporating the lessons of our Aboriginal peoples into our own ecological self-understanding, we have become "addicted to imported myths," which deny our "historical originality."[63] First Nations peoples understand the natural rhythms and processes of *this place*, Turtle Island, and the country will flounder unless it recognizes and embraces this understanding.

Language, Land, and the Residential Schools

Source: Library and Archives Canada/PA-042133

Fort Resolution Residential School, NWT. Forced learning of English was an attempt to sever the connection between First Nations peoples and their land.

In June 2015 the Truth and Reconciliation Commission's report on the residential school system began to be released to the public. Commenting on the report, Chief Justice of the Supreme Court Beverley McLachlin (echoed by former Prime Minister Paul Martin and many others) asserted that the system represented an attempt at "cultural genocide" on the part of the Canadian government towards First Nations' peoples. This is an extremely provocative claim and in this case study we will explore one way of trying to come to grips with it.

Overcoming the status of victim cannot proceed without first understanding the precise nature of the wrongs that have been done. On 11 June 2008, Canadian prime minister Stephen Harper rose in the House of Commons to apologize to Canada's Native population for what was done to them under the auspices of the residential school system. Here is part of his speech:

Continued

Mr. Speaker, I stand before you today to offer an apology to former students of Indian residential schools. The treatment of children in Indian residential schools is a sad chapter in our history. In the 1870s, the federal government, partly in order to meet its obligation to educate aboriginal children, began to play a role in the development and administration of these schools. Two primary objectives of the residential schools system were to remove and isolate children from the influence of their homes, families, traditions and cultures, and to assimilate them into the dominant culture. These objectives were based on the assumption that aboriginal cultures and spiritual beliefs were inferior and unequal. Indeed, some sought, as it was infamously said, "to kill the Indian in the child." Today, we recognize that this policy of assimilation was wrong, has caused great harm, and has no place in our country.

What is the relevance of this apology and the history it revisits for environmental ethics? One of the key objectives of the residential school system was to force First Nations peoples to forget their languages and to learn English instead. But for First Nations peoples, there is no separation between knowledge of the land and the ways they have of naming its features in their languages. To deprive them of the language is to deprive them of the sense of place that has defined them for thousands of years. For example, as Hugh Brody reports, for many northwest coast societies, "the singing of a song and the telling of history are central expressions of a people's rights to, and their management of, territory." And for the Gitxsan and Wet'suwet'en peoples, "inheritance of land, and inheritance of the stories that establish rights to the land, are inseparable." Brody summarizes these connections this way: "[T]he ability to name is hard to separate from the right to have, use and enjoy that which is your own."

You can deprive a people of these connections by making them *ashamed* of their language. Here is Georgina Gregory, a one-time resident of File Hills Indian Residential School in Balcarres, Saskatchewan:

> There were children who would arrive now and then who could not speak English. They were ridiculed and discouraged from speaking their language and had no choice but to speak English. . . . They saw to it that those students forgot their language through humiliation and shame. . . . Why was it necessary to take away the Indian's language? Why was it necessary to make them ashamed of their culture?

Shaming—at least in its most toxic form—works by implanting a sense of deep worthlessness and self-loathing in agents in virtue of some specific practice they engage in or trait they possess. Children in residential schools were not only sexually and physically assaulted, they were also made to hate themselves for who they were. Shame is thus an attack on self-identity. But the identity of these peoples, as we have seen, is rooted in a respect for nature and its processes. Whatever else it is, this imposed self-hatred was therefore a deeply political act on the part of the authorities. It was an attempt to deprive a people of pride of place so as to loosen their attachment to that place. If successful, of course, it would have been easier to continue to deny their rights to lands that were historically theirs. It is difficult to say how the prime minister's apology will play out in the political sphere in years to come. Perhaps someone in government will realize that by apologizing to the First Nations for the residential schools, Canadians are also, implicitly at least, recognizing the legitimacy of unresolved First Nations land claims.

Sources: Brody 2000; Constance Deiter, *From Our Mother's Arms: The Intergenerational Impact of Residential Schools in Saskatchewan* (Toronto: United Church Publishing House, 1999); Prime Minister Stephen Harper in the House of Commons, 11 June 2008, from CTV news online: www.ctv.ca/CTVNews/QPeriod/20080611/harper_text_080611.

Study Questions

1. The 2014 Supreme Court's ruling in the Tsilhquot'in land dispute case seems to have altered profoundly First Nations' relations with the federal government. Effectively, the ruling gives First Nations a veto over development on lands to which they have title (and perhaps even on lands to which they merely claim title). Can you see how the ruling addresses some of the concerns laid out in this chapter's case study?

2. Explain Sioui's notion of the Amerindian idea of creation.

3. Morito presents a very vivid picture of the key elements of some First Nations cultures. What features of these cultures stand out for you and how can such features enhance your own cultural self-understanding?

4. Explain TEK. Do you think it is a useful tool for densely populated, large-scale, technocratic societies seeking to manage their relationship to nature sustainably? Why or why not?

Further Reading

Margaret Atwood. 1972. *Survival: A Thematic Guide to Canadian Literature*. Toronto: McClelland and Stewart.

Hugh Brody. 2000. *The Other Side of Eden: Hunters, Farmers, and the Shaping of the World*. Vancouver: Douglas and MacIntyre.

J. Baird Callicott. 1994. *Earth's Insights: A Survey of Ecological Insights from the Mediterranean Basin to the Australian Outback*. Berkeley: U of California P.

Peter Knudson and David Suzuki. 1992. *Wisdom of the Elders: Native and Scientific Ways of Knowing about Nature*. Vancouver: Greystone Publishers.

John Ralston Saul. 2008. *A Fair Country: Telling Truths about Canada*. Toronto: Viking Canada.

Georges E. Sioui. 1992. *For an Amerindian Autohistory*. Montreal and Kingston: McGill-Queen's UP.

CHAPTER 9

Environmental Virtue Ethics

Ecological Intuition Pump

Imagine you live in a country with a huge and richly biodiverse rainforest where trees are being cut down at an unsustainable rate. The rest of the world recognizes the value of your forest as both a repository of species diversity and a huge carbon sink. They would like your country to stop deforestation, and international negotiations are being conducted to construct a system whereby you can be paid to do so. But the negotiations are bogged down in details, and it is hard to say when they will conclude. Meanwhile, deforestation proceeds apace. Situations like this might force us to wonder to what extent we should allow our morally loaded decisions to be based on calculations about what other people are going to do. According to philosopher Dale Jamieson, one way out of this sort of trap is to develop a set of environmental virtues—call them virtues of sustainability—that dispose us to do right by the environment regardless of what others are doing. These virtuous dispositions give our characters a kind of welcome inflexibility. In the case at hand, they might incline you to refuse to wait on the outcome of the negotiations before deciding to help stop an environmentally destructive practice. You would in effect say that allowing this practice to continue is not what an environmentally virtuous *person* would do, and that we should not be paid to do the right thing in any case. But could such inflexibility lead to moral complacency? We do after all seem to prize flexibility, and being flexible is nearly synonymous with being non-complacent. How would you resolve this tension?

Ⓐ Introduction

Imagine that a powerful timber company wanted to clear-cut large expanses of a lowland tropical forest. In an effort to gain local assent to the project, the company might send its officials out to bribe politicians and tribal headmen. Compared to the massive profits available to the company if the project goes through, the money it would need to spend on these bribes is a pittance. So, in return for the loss of a large portion of an ancient and highly biodiverse forest ecosystem, the local population will receive almost nothing.[1] Quite apart from views we might hold about the blameworthiness of practices like clear-cutting or actions like bribery, stories like this can also cause us to question the *character* of the officials involved. How should we understand a reaction like this?

Virtue ethics helps us explain and justify this kind of reaction. As an approach to normative ethics, it contrasts sharply with both the deontological focus on moral duties or rules and the consequentialist focus on outcomes or consequences. More positively, virtue ethics strives to offer a richer description of our moral lives than its competitors. Its proponents believe it can do this by eschewing the traditional focus on actions or rules, focusing instead (or in addition) on the character of moral agents. Although he was not an avowed virtue ethicist, Aldo Leopold, the founder of the land ethic (examined in Chapter 3), had something like this in mind in claiming that the new environmental ethics he advocated will involve a change in our "intellectual emphasis, loyalties, affections, and convictions."[2] This collection of attributes is meant to highlight the importance of the *whole person* as the locus of moral attention and is encapsulated in the concept of a virtue.

A virtue is a disposition to act and feel in a certain manner as called for in a particular situation. So the compassionate person is not the one who happens to perform this or that compassionate action, moved by a sudden and perhaps fleeting desire to do good (when presented with television images of starving children, for example). Nor is it the person who believes that she will somehow benefit from performing compassionate actions (by gaining a certain reputation for benevolence, for example). Nor, finally, is it the person who performs compassionate actions because, and only because, she believes it is her duty to do so (whether or not she also *wants* to). The genuinely virtuous person is not motivated by sentimentalism, mere prudence, or rule worship.

Rather, when presented with a case of remediable suffering, the entire moral-psychological economy of the virtuously compassionate person will be mobilized. This means that she will, as Aristotle puts it, have the right feelings, at the right times about the right things, toward the right people, for the right end, and in the right way.[3] This formulation underscores a further key feature of the virtuous agent: her possession of practical wisdom or *phronesis*. To be fully virtuous, it is not enough to be well-intentioned. One must also be a skilled reader of moral situations, knowing what sorts of actions and feelings specific situations should elicit.

The idea that the proper focus of attention in moral assessment should be the agent, construed in this comprehensive manner, is what fundamentally separates virtue ethics from its competitor theories. Rather than asking whether or not a proposed course of action accords with a rule or will produce the best consequences (however these are specified), the virtuous agent asks whether the action is just, or courageous, or compassionate, and so on. Assuming she herself has the relevant virtue, this is really just a way of asking a more basic question: "Am I the sort of *person* who could perform this action?" The answer to *that* question, in turn, will likely open up a concrete course of action to the virtuous agent.

Ⓑ Human Excellence and the Environment

Environmental virtue ethics is a very new field. Until quite recently, it was assumed that the problems investigated by environmental ethicists could only be addressed in the traditional framework of deontology and consequentialism (the latter usually in its utilitarian guise). The precise historical point at which the virtue-ethical approach emerged into this field is difficult to pin down precisely, but many philosophers trace it to Thomas E. Hill's article "Ideals of Human Excellence and Preserving Natural Environments," to which we now turn.

IDEALS OF HUMAN EXCELLENCE AND PRESERVING NATURAL ENVIRONMENTS

Thomas E. Hill, Jr

The moral significance of preserving natural environments is not entirely an issue of rights and social utility, for a person's attitude toward nature may be importantly connected with virtues or human excellences. The question is, "What sort of person would destroy the natural environment—or even see its value solely in cost/benefit terms?" The answer I suggest is that willingness to do so may well reveal the absence of traits that are a natural basis for a proper humility, self-acceptance, gratitude, and appreciation of the good in others.

A wealthy eccentric bought a house in a neighbourhood I know. The house was surrounded by a beautiful display of grass, plants, and flowers, and it was shaded by a huge old avocado tree. But the grass required cutting, the flowers needed tending, and the man wanted more sun. So he cut the whole lot down and covered the yard with asphalt. After all, it was his property, and he was not fond of plants. It was a small operation, but it reminded me of the strip mining of large sections of the Appalachians. In both cases, of course, there were reasons for the destruction, and property rights could be cited as justification. But I could not help but wonder, "What sort of person would do a thing like that?" . . .

Incidents like these arouse the indignation of ardent environmentalists and leave even apolitical observers with some degree of moral discomfort. The reasons for these reactions are mostly obvious. Uprooting the natural environment robs both present and future generations of much potential use and enjoyment. Animals too depend on the environment, and even if one does not value animals for their own sakes, their potential utility for us is incalculable.

Plants are needed, of course, to replenish the atmosphere, quite aside from their aesthetic value. These reasons for hesitating to destroy forests and gardens are not only the most obvious ones but also the most persuasive for practical purposes. But, one wonders, is there nothing more behind our discomfort? Are we concerned solely about the potential use and enjoyment of the forests, etc., for ourselves, later generations, and perhaps animals? Is there not something else that disturbs us when we witness the destruction or even listen to those who would defend it in terms of cost/benefit analysis?

Imagine that in each of our examples those who would destroy the environment argue elaborately that, even considering future generations of human beings and animals, there are benefits in "replacing" the natural environment that outweigh the negative utilities that environmentalists cite. No doubt we could press the argument on the facts, trying to show that the destruction is short-sighted and that its defenders have underestimated its potential harm or ignored some pertinent rights or interests. But is this all we could say? Suppose we grant, for a moment, that the utility of destroying the redwoods, forests, and gardens is equal to their potential for use and enjoyment by nature lovers and animals. Suppose, further, that we even grant that the pertinent human rights and animal rights, if any, are evenly divided for and against destruction. Imagine that we also concede, for argument's sake, that the forests contain no potentially useful endangered species of animals and plants. Must we then conclude that there is no further cause for moral concern? Should we then feel morally indifferent when we see the natural environment uprooted?

Suppose we feel that the answer to these questions should be negative. Suppose, in other words, we feel that our moral discomfort when we confront the destroyers of nature is not fully explained by our belief that they have miscalculated the best use of natural resources or violated rights in exploiting them. Suppose, in particular, we sense that part of the problem is that the natural environment is being viewed exclusively as a natural resource. What could be the ground of such a feeling? That is, what is there in our system of normative principles and values that could account for our remaining moral dissatisfaction?

Some may be tempted to seek an explanation by appeal to the interests, or even the rights, of plants. After all, they may argue, we only gradually came to acknowledge the moral importance of all human beings, and it is even more recently that consciences have been aroused to give full weight to the welfare (and rights?) of animals. The next logical step, it may be argued, is to acknowledge a moral requirement to take into account the interests (and rights?) of plants. The problem with the strip miners, redwood cutters, and the like, on this view, is not just that they ignore the welfare and rights of people and animals, they also fail to give due weight to the survival and health of the plants themselves.

The temptation to make such a reply is understandable if one assumes that all moral questions are exclusively concerned with whether acts are right or wrong and that this, in turn, is determined entirely by how the acts impinge on the rights and interests of those directly affected. On this assumption, if there is cause for moral concern, some right or interest has been neglected, and if the rights and interests of human beings and animals have already been taken into account, then there must be some other pertinent interests, for example, those of plants. A little reflection will show that the assumption is mistaken, but in any case, the conclusion that plants have rights or morally relevant interests is surely untenable. We do speak of what is "good for" plants, and they can "thrive" and also be "killed." But this does not imply that they have "interests" in any morally relevant sense. . . .

Early in this century, due largely to the influence of G.E. Moore, another point of view developed that some may find promising.[4] Moore introduced, or at least made popular, the idea that certain states of affairs are intrinsically valuable—not just valued but valuable, and not necessarily because of their effects on sentient beings. Admittedly, Moore came to believe that in fact the only intrinsically valuable things were conscious experiences of various sorts,[5] but this restriction was not inherent in the idea of intrinsic value. The intrinsic goodness of something, he thought, was an objective, non-relational property of the thing, like its texture or colour, but not a property perceivable by sense perception or detectable by scientific instruments. In theory at least, a single tree thriving alone in a universe without sentient beings, and even without God, could be intrinsically valuable. Since, according to Moore, our duty is to maximize intrinsic value, his theory could obviously be used to argue that we have reason not to destroy natural environments independently of how they affect human beings and animals. The survival of a forest might have worth beyond its worth to sentient beings.

This approach . . . may appeal to some but is infested with problems. There are, first, the familiar objections to intuitionism, on which the theory depends. Metaphysical and epistemological doubts about non-natural, intuited properties are hard to suppress, and many have argued that the theory rests on a misunderstanding of the words *good*, *valuable*, and the like.[6] Second, even if we try to set aside these objections and think in Moore's terms, it is far from obvious that everyone would agree that the existence of forests, etc., is intrinsically valuable. The test, says Moore, is what we would say when we imagine a universe with just the thing in question, without any effects or accompaniments, and then we ask, "Would its existence be better than its non-existence?" Be careful, Moore would remind us, not to construe this question as "Would you prefer the existence of that universe to its non-existence?" The question is "Would its existence have the objective, non-relational property, intrinsic goodness?"

Now, even among those who have no worries about whether this really makes sense, we might well get a diversity of answers. Those prone to destroy natural environments will doubtless give one answer, and nature lovers will likely give another. When an issue is as controversial as the one at hand, intuition is a poor arbiter.

The problem, then, is this. We want to understand what underlies our moral uneasiness at the destruction of the redwoods, forests, etc., even apart from the loss of these as resources for human beings and animals. But I find no adequate answer by pursuing the questions: "Are rights or interests of plants neglected?" . . .

[or] "What is the intrinsic value of the existence of a tree or forest?" My suggestion, which is in fact the main point of this paper, is that we look at the problem from a different perspective. That is, let us turn for a while from the effort to find reasons why certain acts destructive of natural environments are morally wrong to the ancient task of articulating our ideals of human excellence. Rather than argue directly with destroyers of the environment who say, "Show me why what I am doing is immoral," I want to ask, "What sort of person would want to do what they propose?" The point is not to skirt the issue with an ad hominem, but to raise a different moral question, for even if there is no convincing way to show that the destructive acts are wrong (independently of human and animal use and enjoyment), we may find that the willingness to indulge in them reflects the absence of human traits that we admire and regard as morally important. . . .

What sort of person, then, would cover his garden with asphalt, strip mine a wooded mountain, or level an irreplaceable redwood grove? Two sorts of answers, though initially appealing, must be ruled out. The first is that persons who would destroy the environment in these ways are either short-sighted, underestimating the harm they do, or else are too little concerned for the well-being of other people. Perhaps, too, they have insufficient regard for animal life. But these considerations have been set aside in order to refine the controversy. Another tempting response might be that we count it a moral virtue, or at least a human ideal, to love nature. Those who value the environment only for its utility must not really love nature and so in this way fall short of an ideal. But such an answer is hardly satisfying in the present context, for what is at issue is why we feel moral discomfort at the activities of those who admittedly value nature only for its utility.

That it is ideal to care for non-sentient nature beyond its possible use is really just another way of expressing the general point that is under controversy.

What is needed is some way of showing that this ideal is connected with other virtues, or human excellences, not in question. To do so is difficult, and my suggestions, accordingly, will be tentative and subject to qualification. The main idea is that though indifference to non-sentient nature does not necessarily reflect the absence of virtues, it often signals the absence of certain traits that we want to encourage because they are, in most cases, a natural basis for the development of certain virtues. It is often thought, for example, that those who would destroy the natural environment must lack a proper appreciation of their place in the natural order and so must either be ignorant or have too little humility. Though I would argue that this is not necessarily so, I suggest that, given certain plausible empirical assumptions, their attitude may well be rooted in ignorance, a narrow perspective, inability to see things as important apart from themselves and the limited groups they associate with, or reluctance to accept themselves as natural beings. Overcoming these deficiencies will not guarantee a proper moral humility, but for most of us it is probably an important psychological preliminary. . . .

Consider first the suggestion that destroyers of the environment lack an appreciation of their place in the universe. Their attention, it seems, must be focused on parochial matters, on what is, relatively speaking, close in space and time. They seem not to understand that we are a speck on the cosmic scene, a brief stage in the evolutionary process, only one among millions of species on Earth, and an episode in the course of human history. Of course, they know that there are stars, fossils, insects, and ancient ruins, but do they have any idea of the complexity of the processes that led to the natural world as we find it? Are they aware how much the forces at work within their own bodies are like those that govern all living things and even how much they have in common with inanimate bodies? Admittedly, scientific knowledge is limited, and no one can master it all, but could one who had a broad and deep understanding of his place in nature really be indifferent to the destruction of the natural environment?

This first suggestion, however, may well provoke a protest from a sophisticated anti-environmentalist. "Perhaps some may be indifferent to nature from ignorance," the critic may object, "but I have studied astronomy, geology, biology, and biochemistry, and I still unashamedly regard the non-sentient environment as simply a resource for our use. It should not be wasted, of course, but what should be preserved is decidable by weighing long-term costs and benefits." "Besides," our critic may continue, "as philosophers you should know the old Humean formula, 'You cannot derive an ought from an is.' All the facts of biology, biochemistry, etc., do not entail that I ought to love nature or want to preserve it. What one understands is one thing; what one values is something else. Just as nature lovers are not necessarily scientists, those indifferent to nature are not necessarily ignorant."

Although the environmentalist may concede the critic's logical point, he may well argue that, as a matter of fact, increased understanding of nature tends to heighten people's concern for its preservation. If so, despite the objection, the suspicion that the destroyers of the environment lack deep understanding of nature is not, in most cases, unwarranted, but the argument need not rest here.

The environmentalist might amplify his original idea as follows: "When I said that the destroyers of nature do not appreciate their place in the universe, I was not speaking of intellectual understanding alone, for, after all, a person can know a catalogue of facts without ever putting them together and seeing vividly the whole picture that they form. To see one-self as just one part of nature is to look at one-self and the world from a certain perspective, which is quite different from being able to recite detailed information from the natural sciences. What the destroyers of nature lack is this per-spective, not particular information."

Again our critic may object, though only after making some concessions: "All right," he may say, "some who are indifferent to nature may lack the cosmic perspective of which you speak, but again there is no necessary connec-tion between this failing, if it is one, and any particular evaluative attitude toward nature. In fact, different people respond quite differ-ently when they move to a wider perspective. When I try to picture myself vividly as a brief, transitory episode in the course of nature, I sim-ply get depressed. Far from inspiring me with a love of nature, the exercise makes me sad and hostile. . . ." In sum, the critic may object, "Even if one should try to see oneself as one small, transitory part of nature, doing so does not dictate any particular normative attitude. Some may come to love nature, but others are moved to live for the moment; some sink into sad resignation; others get depressed or angry. So indifference to nature is not necessarily a sign that a person fails to look at himself from the larger perspective."

The environmentalist might respond to this objection in several ways. He might, for example, argue that even though some people who see themselves as part of the natural order remain indifferent to non-sentient nature, this is not a common reaction. Typically, it may be argued, as we become more and more aware

that we are parts of the larger whole, we come to value the whole independently of its effect on ourselves. Thus, despite the possibilities the critic raises, indifference to non-sentient nature is still in most cases a sign that a person fails to see himself as part of the natural order.

If someone challenges the empirical assump-tion here, the environmentalist might develop the argument along a quite different line. The initial idea, he may remind us, was that those who would destroy the natural environment fail to appreciate their place in the natural order. "Appreciating one's place" is not simply an intellectual appreciation. It is also an attitude, reflecting what one values as well as what one knows. When we say, for example, that both the servile and the arrogant person fail to appreci-ate their place in a society of equals, we do not mean simply that they are ignorant of certain empirical facts but rather that they have certain objectionable attitudes about their importance relative to other people. Similarly, to fail to appreciate one's place in nature is not merely to lack knowledge or breadth of perspective but to take a certain attitude about what matters. A person who understands his place in nature but still views non-sentient nature merely as a resource takes the attitude that nothing is important but human beings and animals. . . .

So construed, the argument appeals to the common idea that awareness of nature typically has, and should have, a humbling effect. The Alps, a storm at sea, the Grand Canyon, towering redwoods, and "the starry heavens above" move many a person to remark on the comparative insignificance of our daily concerns and even of our species, and this is generally taken to be a quite fitting response. What seems to be missing, then, in those who understand nature but remain unmoved is a proper humility. Absence of proper humility is not the same as selfishness or ego-ism, for one can be devoted to self-interest while still viewing one's own pleasures and projects as trivial and unimportant. And one can have an exaggerated view of one's own importance while grandly sacrificing for those one views as inferior. Nor is the lack of humility identical with belief that one has power and influence, for a person can be quite puffed up about himself while believing that the foolish world will never acknowledge him. The humility we miss seems not so much a belief about one's relative effectiveness and recognition as an attitude that measures the importance of things independently of their relation to oneself or to some narrow group with which one identifies. . . . The suspicion about those who would destroy the environment, then, is that what they count important is too narrowly confined insofar as it encompasses only what affects beings who, like us, are capable of feeling.

But however intuitively appealing, the idea will surely arouse objections from our non-environmentalist critic. "Why," he will ask, "do you suppose that the sort of humility I should have requires me to acknowledge the importance of non-sentient nature aside from its utility? You cannot, by your own admission, argue that non-sentient nature is important, appealing to . . . intuitionist grounds. And simply to assert, without further argument, that an ideal humility requires us to view non-sentient nature as important for its own sake begs the question at issue. If proper humility is acknowledging the relative importance of things as one should, then to show that I must lack this, you must first establish that one should acknowledge the importance of non-sentient nature."

Though some may wish to accept this challenge, there are other ways to pursue the connection between humility and response to non-sentient nature. For example, suppose we grant that proper humility requires only acknowledging a due status to sentient beings. We must admit, then, that it is logically possible for a person to be properly humble even though he views all non-sentient nature simply as a resource. But this logical possibility may be a psychological rarity. It may be that, given the sort of beings we are, we would never learn humility before persons without developing the general capacity to cherish, and regard important, many things for their own sakes. The major obstacle to humility before persons is self-importance, a tendency to measure the significance of everything by its relation to oneself and those with whom one identifies. The processes by which we overcome self-importance are doubtless many and complex, but it seems unlikely that they are exclusively concerned with how we relate to other people and animals. . . .

This last argument, unfortunately, has its limits. It presupposes an empirical connection between experiencing nature and overcoming self-importance, and this may be challenged. Even if experiencing nature promotes humility before others, there may be other ways people can develop such humility in a world of concrete, glass, and plastic. If not, perhaps all that is needed is limited experience of nature in one's early, developing years; mature adults, having overcome youthful self-importance, may live well enough in artificial surroundings. More importantly, the argument does not fully capture the spirit of the intuition that an ideal person stands humbly before nature. That idea is not simply that experiencing nature tends to foster proper humility before other people; it is, in part, that natural surroundings encourage and are appropriate to an ideal sense of oneself as part of the natural world. Standing alone in the forest, after months in the city, is not merely good as a means of curbing one's arrogance before others; it reinforces and fittingly expresses one's acceptance of oneself as a natural being.

Previously, we considered only one aspect of proper humility, namely, a sense of one's relative importance with respect to other human beings. Another aspect, I think, is a kind of self-acceptance. This involves acknowledging, in more than a merely intellectual way, that we are the sort of creatures that we are. Whether one is self-accepting is not so much a matter of how one attributes importance comparatively to oneself, other people, animals, plants, and other things as it is a matter of understanding, facing squarely, and responding appropriately to who and what one is—for example, one's powers and limits, one's affinities with other beings and differences from them, one's unalterable nature, and one's freedom to change. Self-acceptance is not merely intellectual awareness, for one can be intellectually aware that one is growing old and will eventually die while nevertheless behaving in a thousand foolish ways that reflect a refusal to acknowledge these facts.

One fails to accept oneself when . . . [one's] patterns of behaviour and emotion are rooted in a desire to disown and deny features of oneself, to pretend to oneself that they are not there. This is not to say that a self-accepting person makes no value judgments about himself, that he likes all facts about himself, wants equally to develop and display them; he can, and should, feel remorse for his past misdeeds and strive to change his current vices. The point is that he does not disown them, pretend that they do not exist or are facts about something other than himself. Such pretence is incompatible with proper humility because it is seeing oneself as better than one is.

Self-acceptance of this sort has long been considered a human excellence, under various names, but what has it to do with preserving nature? There is, I think, the following connection. As human beings we are part of nature, living, growing, declining, and dying by natural laws similar to those governing other living beings; despite our awesomely distinctive human powers, we share many of the needs, limits, and liabilities of animals and plants. These facts are neither good nor bad in themselves, aside from personal preference and varying conventional values. To say this is to utter a truism that few will deny, but to accept these facts, as facts about oneself, is not so easy—or so common. Much of what naturalists deplore about our increasingly artificial world reflects, and encourages, a denial of these facts, an unwillingness to avow them with equanimity.

My suggestion is not merely that experiencing nature causally promotes such self-acceptance but also that those who fully accept themselves as part of the natural world lack the common drive to disassociate themselves from nature by replacing natural environments with artificial ones. A storm in the wilds helps us to appreciate our animal vulnerability, but, equally important, the reluctance to experience it may reflect an unwillingness to accept this aspect of ourselves. The person who is too ready to destroy the ancient redwoods may lack humility, not so much in the sense that he exaggerates his importance relative to others but rather in the sense that he tries to avoid seeing himself as one among many natural creatures.

My suggestion so far has been that, though indifference to non-sentient nature is not itself a moral vice, it is likely to reflect either ignorance, a self-importance, or a lack of self-acceptance, which we must overcome to have proper humility. A similar idea might be developed connecting attitudes toward non-sentient nature with other human excellences. For example, one might argue that indifference to nature reveals a lack of either an aesthetic sense or some of the natural roots of gratitude. . . .

The anti-environmentalist, however, may refuse to accept the charge that he lacks aesthetic sensibility. If he claims to appreciate seventeenth-century miniature portraits but to abhor natural wildernesses, he will hardly be convincing. Tastes vary, but aesthetic sense is not that selective. He may, instead, insist that he does appreciate natural beauty. He spends his vacations, let us suppose, hiking in the Sierras, photographing wildflowers, and so on. He might press his argument as follows: "I enjoy natural beauty as much as anyone, but I fail to see what this has to do with preserving the environment independently of human enjoyment and use. Non-sentient nature is a resource, but one of its best uses is to give us pleasure. I take this into account when I calculate the costs and benefits of preserving a park, planting a garden, and so on. But the problem you raised explicitly set aside the desire to preserve nature as a means to enjoyment. I say, let us enjoy nature fully while we can, but if all sentient beings were to die tomorrow, we might as well blow up all plant life as well. A redwood grove that no one can use or enjoy is utterly worthless."

The attitude expressed here, I suspect, is not a common one, but it represents a philosophical challenge. The beginnings of a reply may be found in the following. When a person takes joy in something, it is a common (and perhaps natural) response to come to cherish it. To cherish something is not simply to be happy with it at the moment but to care for it for its own sake. This is not to say that one necessarily sees it as having feelings and so wants it to feel good; nor does it imply that one judges the thing to have Moore's intrinsic value. One simply wants the thing to survive and (when appropriate) to thrive, and not simply for its utility. We see this attitude repeatedly regarding mementos. They are not simply valued as a means to remind us of happy occasions; they come to be valued for their own sake. Thus, if someone really took joy in the natural environment but was prepared to blow it up as soon as sentient life ended, he would lack this common human tendency to cherish what enriches our lives.

While this response is not itself a moral virtue, it may be a natural basis of the virtue we call "gratitude." People who have no tendency to cherish things that give them pleasure may be poorly disposed to respond gratefully to persons who are good to them. Again, the connection is not one of logical necessity, but it may nevertheless be important. A non-religious person unable to "thank" anyone for the beauties of nature may nevertheless feel "grateful" in a sense, and I suspect that the person who feels no such "gratitude" toward nature is unlikely to show proper gratitude toward people.

Suppose these conjectures prove to be true. One may wonder what is the point of considering them. Is it to disparage all those who view nature merely as a resource? To do so, it seems, would be unfair, for even if this attitude typically stems from deficiencies that affect one's attitudes toward sentient beings, there may be exceptions and we have not shown that their view of non-sentient nature is itself blameworthy. But when we set aside questions of blame and inquire what sorts of human traits we want to encourage, our reflections become relevant in a more positive way. The point is not to insinuate that all anti-environmentalists are defective but to see that those who value such traits as humility, gratitude, and sensitivity to others have reason to promote the love of nature.

Suppose you were a governmental policy-maker forced to choose between the preservation or the destruction of a certain natural landscape. Perhaps it contains an economically valuable lode of minerals but is also home to a rich diversity of species whose existence would be threatened by development. How would you choose? Preservation might often seem like the best course, even though it may be difficult to justify purely in terms of cost-benefit analysis (see Chapter 4 for more on this concept). Hill claims that we should in many of these cases opt for preservation but not because what we propose to preserve has "rights" or intrinsically valuable properties that we can somehow "intuit" (see Chapter 5 for a critique of the notion of intrinsic value). The preservationist idea that non-sentient nature is important for its own sake is a good one, but these are philosophically unacceptable ways of justifying it. Instead, Hill's proposal is that valuing nature in the right way is connected to other *human excellences* in certain ways such that to disvalue nature, by destroying it, is to lack these excellences. What are these excellences and their corresponding shortcomings? Two of them stand out.

First, Hill argues that the non-environmentalist—that is, the person who sees nature strictly as a resource to be used as we see fit, even if this involves massive destruction of the natural environment—lacks *proper humility*. Although it is logically possible for this type of person to value non-sentient nature for its own sake, Hill claims that this is a "psychological rarity." More typically, such a person will be objectionably self-important. Closely connected to this idea is the thought that the non-environmentalist also fails properly to accept his *own* finitude, needs, and limitations as a "living, growing, declining, and dying" being. How are these psychological shortcomings related to appreciation of nature for its own sake? Hill's central claim here is that such appreciation (a) contributes causally to being properly humble and self-accepting and (b) prevents people from engaging in actions that destroy non-sentient nature.

Although Hill does not talk about it directly, the vice that corresponds to the virtue of humility is *hubris*, a Greek term meaning improper pride or arrogance. The environmental literature is replete with references to this vice and to the corresponding virtue, an indication that Hill has hit on something central to our way of viewing our ideal relation to the natural world. For example, throughout his career, Canadian environmentalist and scientist David Suzuki has been an outspoken critic of what he sees as our excessive trust in the scientific approach to nature. Science has provided us with many benefits and insights, but the over-reliance on it has blocked alternative ways of knowing nature, especially those contained in the wise practices of the world's Aboriginal communities. Were we to attend more carefully to these ways of knowing nature, we might become more humble:

> Science . . . allows us to extract great detail by sacrificing that sense of nature's vast breadth and immeasurable complexity. In the Monteverde cloud forest, one is overwhelmed by the immensity of our ignorance, a sense of humility about our abilities, and a reverence for nature, which put our sense of achievement into perspective.[7]

For Aldo Leopold, the ability to "see the cultural value of wilderness boils down, in the last analysis, to a question of intellectual humility."[8] And in her environmental classic, *Silent*

Spring, which brought the dangers of chemical insecticides like DDT to public awareness for the first time, Rachel Carson sounds a similar note:

> The "control of nature" is a phrase conceived in arrogance, born of the Neanderthal age of biology and philosophy, when it was supposed that nature exists for the convenience of man.... [The] extraordinary capacities of life have been ignored by the practitioners of chemical control who have brought to their task ... no humility before the vast forces with which they tamper.[9]

If Hill, Suzuki, Leopold, and Carson are right, then, when we object to the human destruction of non-sentient nature—think again of those hypothetical timber company officials—a significant aspect of our objection is directed at the hubristic character of the destroyer herself. So virtue ethics is meant to explain a certain type of reaction to moral failing. The virtue ethicist believes that we are repelled by the havoc wreaked on natural landscapes by actions like this in large measure *because* we think that those who are responsible for them are, among other things, lacking in proper humility toward nature. The moral assessment of character in a statement like this need not be subordinate to the analysis of negative consequences or the discovery that some moral rule has been breached. It is normatively primary.

The second excellence Hill isolates is *gratitude*. As he puts it, "if someone really took joy in the natural environment but was prepared to blow it up as soon as sentient life ended, he would lack [the] common human tendency to cherish what enriches our lives." There is an incompatibility between an attitude of respect for non-sentient nature and the willingness to think of the latter as a mere resource. Instead, such respect should engender the desire to see the respected object "survive ... and thrive." Here, in the abstract, we are presented with an agent who seems to have at least some of the proper attitudes toward nature. There is, after all, the recognition on the part of this agent that pristine nature can "enrich" or bring "joy" to our lives. And yet the agent performs actions that destroy nature. What has gone wrong here? Here is an example that might help us make sense of this.

In the Yakoun River Valley of British Columbia's Haida Gwaii (the Queen Charlotte Islands), there once stood a magnificent Sitka spruce tree, 16 storeys tall and more than six metres around. But it was not its size that distinguished this tree from the others around it; it was its colour. The tree was golden. This was so rare that it received its own name: *Picea sitchensis aurea*, the Latin name for the Sitka spruce plus the addition "aurea," Latin for "golden" or "gleaming like gold" (but also "beautiful" or "splendid"). This "arboreal unicorn" had stood since approximately 1700, achieving during that time a central place in the mythology of the local Aboriginal group, the Haida Nation.[10] They too had their own word for the tree, K'iid K'iyaas (Elder Spruce Tree). The tree was so iridescent that it could be seen from several kilometres away.

On a winter night in 1997, Grant Hadwin, a disillusioned timber prospector, cut the Golden Spruce down. In his own words, he "butchered" this "magnificent old plant."[11] He did this as a form of protest against what he saw as a predatory forestry industry and also the "university trained professionals" who worked for and advised big logging companies like MacMillan

Bloedel. Hadwin believed that these companies treated the Golden Spruce as a "pet plant." By advertising their refusal to cut down this particular tree, MacMillan Bloedel could paint itself—falsely, in Hadwin's view—as a wise and sensitive steward of the forest and its resources. Of course, very few people were sympathetically moved by Hadwin's "message." Most were profoundly shocked and disgusted by it. In a press release, the Council of the Haida Nation described the death of K'iid K'iyaas as "a deliberate violation of our cultural history."[12] Although it is possible to assess Hadwin's act by looking at its negative consequences (especially for the Haida Nation) or by citing some rule he might have breached in cutting down the tree, it might be more natural and illuminating to ask Hill's question: "What kind of *person* would do a thing like this?"

At first glance, it would seem that this case does not sit particularly well with Hill's analysis of environmental excellences. After all, Hill's target is the person who conceives of the natural world strictly as a resource. Indeed, we could personify Hill's agent as a spokesperson for MacMillan Bloedel, someone with the same sort of mindset at which Hadwin's destructive act was symbolically aimed. But there is a deeper similarity between Hill's non-environmentalist and Hadwin. Is it not the case that our reaction to Hadwin is one of moral disgust or outrage? If it is and we ask what grounds this reaction, we will be directed to Hadwin's moral character. At this point, we will likely want to say that in destroying a piece of non-sentient nature of surpassing beauty and cultural significance, Hadwin was above all *ungrateful* to a natural world he professed to cherish. If so, then the most appropriate way to assess Hadwin morally may be to say that he was environmentally vicious: he was ungrateful, egotistical, and arrogant.

But one of the appealing features of virtue ethics is that it offers us another option in cases like this. It may be that we want to describe Hadwin as lacking *phronesis*. Although he really did believe that nature ought to be respected, perhaps he simply failed to understand how best to translate this respect into action, so he behaved like a well-intentioned but impulsive child. This is clearly different from being motivated by disrespect for nature or the Haida (or both) in cutting down the tree. The analogy with the child is apt as far as it goes, but we must be careful employing it. The key difference between such a child and an adult like Hadwin is that in his case we likely want to say that he *should have* known better and is thus fully blameworthy for what he did (unlike the child). We would require more detail about the case in order to decide how best to describe and assess Hadwin's action. But however we decide this issue, at the level of character assessment there will be no very dramatic difference between Hadwin and the officials he so despised. Both have blameworthy dispositions regarding non-sentient nature.

These considerations reveal the promise of the virtue ethical approach to environmental issues. Although, as Hill shows, it provides an effective way to counter the narrow nature-as-resource view, it also gives us insight into what sorts of positive attitudes and dispositions we might cultivate toward sentient and non-sentient nature. Furthermore, Hadwin's case shows us that we should be alert to the possibility that the best way to describe a particular moral failing *might* need to make reference to an agent's (often blameworthy) lack of practical wisdom rather than, or in addition to, his vicious dispositions. As far as cultivating the proper set of dispositions toward nature is concerned, the story of the Golden Spruce suggests that the Haida Nation might have it right. Both K'iid K'iyaas in particular and nature generally are regarded by them with humility, respect, and the kind of gratitude one displays to beings that provide crucial physical and cultural sustenance.

Ⓒ Epistemic Environmental Virtues

So far, we have considered only moral virtues, those having to do ultimately with the moral-psychological springs of action. But philosophers have been turning increasingly to the analysis of "intellectual" or "epistemic" virtues as well. These have to do with what traits or dispositions are best suited to the ideal organization of our beliefs about the world. Since we have many beliefs about the environment, the new focus by epistemologists on epistemic virtue and vice should apply nicely to this area of our thinking. Our next selection is the first attempt in the philosophical literature to make the connection.

EPISTEMIC VIRTUE AND THE ECOLOGICAL CRISIS

Byron Williston

I. Introduction

When Oklahoma Senator James Inhofe declared on the United States Senate floor that "climate change is the greatest hoax ever perpetrated on the American people,"[13] many of us were inclined to wonder, "What sort of person could believe a thing like that?" Thomas E. Hill Jr provides the original motivation for this kind of question applied to our environmental concerns and for the general move to a virtue-oriented approach to environmental ethics. Hill points out that sometimes, when contemplating the blameworthy actions of others (or ourselves for that matter), we are tempted to ask the question, "What sort of person would do a thing like that?" The question is meant in the first place to refocus the question of moral wrongness. Whereas traditionally we focus on the action (asking whether anyone's rights were violated in its performance, for instance), we are now asking a question about the moral character of the agent who performed the action. By analogy, our reaction to Inhofe's egregiously stupid comment suggests that we are sometimes concerned not only with the propositions to which a per-

son assents—his beliefs—but also, and perhaps more pointedly, with that person's underlying epistemic character. This paper enlarges on this notion in the context of our ecological crisis.

II. Motivating the Focus on Epistemic Virtue

Some epistemologists have recently criticized their subdiscipline for its exclusive and narrow focus on the properties of propositions—truth, warrant, and justification. As an alternative, or supplement, to this traditional focus they suggest looking to the agents who entertain propositions, put them forward as putatively accurate depictions of reality, debate them with others, and so on.[14] Here is the thumbnail version of what is behind this new approach, applied to a person whose epistemic house appears to be in order. There are many people who have more true beliefs than false ones but who pick up the former and avoid the latter quite haphazardly. Think, by contrast, of a person who displays both a dogged love of truth and an unusually high level of competence in ordering her beliefs in accordance with accepted standards of evidence and logic. This person seems to care more than most about both getting things right and disseminating her beliefs responsibly and respectfully. She fights for the truth, is always willing to debate important topics, gives her assent to propositions even when it makes her unpopular to do so, and is open-minded to and respectful of challenges to her claims (up to a point).

Many details could be added to this picture. For instance, this person also likely possesses moral courage as well as the proper balance between humility and autonomy. Note also that the same general picture could be drawn about the opposite sort of person, someone who didn't much care about discovering and disseminating the truth. Here too we would need supplementary detail: such a person might be lazy or arrogant, for instance. The point is that to the extent that we make rough and ready distinctions among people on the basis of the efforts they make to get their beliefs to match the world, we should have little trouble in endorsing the notion that there are epistemic agents and that their traits or dispositions are proper objects of moral evaluation. The approach is illuminating for all epistemic performances but especially for cases of either superlative epistemic achievement or abysmal epistemic failure. Inhofe certainly fits the latter category, but the rest of this paper is going to focus on a somewhat less dramatic (though no less worrying) type of case.

My aim here is to employ the theoretical model just sketched to help us understand and, hopefully, move some distance toward overcoming our ecological crisis. The most important general feature of this crisis is that it is the product of multiple, and often overlapping, collective action problems. The collectives that concern us come in all sizes and degrees of cohesion—from a small group managing an inland fishery to multiple individuals consuming competitively and nations negotiating greenhouse gas emissions targets—but they have a recurring structure. A group of agents must cooperate to achieve an optimal outcome, but each agent, pursuing her perceived self-interest, acts in a way that—when all or even a critical mass act this way—results in the nonachievement of this outcome. The only reliable way to avoid this outcome is for all to agree to a system in which everyone *constrains* himself or herself in accordance with the demands of the optimal outcome.

This is extraordinarily difficult to achieve in practice but "suboptimal" results are not inevitable. Nobel Prize–winning economist Elinor Ostrom has shown that in some cases—she discusses certain inland fisheries, for example—agents have a good chance of avoiding it. In these cases, the agents in question have devised strategies of cooperation that seem to work over relatively long periods of time.[15] But where larger and less cohesive collectives are involved, suboptimal outcomes appear to be the norm. This is rarely intended by any of the individual agents involved in decision-making processes, so how does it arise? Unfortunately, these structures invite knavery (free-riding). Because any agent will benefit if everyone else constrains while she does not, and everyone knows this, all are on the lookout for knaves or are tempted by knavery themselves. These structures are thus permeated by decisional anxiety. While caught up in them, we are prompted mainly by the desire to get ahead of the next guy and the fear that he will get ahead of us.

Such psychological forces are, in turn, abetted by a reactive and calculating cognitive structure.[16] Rather than defining ourselves as just or compassionate people and letting these moral dispositions guide our choices, we wait to see how the other players are going to act and make our decisions on this basis. When everyone operates this way, the result is bad for everyone. Now the ultimate "loser" in many of these scenarios is the environment. The forces just analyzed are the main cause of the degradation of the atmosphere, our fisheries, soils, waterways, etc. For example, countries are reluctant to sign on to aggressive measures to constrain their carbon emissions for fear that in doing so they will lose one more year (or other fixed period) of economic growth while other countries increase (or don't decrease) their emissions. The result? Nobody reduces and the atmospheric (as well as oceanic) commons is degraded. So, as we examine how to structure our beliefs in the age of climate change, massive biodiversity loss, the steady depletion of freshwater resources, the acidification of our oceans, and the poisoning of agricultural lands, we must not lose sight of this social-structural backdrop (i.e., the problem of collective action).

We are, as it stands, quite incompetent at acting in large groups, and yet if we are going to place morally responsible controls on our large-scale technological interventions into the natural world, we cannot avoid collective action. How should we reconceive ourselves so as to help produce better outcomes to such challenges? One recent suggestion is that we should cultivate certain virtues of cooperativeness that impart a measure of firmness to our characters. This way, when we find ourselves in collective action scenarios, we won't be tempted to engage in the sort of calculative, wait-and-see behaviour that produces suboptimal outcomes. For example, by definition a just person recognizes the legitimate claims of others to their fair share of scarce resources. And she will not act in ways that run counter to the demands of this virtue, *no matter what other agents in the collective do.*[17]

Since from an epistemic standpoint we are interested in what sorts of beliefs are typically operative in collective action problems, we

should, by analogy, examine what a virtue epistemology for such problems might look like. Collective action scenarios are dense with information. The fishermen studied by Ostrom need to have accurate and timely access to knowledge both about the physical resource (fish stocks, etc.) and about the unique perspectives of other agents in the enterprise (maybe someone fishing a far shore has found evidence of an invasive species, for example). They also need to have accurate beliefs about whether or not the institutions and rules they have crafted for organizing the collective are up to the task. If they are not, the agents should be open to reforming those institutions and rules. Finally, it is crucial that each agent not allow self-serving bias to cloud his or her thinking in any of these areas, and that information can be readily pooled for common use. We need epistemic virtues *of* cooperation in order to achieve these things.

Now, the most environmentally challenging problems we face arise because of the way in which our collective decisions and actions are constructed through and by our high-tech, high-energy, neo-liberal capitalist economy. So I'm going to suggest that the foundational epistemic issue has to do with the beliefs we have in this broad area. Thus, following on the claims made in the previous paragraph about the importance of accurate information, I'll focus on beliefs about (a) how science and the economy should interact; (b) whether our current political and economic institutions are adequate to our challenges; and (c) the extent to which it is rational to trust other agents about what to believe.

III. Epistemic Character in the Capitalist Economy

Over the years, I've had discussions with hundreds of business students and professionals and while many of them are perfectly charming people, I sometimes wonder about the world view they are absorbing in their educational programs and workplaces. They appear to admire entrepreneurial success above all other goods and assume that in the competition for social rewards anything they can get away with goes. They think it is much more important to be "innovative" than good or truthful. In fact, morality seems at best an afterthought for them, providing a sometimes useful vocabulary for dressing up their actions in socially acceptable garb. They appear most concerned not with who they are but with how others perceive them, and so they are constantly massaging their self-images to please this or that crowd. They are both unapologetically aggressive in pursuit of material and reputational success and unreflectively conformist regarding the economic values of the larger social world. Because we can safely generalize this description beyond these people, this then becomes an issue about the goals and challenges of moral education in the context of a globalized and hypertrophied business-first culture.

From this perspective it makes sense to ask whether or not we *should* be educating people this way in the age of ecological crisis. To focus the problem, let's look at Joel Kovel's portrait of what it takes to get ahead in the capitalist economy. This strikes me as a pretty good, if incomplete, description of the type of agent under consideration:

> To succeed in the market place and to rise to the top, one needs a hard, cold, calculating mentality, the ability to sell oneself, and a hefty dose of the will to power.[18]

Think of a person taking this recipe for success to heart. Let's call him Dave. Recalling questions (a)–(c) above and extrapolating from Kovel's description, we can speculate about what sorts of propositions Dave likely endorses. If pressed on what he thinks about the relation between science and the economy (a), he would likely believe the following:

> B1: Science is at best a servant of commerce. Beyond its applications in practical fields like engineering, for example, science is irrelevant to the larger workings of the business world.

Asked about the advisability of addressing crises through radical reforms to the current economic and political order (b), Dave will surely be ultra-cautious:

> B2: The economic status quo gets things largely right, so there is no good reason to challenge it. The goal is to get ahead in the marketplace, not to overturn its fundamental assumptions.

Finally, we might wonder about some of Dave's meta-beliefs, those concerning the most rational way of going about getting beliefs. What, for instance, does Dave think about the extent of our reliance on other people in the belief-formation process (c)? This is a very important question to ask because collective action problems have centrally to do with how we interact with other agents. There are inevitably lots of competing beliefs that have to be managed rationally in these situations. Intelligent trust is thus crucial. Noting his very individualistic "will to power," we might be tempted to ascribe this meta-belief to Dave:

> B3: It is irrational to place significant trust in others in the process of belief formation. This is because everyone

is, ultimately, out to assert his or her power over others and trusting makes us too vulnerable.

What dispositions—epistemic traits—typically underlie beliefs like (B1)–(B3)? B1 means that Dave is at best selectively truthful. Because of B2 we can say that he is epistemically conservative. He is relatively closed to radically new and challenging information. And because of B3 we can say that he is epistemically independent, and that he thinks of trusting as a form of gullibility. These are not intrinsically objectionable epistemic traits to have. We are all inundated by information, so selective truthfulness seems warranted. Nor can we believe every new idea that comes our way, which makes a certain kind of epistemic conservatism look attractive. And because autonomy is a deeply held value in our culture, it is difficult to be wholly critical of the sort of epistemic independence Dave espouses.

Traits—moral or epistemic—become virtues or vices when their exercise is, respectively, either morally beneficial or harmful. This is a highly context-dependent matter. Moral or epistemic independence, for example, might be a vice in one context, a virtue in another, and morally neutral in a third. So the question we need to ask about Dave is what moral status his three epistemic traits have, and in asking this, we need to specify the context in which they are exercised. I'll come back to this question toward the end of this paper, but first I want to set up a contrast between Dave and an altogether different sort of epistemic agent. Proceeding the way we did with Dave, if we try to understand this agent in terms of questions (a) and (b), these two beliefs stand out:

B4: The findings of scientists can be relevant to our public policy and the shape of our institutions, especially in times of crisis.

B5: We should be open to the possibility that many of our institutions, including economic institutions, will need to be substantially altered to deal with our problems. The status quo is not authoritative.

And again, with regard to question (c), our virtuous agent will have at least this meta-belief:

B6: A fully justified belief-set is partly constructed through multiple social dependencies. Because it is a basic feature of cooperative activity, far-reaching epistemic trust is therefore rational.

Obviously, B4–B6 are directly antithetical to B1–B3, as are the epistemic traits each set expresses. In place of selective truthfulness, conservatism, and independence, we get robust truthfulness, openness to (possibly radical) social reform, and the capacity for wide-ranging trust (I take this to include trust*worthiness*). Again, without a context of application, we can't say whether or not these traits are virtues or vices. But, as it happens, they are key epistemic virtues given the need for cooperation in an ecological crisis. Let's look at each of them in a little more detail.

By robust truthfulness I mean the quality of will that keeps an agent attuned to eudaimonistically significant propositions—that is, propositions that purport to capture aspects of reality the knowledge of which is necessary for achieving optimal outcomes in collective action scenarios.[19] For example, if we want to manage the planet's carbon sinks prudently—atmosphere, forests, soil, oceans—it is important that we understand how the carbon cycle works and how anthropogenic interference in this cycle can cause dangerous imbalances in it. To the extent that we discover that some of our activities *are* disrupting this cycle in a way or to a degree that threatens the long-term ability of our species (or other species) to flourish, the robustly truthful agent believes we should alter our behaviour. This point is connected to the second virtue, openness to possibly radical social reform. Wijkman and Roström have argued persuasively that the sources of the ecological crisis are so interwoven through the fabric of our cultures that in order to stop the mounting damage, we require "nothing less than a revolution, both in attitudes and in social and economic organization."[20] They are by no means alone in this assessment.

This brings us to meta-belief B6. We need to recognize our epistemic dependency in two distinct, though related, areas. First, most of us are non-experts in environmental science, so we have no option but to defer to—that is, to *trust*—experts on scientific matters that have important social implications. Can this be justified? Take the Intergovernmental Panel on Climate Change (IPCC) as an example. One reason we can justify trusting the IPCC is the extraordinarily rigorous process of peer review through which its reports go. These reports are not flawless, and subsequent peer-reviewed research is useful in pointing out their mistakes, but they are remarkably thorough. The general point is that we can justify believing the conclusions issued by bodies like the IPCC on the basis of the soundness and transparency of the *procedures* leading to those conclusions.

Second, we must embrace the deep implications of the banal truth that each of us is just one person among many and that solutions to our large-scale problems require both compromise and constraint on the part of all. It is simply not true in this setting that only our individual perspective is important. The beliefs of other

agents in the cooperative enterprise must be taken seriously. These beliefs generally spring from unique social and psychological circumstances that the other agent knows more about than we do, and so again there is an issue of trust here. The same point could be made at the level of nations negotiating environmental treaties. In short, in the process of constructing that subset of our beliefs concerning what to do in cooperative enterprises—specifically, the sort involved in managing environmental goods—we are irreducibly epistemically dependent on both scientific experts and all the other ordinary people making decisions with us. We either trust them—intelligently, not blindly—or fail to obtain the goods of our many cooperative enterprises.

Since the epistemic traits I have just examined—robust truthfulness, openness to (possibly radical) social reform, and the capacity for wide-ranging trust—are morally salutary, it is proper to call them virtues. In fact, I submit that they are, together, the cardinal epistemic virtues of cooperation in the age of deep ecological crisis. They should be cultivated *along with* the appropriate set of moral virtues: justice, compassion, courage, etc.

By contrast, the qualities that make Dave a likely candidate for success in the capitalist economy are precisely those that make him destructive and unreliable in our new context. Kovel's description of the "hard, cold, calculating mentality" continues this way:

> None of these traits is at all correlated with ecological sensibility or caring, and all are induced by the same force field that shapes investment decisions.

In collective action scenarios Dave is clearly the sort of player whose actions, based on his beliefs, are instrumental in bringing about suboptimal outcomes. Because he is always looking to play the knave, he is subject to a self-serving bias in his treatment of data and will readily resort to deception if he thinks it serves his needs. Dave's traits are therefore epistemic vices. He is, we now must say, willfully *ignorant* in setting aside the clear warnings of scientific experts about the dangers of our current way of doing things. He is also epistemically *reckless* where some degree of caution is called for. This recklessness is, perhaps paradoxically, paired with an objectionable *timidity* when it comes to the economic status quo. However, the appearance of paradox here is diminished somewhat when we note that the latter two traits—recklessness and timidity—are the dispositional springs of, respectively, the aggressiveness and conformism about which I spoke earlier. Finally, Dave is *arrogant* in both his steadfast refusal to trust others and his willingness to deceive them.

IV. Conclusion

Can any decent or self-respecting person remain unruffled while being labelled ignorant, reckless, timid, and arrogant? One hopes not, but *if* not, then the preceding exercise has an important motivational component to it. Dave's way of seeing things is clearly celebrated in our culture. I don't mean this univocally. We would not approve of someone who thought it fit to exercise his unfettered will to power in his interactions with his grandmother or his pet gerbil. But people like Dave do undeniably tend to get ahead in our public culture, both materially and reputationally. Recognizing this, and internalizing the thought that he is epistemically vicious, we might reconsider some of the assumptions in our system of social rewards, or at least seek to produce fewer Daves.

We should also strive to avoid a certain mischievous complacency here. It's all too easy to consider ourselves off the hook morally because although we endorse one or two of the offending beliefs upheld by Dave, we don't endorse all three, or at least not in the *exact* form in which they are presented here. In an ecological crisis as deep as the one we are currently experiencing, we cannot afford to indulge any of these beliefs or their permutations. My claim here has been that we should focus our critique on the traits supporting the beliefs. Unfortunately, we are all caught up in a set of economic institutions that privileges this psychic support. We have pressing reasons to question the fundamental assumptions and organizational prerogatives of any system that normalizes this model of epistemic agency.

D Two Objections and Responses

Together, Hill and Williston give us a robust picture of the environmentally virtuous agent, one whose emotions *and* beliefs are properly attuned to the environment. But are there flaws in the virtue ethical approach to the environment? Though a recent innovation in applied ethics, the theory has already received a good deal of criticism from philosophers. Numerous objections to it have been made, two of which are worth highlighting: the claim that environmental virtue ethics is unacceptably anthropocentric and the claim that it is not sufficiently action-guiding. Let's look at these criticisms in order.

The charge of unacceptable anthropocentrism has been made by Holmes Rolston III, among others, and can apply to the emphasis on either moral virtue or epistemic virtue. Rolston III thinks that the virtue ethical approach to environmental issues distorts our proper relation to nature by placing primary value in the agent rather than in those entities of relevant moral concern *external* to the agent:

> Caring needs to be elicited by the properties of what is cared for. . . . We may say, before callous destruction of passenger pigeons, or desert fish: "No *self-respecting* person would do that." Yes, but the reason is that my *respect for the other*, which ought to be realized and respected in myself, is diminished, not that my self-respect *per se* has been tarnished. It is virtuous to recognize the rights of other persons, but their motivating force is their rights that I appreciate, not my self-respect.[21]

The idea is that virtue ethics puts the philosophical cart before the horse. Although it is good to possess and cultivate certain virtues, this is only *because* doing so is a response to the value one has independently discovered in nature. Virtue ethics, by contrast, would have us *derive* the value of nature from the value of being a virtuous agent.

This charge can be parried. Although the focus of our moral assessments is on the dispositions of agents, it does not follow that only such agents have moral standing. Indeed, the environmentally virtuous agent is constitutively—that is, *essentially*—attuned to the objective interests of non-human and even non-sentient beings. To fail to appreciate and be motivated by such interests is to be prey to one or more of the environmental vices and thus simply not to be an environmentally virtuous agent at all. Philosopher Ronald Sandler has put this point well. It is *not* the case, he argues, that environmental virtue ethics is committed to the view that

> the only demands of the world are agent or human flourishing; that humans are afforded a special, privileged place within nature . . . ; or that special moral standing is attributed merely for membership in the species *Homo sapiens*. In denying each of these, the approach avoids what most environmental ethicists rightly find objectionable about some forms of anthropocentrism: unjustified bias toward humans to the detriment of non-humans.[22]

With regard to the second claim—that environmental virtue ethics is not sufficiently action-guiding—the idea here is that although the approach can tell us valuable things about the ideal dispositional qualities of agents, it cannot provide concrete guidance on tough environmental issues. What specific *actions* does humility (Hill) or robust truthfulness (Williston), for example, counsel me to perform? How does possession of these virtues allow me to make decisions when I am confronted by conflicting demands? An approach that was incapable of providing clear answers to these sorts of questions would be deficient.

However, in his discussion of epistemic virtue, Williston insists on the notion that right action in the context of collective action scenarios is dependent on specific beliefs we have about science, our institutions, and other agents in the cooperative enterprise. There is no way to separate our beliefs, and the traits that support them, from the things we do. Similarly, Ronald Sandler has recently argued that the virtue ethical approach is indeed action-guiding. He focuses on the issue of **genetically modified (GM) crops**. His claim is that given current practices, while it is environmentally virtuous to cultivate some GM crops, this is not the case with others. The criteria for making this distinction are whether or not the crop would (a) increase monocultural and chemical agriculture and (b) be likely to disperse beyond its intended fields.[23]

Either of these outcomes would be environmentally damaging. Monocultures are both more susceptible to pests than diverse plantations and less able to support pollinators and other organisms that act as natural pest controls.[24] This is why they require intensive applica- tion of chemicals. And the dispersal of GM seeds beyond their intended fields can ultimately pollute the genetic makeup of many other plant species. On these criteria, only the beta carotene–enhanced rice dubbed "golden rice" passes the test. The vast majority of GM crops currently being grown—mainly soybeans, corn, cotton, and canola—do not. For these reasons and others, the cultivation of most GM crops is contrary to basic principles of sustainability. Since sustainability is about securing environmental goods and services indefinitely into the future, these practices are contrary to what Sandler calls "virtues of sustainability"—such as temperance, frugality, far-sightedness, and attunement—as well as virtues of steward- ship like benevolence.[25] A similar example has to do with our small but growing reliance on biofuels like ethanol and biodiesel. Biodiesels are made from vegetable oils such as soy and canola, ethanol from grains, corn, and potatoes.[26] Canada is not a large producer of biofuels, currently producing about 600 million litres of ethanol annually (compare this to Brazil and the United States, which together produce 44.7 billion litres annually). But the federal government may want to change this. It introduced a $200-million four-year program (end- ing in 2011) called the ecoAGRICULTURE Biofuels Capital Initiative (ecoABC). The goal of this program is to stimulate production of biofuels by providing rebates to those who invest in biofuel infrastructure development.[27]

Biofuels have clear environmental benefits. Because of their relatively high oxygen con- tent, they combust much more efficiently than conventional fuels, thus reducing emissions of greenhouse gases like carbon dioxide and pollutants like carbon monoxide, and they are renewable, biodegradable, and non-toxic.[28] Still, there are potential problems associated with the push to convert agricultural lands on a large scale to the production of biofuels.

First, conventional fuel use *does* still figure heavily in the production of biofuels. As with all intensive agriculture, the plants are grown through heavy application of fertilizers (the production of which requires large amounts of conventional fuels, especially natural gas) and with the use of agricultural machinery. Second, the crops—especially corn for ethanol—are grown largely in monocultures and are thus subject to the problems noted above in connection with GM crops. Finally, although it is not an environmental issue per se, there is also a social and political problem associated with biofuels. Since there is often an increased profit to be made by converting land from the production of food sources to that of plants for biofuels, less land is available for growing food. This could drive food prices up, leading to significant social unrest.[29]

For these reasons, an environmentally (and socially) virtuous agent might be inclined to oppose the development of biofuels. But scientists are now working on developing biofuels from non-food sources such as wood wastes, algae, and some grasses. These are sometimes referred to as "cellulosic" fuel sources. If successful, this technology could avoid all three of the problems just mentioned. Converting these materials to fuels relies less on conventional fuel use than is the case for food-based material conversion; they are not best grown in monocultures; and many of them are grown in semi-arid areas not suitable for crops.[30] Sandler's virtues of sustainability and stewardship therefore favour the development of non-food–source biofuels over both conventional fuels and food-source biofuels. If one wants to know how to choose virtuously among these three options, these considerations should suffice to guide one's actions. And one *can* act here, either by putting pressure on the Canadian government to include non-food–source biofuels in its ecoABC program (or by pushing them to favour the development of such fuels in the program) or by investing in (or otherwise supporting) companies like Ottawa-based Iogen, which is developing the technology to convert these materials into useful fuels.

A similar analysis could be made with respect to other environmental issues: land-use decisions, the choice between organic and conventionally grown foods, family-planning decisions, specific consumption choices, and so on. With all such issues, environmental virtue ethics claims that to the extent that we have the right set of virtues—both moral and intellectual (or epistemic)—we will be strongly inclined to behave in an environmentally sound manner. If this is correct, then environmental virtue ethics cannot be dismissed on the grounds that it fails to be action-guiding.

Ⓔ Conclusion

Despite its ability to answer the two criticisms just considered, questions remain about the viability of a virtue ethical approach to the environment. For instance, although we have seen that the theory is action-guiding, we might wonder whether it is, among the normative theories on offer, action-guiding in a unique way. That is, with respect to issues like the development of biofuels or the cultivation of GM crops, does it guide us in a manner different from the way consequentialism or deontology would? Couldn't we arrive at the very same practical conclusions about these two cases by employing one of the two competitor theories? If not, then there is much comparative work to be done showing *how* virtue ethics is uniquely capable of giving us the right answers in cases like this. But if so, then more needs to be said about why we should take the virtue ethical approach at all rather than one of the competitor approaches. Perhaps virtue ethics can present the right choice more vividly than the others can. Or perhaps it accords with our intuitions better. These possibilities need to be explored.

Also, much work remains to be done in describing the details of the various environmental virtues. Sandler has provided a very impressive typology of environmental virtues. It includes love, considerateness, gratitude, wonder, frugality, non-maleficence, optimism, creativity, diligence, perseverance, cooperativeness, loyalty, and honesty, among others.[31] But neither he nor anyone else has so far given us a comprehensive account of how each of these virtues ought to figure among the dispositions of environmentally virtuous agents. Such a project would need to take account of many factors: what specific environmental actions each virtue would incline agents to perform; whether or not the virtues can come into conflict with one another and, if they do, what theoretical mechanism is available for resolving such conflicts; to what extent a focus on these virtues needs to be supplemented by a consideration of consequences or rules; the impact that cultivation of the whole set of environmental virtues will have on agents' ethical relations with other humans; and what vices correspond to the list of virtues. Finally, while Williston has paved the way for an epistemically focused environmental virtue ethic, much work obviously needs to be done in drawing out this type of analysis.

CASE STUDY

Three Canadian Environmentalist Exemplars

One puzzle for virtue ethics is how to explain the process by which we become virtuous. As we have seen, virtue ethics places much less stress on the role of rules in moral education than the other normative theories do. So much about sound moral judgment is, for virtue ethics, about responding well to situational nuance. What is courage? Well, according to Aristotle, it is about striking a balance between recklessness and cowardice, and it involves thinking, feeling, and acting in ways that are highly responsive to facts, including facts about ourselves. If my extremely athletic friend rushes into a burning building and saves a person from the flames, I might be inclined to praise her courage. But given my physical frailty, the same action performed by me would be foolhardy and reckless.

Sources (left to right): The Canadian Press/Adrian Wyld; Clinton Global Initiative/Adam Schultz; Peter Power/GetStock.com

Elizabeth May, Maude Barlow, and David Suzuki. Moral education can be largely about learning from other virtuous agents.

Nonetheless, because of its focus on the whole person and the way she lives her life, virtue ethicists do think we can learn from the moral examples of others. Sometimes the best way

Continued

to learn how to become a courageous person is simply to study one over time. Some ethicists—Alasdair MacIntyre, for one—have elucidated the idea by pointing to the relationship between a master and an apprentice. If you want to become a good guitar player, for instance, you should find a guitar master and study the instrument with this person, not just read a book then try to apply what you've absorbed. The ideal outcome of this sort of arrangement is not that you become your master's clone but that you adapt his guitar-playing wisdom and know-how to the peculiarities of your own abilities and limitations. Nor does it mean that you should canonize or idealize the master. An exemplar is not a saint and is subject to many of the same foibles as the rest of us. The same point goes for the virtues. If you want to learn how to be kind, or temperate, or just, seek out and study the actions of kind, temperate, or just people. This is the **exemplar theory of moral education**.

In David Suzuki—broadcaster, activist, scientist—Canadians have a very good environmental exemplar ready to hand. Suzuki began his career as a geneticist and has always been a staunch advocate for environmental policy that is both science-based and rooted in more traditional ways of knowing—those of the world's Aboriginal communities, for example. He has been a tireless advocate for various environmental issues for more than 30 years. Suzuki has received many awards for his environmental work, including the UNESCO Science Prize, the Order of Canada, and the UN Environment Programme Medal. He has 24 honorary degrees from universities all over the world, and he is the author of 47 books, 17 of which are for children. He is co-founder of the David Suzuki Foundation, which has as its goals finding solutions to global warming, protecting human health, conserving oceans, promoting global conservation, and building a sustainable economy.

There are other good exemplars on the Canadian environmental scene as well. Think, for example, of Elizabeth May, current leader of the federal Green Party. May once sat on the steps of Parliament for two weeks on a hunger strike to protest the federal and Nova Scotia governments' insufficient response to the toxic mess that is the Sydney Tar Ponds (see the Introduction in Chapter 14 for more on this). May had been working for 15 years in the interests of the local residents of Sydney, many of whom have suffered increased levels of cancer and other illnesses as a result of exposure to the ponds. Or think of Maude Barlow, chairperson of the Council of Canadians, who through her political activism and books has probably done more than anyone in the world to alert us to the critical state of our freshwater resources. Here is how she sums up the life of an environmental activist:

> The life of an activist is a good life because you get up in the morning caring about more than just yourself or how to make money. A life of activism gives hope, which is a moral imperative in this work and in this world. It gives us energy and it gives us direction. You meet the nicest people, you help transform ideas and systems and you commit to leaving the earth in at least as whole a condition as you inherited it.

Think of these goals and attitudes in the context of what Sandler calls virtues of sustainability: temperance, frugality, far-sightedness, and ecological attunement. It is difficult to imagine how a person could work consistently and *effectively* toward the goal of building a sustainable economy, for example, if that person were not himself or herself disposed to act in accordance with these virtues. Do you want to develop these virtues in yourself? If so, begin by studying the life and work of Suzuki, May, or Barlow. Doing so will expose you to people totally devoted to the goal of promoting environmentally sustainable policy and practices in Canada. This should inform and deepen your own commitment to the key environmental issues of our time through the development of a strong set of dispositions to act, think, and feel in ways that are environmentally sound.

Sources: Draper and Reed 2009, 58; David Suzuki Foundation: www.davidsuzuki.org; Council of Canadians: www.canadians.org/about/Maude_Barlow.

Study Questions

1. What do you think of the exemplar theory of moral education? Does it really present a clear and viable alternative to other models of how we learn what it means to be moral (e.g., learning a list of dos and don'ts)?
2. Do you think it sometimes makes sense to turn our moral attention to agents' characters as we try to understand wrongdoing? Is this approach sufficient as a way of coming to terms with the damages we are inflicting on the natural world?
3. Is Williston correct to extend the virtue ethical analysis of our environmental challenges to the epistemic arena? Why or why not?
4. Can you think of virtues other than those discussed in this chapter that we might need to cultivate in the context of our ecological crisis?

Further Reading

Dale Jamieson. 2007. "When Utilitarians Should Be Virtue Ethicists." *Utilitas* 19 (2): 160–83.

Alasdair MacIntyre. 1984. *After Virtue.* Notre Dame: U of Notre Dame P.

Ronald Sandler. 2007. *Character and Environment.* New York: Columbia UP.

Ronald Sandler and Philip Cafaro, eds. 2005. *Environmental Virtue Ethics.* New York: Rowman and Littlefield.

David Suzuki. 2004. "Leaders, Role Models, and Success Stories." *The David Suzuki Reader.* Vancouver: Greystone. 319–76.

Louke van Wensveen. 1999. *Dirty Virtues.* Amherst, NY: Humanity Books.

CHAPTER | 10

Social Ecology and Environmental Activism

Ecological Intuition Pump

Biomimesis is the idea that we have much to learn from nature about how to organize our societies and the various social practices that make them up. For example, in her book *Biomimicry*, Janine Benyus argues persuasively that our farming practices would benefit immensely from looking more carefully at the natural organization of prairies, that we can gain important insights about how to harness energy from studying photosynthesis, that businesses could learn a thing or two about internal organization from a redwood forest, and much more. The key to all the examples is that the natural systems appealed to do not achieve their remarkable results through the heavy-handed imposition of some form of control on the internal processes. The results are rather the product of symbiosis worked out over the ages among the elements of the system. We humans have been living together in large groups for a very short time by comparison with the natural systems Benyus is talking about. Ever since the agricultural revolution we have been organizing ourselves into hierarchies. Because of this, we project hierarchy onto the rest of nature, assuming that there is much more hierarchy out there than there really is. It is easy to move from this projection to the idea that since hierarchy is the norm, we should be at the top. And here is where all our problems with nature enter the picture. What would happen to our societies if we rejected this false and destructive model and instead became more thoroughly biomimetic in our social relations? Would we not then begin to *see* nature's systems more and more clearly?

Ⓐ Introduction

In 2011 hundreds of protesters "occupied" Zuccotti Park in New York in what became known as the Occupy Wall Street movement. The protests spread around the world, including to many Canadian cities. One of the key messages of these groups was contained in the phrase, "We are the 99%!" Beyond this theme, however, the aims of the protesters were less focused. As Ross Jackson observes:

The demonstrators were quickly criticized by some observers for the wide range of issues they bring to the table and for their lack of any coherent alternative. This should not surprise anyone for the issues are indeed complex, interrelated and actually quite a bit broader than most of the demonstrators and their critics even realize.[1]

The lazy, knee-jerk reaction of media pundits like the *Globe and Mail*'s Margaret Wente or the CBC's Rex Murphy to events like this is always to criticize the protestors' lack of focus. How many times have we heard such commentators lament the fact that while we know what these protestors are *against* (everything!) we have no idea what are they are *for*? Down the ages, this is how all social conservatives have reacted to any fundamental challenge to the social order. The strategy in the case of the Occupy movement is to dismiss the protests as the juvenile rantings of a few hotheads beguiled by charismatic philosophers like Slavoj Žižec. The protestors should grow up, take a shower, and get a job. However, Jackson's point is that *in their very lack of cohesion* the protesters were on to something important. The breadth of their concerns expressed an appreciation of the interrelatedness of our social problems. This chapter takes this insight to heart through an analysis of **social ecology**, which is a response to the recognition that our environmental woes are in large part a result of the failure to organize the rest of society in a just or rational way. A very large part of the problem is that we have allowed the system of global capitalism to set the terms of debate about what can and cannot be done to affect the shapes our societies take. The system has become a force over which we claim to have no fundamental control. But it is difficult to dispute the fact that modern capitalism has created vast inequalities both within nations and across them. More to the point, it has allowed us to internalize a conception of our relation to the natural world that is destructively extractive. The Occupy protests are instructive because they made us look at the current economy in a way that stresses its inequality—the domination of the 1 per cent over everyone else, the way it perpetuates a hierarchy of wealth among us. Social ecologists, as we will see, think that this very mindset is also the principal cause of our unhealthy relation to nature.

Ⓑ Bookchin's Social Ecology

Murray Bookchin (1921–2006), the founder of social ecology, was an anarchist and libertarian social and environmental philosopher. That's a pretty loaded description of any one thinker, and Bookchin is indeed a complex and challenging figure, one difficult to pigeonhole. In this section I will present the outlines of his environmental ethics and also say something about his influence on the field. A good place to start is with an extended quote from one of Bookchin's intellectual heroes, the Russian anarchist philosopher Peter Kropotkin (1842–1921):

We are enabled to conclude that the lesson which man derives from both the study of Nature and his own history is the permanent presence of a double tendency: towards a greater development on the one side of sociality, and, on the other side, of a consequent increase in the intensity of life.... This double tendency is a distinctive characteristic of life in general. It is always present and belongs to life as one of its attributes, whatever aspects life may take on our planet or elsewhere. And this is not a metaphysical assertion of the "universality of the moral laws," or a mere supposition. Without the continual growth of sociality, and consequently of the intensity and variety of sensations, life is impossible.[2]

There's much to unpack here but doing so carefully will pay dividends by helping to illuminate the key features of Bookchin's thinking. Let's isolate three elements. The first is the claim that the study of humans, and especially of our sociality, must be embedded in the larger context of a study of nature. For Bookchin this means seeing the human capacity for sociality as the upshot of distinctively evolutionary processes. The broad claim is that through the process of evolution, nature produces more and more complex structures. This process culminates in the ecosystemic wholes we see everywhere in the biosphere and also the social wholes that our species has evolved over the course of our history.

Second, increased sociality or organizational complexity brings with it an increase in the "intensity" of experiences individual members of collectives have available to them. This claim is more vague than the first one. What could it mean? One way to put it, with reference to non-sentient systems, is that the more complex such systems are, the greater the number of causal relations in which individual members of that system are caught up. Think here of the bewildering biodiversity of rainforests as an example. A single beetle or bird species in the Amazon is involved in thousands of causal relations with other species of flora and fauna. The same sort of thing is true of human collectives, and here the reference to intensity of experience becomes much easier to apply (because of our sentience). One of the reasons teenagers are so keen to flee the nest at 18 or 19 is that the intensity of experiences within the family collective has become too simple, too restrictive. Young adults are, among other things, seeking to intensify the sensations and experiences available to them when they leave home, and they do this for the most part by inserting themselves into larger collectives. So Kropotkin and Bookchin seem to be pointing to something general here.

However, and this is the third point, not all human social collectives are created equal. Bookchin's main concern is that the sorts of societies we have allowed to flourish since the agricultural revolution some 10,000 to 12,000 years ago are in fact a threat to full human development. These social constructs have arrested the full potential of the species to achieve the maximal intensity of experience of which it is capable. Since this third point gets to the nub of social ecology, we will spend some time going over its details. The chief problem with current social organizations is that they are based on hierarchy, the dominance of one group—a class, a gender, a race, etc.—over another. Bookchin's guiding thought is that our reckless attempts to dominate nature spring from these social forms of domination. Social ecology is the thesis that "the very notion of the domination of nature by man stems from the very real domination of human by human."[3]

In his magnum opus, *The Ecology of Freedom*, Bookchin sets up the key contrast between all such societies and the alternative form this way:

> I would like to emphasize that this book is structured around contrasts between pre-literate, nonhierarchical societies—their outlooks, technics and forms of thinking—and "civilizations" based on hierarchy and domination.[4]

Many of those working to emphasize the importance of Aboriginal ways of thinking in the environmental crisis have sought inspiration in Bookchin's writing. Think back to some of the themes from Chapter 8. One way to think about the issues here has to do with the connection between non-hierarchical organization and respect for "Otherness." It has been a central tenet of most moral and political philosophy in the Western tradition that an ethics of respect among members of some group—the nation, the species, and so on—must be grounded in a conception of the fundamental equality of the members of the group. A potential problem for any hierarchy, then, is that those on the lower rungs of the social order are systematically disrespected by those higher up. Kant thinks there is an intimate connection between moral equality and respect. He argues that all humans are worthy of respect by all others precisely because we are all equally possessed of a rational nature. Similarly, one way of thinking of environmental ethics generally is to see it as attempting to pick out those things that are worthy of respect by pointing to a feature that they all share (e.g., sentience or life).

One of the signal features of the First Nations world view is the idea that the land should be respected:

> For traditionally schooled aboriginal people in many regions the environment is seen as a whole, all the parts are interconnected in a seamless web of causes and effects, actions and outcomes, behaviours and consequences. People, animals, plants, natural objects, and supernatural entities are not separate and distinct. Rather, they are all linked to each other and to the places where they reside through cultural traditions and interactive, reciprocal relationships. Because of the integration of the secular with the spiritual, of the past with the present, and of all parts of the living universe, people have a sense of spiritual and practical *respect* for their lands, waters, and all the environmental components that they recognize.[5]

Respect flows from a recognition of equality, which itself is supported by an insight into interconnectedness and interdependency. What we need to explore now is the sense in which this insight is the product of a very intimate and localized knowledge of a particular place.

Often, the best way to illuminate a concept or idea is to see what it is in sharpest contrast with. Canadian writer and anthropologist Hugh Brody has advanced a provocative thesis about the contrast between hunter-gatherer societies and agricultural societies that might help us gain some traction here. Euroamericans are agriculturalists, even those of us who live in cities, since people in cities are fed mostly by farm produce. But many—not all—of the First Nations were, and are, hunter-gatherers. So when the two civilizations came into contact, what clashed were two fundamentally distinct, and in many ways opposed, ways of thinking about how to manage land. Let's begin by quoting Brody at some length to get a sense of the contrast:

The distinction between respect and control is of immense importance to an understanding of how agriculturalists approach hunter-gatherers. The skills of farmers are centred not on their relationship to the world but on their ability to change it. Technical and intellectual systems are developed to achieve and maintain this as completely as possible. Farmers carry with them systems of control as well as crucial seeds and livestock. . . . Their thinking makes use of analytic categories that are independent of any particular geography.[6]

Hunter-gatherers, by contrast,

rely on a relative absence of exact or abstract categories that transcend geography and specific facts. Their knowledge is compounded of many specifics. . . . At the most general level failure to take care of the land is a moral risk that brings spiritual danger. It also involves spiritual realities, and the relation between humans and the spirits of the places where both humans and spirits live. These places cannot be controlled and must not be reshaped.[7]

Here are three examples of the sort of respect for the land that is implied by a humble acceptance of the fact that one does not fully control it. (1) Some Inuit place fresh water in the mouth of freshly killed seals to signal to other animals that the hunter respected the willingness of the seal to be hunted. (2) Thanks are given by the Gitxsan to the salmon they harvest each year in recognition of the willingness of the salmon to provide sustenance to members of the tribe. (3) Bones of bears they have killed are placed in trees by Algonquian hunters to mark their respect for the bear.[8] Implicit in each case—and there are many more examples like them—is the idea that objective moral imperatives pervade the relationship between the hunter and the land. These imperatives place specific constraints on the hunter to treat the parts of the interconnected natural world with appropriate respect. And the constraints are tailored to the specificities of the situation. The hunter asks, "Now that I have killed it, what am I required to do to—and for—*this* animal and its kin?"

By contrast, the agricultural mind approaches the land as a resource. Since it is inanimate, it can simply be harvested, forced to give up whatever it contains that is useful for human purposes. And it can be radically reshaped for this purpose. Recall the case study in Chapter 3 dealing with the tallgrass prairie that originally dominated much of the Prairie provinces. It was a stunningly beautiful and biodiverse place at one time but has since been converted into a vast, virtually monocultural food factory synthetically powered by fossil fuels, pesticides, and fertilizers. It has been fully subdued, controlled. Brody's point is that such subjugation and control has, in a sense, deprived the place of its original and highly specific *identity*. The specific mix of species on the tallgrass prairie made it one of the planet's most unique bioregions. But one wheat field is really the same as any other. With regard to our knowledge of the land, what is valued is not the cache of details defining intricate ecosystems but, for example, calculations of how many bushels per hectare can be extracted from the land, with the assumption that the more we can make it give up, the better. Mathematics has become the tool for understanding natural processes.

Again, Brody's thesis is best understood as an attempt to distinguish between two modes of land management. Often, the image of the circle is contrasted with that of the straight line and the notion of linear development that accompanies it. The agricultural mind was forged by the Western, monotheistic idea that the world began at some point in the deep past and is moving toward some divinely ordained goal. The human enterprise is to understand and aid in this movement. The key here is that the world must be changed because, ideally at least, it can be made better than it is right now. Of course, this claim is itself the upshot of the myth that we have in fact fallen from a state of grace, some Edenic place or state of perfection. The biblical command to go forth, multiply, and establish dominion over the Earth is a call to transform it by sheer force of will. And beyond the myth there is indeed a direct connection between the development of agriculture, the expansion of the human population, and the radical transformation of the Earth that accompanied these events.

By contrast, First Nations peoples do not believe that the world must be—or indeed *can* be—transformed for the better:

> [T]he central preoccupation of the hunter-gatherer economic and spiritual systems is the maintenance of the natural world as it is. The assumption held deep within this point of view is that the place where a people lives is ideal: therefore change is for the worse.[9]

This goes to the heart of the contrast between respect and control. To respect a thing is, in a way, to *leave it alone*. Think again of Kant's notion of respect: there are some things we may not do to other rational beings, because to be rational is to be self-determining or autonomous. This is what makes coercion and deception so wrong: they interfere with an agent's capacity to act on and for her own reasons. But respect for the other does not entail disengagement from the other. It only means that in engaging with her, one may not bypass her rationality in an attempt to manipulate her for one's own ends. The First Nations peoples are saying something similar about the land. To respect it is to see that it has a reality independent of our desires and purposes, a reality that imposes moral constraints on us. This means that it is impermissible for us to engage in actions and practices that degrade it. We must seek to live in balance with it.

Let's bring this back to Bookchin. Here, quoted at length, is his description of the connection between balance and non-domination in social relations on the one hand and a balanced relation with nature on the other:

> Yet the thought lingers that, at the dawn of history, a village society had emerged in which life seem to be unified by a communal disposition of work and its products; by a procreative relationship with the natural world, one that found overt expression in fertility rites; by a pacification of the relationships between humans and the world around them. The hunter-gatherers may have left the world virtually untouched aside from the grasslands they cleared for the great herds, but such an achievement is safely marked by its absence of activity. . . . [T]he matricentric horticulturists managed to touch the earth and change it, but with a grace, delicacy

and feeling that may be regarded as evolution's own harvest. Their archaeology is an expression of human artfulness and human fulfillment. Neolithic artifacts seem to reflect a communion of humanity and nature that patently expressed the communion of humans with each other: a solidarity of the community with the world of life that articulated an intense solidarity with the community itself. As long as this internal solidarity persisted, nature was its beneficiary. When it began to decay, the surrounding world began to decay with it—and thence came the long wintertime of domination and oppression we normally call "civilization."[10]

ⓒ The Importance of Activism

I have been suggesting that this older world view is deeply connected to an ethics of respect for Otherness (both nature and other members of the collective). Presumably, the corollary is that hierarchy as such is incompatible with such respect. There are probably good reasons to doubt this claim. That is, there may be certain forms of "domination" that are compatible with respect for Otherness. I'll return to that issue in the next section of the chapter. For now, it is crucial to emphasize that social ecology's key claim—that our improper domination of nature is rooted in improper social relations—is, among other things, a clear call to environmentalists to focus on political reform. This is what makes environmental activism so important. Our first article in this chapter explores this issue.

ENVIRONMENTAL POLITICAL ACTIVISM

Nir Barak and Avner de-Shalit

The environmental movement has a rich history, many founders, and a widely divergent heritage of activism. The kind of activism undertaken and its goals gradually evolved as the understanding of environmental issues developed. Its different phases include the formation of environmental and wildlife protection agencies, direct action against environmental destruction, protests against nuclear power and use of pesticides, and environmental justice campaigns that also address human rights. In recent years, environmental action is becoming ubiquitous in business corporations, in the promotion of an urban environmental policy, and in a general sense of "greening" of everyday life. This prevalent "green" attitude is of course a welcome trend, though it blurs the lines and complicates the ability to differentiate the different types of activism and the causes advocated for. Thus, this paper offers some analytic tools for students of environmental ethics and/or environmental politics in order to address environmental activism in a relevant manner. Our central thesis is that in order for an environmental campaign to be successful you should know exactly what you're fighting over, but be prepared to compromise.

I. Modes of Environmental Activism

The epic depiction of "eco-warriors" holding campaigns against whale hunting, against radiation experimentation, and in support of wildlife protection is somewhat passé, not because these issues received a laudable solution, but rather because environmental activism has changed dramatically over the last decades. What was once the dominion of a minority of nature-loving "tree-huggers" has grown to become a vast field of activists of many modes. At least eight different modes of environmental activism may be discerned:

- Classic—carrying out protests, sit-ins and other non-violent means of action in association with protection of open spaces, biodiversity, endangered species, and natural habitats.
- Radical—practising aggressive direct action, namely somewhat violent action that a group takes in order to either obtain demands from the government or demonstrate a problem or a possible solution. The action is termed "direct" because it is not mediated through political institutions. Under this category we include eco-terrorism, applied either against persons or against property. Eco-terrorism is often unlawful, although it is justified by those committing it as an act against those who have already broken the law. This mode is becoming less common.[11]
- Economic—associating environmental campaigns with issues of global, social, or environmental justice. This might take the form of protests against a specific industrial plant or a more internationally oriented protest against global economic summits.
- Technological—developing and promoting safer technologies for the ecosphere and humans alike. This kind of activism can be used by professionals only.
- Legalistic—lobbying and leading public campaigns in order to effect legislation

in issues regarding the rights of nature or the regulation of environmental standards.
- Communication—involves the various means that are utilized for communicating environmental messages. Web activism is a rising type of activism, and it implies using social media for echoing environmental concerns. Journalist activism is the ongoing media coverage of environmental issues about the non-human world, scientific discoveries, and environmental advocacy. Making documentary films, fiction movies, or television programs has proven to be a very powerful means for bringing environmental issues to the public agenda.
- Education—teaching environmentally benign behaviour and lifestyle or educating the next generation of environmental leaders. It often comes along with outdoor education emphasizing experiential aspects of nature.[12]
- Private, or *at home and at work* activism—adopting a "green" lifestyle that may include a vegan or vegetarian diet, "simple living" (e.g., always preferring public transportation to private cars or giving up food that was transported from far away), or a general following of the maxim "Be the change you want to be."

These various ways of intervention and countless numbers of ongoing environmental campaigns make it difficult to recognize what the unified meta-goals of each specific instance are. Of course they are all associated with promoting environmental objectives, but what exactly the comprehensive world view is that stands behind these actions is not always clear. In any campaign it is extremely important to know exactly what you are fighting over, since the various messages and rhetoric that are utilized may be confusing. In other words, sometimes for tactical reasons there's a difference between the arguments that are being sent to the public and the private reasoning that motivates political action. This issue will be discussed in the last section. At the moment, let us emphasize that by getting engaged in political activism many activists wish to scrutinize the ideas that they represent in a clear and distinct manner. Therefore, the following section offers two ideal types of environmental political activism to assist students and activists in their introspection.

II. Typology of Environmental Political Activism

In environmental political activism, we refer to the act of active participation or engagement in a particular sphere of activity in order to foster political or social change that exceeds present conditions. Environmental political activism differs from routine involvement and participation in everyday management of environmental issues in that it has a visionary aspect that seeks to redefine the manner in which these issues are dealt with. This does not imply that daily management lacks active participation, although this type of engagement differs from activism, since an essential part of activism is to stretch and challenge the routine and to suggest possible political and social alternatives. Thus, the analysis is focused on these visionary approaches.

The two main trends in environmental political activism are sustainability and ecologism, and they may be distinguished by the intensity of the suggested political transformation.[13] The concept of sustainability has come a long way and experienced many theoretical developments since it was first conceived in the Brundtland report (1987); yet it is still

imaginative in its suggested world view. It seeks to resolve immanent conflicts between social, environmental, and economic considerations. While sustainability attempts to bring about significant changes in the manner in which economic development is understood and fostered, it tries to do so in a reformative method without fundamental changes in present values.[14] Ecologism, on the other hand, is a radical approach that primarily challenges the manner in which the relationship between humans and nature is conceived and valued.[15] Like sustainability, it seeks to reconcile the tensions associated with development, though unlike sustainability it also calls for critical assessments of present values and radical changes in our political and social lives.[16] Each of the above holds a comprehensive world view that conveys care for the environment, yet they are significantly different. The following compares different goals and motivations for engagement in activism. The analysis is abstracted and somewhat exaggerated in order to yield ideal types and provide clear distinctions that in practice are naturally somewhat vague.

The first criteria in which these two approaches are differentiated concerns how the Earth is conceived. A key metaphor for understanding how sustainability activists relate to the Earth is that of a *life support system*. The abuse of nature, pollution, and over-extraction associated with economic development are detrimental primarily because they reciprocally destroy the ecosystems that enable life on earth. Much emphasis is put on the counter-effects of the abuse of the atmosphere and biosphere on human health and well-being. The Earth is valued primarily in utilitarian and instrumental measures. Ecologism's valuing of Earth, on the other hand, is primary, as it is intrinsically valuable for itself, regardless of its instrumental benefits for humanity. Contemporary modes of

TABLE 10. 1 | Comparison of Environmental Political Activism

	SUSTAINABILITY	ECOLOGISM
Earth	Life support system	Intrinsic value
Nature	"resource" Conservation	"nature" Preservation
Other species	Biodiversity	Ecosystem integrity
Consciousness	Environmentally benign	Holistic
Development	Improved management and technology	Questioning consumption and production
Change	Reform and mainly improved and more efficient management	Revolution or significant change of current political and social lifestyle
Citizenship	Current (national) citizenship	Environmental citizenship
Intersections	Environmental justice	Feminism, anarchism, social justice

production are opposed, since they destroy valuable livelihoods; the effects on human health and well-being are significant yet a secondary consideration. For example, sustainability activists would protest against the use of chemical pesticides owing to their effect on human health; ecologism activists would oppose the same policy because of its perilous and unjustified influences on wildlife, plants, and livelihoods.

The second category is the conception of what nature is. Sustainability activists debate over natural resources, while ecologism activists discuss natural treasures. This may seem like a semantic difference, yet the terminology is most significant, as it yields different justifications for the protection of nature and therefore also (sometimes) different modes of action. Sustainability is more closely associated with the legacy of resource conservation. Gifford Pinchot, the founder of the US Forest Service and its head (1905–10), is a good example of such an activist, promoting conservation and long-term careful use of forests through planning based on profits and commercial use on the one hand and respect for ecosystems on the other.[17] This is a managerial approach that seeks to promote the *proper and efficient* use of nature. Nature is interpreted as resource for human use

(consumption or leisure) and is managed and conserved in order to prevent wasteful utilization. Ecologism is more closely related to the legacy of nature preservation. Nature is protected and preserved regardless of present or future benefits. John Muir's initiative to establish national parks and by this to preserve wilderness, the Sierra Club's campaigns in the United States, and Nature Canada's activities in Canada are good examples. They seek to protect nature often for its spiritual value, or its intrinsic value, even in cases where this seems economically costly.[18] The Sierra Club seeks to establish nature reserves in order to preserve some part of nature in a pristine state or at least as unaffected as possible and to protect it from any future utilization. This is why the Sierra Club, for instance, supported transferring Yosemite Park from state to federal hands, to guarantee that it would not be used and that its land would not be taken by the state of California.

Within the third category are other species and justifications for protecting them.[19] Sustainability activists put great emphasis on biodiversity and the significance of it for human purposes. Agriculture relies on biodiversity in order to guarantee food security and to offer a wide range of choices for consumption. Medical research depends on biodiversity for research and development of safer and more efficient medicine. Industry is dependent on biodiversity because nature is the source of many industrial materials. These and other reasons provide sustainability activists with solid justification for protecting biodiversity. With respect to other species from this perspective, the emphasis of these activists is mostly on rare, endangered, and threatened species and biodiversity *hotspots*. Conversely, deriving human utility is not the leading justification that ecologism activists have for protecting other species. They are viewed as an integral part of the ecosystem, and preservation of ecosystem integrity is the key justification. As initially argued in Leopold's *A Sand County Almanac,* "A thing is right when it tends to preserve the integrity, stability, and beauty of the biotic community. It is wrong when it tends otherwise."[20] In addition, ecologism activists frequently accept notions of biospheric egalitarianism, which means that all things in the biosphere have an equal right to live and blossom, and to reach their own individual forms of "self-realization in the world."[21] However, due to pragmatic reasoning, while this concept is accepted "in principle," ecologism activists put greater emphasis on rare, endangered and threatened species.

The three following categories will be analyzed in unity: the promoted change of public consciousness, the endorsement of economic development, and the intensity of the suggested change. The watchword of sustainability activists is "small changes, big impact." They put a lot of effort into the greening of contemporary lifestyle. They bring environmental considerations into public debate and offer accessible solutions that require some gradual changes without challenging current values and standards of life. Often anti-road activists do just this. They do not oppose using

the car, they do not oppose having roads, but they want our roads to be as minimal as possible, and as environment friendly as possible, so that they don't spoil pleasing landscapes or cut through ecosystems where rare birds nest. In general, the attitude of sustainability activists toward development and technology may be characterized as critical endorsement. On the one hand, they are far from having blind faith in economic development and technological fixes, and on the other, they argue that safer and more efficient technologies can support sustainable development.[22] Thus, the intensity of the change advocated by sustainability activists is accessible and reformative in essence yet dramatic and significant in its environmental impact.

The motto of ecologism activists is "More of the same is not enough." They promote a holistic consciousness that implies that environmental, political, economic, social, and cultural issues cannot be separated. On this basis, they encourage critical assessments of present institutions, economic systems, and moral values and suggest radical changes in our political and social lives. Their attitude toward development and technology may be characterized as critical scepticism. On the one hand, they are not technophobes and do not reject economic development, and on the other hand, they radically question contemporary patterns of production and especially of consumption.[23] So, the intensity of the change advocated by ecologism activists is radical and very challenging in comparison to contemporary trends. Their anti-road campaigns often include a call for a switch from roads to railroads and from private to public transportation. They are intended to have a significant environmental impact and usher in significant cultural transformations.

The intensity of the changes promoted by environmental activists is apparent in the category of political citizenship. Sustainability activists seek to promote significant environmental changes within present political and social frameworks. This means that although their vision is environmental and global, they rely on nation-states and international organizations to promote and adopt their suggested policies and world view. While their ideas are cosmopolitan, the institutions that carry them out are national or based on national institutions and cooperation. Ecologism activists promote political notions of ecological citizenship that challenge national citizenship.[24] The pith of ecological citizenship comprehends political space in non-territorial terms and is located in individual ecological footprints. The non-territorial aspect takes into consideration the asymmetric impact of some nations on others and seeks to distribute environmental goods and bads in a more egalitarian and just manner across the globe. Sustainability activists use ecological footprints as a measure for assessing national and international policy; ecologism activists use them as the symbol and practice of "post-national" ecological citizenship. In addition to this challenge to national citizenship, contemporarily it is also being challenged from sub-state entities like cities. Urban ecological citizenship has been posed as an alternative form of active environmental citizenship.[25] So, the intensity (and accessibility) of the changes suggested by activists is also manifested in their conceptions of citizenship. Sustainability activists work in the spirit of gradual reforms in existing political institutions; ecologism activists take the radical attitude in political citizenship as well.

The seventh category consists of intersections of goals and causes of other (not necessarily environmental) political activists. Environmental activism shares various initiatives regarding human rights that are manifested especially in environmental justice. This implies a close relationship between environmental activists and political activists with respect to issues like discrimination in access to environmental goods and healthy livelihoods, mitigation of poverty, and empowerment of disadvantaged population. There are significant correlations between each of these socio-political factors and environmental degradation, and each factor intersects with environmental activism. Sustainability activists focus on these issues, and most of them frequently try to resolve them or at least promote an adequate policy within the general framework of liberal-capitalistic democracies. This marks a most significant cleavage between sustainability and ecologism activists. The latter are more radical in socio-political terms and constantly criticize and challenge capitalism as a crucial source of environmental degradation and environmental injustice. As phrased by Bookchin, the environmental movement can "become institutionalized as an appendage of the very system whose structure and methods it professes to oppose."[26] Stated clearly, liberal-capitalism is the problem, it cannot be the solution. Thus, ecologism activists have wider intersections with socio-political activists. They share common goals with feminists, with some strands of anarchists, but mostly with activists of social justice. Sustainability activists address issues of environmental justice within existing political institutions; ecologism activists focus on socio-economic change and social justice.

In summary, this section introduces two ideal types of environmental political activism. Each type encompasses a comprehensive world view that has specific goals and an alternative social and political vision. The typology presented above was intended to present the dominant trends in environmental political activism in a clear and distinct manner. However, it is important to notice that these ideal types are what we recognize as the *private face* of environmental political activism. Although these two types are significantly different in their justifications and motivations for political action, these differences are sometimes intentionally blurred in order to make the underlying ideas more accessible to policy-makers and the general public. This simplification is what we recognize as the *public face* of environmental political activism. The interaction between the public and private face of environmental political activism and practical reasoning in activism will be analyzed in the subsequent section.

III. Activism in Practice

Frequently, activists utilize arguments that differ from their original standpoint in order to persuade and recruit. Let's take vegetarianism as an example. The subject can be approached from an ecological perspective, animal rights perspective, or individual health. Among activists who support vegetarianism, the *private* discussion is frequently on issues concerning the ecological and moral implications of meat consumption; however, the *public* face that these activists present is very different. They would start their argument by addressing the implications of meat consumption on health and the various dangers and risks involved. Only when the activist has grabbed the attention of the listener will arguments regarding environmental and moral considerations, such as animal rights, be made. Even then, those arguments would most frequently be addressed as a form of enlightened self-interest. However, it is important to note that whether a person chooses to adopt vegetarianism for moral reasons or decreases or refrains from consuming meat because of health issues, they are both adopting the same policy.

The practical reasoning for adopting a *public* face is that the same policy may be justified in different ways. As in the example above, a person who chooses to adopt vegetarianism for moral reasons and one who chooses to decrease or refrain from consuming meat because of health issues are both carrying through the same policy. They both have their own *private* justification, though in practice they both do not eat meat products. Utilizing arguments that are more likely to be accepted by the public is a practical method of making ends meet and promoting a policy despite the different methods of justification. In this way, ecologism activists may publicly "dress up" as sustainability activists and vice-versa if doing so supports the goals of the campaign.

This leads us to say something about compromise. Whether you are a sustainability or an ecologism activist, in reference to the status quo both positions are quite extreme. But the ethical question at stake derives from the fact that the more extreme the change, the less likely it is to be achieved within a foreseeable future. Some people think that compromise is a sign of weakness and may be depicted in terms of disloyalty to one's values. However, it has previously been suggested that an all-or-nothing approach would necessarily lead to loss of the campaign.[27] Maybe, as Avishai Margalit argues, what counts about a person's morality is not her initial standpoint and how true to herself she is being, but rather what she is ready to compromise about and how far she is ready to compromise.[28] We are not saying you should be ready to compromise as if giving up all your positions, but rather that compromise is a more practical method and more likely to be successful. This might appear, especially to young readers, to be a disappointing conclusion for a chapter on environmental activism. However, don't forget that activism is a form of political action, that politics is the art of achieving what is possible, and good practical activists know that "first we take Manhattan, then we take Berlin."

The first thing to notice about this marvellously lucid exposition is that there is not much *direct* reference in it to Bookchin's idea that what we require is a necessarily radical politics, one directed at overturning all social hierarchy. But notice that the "ecologism" option discussed by de Shalit and Barak does, as they point out, have "revolutionary" implications for our current social order. These authors do not argue, as Bookchin and his followers certainly would, that this is the best approach to take to our environmental problems. Instead, de Shalit and Barak suggest, more cautiously, that we should at least be aware of what our activist goals are and seek to communicate these as clearly as possible to others. But let's suppose that the "sustainability" option is sound. Obviously, it does not entail the sort of social upheaval ecologism warrants. Is this to license unjust domination, even as we seek solutions to our environmental problems? If so, that would be unfortunate, because one of social ecology's key insights is that improper forms of social domination are the causes of our reckless domination of nature.

D The Problem of Domination

The problem of domination is a key theme of ecofeminism too, as we have seen. It is worthwhile to consider social ecology and ecofeminism together here. Indeed, Bookchin argues that patriarchy is one of the earliest and most insidious forms of social domination, one that paved the way for many of the others:

> Even as the Mother Goddess continued to occupy a foremost place in mythology . . . women began to lose whatever parity they had with men—a change that occurred not only in their social status but in the very view they held of themselves. Both in home and in economy, the social division of labour shed its traditional egalitarian features and acquired an increasingly hierarchical form. Man staked out a claim for the superiority of his work over woman's; later, the craftsman asserted his superiority over the food cultivator; finally the thinker affirmed his sovereignty over the workers.[29]

Recall, in this connection, Karen Warren's key claim (see Chapter 6): there is a "logic of domination" governing men's relation to both nature and women. This logic constitutes an "oppressive conceptual framework," the purpose of which is—since it is tied to specific normative assessments—to provide a *justification* for subordination. Social ecologists and ecofeminists seek to challenge this pattern of thought and thus overturn the justification. So the point must be that domination as such is a bad thing, that it is impermissible for us to dominate other people or bits of non-human nature (perhaps even whole systems). Leaving aside the morality of dominating other people—which is relatively unproblematic—is this claim correct?

It is usually ill-advised to begin the analysis of a concept with a definition, but sometimes this approach can be helpful. The *Oxford English Dictionary* defines "domination" as "command" or "control." The question we need to examine here is whether or not domination of nature, in the sense of commanding and controlling it, is always impermissible. And surely we want to deny that this is the case. Any Canadian who is an avid gardener will have come across plant species like the trumpet vine. This vine can look great splayed across the side of a brick wall or a fence, but it is terribly invasive and aggressive. Left to its own devices, it will quickly overwhelm and choke off other plant species. As a result, it needs constant pruning.

Now, this may sound like a trivial example, but surely we want to say that pruning the vine—which is undoubtedly an act of command or control over a part of the natural world—is an entirely *permissible* action on the part of the gardener who does not want his garden overrun with such an aggressive species. But we can go even further. If the plant threatens to jump his fence and ruin his neighbour's carefully tended rosebush, then he also has a *duty* to prune it—that is, to dominate it. Apparently we are often permitted and sometimes obligated to dominate nature.

The point can be generalized. We have permissions and/or duties to dominate nature in many different areas: from training domestic animals to ploughing and fertilizing fields and even, possibly, to polluting in responsible or "optimal" ways. Even getting a regular haircut is a way of dominating nature. If this is right, there can be no blanket prohibition on dominating nature. In response, ecofeminists and social ecologists would no doubt point out that they are only talking about *unjustified* forms of domination. The problem with this response is that in the original argument the fact that an act involves domination is the basis for saying that it is morally unjustified. Now, it would seem that we have to say that some acts of domination are morally justified, some not.

How can we do this? We cannot invoke any of the concepts that pop up so often in the literature—such as subordination and oppression—to buttress the distinction we are after. To say that only those forms of domination that do not involve subordination or oppression are justified is to argue in a circle, because these terms, at least as generally used by ecofeminists and social ecologists, are basically synonyms for domination. It therefore seems that we must go back to the drawing board of environmental ethics. That is, we need to make an argument about which sorts of things have moral standing and why. If we decide that non-human animals, or non-human living things, or ecosystems have moral standing, then we have a *prima facie* case for saying that it is not permissible to dominate them. However, if we do this, then the philosophical action is really with the traditional forms of environmental ethics—for example, animal welfarism, biocentrism, the land ethic—and it is difficult to see what *unique* contribution ecofeminism or social ecology can make to this project.

So the ecofeminist or social ecologist is left to argue either (1) that all forms of domination are impermissible or (2) that some are and some are not. But (1) is deeply implausible, whereas (2) does not require appeal to specifically social ecological or ecofeminist doctrines.

But there *is* much of value in social ecology in spite of these points. For one thing, social ecology makes us notice the connection that often obtains between our ways of thinking about social relations generally and our ways of thinking about nature. Here's an example, which again brings us back to the male domination of women. Francis Bacon is one of the founders of the modern scientific method. He claims that science is useful because it enslaves nature for our use, in the same way that marriage is useful because it enslaves women for the use of men. Claims like this alert us to the fact that sometimes we do apply something like a logic of domination to distinct "objects" and that it is wrong to do so. Where ecofeminism goes wrong is in pointing the finger of blame at the fact of domination rather than at the reasons we have for thinking that there are justified constraints on the *extent to which* we can dominate nature.

To enslave someone or something is to assert morally unconstrained control over it. The sort of attitude Bacon endorses might therefore incline us to believe that we are permitted to treat non-human nature any way we like, whereas many of us think this is just not true. In other words, what shocks us about Bacon's idea is that in one case—that of women—he does not recognize our duties to refrain altogether from domination, while in the other—that of nature—he does not recognize that there are limits to how much we may dominate. For example, we are, we might think, fully justified in dominating nature as long as we do so in sustainable ways. But again, although ecofeminists and social ecologists can lead us—and have led us—to notice such problematic patterns in our thinking, it appears that deciding where to draw the line between justified and unjustified forms of domination over non-human nature will have to appeal to arguments about who or what has moral standing.

Ⓔ Integrity and Stewardship

It is easy to misunderstand Bookchin in one crucial respect. He is not critical of contemporary culture and society simply in view of its hierarchically organized *complexity*. He does *not* pine for simplicity. Recall the quote from Kropotkin (section A, above). It stresses the idea that the higher achievements of the evolutionary process are magnificent precisely in view of their internal complexity. This should remind us of the concept of "dynamic equilibrium," first introduced in Chapter 3. Healthy ecosystems, recall, are both internally complex and unified, they display a unity in diversity. Another concept used to signal this achievement is, of course, that of "integrity." To see how it might apply in the present context, we turn now to this chapter's second article.

ECOLOGICAL INTEGRITY: BETWEEN ETHICS AND LAW

Laura Westra

Ecological or biological integrity originated as an ethical concept in the wake of Aldo Leopold and has been present in the law, both domestic and international, and part of public policy since its appearance in the 1972 US Clean Water Act (CWA). Ecological integrity has also filtered into the language of a great number of mission and vision statements internationally, as well as being clearly present in the Great Lakes Water Quality Agreement between the United States and Canada, which was ratified in 1988.

The generic concept of integrity connotes a valuable whole, the state of being whole or undiminished, unimpaired, or in perfect condition. Integrity in common usage is thus an umbrella concept that encompasses a variety of other notions. Although integrity may be developed in other contexts, wild nature provides paradigmatic examples for applied reflection and research.

In my work, I consistently base my arguments on the scientific understanding of ecological integrity, because to accept ecological integrity, as some have done, as a socially definable notion, meant to lose it as a firm point of reference. In that case, it could not be understood as a firm starting point if it were open to a variety of opinions. I proposed the principle of integrity as a solid principle of a new ethics,[30] reaching beyond a simple environmental ethic to a moral principle that acknowledges the primacy of a consideration for human rights and the human good, even the right to life; all must start with the protection for the habitat of mankind that also—at the same time—ensures respect for the biological integrity of all human beings.

That understanding, based on science, was eventually fleshed out in its full meaning and connotations by the members of the Global Ecological Integrity Group (GEIG), initially funded by the Social Sciences and Humanities Research Council (SSHRC; 1992–9). Since then it has been meeting every year, starting with the 2000 meeting funded by NATO to discuss a 50 per cent Eastern Europe participation, which took place in Budapest. The final collectively agreed upon definition was published combining the work of ethicists, conservation biologists, ecologists, and other scholars.[31]

Essentially, the Global Ecological Integrity Project (as it was known at the start) has been guided by two complementary policy imperatives: conserve integrity and live sustainably.[32] I have defined sustainability as a system's capacity to retain its specific functions, that is, its critical life-support processes as well as its parts or components.[33] The emphasis on the scientific meaning of integrity was never lost by the GEIG, but after 2000, when the final definition was reached, our focus became based primarily on the work of James Karr and the development of his Index of Biotic Integrity.[34] Fidelity to the scientific consensus reached remained constant.

This starting point was established against the arguments of philosophers of science,[35] but also against the practical resistance of globalized corporate interests, strongest at Great Lakes Meetings and other venues. Their mantra of "development" at all costs contradicts the very basis of strong sustainability my group and I have supported,[36] which renders ecological integrity both hard to implement and unpopular.

As well, aside from the hostile responses to the primacy of integrity on the part of others, from various standpoints, there is also the somewhat "despairing" position of one of our founding members, William Rees:

> Ecological footprint analysis, developed by William Rees and his students, is another important tool for diagnosing unsustainability in relation to off-site impacts. It is not enough that one's immediate habitat or environment is stabilized. . . . We need to ensure that the distant terrestrial and aquatic ecosystems we "appropriate" through trade and by exploitation of the global commons . . . also remain in a productively healthy state.[37]

But Rees also identifies a further grave difficulty, for which we cannot blame only external factors and economic interests: our own propensity to accept "social myths" while totally ignoring the "ecological reality that surrounds us," which makes us prey to "a collection of shared illusions."[38] Thus, one of the foremost names in Canadian ecological economics and regional planning, William Rees, is co-author of the introductory chapter of a law textbook currently used to teach environmental law. This is specific to Canada and would not likely be found in the US, for instance.

Another unique Canadian trait, as we shall see, is the frequent reference one finds in law to Canadian First Nations, particularly in legal scholarly work regarding the environment. At any rate, in 1998 I proposed eight "secondary order principles" (SOPs) to supplement the principle of integrity and to expand on it, attempting a response to the question: "How can we conserve integrity and live sustainably, with concern for environmental justice and the rights of those living in the third world?" The necessary step from theory to public policy suggested these principles for practical guidance:

SOP 1   In order to protect and defend ecological integrity, we must start by designing policies that embrace complexity.

SOP 2   We should not engage in activities that are potentially harmful to natural systems and to life in general. Judgments about potential harms should be based on the approach of "post-normal" science.

SOP 3   Human activities ought to be limited by the requirements of the precautionary principle.

SOP 4   We must accept an "ecological worldview" and thus reject our present "expansionist worldview" and reduce our ecological footprint.

SOP 5   It is imperative to eliminate many of our present practices and choices as well as the current emphasis on "technical maximality" and on environmentally hazardous or wasteful individual rights.

SOP 6   It is necessary for humanity to learn to live as in a "buffer." Zoning restraints are necessary to impose limits both on the quality of our activities, but also on their quantity. Two corollary principles follow: (a) we must respect and protect "core"/wild areas; (b) we must view all our activities as taking place within a "buffer" zone. This is the essential meaning of the ethics of integrity.

SOP 7   We must respect the individual integrity of single organisms (or microintegrity), in order to be consistent in our respect for integrity and also to respect and protect individual functions and their contribution to the systemic whole.

SOP 8   Given the uncertainties embedded in SOPs 1, 2, and 3, the "Risk Thesis" must be accepted, for uncertainties referring to the near future. We must also accept the "Potency Thesis" for the protection of individuals and whole in the long term.[39]

No doubt, philosophers may still be cringing at my use of an "is" to generate an "ought." But *pace* Hume, Kant's support of the infinite value of life is closer to the approach I have been taking. Onora O'Neill discusses the difference between globalization and cosmopolitanism. While the former is primarily *procedural* in its structures, and primarily influenced by powerful, market-oriented powers, the latter is based primarily on *substantive* moral principles of justice that include but also transcend the economic realm and rely on Kantian principles. States may or may not be fully just within their own borders, but even at best, they may well injure those outside their borders by exclusionary practices, and these are *direct* injuries.[40] The practices we don't accept provide *indirect* injuries instead. This is a form of indirect injustice, as "destroying parts of natural and manmade environments injure those whose lives depend on them." In addition,

> [T]he principles of destroying natural and manmade environments, in the sense of destroying their reproductive and regenerative powers, is not universalizable.[41]

Ecological and biological integrity is precisely what O'Neill terms "regenerative and reproductive powers," or true sustainability:

> Environmental justice is therefore a matter of transforming natural and manmade systems only in ways that do not systematically or gratuitously destroy the reproductive and regenerative powers of the natural world, so do not inflict indirect injury.[42]

This is what "has been argued from a scientific and a moral point of view in the work of the 'Global Ecological Integrity Project.'"[43] In O'Neill's terms, moral principles represent the "blueprint" and the "specifications," which define the "product" to be eventually produced. In a similar sense, strategies based upon principles are not, as such, the strategic tools to use in order to achieve just aims, but they define what forms such tools might take. O'Neill says:

> The move from abstract and inconclusive principles of justice toward just institutions, policies and practices is analogous to moves from design specification towards finished product.[44]

Cosmopolitanism, based on Kantianism, can supply the principles and also the guidelines that are largely absent from even the best among the advocates of liberal democracy, as the roots of injustice are seldom sought out by these thinkers:

> The idea that our economic policies and the global economic institutions we impose make us causally and morally responsible for the perpetuation and even aggravation of world hunger, by contrast, is an idea rarely taken seriously by established intellectuals and politicians in the developed world.[45]

O'Neill's focus is on the role of conservation and respect for natural systems, as she acknowledges the interface between these and human rights. Following along these lines, we can add the insights Alan Gewirth proposes. The foundational arguments proposed by Gewirth help to shed light on that basic connection between humans and their habitats. Gewirth argues that human rights are not based primarily on human dignity,[46] but that this Kantian principle is only partially right. He prefers to base "human rights on the necessary conditions of human action,"[47] as morality is intended to give rise to moral action. Gewirth adds that "human rights are the equivalent to 'natural' rights, in that they pertain to all humans by virtue of their nature as actual or prospective agents."[48] He cites five reasons in support of his claim: (1) "the supreme importance, of the conditions of human actions" (and we will return to this point below); (2) action is "the common subject matter of all moralities"; (3) "action" is more specific and less vague than "dignity" or "flourishing"; (4) thus "action" ultimately secures "fundamental moral status" for persons; (5) "action's necessary conditions provide justification for human rights—as every agent must hold that he has a right to freedom and well-being as the necessary conditions of his actions."[49]

Beyleveld, Deryck, and Brownsword argue that the "basic" or "generic needs" that represent the preconditions of all action, including moral action, are "freedom or voluntariness" and "well-being or purposiveness," where the former are procedural and the latter "substantive,"[50] and they view freedom as instrumental to well-being. I want to propose inverting this order. Life, health, and the mental ability to comprehend and choose precede the exercise of voluntariness and are not only necessary for it, but sufficient, when all these conditions are in fact present.

In essence, this has been the argument of the previous section: "basic rights"[51] represent the minimum all humans are entitled to, and they are prior to all other rights, both conceptually and temporally. For Gewirth as well, life and the capacities named above can be "threatened or interfered with."[52] Thus, to say we have rights is to say equally that the preconditions of these rights represent something we are entitled to have, not only in morality, but also in the law. In other words, any legal instrument that supports the existence of human rights, *ipso facto,* ought to proclaim the requirement that their preconditions be equally supported and respected.

Some argue that the dignity of human beings is only partially the ground of human rights and that dignity itself is based on agency; still, the argument allows the introduction of at least a further point in favour of extending human rights to life and health. The introduction of "preconditions" means the introduction of conditions that are not only conceptually but temporally prior to agency; hence the protection of these preconditions entails the acceptance of potential consequences in the protection of agency. Thus, not only does the scientific approach and the ensuing definition of integrity foster and support the interface between ecological integrity and human rights, but that

connection may also be found in the thought of moral philosophers, although the explicit reference to ecological integrity, a recent scientific concept, is clearly missing.

However, the most significant aspect of ecological integrity relating to Canada can be found in the multiple areas where biological or ecological integrity figures prominently in the law. It is present in the Great Lakes Water Quality Agreement (1978, rat. 1988). It is sad to acknowledge that its mandates, however, are only paper tigers, as they are largely ignored in a series of annual or biennial meetings, where a minor advancement is described as a great success, while, except for the notable cleanup of Lake Erie, for the rest, it is mostly business as usual.

In contrast, there are two other areas of law where ecological integrity or its equivalent takes centre stage, at least on paper: first, the environmental legal regimes regulating Canadian parks, and second, the extensive jurisprudence past and present regarding the rights of First Nations. The role of ecological integrity in Canadian law is explained by Shaun Fluker:

> [H]uman activity necessarily impairs ecological integrity and thus paradigm ecological integrity is found in ecosystems protected from human disturbance. . . . It thus comes as little surprise that the norm of ecological integrity figures prominently in the management of Canada's national parks.[53]

After a study mandated by the minister of Canadian heritage,[54] the appointed scientists confirmed what Canadians know well, especially our environmentalists, that is, that ecological integrity was not central to park management, despite the stated goal of maintaining ecological integrity. But such first priority has often been lost to other interests proposing the primacy of human activities in the parks.[55] Hence, two sections were added to Canadian legislation in 2001:

Section 2(1) – Definitions

"ecological integrity" means, with respect to a park, a condition that is determined to be characteristic of its natural region and likely to persist, including abiotic components and the composition and abundance of native species and biological communities, rates of change and supporting processes.

Section 8(2) – Ecological Integrity

Maintenance or restoration of ecological integrity, through the protection of natural resources and natural processes, shall be the first priority of the Minister when considering all aspects of the management of parks.

However strong the language of these mandates for Canadian parks, the facts on the ground contradict the high-sounding environmental concerns expressed here. Nowhere is this contradiction clearer than in cases where there is a conflict between a First Nation's interest and that of a natural park adjacent to their lands—for example, the case regarding Wood Buffalo National Park.[56] This park straddles the province of Alberta and the Northwest Territories. In 1998 the municipality of Fort Smith submitted an application to Parks Canada, "seeking approval to construct and operate a road crossing the park from east to west along the Peace River."[57] The park had been originally set aside

in 1922 to protect the Wood Buffalo, and the proposed road was not found to serve any park purpose, but it would advance economic interests of others outside the park instead. Parks Canada approved the road's construction in 2001, with no reference to ecological integrity, and Justice Gibson ruled, in a requested judicial review, that Parks Canada's lack of consideration for ecological integrity did not invalidate the decision.

In contrast, there was at the same time another lawsuit pending, one based on the legal rights of the Mikisew First Nation instead, which was eventually heard by the Supreme Court of Canada. In that case,[58] Madam Justice Hansen stated that although ecological integrity was perhaps not the first priority in reaching a decision, balancing it with the interests of those living near the park was paramount. Therefore, the First Nations' rights to Aboriginal hunting and their traditional lifestyle would prevail.[59]

Thus, despite the explicit wording added to Canadian parks legislation, the rights of First Nations to their unspoiled habitat, one that maintained wildlife, take precedence. The connection between the natural environment wherein they reside and its ecological integrity is inescapably tied to their cultural integrity, their rights to their religious practices, and, in general, their traditional lifestyle.[60]

In conclusion, whether explicitly cited or not, ecological integrity is more than a foundational necessity for sound moral principles in defence of human rights. It also acquires a specific Canadian flavour, not only because of the amount of research funded by Canadian sources (SSHRCC and Health Canada), but also because of the legal rights to the "commons,"[61] which is expressed in Canadian regulatory regimes, whether directly, as in Parks Canada's explicit language, or indirectly, as a substantive aspect of the rights of Canadian First Nations.

Westra does a lovely job of tying together some of this book's key themes, especially the connection between the concept of ecological integrity and how we think about the ideal shape our laws and social practices should take. In so doing, she advances the key thrust of social ecology: to show that there are important synergies between social organization and the organization of the natural world. Are there other fruitful connections we can make in this area? One possibility is to notice that the concept of "stewardship," so essential to environmental ethics, is also an implicit feature of social ecological thinking. After all, the injunction to end unjust domination of nature is virtually equivalent to the idea that we should learn better how to care for nature. Further, the idea of a good environmental steward gives us clues about the sorts of people we want, ideally, to be, and in this way we can connect themes in social ecology to environmental virtue ethics (explored in detail in Chapter 9). Let's see how we might begin to put all of this together.

What does it mean to be a good steward of the land? A steward is someone who manages or supervises something, very often with the implication that he or she may not strictly own that thing. Historically, for instance, a steward was an officer of the British royal household, charged with the administration of Crown lands. Although the idea of non-ownership is not essential to the concept of stewardship, for our purposes it nevertheless underlines one of its more important aspects, because it suggests that the good of the thing cared for cannot be determined solely by reference to the interests of the steward. Of course, in determining the nature of that good, we might ultimately need to consult the interests of some *other* human agent or agents. But it might also be necessary to think entirely independently of human interests. According to Geoffrey Frasz, the best way to do this is to cultivate the virtue of environmental benevolence and its offshoots.[62]

What is benevolence more generally? It has three central features: (1) an imaginative reconstruction of the life of the other; (2) an attempt to determine what is in the best interests of the other; and (3) the motivation to act with the interest of the other in mind. Let's see how the features might figure into the concept of the environmentally benevolent steward. I have indicated that feature (2) might already be contained in the concept of the competent steward. Furthermore, to the extent that entering imaginatively into the life of the other is a necessary condition for finding out what its peculiar interests are, so is feature (1). But feature (3) certainly adds something new, since it refers to the agent's *motivation* or willingness to promote the interests thus discovered. The etymology of benevolence highlights this. The word is derived from two Latin words: *bene* or "good" and *volens* or "willing," giving us "good-willing." Features (1) and (2) do not yet describe a virtue, just a skill. However, with the addition of (3) we do have a full-blown virtue: a disposition on the part of the environmental steward to seek out and *promote* the interests of the land.

As this reference to the land indicates, a crucial aspect of the account is the claim that our benevolence ought to extend beyond sentient nature to non-sentient nature. To make this move, it is crucial to divorce benevolence from compassion. Compassion is, literally, co-suffering. As such, although it is appropriate to our relations with sentient beings—since they are by definition capable of suffering and their interests can in large measure be determined by reference to this fact—it is clearly not appropriate to our dealings with non-sentient beings. Nevertheless we can, with Leopold and Westra, speak accurately of the "integrity" or overall well-being of such entities. Thus, it is possible to be benevolent stewards of forests, ecosystems, species, and so on.

Frasz shows that corresponding to this virtuous extension of our moral concern are specific vices: for example, the arrogance and chauvinism that cause us to view non-human nature only in terms of human interests. Frasz also claims (plausibly) that there is a cluster of vices involved both in our failure to adequately grasp the interests of non-human nature and in our unwillingness to act on this knowledge if we do obtain it: laziness, sloth, lack of curiosity, impatience, and cynicism. My suggestion here is that we can make useful connection between the social ecologist and the environmental virtue ethicist if we are willing to talk about the virtues and vices through the lens of social ecology's push to end unjust forms of domination.

One reason to be hopeful that the appropriate connections can be forged is that virtues are relatively stable dispositions, that is, they are ideally applicable across distinct contexts. For example, if I am brave on the battlefield, it's safe to assume that I'm also brave on the rugby pitch or when facing the threats of a colleague. Or, if I fail in one or another of these areas, that I *should* seek to change precisely insofar as I think of myself as a brave person. So, if I've got the virtues of a good environmental steward and have avoided the relevant vices, then not only do I have morally praiseworthy relations with nature, I'm also likely to have the same praiseworthy relations with other people. Indeed, the social ecologist may be correct to insist that I should seek to cultivate these virtues *first* in my social relations and that a healthier relation with the rest of the biosphere would follow. If the key virtues help us avoid unjust domination in one sphere they should help us do so in the other as well.

Ⓕ Conclusion

One element of Bookchin's thinking not touched on here is his anarchism. We should conclude with a brief look at this concept. The key principle of anarchism is that there is a fundamental tension between authority and autonomy. Authority demands submission, while autonomy demands that all submission be removed from social relations. As Robert Paul Wolf argues, if autonomy and authority are genuinely at odds, then we must either (a) embrace anarchism and "treat all governments as non-legitimate bodies whose commands must be judged and evaluated in each instance;" or (b) give up on full autonomy. Wolf thinks that autonomy is non-negotiable—to think otherwise is to commit the "sin of willful heteronomy"—leaving only option (a).[63]

Wolf also believes that our alienation from each other is the product of unjust domination: "man also confronts a social world which appears other, which appears to stand over against him, at least partially independent of his will and frequently capricious in its frustration of his desires."[64] What is "alienation"? The concept has to do with *estrangement*. Because of the fact of unjust social domination we sometimes feel as though social institutions are forces set against us, that they have a logic all their own that we cannot control. These are constructs that we believe ought to be expressions of our own socially creative powers. But we also feel that they have gotten away from us. Think of the way the demands of "the economy" are typically presented to us: they are brute facts about the need for perpetual growth, the way we must respect the global flow of capital, create the right "conditions for investment" in our societies (by keeping corporate tax rates low, for example), and so on. And we, so goes the argument, must simply submit to this force in the design of our societies. One upshot of this way of thinking is that we are encouraged to think of everybody else as competitors with us for the world's scarce resources. Consumption is, at its heart, a competitive activity. When you buy this or that item, you may feel as though you are only indulging in an innocent pleasure, but you are also displaying your status to others. So in capitalist societies we are estranged—alienated—both from this thing called "the economy" and, just as importantly, from each other.

Governments and business elites talk and behave as though they cannot question the fundamental prerogatives of the global capitalist economy. What is lost in this way of thinking is the truth that the economy is a social construct. We made it! In principle, we can therefore give it whatever shape we want, and that means that if we wish to have an economy that is a better respecter of marginalized people and the natural world, we can do so. Anarchists think that ending alienation—from nature, from each other—is all about reclaiming our freedom. It is no accident that Bookchin's major work is called *The Ecology of Freedom*. Wolf concludes that "rational people of good will can in principle eliminate social estrangement."[65] Social ecologists would add that our current estrangement from nature would also melt away were we to achieve this. What our politics will look like after this is not clear. Do we need the full resources of political anarchism—the replacement of state structures with voluntary communal organizations of various kinds—in order to accomplish it?

CASE STUDY

Fighting Hierarchy through Access to Knowledge

One of the most momentous events in the history of Western culture occurred in the early-modern period with the rise of Protestantism. This is because one of the things reformers like Luther and Calvin insisted on was that the message of the Bible ought to be directly available to ordinary people. "Directly" here means that such people did not require an "expert" to read and interpret the book's contents. It is difficult now to appreciate that a political upheaval could be caused by the suggestion that a book should be translated from Latin into the various languages of Europe. The incident highlights a key general truth: that knowledge is indeed power. When ordinary people could read the Bible for themselves, a great deal of power was transferred from Rome to the peoples of Europe. Hierarchies of knowledge acquisition often abet cruder forms of social domination. If I have access to the truth about what measures are required to save your eternal soul from damnation, and I can get you to believe this is the case, then you had better listen to my recommendations about exactly how you should live your life.

Continued

The Death of Evidence March, Ottawa 2012. Is democracy threatened when governments suppress important scientific information about the effects of our industrial activities on the environment?

Today we rely on knowledge just as much as people did back then, though the ways in which knowledge is accumulated and organized are more difficult to analyze. Let's explore the organization and dissemination of knowledge about the environment in Canada. In 2012, and then again in 2015, hundreds of Canadian scientists and their supporters marched on Parliament Hill (and elsewhere) in an effort to bring wider public attention to the Conservative government's attack on environmental science as well as the environment itself. This was dubbed the "Death of Evidence March," aimed at Bill C-38, the controversial piece of legislation introduced by Stephen Harper's government in that year. The bill amended 60 separate environmental acts and repealed 12 more. An op-ed piece in the *Guelph Mercury* noted that "the federal government is taking a sledgehammer to environmental protection across Canada." This legislation and the broader assault on environmental protection in Canada have implications well beyond the shelf-life of the Conservative government, for they will tie the hands of governments for generations to come. We should therefore look at the government's actions more closely. One of its most extraordinary moves was to eliminate federal funding for the Experimental Lakes Area (ELA) in northern Ontario. The government's excuse for doing this—to save money—is risible, since the ELA costs only about $1 million per year, a microscopic fraction of a drop in the bucket of available federal funds. So why did they cease to fund the project? The ELA is a large-scale outdoor biological laboratory. In fact, as Chris Turner notes, scientists have created a whole "biosphere" up there for experimental purposes:

The ELA is not a lab in any conventional sense but rather a sort of contained biosphere where contained experiments sometimes involve altering the fundamental biology and chemistry of a whole lake—or several lakes—for years at a stretch. It may be the most important freshwater research facility on the planet, and its researchers . . . had made discoveries of global import, uncovering the mechanisms by which acid rain poisons aquatic ecosystems and industrial phosphorus runoff damages freshwater chemistry.

The ELA also examined the effects of climate change on freshwater resources and the species they contain. Now put these developments together with another one. The Harper government has been criticized internationally for preventing its environmental scientists from communicating their findings freely with the public. Everything these scientists want to say first has to go through the party's press officer to ensure that it is not at odds with the official government view. It is now the accepted view of this government that Canada should be devoted to unfettered resource extraction. That is our chief role in the global economy: we have an abundance of resources—oil, minerals, freshwater, etc.—that we will sell to anyone willing to pay for them.

But a problem arises when aggressive resource extraction begins to damage the environment on a dangerous scale. Where this happens, you can do one of two things as a government: (1) seek a better balance between environmental care and resource development; and (2) continue to extract resources ever more aggressively but cover up the damage this is doing. Unfortunately, our government has clearly opted for (2). As noted, the approach is two-pronged: first, eliminate as much of the uncomfortable data as possible by closing down research institutes like the ELA; and second, where the data gets through to the scientists working for the government, muzzle these scientists so that they cannot report it fully to the public.

The effect is much the same as it was in the Middle Ages with the concentration of knowledge in the hands of a few Catholic clergy: those in positions of power control what is known about a matter of vital concern to ordinary citizens or subjects. Of course, the big difference between now and then is that knowledge is much more widely available now. It is difficult to control it the way Harper has tried to. But even if the control from above is not total, it is difficult to deny that it has an effect on the public perception of risks associated with our currently unconstrained industrial assault on the environment. If we want to follow the social ecologists in reducing the harmful effects—on society and nature—of pernicious hierarchy, we would do well to focus on the issue of access to information about what we are doing to the environment.

Source: Chris Turner, *The War on Science: Muzzled Scientists and Willful Blindness in Stephen Harper's Canada* (Vancouver: Greystone Books, 2013).

Study Questions

1. The fundamental claim of social ecology is that the domination of nature is a product of human domination. Do you agree with this claim? Why or why not?
2. Westra connects the concept of ecological integrity with how the law works in Canada. Explain this connection.

3. What is the main message about environmental activism put forward by de Shalit and Barak. Do you agree with it? Why or why not?

4. What do you think about the Occupy movement? Was it a one-time protest of the disgruntled and marginalized against the powerful or is it likely to have repercussions in the future? What tensions and conflicts in the social order might cause a similar movement to resurface? How does all of this connect with our environmental concerns?

Further Reading

Janine M. Benyus. 2012. *Biomimicry: Innovation Inspired by Nature*. New York: HarperCollins.

Murray Bookchin. 2004. *Post-Scarcity Anarchism*. Edinburgh: AK Press.

———. 2005. *The Ecology of Freedom: The Emergence and Dissolution of Hierarchy*. Edinburgh: AK Press.

Ross Jackson. 2012. *Occupy World Street: A Global Roadmap for Radical Economic and Political Reform*. Totnes, UK: Green Books.

Andrew Light. 1989. *Social Ecology after Bookchin*. London: Guilford Press.

Robert Paul Wolf. 1970. *In Defence of Anarchism*. Berkeley: University of California Press.

Part  III

Environmental Issues

In Part III, Environmental Issues, we examine four key environmental problems. The first (Chapter 11) is climate change—placed first because it is in many ways the biggest and most complex environmental issue of our times. Trying to solve this problem will test our ability to cooperate on a global scale, our courage to enact tough and far-reaching policy measures, and our compassion for the people who will suffer because of it. In this chapter, we look at two broad issues: who is responsible for climate change and how we can explain and possibly overcome the political inertia in which we seem mired at the moment. The next chapter (Chapter 12) is about two closely related problems: our population and how much we consume. The central question of the chapter is how to balance the need to bring people out of poverty—especially those whom Paul Collier calls "the bottom billion"—with the need to drastically reduce the environmental impact of our presence on the planet. It may be that we are too sanguine about our ability to do both of these things, in which case we may need to make hard choices about which goal is more important to us and why. Next, we look at the biodiversity crisis (Chapter 13). Humans are having a massive impact on the ability of other species to survive. Indeed, our activities are causing the sixth great extinction event in the history of life on Earth. The chapter investigates the nature and value of species biodiversity to help us better appreciate the extent of the crisis we face. How did we come to this state of affairs, and how can we repair the damage we have done, or at least slow down our destruction of life? Finally, we look at the contested concept of sustainability (Chapter 14). It crops up everywhere in our culture, but it is not obvious that sustainability is a helpful or even meaningful concept. The chapter argues that the concept is, in fact, useful, but it needs to be defined carefully. It is appropriate to place this chapter last because of the sense that all of environmental ethics is ultimately aimed at laying out the conditions under which we can live more sustainably.

CHAPTER | **11**

Climate Change

Ecological Intuition Pump

Much of environmental ethics concerns the world we are leaving for posterity. We wonder whether they might despise us for failing to bequeath a richly biodiverse and generally healthy planet to them. This is how many philosophers urge us to think about climate change, the impacts of which will be visited mainly on future people. But philosopher Derek Parfit has famously asked what basis posterity could possibly have for taking this critical view of us. Suppose that the identity of future people is radically contingent on various decisions taken by prior people. You, for example, would not have been born at all had your parents not conceived you just when they did, and this would not have happened had they not met exactly when they did, which would not have happened had the coffee shop in which they met been closed that Sunday, and so on. Similarly, if we adopt policies that are toxic to the environment, this may cause different future people to be born than would have been born had we adopted greener policies. Suppose we choose this toxic future world. On the assumption that life is still worth living in that world, doesn't it follow that those born into it have no cause to blame us for the state of their world, since *they* would not have been born had we not adopted our toxic policies? Still, even if there is no way for us to say that those future people have been harmed by us, can we not say that we are blameworthy simply for having failed to produce the best world, among the options available to us?

Ⓐ Introduction

In October 2009, 76 Tamil refugees from Sri Lanka arrived in Vancouver harbour aboard the *Ocean Lady*. Then in August 2010, nearly 500 more Tamils—passengers on the *Sun Sea*—landed, this time in Esquimalt, BC. The arrival of all these people, and their subsequent attempt to remain in Canada as refugees from their war-ravaged country, sparked a sometimes heated debate among Canadians about how we should deal with this sort of situation. This event can be seen as a small precursor to the reality that likely awaits us in a world in which the effects of climate change become increasingly apparent and more dramatic. According to some estimates, in the next 50 years the world could see anywhere between 50 million and 1 billion climate refugees, people driven from their homelands by severe climatic disruptions. And since high-latitude countries like Canada will probably not be subject to the sorts of ravages that will befall mid-latitude countries (especially those in sub-Saharan Africa and Southeast Asia), they will surely become very attractive destinations for climate refugees. How will we react when these desperate people, in their tens or hundreds of thousands, turn up on Canadian shores?

This is one among many wrenching questions with which we will undoubtedly have to grapple in a world transformed by climate change. For according to the best science we have, our world is about to become significantly warmer, a fact some believe could threaten human civilization. How did our species come to this dangerous point? **Anthropogenic climate change** results from human activities increasing the atmospheric concentration of greenhouse gases (GHGs), a process that alters the energy balance of the climate system. Carbon dioxide is the most important greenhouse gas, although methane, nitrous oxide, and water vapour are also important. These gases work as a kind of blanket over the Earth: they allow incoming radiation from the sun but absorb, and therefore trap, some of the infrared heat radiating back from the Earth. This effect is what allows for life on Earth: our planet possesses a delicately poised concentration of greenhouse gases in the atmosphere. Too great a concentration of greenhouse gases and the planet would be too hot to support life; too little and it would be too cold to do so.

Ever since the work of the nineteenth-century scientists Jean-Baptiste Fourier and John Tyndall, we have known about the absorptive power of greenhouse gases. But in 1938, the British engineer Guy Callendar showed that there is a causal connection between the combustion of fossil fuels and the atmospheric concentration of greenhouse gases. Fossil fuels are rich repositories of carbon. They are the remnants of plant and animal life on Earth, compressed and stored beneath the surface of the Earth over millions of years. When we haul them out and burn them, we release their stored carbon into the atmosphere, where it acts as a warming agent. Since Callendar's time, our understanding of this causal relation has improved dramatically. In its fifth and latest report on climate change, the **Intergovernmental Panel on Climate Change (IPCC)** notes that "it is extremely likely that human influence has been the dominant cause of the observed warming since the mid-20th century. The evidence for this has grown, thanks to more and better observations, an improved understanding of the climate system response and improved climate models."[1] Whereas the atmospheric concentration of carbon dioxide was about 280 parts per million (ppm) in pre-industrial times, it stood at 379 ppm in 2005 and has now exceeded 400 ppm. Human-induced climate forcing comes primarily from the coal, oil, and natural gas we burn for three broad purposes: electricity and heat, transportation, and industry. Together, these uses comprise nearly 55 per cent of the CO_2 we emit. A good deal of such emissions is also produced—in the form of both CO_2 and the very powerful greenhouse gas methane (CH_4)—from land-use changes (chiefly deforestation), industrial processes, and, especially, agricultural production. The result of all of this activity is what the IPCC calls "unequivocal" warming of the climate system, evidenced by increases in global average surface temperatures and ocean temperatures, melting of snow and ice, and rising global average sea level.

We are now going to see a rise in temperature of 1.1 to 6.4 degrees (Celsius) this century relative to the pre-industrial baseline (and we seem headed for the upper end of this range). This may sound small, especially to those of us living in a cold climate like Canada's. But think of it this way. The difference between global average temperatures in the midst of the last ice age and now is about five degrees, and there is of course a profound difference between these two climates with respect to human habitability. We have set in motion equally profound changes in the opposite direction, changes that are now very likely to occur with astonishing rapidity. What harms might be caused by a temperature rise like this? The phenomena of most concern to scientists are warmer and more frequent hot days and nights over most land areas, an increase in the frequency and severity of heat waves, heavy precipitation events, an increase in drought-affected areas, an increase in the intensity of tropical cyclones and hurricanes, and an increased incidence of extreme high sea levels.

The impact on human societies will be profound. There will be more landslides, avalanches, and mudslides in some areas where rain will increase and more drought and attendant crop failure where rain has diminished. Indeed, there will likely be increased desertification around the world, especially in the Mediterranean area and the American southwest. A rise in sea level of between 10 centimetres and one metre would cause the displacement of millions of people living in coastal areas. Whole island nations—like the Maldives and the Solomon Islands—would disappear. The loss of significant portions of habitable land in a low-lying country like Bangladesh—where 20 per cent of the habitable land is just above sea level—could force millions of refugees to pour across India's eastern border with that country. Since India may itself be struggling to cope with water shortages caused by the drying up of the glaciers that feed rivers like the Brahmaputra, it will simply not be able to cope with this influx. Regional conflicts like this will flare up all over the world and could in some cases lead to war, possibly between countries equipped with nuclear weapons. These scenarios only scratch the surface of the harms that are in store for us. In short, as philosopher James Garvey puts it, because of climate change "there is going to be a lot of death in the future."[2]

Ⓑ Confronting Climate Change Denial

Before getting to the grave ethical challenges climate change poses, it is probably a good idea to confront the arguments of those who, in one way or another, deny that there *is* a climate change problem. In 1988, NASA's James Hansen testified before a US Senate committee that anthropogenic climate change is an indisputable fact and that we had better start doing something about it. This testimony was evidently a wake-up call for those with vested political and economic interests in maintaining the status quo. Since then climate change denial has taken wing, assuming many forms over the years. In this section, we will focus on three of these forms and show why each is misguided.

The first denialist claim is that although it may have looked as though there was significant global warming over the past half-century, the warming has now stopped. Indeed, average global surface temperatures have been falling since 1998, proof that the planet is now in a cooling phase. The error here involves what is often called "framing" or "cherry-picking." It is no accident that those making the claim cite 1998 as the starting date of the alleged cooling per-iod. That year saw an extreme upward spike in average global temperatures, smashing many records across the globe. So, of course, the years that follow this unusually hot year will show a downward trend. However, this only means that to get an accurate record of temperature change over a long period of time, we need to smooth out the data in an effort to detect the climate signal amidst all the weather noise. When we do this, the data clearly show a steady upward curve in temperature. No overall cooling trend is evident. Here's another example of the same error. Deniers sometimes point to the fact that in 2009 there was a 9.4 per cent *gain* in Arctic sea ice compared to the previous year. They then draw the immediate conclusion that there is no climate change or at least that the climate change threat has been overblown. Again, this sort of cherry-picking ignores long-term trends and draws hasty conclusions from insufficient data. In the case of Arctic sea ice, that trend is now clear: the Arctic could be ice-free in the summer as early as 2025 or 2030.

The second denialist claim is that climate change is real but is not anthropogenic in origin. On this view, what climate change we are seeing is the product of purely natural climatic variations of the sort that have occurred often in the history of the planet. For example, the increase in average global surface temperatures over the past 30 years or so can be attributed to changes in solar output. There are two responses to make to this challenge. The first is directed at the particular claim about changes in solar output. The problem with this view is that if it were correct, the warming in the stratosphere (upper atmosphere) should be equal to or greater than that in the troposphere (lower atmosphere) in the relevant time period. After all, the stratosphere is closer to the sun. But models and observations show *exactly the opposite*, from which it follows that the increased warming is coming, as it were, from the ground up. Second, and more generally, IPCC data show conclusively that the observed warming cannot be explained with sole reference either to anthropogenic causes or to purely natural causes but only by combining the two causes. Moreover, of the two causes the anthropogenic is clearly dominant.

The larger point here is that the greenhouse effect is a well-established and elegant explana-tion for the warming that we've seen and it is simply indisputable that human activities have enhanced this effect. Here is how climatologist David Archer puts the point:

> Shifting the blame to something else would require an explanation of why the CO_2 would not be trapping the heat as we expect it would be doing. Think of it like a murder mystery. The butler (CO_2) was caught with a smoking gun in his hand in the room with the dead guy. . . . Yes, the bullets came from the gun. . . . Yes, the gun was purchased by the butler. Everything checks out. But now your partner, Bob . . . argues that it was really the chauffeur [who] did it. Actually you find out that the chauffeur was at his sister's wedding on the other side of town for the whole time and lots of people saw him. . . . [I]f Bob is going to convict the chauffeur, he has to think of a way to unconvict the butler. He would have to come up with an innocent explanation for the butler's smoking gun and the bullets, and all that.[3]

Similarly, those who say that there must be some other cause for climate change must not only find the guilty source but also explain how we should go about *unconvicting* the obvious culprit, human emissions of CO_2. That, as Archer understatedly puts it, is a "tall order."

The responses to the first two challenges show (a) that the climate change observed since the industrial period and especially over the past half-century is real and continuing and (b) that it is chiefly anthropogenic in origin. The final denialist claim is different. It says that although anthropogenic climate change is undeniable, it is not as big a threat as it is sometimes made out to be. So although this isn't really denialism in the technical sense, it makes sense to consider it here anyway, because just like the first two claims, it implies that we should not take the crisis as seriously as most climatologists believe we should. If adopted, this stance would have profound repercussions on public policy. Indeed, according to the view's most outspoken advocate, Bjorn Lomborg, spending significantly to fight climate change is wasteful in view of the other important things we could and should be doing with our scarce resources (fighting AIDS, reducing poverty worldwide, combatting malaria, and so on).

Lomborg's views get quite a lot of attention these days, but it is not obvious that they should. His central claim, that we should not spend very much to combat climate change, is itself based on the supposition that global warming is "no catastrophe." That is, he believes that the dangers of global warming have been exaggerated by those arguing on the other side of the issue. His chief piece of "evidence" for this claim, however, is that global warming will probably result in lower death rates, since many more people currently die of cold than of heat. Turn up the heat, and although some will surely die of it, many more will be saved from death by cold. There are two responses to this.

First, Howard Friel has demonstrated that Lomborg's data are extremely suspect. For example, Lomborg claims that about 1.5 million Europeans die each year from excess cold and that this exceeds the number of those dying from excess heat by seven times. However, as Friel points out,

> Lomborg's only referenced source for these figures—a chart in the statistical annex of a 2004 World Health Organization report—contains no data on human mortality due to excess heat or cold. In fact the words "excess heat" and "excess cold" make no appearance in the WHO document.[4]

Friel details the manner in which almost all of Lomborg's key statistical claims are similarly unfounded. The result is clear. The normative claim that we should do nothing about climate change is based on the factual claim that climate change will have net health benefits for us. But since the factual claim is not grounded in reliable data, we should reject the normative claim too.

But suppose Lomborg did have the data to back up his factual claim. Should we then take the normative claim seriously? This question brings us to the second point about his analysis. The problem with it is that it fails to consider the indirect health effects of increased temperatures, those that go beyond death by heatstroke. The IPCC is clear that increased heat is likely to lead to increases in drought and consequent starvation, incidences of vector-borne diseases like malaria, cardio-respiratory morbidity, and mortality associated with ground-level ozone, more intense storms and the destruction they bring, and so on. Hundreds of millions, perhaps billions, of people will be killed or have their health adversely affected by such events. And they are all related to increased temperatures. Unlike Lomborg, the IPCC actually provides a balanced view of the matter:

> Overall, climate change is projected to have some health benefits, including reduced cold-related mortality, reductions in some pollutant-related mortality, and restricted distribution of diseases where temperatures or rainfall exceed upper thresholds for vectors or parasites. However, the balance of impacts will be overwhelmingly negative. . . . The analyses suggest that . . . benefits will be greatly outweighed by increased rates of . . . infectious diseases and malnutrition in low-income countries.[5]

We should accept the conclusion of the overwhelming majority of informed scientists that climate change is real, that it is anthropogenic in origin, and that it poses serious threats to the lives, health, and security of present and (especially) future generations, most notably those in low-income countries. These are clear harms for which someone is morally accountable. But who?

ⓒ Responsibility for Climate Change

Where we invoke harms, we need to ask about blame and responsibility. In a catastrophe of this magnitude, is it possible to assess responsibility in an accurate manner? It had better be, because before we can even begin to clean up the climate mess, we need a clear picture of exactly who made it or is making it. A fruitful way of thinking of the climate crisis is by reference to Garrett Hardin's notion of a **tragedy of the commons**. Hardin asks us to picture a pasture freely open to a defined group of herdsmen. Every one of these herdsmen will, in rational pursuit of his own interest, try to graze as many cows on the commons as he can. At any point in the process of adding animals, each herdsman might ponder this question: Should I add one more animal to the commons? And it would seem that the answer will always be yes. The addition of one animal increases the wealth of the herdsman incrementally and does not ruin the land. More specifically, the herdsman will reason that the cost of whatever damage is done to the land will be shared by all the herdsmen, while he alone will reap the benefits of the extra cow. Simple cost-benefit analysis indicates that he should add that cow. However, since each herdsman makes the same calculation, the result is that the commons is overgrazed. What looks rational from an individual standpoint is irrational from the collective point of view. Thus, "freedom in a commons brings ruin to all."[6]

What does all of this have to do with climate change? A great deal, because the atmosphere is a commons. It is a *sink* whose capacity to absorb the carbon we emit is essentially finite. This does not mean that at some point the atmosphere will simply stop taking that carbon, only that there is an upper boundary on the amount we can put up there if we want to avoid dangerous interference with the climate system. But each country is just like one of Hardin's herdsman. Our reasoning has two components, not always explicitly separated. First, we think that just one more year of unrestrained economic growth powered by the consumption of fossil fuels will not *by itself* wreak the climate system. After all, we are just one country, perhaps a relatively small one at that. Second, we calculate that if we do restrain ourselves, others likely will not, which puts us at a comparative economic disadvantage. The result is predictable. Since every country has these thoughts, or behaves as though it does, the atmospheric commons is ruined.

But these thoughts make it seem as though we are all equally to blame for climate change, and this is surely incorrect. For an intriguing picture of proportional responsibility for this crisis, let's turn to the first article of the chapter.

AMERICAN DISENLIGHTENMENT, OR CLIMATE CHANGE MADE IN THE USA

Martin Schönfeld

Climate change is a universal harm.[7] Some consequences are well known, such as the random impacts of more frequent droughts, floods, and storms, or the challenges to public health through more heat weaves, elevated pollen counts, and tropical diseases spreading northward. For non-human life, climate change means a reduction of biodiversity, both on land (via global warming) and in the seas (by marine acidification). For human life, and ultimately for everyone, the impacts of climate change on security, water, and food are insidious.

All the damage being done makes one wonder who's to blame. Generally, this is easy to answer: climate change is largely made in the United States. It is the planetary legacy of the American century. Another and perhaps more vexing issue is the question of why the US finds itself now on the wrong side of history. Why did this highly literate and democratic nation make this happen? What turns climate change into a cognitive and moral blind spot for the US, at this juncture of history, in its ideology and in its overall culture?

Section I is a review of how the US perpetration of climate change unfolded in the first decade after the turn of the millennium. Section II is an account of how affairs evolved more recently, with the US now being polarized domestically and isolated abroad in its environmental policies. Climate change is neither a scientific puzzle nor a technological challenge. It is a cultural problem. The causal structure of climate change is well understood. Scientists know what's happening and tell us what to do. The hardware for a post-carbon switchover to a sustainable civilization is available. Engineers have given us the technology we need. So we know what's going on, we know what to do, we have the tools we need—but in the United States at least, we just don't do it. Section III is an exploration of the cultural foundation of climate change. What might explain this strange combination of perpetrating climate change and denying that it is happening at all? What, in this regard, is the nature of the "American Disenlightenment"?

I. 2000–2009: The Stupid Decade

The first decade of this century should be called "The Age of Stupid." This is also the title of a

2009 sci-fi movie about an old man in a future blighted by climate change.[8] The old man wonders why we, the generation before him, didn't stop global warming when we still had the chance. The crucial tool for doing something about it, for curbing GHG emissions and mitigating climate change, would have been a global treaty to follow the Kyoto Protocol.

Blame spreads collectively, but one nation bore the most responsibility for bringing the Age of Stupid about: the United States. By 2009, the US had emerged as the main perpetrator of climate change. This is no polemic, because climate forcing can be gauged by its primary chemical driver, the atmospheric concentration of CO_2, and because the responsibility for climate change differentiates into the proportionate fault of the "global villagers"—the nation-states that make up world civilization. The UN Statistics Division assesses human-made CO_2-emissions by year and country; by year, country, and the nation's area; and by the change of national annual emissions from a country's 1990 baseline.[9] In the past century and in the early 2000s, the US had been the largest national emitter. At the end of the stupid decade, a comparison of countries in terms of annual emission showed China coming first, with the US second.[10] Comparing countries in terms of per capita emissions showed the small Middle-Eastern nation of Qatar coming first.[11]

China has a thousand times more people than Qatar, and its total annual emissions are a hundred times more than those of Qatar. By 2006, when China overtook the US in emissions, it made the biggest carbon footprint of all nations[12]—as it well should, one could argue, since one-fifth of humankind lives in China. Justice involves the idea that everyone gets their fair share. This suggests that everyone would be entitled to the same share of a commons like the climate system. It seems fair that two humans can have twice and four humans four times what one human can have. By this argument, China would have the right to create one-fifth of humankind's total emissions, which, coincidentally, is rather close to its actual CO_2 output. This exonerates China.

Qatar has the biggest per capita carbon footprint. But nations differ in size and in function. Qatar's footprint is outsized because it is basically one big factory for processing natural gas before feeding it into the world market. Blaming this factory for climate change is like blaming the village gas station for the air pollution caused by all the drivers in the village. This exonerates Qatar, too.

Combining the national and per capita emissions brings us closer to the actual metrics of climate forcing. The 2009 UN data showed the following top 10 countries ranked by total emissions: China (1), USA (2), Russia (3), India (4), Japan (5), Germany (6), Canada (7), UK (8), South Korea (9), and Mexico (10). The top 10 countries ranked by per capita emissions were Qatar (1), United Arab Emirates (2), Kuwait (3), Bahrain (4), Luxembourg (5), Trinidad and Tobago (6), Dutch Antilles (7), Aruba (8), USA (9), and Australia (10).

The United States figures on both rolls. On the per capita list, it is the only country that is large (not a small island or a minor sovereign entity) and consumes (not merely produces) fossil fuels. The top three are OPEC members. The fourth, Bahrain, earns revenues through hydrocarbon industries such as petroleum production. Refineries are emissions-intensive: in a thinly populated country, they translate into giant per capita carbon footprints. The top three plus Bahrain serve as gas stations in the global village: they wouldn't have such emissions were it not for strong consumer demand for their products elsewhere.

The next four on the list, five through eight, are mere specks on the map. Three are Caribbean islands; one is a European duchy. With 100,000 to 500,000 citizens each and up to 20 times the population density of the US, they resemble cities more than countries. The third UN assessment metric—CO_2 emissions by square kilometre—illustrates this structure: because of the spatial compression of small, urbanized countries, their emissions per square kilometre tend to be one order of magnitude greater than those of other, larger countries.[13]

This pushes the US to the per capita top, inviting comparison with China, the total emissions leader. China, as we have seen, has a demographic reason for its huge CO_2 output: with 1.4 billion people by the end of the first decade of the twenty-first century, it is a whale among nations.[14] The US CO_2 output of 5.975 billion metric tons in 2009 was only slightly less than China's 6.1 billion metric tons. But the US had only 300 million people by 2009.[15] Producing as much CO_2 as one-fifth of humankind, while amounting to less than 5 per cent of the world population, seems unfair. The near-equality of Chinese and American emissions means that Americans, on average, had four and a half times the carbon footprint of Chinese citizens at that time. In total annual emissions, the US ranked second but lacked China's demographic justification. It also matters that much of China's growing carbon footprint in the stupid decade was due to the manufacture of consumer goods for the United States.

The real perpetrators of the climate crisis are the *cumulative perpetrators*—the highly developed nations. Great quantities of coal, oil, and gas have been burned far longer there than anywhere else. From 1850 to 2009, developed nations injected 76 per cent of the total CO_2 emissions into the Earth system. The other 24 per cent originated in developing nations such as China, all newcomers to the GHG club.[16] According to recent calculations (2013) of historically accumulated carbon footprints, the US is number one on the cumulative list: from 1850 to 2010, in terms of all GHG emissions, the US contributed 18.6 per cent to the total (all 27 countries of the more populous European Union contributed 17.1 per cent); in terms of energy CO_2 emissions, the US contributed 29.7 per cent of energy CO_2 emissions (24.8 per cent for the EU).[17]

Overall, the United States, whose share in the world population was 4.5 per cent in 2010, has caused a quarter of cumulative GHG emissions and nearly one-third of cumulative CO_2 emissions. This means that the US is responsible for more climate forcing than South America, Africa, the Middle East, Australia, Japan, and Asia—all put together.[18]

II. 2010–2014ff.: The Polarized Decade

Halfway through the second decade of the twenty-first century, the US has become a divided society, and the global village is a multipolar world. The polarization of US society is an economic and political phenomenon. There is now more poverty and more extreme wealth; the middle class has contracted, and young Americans have less money and fewer well-paying job prospects than their parents did when they were young. Economic stratification goes hand in hand with political polarization. The American poor, scared and angry, are easy prey for evangelical fundamentalists and right-wing talk radio hosts; the American rich, seeing the chance of a lifetime, lobby their politicians to lower taxes, to reduce social services, and to aid and abet a freely capitalistic oligarchy.

The polarization mirrors a radicalization of American conservatives. The election of the Democrat Barack H. Obama as first black president of the United States in 2008 triggered a backlash—the rise of the libertarian and predominantly white Tea Party movement, which arrived on the national scene in early 2009. Republicans embraced Tea Party ideas, especially its market ideology and rejection of climate science. Climate denial, in particular, defines Republican policy at federal and state levels. Democrats are on the other side of this divide, and here climate denial is marginal. Barack Obama stated his desire to revise US climate policy in the 2008 presidential election; in 2014, his administration made a new push, the Climate Change Action Plan.[19] In May 2014, the House of Representatives passed an amendment to the National Defense Authorization bill to prohibit the Pentagon from using any funding set aside by the White House for security impacts of climate change. The goal of the amendment is "to block the Department of Defense from taking any significant action related to climate change or its potential consequences."[20]

Analysis of the 2012 dataset by Oakridge National Laboratories (US) and the Netherlands Environmental Assessment Analysis (EU) puts the national contributions of global villagers to climate forcing in context. The proportionate responsibility of top planetary polluters differentiates according to (a) total national emissions per year of CO_2 from fossil fuel use and cement production; (b) national per capita emissions per year of CO_2 from fossil fuel use and cement production; (c) national per capita emissions per year of GHG emissions in general; (d) cumulative national contributions to climate forcing according to current data; and (e) the likely cumulative contributions of individual nations projected to 2030.

The United States figures on all of these lists. First, the US is the biggest national emitter of CO_2 after China. But in the polarized decade,

too, China remains excused: it has four times as many people; it is in the process of completing a First World infrastructure (a construction effort that requires steelmaking and concrete production, both of which add 10–20 per cent to its carbon footprint); and, like Germany, it is creating the world's boldest and most visionary post-carbon economy. China is the world's largest producer of photovoltaics. Its investments in renewables is so massive that it became the driver of global growth in solar installations in 2014.[21] By 2017, China plans to triple currently installed solar capacity to 70 gigawatt (GW) and to boost wind power to 150 GW.[22] (By comparison, a nuclear power plant produces 0.5-1.3 GW on average.) The US, as the number two national emitter, cannot boast of any of this.

Second, the US has the second-largest per capita CO_2 emissions. Number one is Australia. Unlike China, Australia has absolutely no excuse and also has nothing to show for itself: with its national CO_2 emissions rising every year by more than 2 per cent and now being more than 60 per cent higher than they had been in 1990, the Australian free-market government scrapped the carbon price, allowed a national coal seam gas industry to expand, and couldn't be bothered to send a minister to the COP-19 climate talks in Warsaw in 2014.[23] Like the US, Australia has become a problem and is not yet part of global solutions.

The US also has the largest per capita GHG emissions of any country that is neither a city-state (like Aruba or Luxembourg) nor a gas station for the global village (like Qatar or Kuwait). In this metric, the US is worse than Saudi Arabia. It is also worse than Australia: unlike the polluter down under, the US tops both per capita lists.

Finally, the US continues to be number one on the cumulative GHG emissions list. Historically and morally, it is this list that is crucial. As soon as the international blame game begins (which isn't a question of if, but a question of when), all fingers of the human society will point in one direction. Thus, even today, after the end of the stupid decade and well into the polarized decade, global climate change is made in USA.

III. The American Disenlightenment

Kevin Phillips coined the phrase "American Disenlightenment." He means by this "the perils and politics of radical religion, oil, and borrowed money." Disenlightenment is a cultural phenomenon, and if it lasts long enough, it carries the risk of national decline, especially of nation-states with imperial ambitions, such as the US.[24] There is a complex relation between decline and disenlightenment. Decline, as a process of economic, political, and imperial dissolution, is the outcome of prolonged cultural disenlightenment, and in that sense disenlightenment prepares the ground for decline. But disenlightenment is not only a driver, but also a symptom, and thus an effect, of decline. This suggests a negative feedback, which makes disenlightenment hard to treat. People sometimes only wake up when chastened by a downfall. The peril is then for neighbors and for future generations—how much collateral damage will the downfall do?

Disenlightenment is the cultural expression of a cognitive disorder: a collective disconnect from reality in a social setting that, for whatever reason, encourages, endorses, and demands such a disconnect. A reality-disconnect is a willful disregard for objective (scientific, peer-reviewed) information for the sake of ideological fantasizing and wishful thinking. A disconnect is the causal link of disenlightenment to decline. Societies suffering from reality-disconnect risk catastrophic failure—think of the Easter Islanders of ancient times or of Nazi Germany more recently.

What the opposite, enlightenment, means is ambiguous. In the West, this term, together with the definite article and capitalized—*the* Enlightenment—denotes the period from the late 1600s through the 1700s, also known as the Age of Reason. The Enlightenment was a progressive time. The lives of central Europeans became safer, with a tax-revenue system in Prussia to benefit the commonwealth and to protect the poor. Science came into its own, especially physics, mathematics, and chemistry. Real political advances happened: church and state were divided; authority was subjected to a system of checks and balances; kingdoms became republics (as in France), and colonies became democracies (as in America).

In the East, enlightenment, not capitalized and without an article, describes a heightened cognitive state, a spiritual mindfulness. In Zen Buddhism, this is known as "seeing the nature" (*guan ziran* 觀自然 in Chinese) or as "awakening" (*satori* 悟 in Japanese).

A third sense of enlightenment is a secular social ideal, suggested by the German thinker Kant. For Kant, who lived in the Enlightenment and whose predecessors, Leibniz and Wolff, were influenced by Chinese enlightenment via Confucianism, the "spirit" of enlightenment can be expressed in two calls to action: one, *have the courage to use your own reason!* and two, *dare to become wise!*

This third and Kantian sense ties the Western and Eastern variants together. The link is on a cognitive level. The Enlightenment, in the first sense, was a time of advances, both in theory, with an unprecedented progression in the natural and exact sciences, and in practice, with the ideas of human rights and democratic constitutions taking Europe and the American colonies over by storm, which laid the foundation for the EU and the US of today. The Enlightenment was what it was *because information mattered.* Science ruled—and so did common sense, social justice, and compassion. Now, enlightenment, in the second sense, turns on the notion that a well-ordered undercurrent structures the material world. This structure or undercurrent (called the "Way" or Dao 道) gives us the parameters of existence; it interacts with our spirit, the "heart-mind" (心, Chin. *xin*, Jap. *kokoro*), and it is assumed to be compatible with physical reality as known to science. The "awakening" in Eastern enlightenment is a disciplined willingness to see a bigger picture, the "seeing the nature." Both variants, West and East, hinge on the idea of paying attention to reality. What reality is, depends on context. For the Western Enlightenment, it was the material world or physical reality as described by science and accessible to reason. For Eastern enlightenment, it is the deeper structure of the world. In either meaning, reality is a mind-independent, orderly matrix. Opening one's eyes to it is the opposite from sticking one's head in the sand or from staying in denial. So one could say, with Kant, that an enlightened approach to reality is the willingness to revise beliefs when data demand it, to heed the facts, and to contemplate a bigger picture, with clarity and compassion, especially if and when it is inconvenient or unsettling to do so.

This pulls the American disenlightenment into focus. Part of the explanation for it is certainly America's addiction to fossil fuels. The carbon commitment is also tradition to such an extent that it has created an infrastructural lock-in. Just as the empires of the Dutch, the Portuguese, and the Spanish had been built on the smartest use of wind power in their day, and just as the British Empire was built on coal, the American empire was made possible by an abundance of domestic oil. To this day, three of the world's largest oil multinationals—ExxonMobil, Chevron, and ConocoPhillips—

hail from the US. Oil is the basis of American military might and the foundation of American prosperity. In a way, then, oil *is* America. In this situation, the news that a universal harm such as climate change is brought on by the combustion of oil and other fossil fuels is just about the worst possible turn of events for Americans. No wonder it catches US conservatives with their pants down, forcing them into denial.

Let's look at American culture from a distance. What are its features? In asking this, we are not required to consider only features that are mutually consistent with one another. Cultures, like people, are not rational structures. A cultural character, like a personality, is hardly ever consistent. Contradictions weaken a scientific theory. But they can strengthen a character, creating dynamic tensions like the "ornery but lovable uncle" or the "gritty cowboy with a soft heart" that make the blend of traits all the more memorably and, paradoxically, coherent.

So how does the US differ from other societies with a certain cultural cohesiveness, such as Canada, China, or Germany? Four features come to mind. First, the religiosity of Americans is striking. In contrast to these other cultural regions, the US prides itself on being "God's own country," with "in God we trust" minted and printed on its currency. American religiosity is typically monotheistic, usually Judeo-Christian, and predominantly Protestant. Compared to other Christian creeds (such as Catholicism, Lutheranism, or Unitarianism), the fundamentalist worship of Jesus in Protestantism is the most uncompromising of all.

US faith in the free market is equally striking. Capitalism dominates the US as it does Canada, China, and Germany, but only in America does it take on a quasi-religious form. Moreover, nowhere at present is capitalism as unregulated as there. Adam Smith is the prophet of this culture.

Individualism is a third feature. "Liberty" and "independence" are central to US normalcy. In the United States, communitarian ideas seem foreign. US conservatives are libertarians, advocating the freedom of the market. US progressives are liberals, advocating the freedom of lifestyle choices. Both libertarians and liberals stress the self or the individual over others and the community. And the one philosopher ordinary Americans tend to be familiar with is the best-selling ethical egoist Ayn Rand.

There's a fourth feature, one that points beyond the lure of denial to the core of the problem. If we personify the first three traits with (a vengeful, fundamentalist, Tea Party) Jesus, with Adam Smith, and with Ayn Rand, then this final trait of the American disenlightenment evokes the empiricist philosopher and sceptic David Hume. Empiricism is characterized by the attitude that something counts as real only if it's right "there"—if I can point to it. Weather fits the bill but climate does not. Weather is tangible, localized, and observable. But climate is weather in an area over a long time, the minimum interval unit being 30 years.

This, then, is the cognitive square that defines the culture of the American disenlightenment: fundamentalist religiosity, libertarian capitalism, egotistical individualism, and sceptical empiricism. Any culture informed by these traits is bound to resist climate information. Tragically, this very resistance has turned not only climate change into the massive problem it is now, but also the United States of America into the primary, cumulative, and proportional perpetrator.

However, this cognitive square of disenlightenment doesn't really describe US normalcy anymore (as it did in the stupid decade). Rather, it describes its *conservative* version. Normalcy has fragmented into opposing subcultures. This is the polarized decade, and there are two Americas now.

And this allows us to end on a hopeful note. Future generations, who will have to make do in a tougher world, may well judge the conservative normalcy of the US disenlightenment as the great scourge of global civilization in this century.

But chances are, they won't identify this scourge with America.

One of the virtues of Schönfeld's striking analysis is that he forces us to grapple with the possibility that the climate crisis was not inevitable. Rather, it is largely the product of a single culture's world view and the principles and ideals that underlie that world view. In particular, Schönfeld sees four modes in America's "flawed cognition" vis-à-vis climate change. They are flawed because the fact of climate change proves them all indefensible.

Were these four ideals—religiosity, individualism, capitalism, and scepticism—not so firmly entrenched in American culture, we might not have arrived at such an impasse. But here we are, so where do we go from here? Pointing the finger of blame is not an end in itself. We should do it only insofar as it helps us arrive at a philosophically defensible way of assigning duties. Here, James Garvey is helpful. He has argued that we should rely on three criteria of moral adequacy for any proposal about how to tackle climate change. The first concerns historical responsibilities. Since some countries—the "cumulative culprits," as Schönfeld calls them—are more historically responsible than others for the climate crisis, they ought to bear a larger share of the cleanup burden. Second, there is the fact of current capacities. If a country has benefited historically from greenhouse gas emissions, then, other things being equal, it is probably relatively prosperous now and can therefore afford more in the way of cleanup than a country without this history. The third criterion has to do with sustainability. Any decision about how to tackle the climate crisis must be constrained by how the legitimate interests of future generations will be affected by the decision. It is helpful to think of these three criteria in temporal terms: the past (historical responsibilities), the present (current capacities), and the future (sustainability).

It seems natural to suppose that all humans have an equal right to use whatever is left of the carbon sink. As Garvey puts it, "a finite and precious resource should be distributed equally unless we have some morally relevant criteria for departing from equality."[25] Let's assume that no such criteria are forthcoming. With the goal of emissions equality in mind, a fair application of Garvey's first two criteria of moral adequacy (just discussed) indicates that industrialized Western countries, led by the United States, ought to bear the lion's share of the financial burden of addressing climate change. Developed countries have become rich on the back of greenhouse gas emissions. This is a fact about both their historical responsibility and their current capacities. So even though we have in view an equal distribution of the atmospheric resource, developed countries will have to absorb more of the cleanup cost than developing countries. Perhaps, some have suggested, all of it.

However, this is not the end of the story, because Garvey's third criterion is meant to act as a constraint on the first two. Suppose we calculate that the developed world is responsible for all of the cleanup, or nearly all of it. This conclusion might make them reluctant to do anything at all about the problem. Indeed, this kind of thinking has been present in the public pronouncements of many Western leaders, including Canada's Stephen Harper, who always opposed the Kyoto Protocol (and who pulled Canada out of the treaty after the country originally ratified it). Kyoto made no requirements on developing countries. Harper's response, in effect, was that "we won't make any cuts if they don't." Setting aside the dubious principles underlying this attitude, since it would result in widespread political inertia, its effect is likely to be disastrous for future generations. And this violates the sustainability criterion of moral adequacy. An effective solution to the climate crisis must clearly involve *all* countries, even if some must shoulder a greater financial load than others.

⒟ The Problem of Political Inertia

Still, even if we could agree on the idea that all countries must participate in the climate cleanup, we seem to have made little progress toward getting anything meaningful done. Why is this? The science is in, developing countries like China and India are at least talking about doing their fair share to help, and yet we appear to be stuck. One possible explanation is that for many of us climate change is still somehow unreal. It's largely a problem for the future, and it is difficult to constrain present consumption significantly for the benefit of people who are not even born yet. Responding to this sort of perception, politicians (including Canadian prime ministers) promote business as usual, carefully packaged as wise and "balanced" reform. They offer subsidies to public transit users, propose voluntary targets for industry, vow to reduce the amount of carbon per output of GDP (so-called intensity targets), set emissions targets for future dates with no specific policy map to get us there, and generally paint themselves green. With respect to the sort of policies required to avoid climate change's more catastrophic scenarios, as laid out by mainstream scientists, these measures are all basically useless.

Let's be more specific. In May 2015, the Canadian government announced its intention to reduce emissions 30 per cent below 2005 levels by 2030. There are two things to say about this. First, it is an extremely unambitious target, well below that of the US or what was called for in the Kyoto Protocol. Second, there is in any case no reason to believe that the government will enact policies allowing us to reach it. This is because (a) the government refuses to place restrictions on the growth of our biggest single source of emissions, the tar sands; and (b) the government's own data show that we are already in the process of failing abysmally to meet our previous target: 17 per cent below 2005 levels by 2020.[26] Climate change may indeed be largely "made in the USA," but countries like Canada are evidently doing nothing to blunt its future impacts.

Meanwhile, temperatures and sea levels just keep rising. In light of all this, it might be suggested that what we may need is a good disaster, something to shake us out of our complacency and torpor. Even better, perhaps it would benefit us—in a perverse sort of way—to accept the fact, if it is one, that catastrophic climate events may soon be the norm. Not so fast, says Stephen M. Gardiner.

SAVED BY DISASTER? ABRUPT CLIMATE CHANGE, POLITICAL INERTIA, AND THE POSSIBILITY OF AN INTERGENERATIONAL ARMS RACE

Stephen M. Gardiner

In recent years, scientific discussion of climate change has taken a turn for the worse. Traditional concern for the gradual, incremental effects of global warming remains, but now greater attention is being paid to the possibility of encountering major threshold phenomena in the climate system, where breaching such thresholds may have catastrophic consequences. As recently as the 2001 report of the . . . IPCC, such events were treated as unlikely, at least during the current century. But the work of the last six years tends to suggest that these projections are shaky at best. . . .

In this paper, I want to explore some ways in which this paradigm shift may make a difference to how we understand the moral and political challenges posed by climate change and in particular the current problem of political inertia. I will examine two suggestions. The first is that abrupt climate change undermines political inertia. . . . The second suggestion is that this shift is in one respect beneficial: the focus on abrupt, as opposed to gradual, climate change actually helps us to act. On the one hand, it supplies strong motives to the current generation to do what is necessary to tackle the climate problem on behalf of both itself and future generations; on the other hand, failing this, it acts as a kind of fail-safe device, which at least limits how bad the problem can, ultimately, become.

My thesis is that these suggestions are largely mistaken, for two reasons. First, the possibility of abrupt change tends only to reshape rather than undermine the usual concerns; hence, the root causes of moral corruption remain. Second, the possibility may make appropriate action more, rather than less, difficult and exacerbate, rather than limit, the severity of the problem. Worst of all, it may provoke the equivalent of an intergenerational arms race.

I. Abrupt Climate Change and Political Inertia

Until recently, scientific discussion of climate change has been dominated by what I shall call "the gradualist paradigm." Researchers tended to assume that the response of natural phenomena to increases in greenhouse gas concentrations would be mainly linear and incremental, and this assumption tended to result in analogous claims about likely impacts on human and non-human systems. Hence, for example, the original IPCC report projected a rise in global temperature at an average of 0.3°C per decade in the twenty-first century, and typical estimates of the economic costs of impacts ran at around 1.5–2 per cent of gross world product. Such results are hardly to be taken lightly, and much of the first three IPCC reports was taken up with showing how and why they are matters of serious concern.

But recent research suggests that they may underestimate the problem. This is because there is increasing evidence that the climate system is much less regular than the gradualist paradigm suggests. In particular, there may be major threshold phenomena, and crossing the relevant thresholds may have catastrophic consequences. Scientists have been aware of the possibility of such thresholds for some time. But recent work suggests that the mechanisms governing them are much less robust, and the thresholds themselves much closer to where we are now, than previously thought. This suggests that we need an additional way of understanding the threat posed by climate change. Let us call this "the abrupt paradigm." . . .

Still, perhaps the news is not all grim. For it has been suggested that the possibility of abrupt change may help us out of our current problem of political inertia. Let me first sketch this problem and then suggest why it might be thought that the threat of abrupt change might help.

The fact that climate change poses a serious threat has been known for some time. . . . The IPCC's main conclusions have been endorsed by all major scientific bodies, including the national Academy of Sciences, the American Meteorological Society, the American Geophysical Union, and the American Association for the Advancement of Science. This consensus appears to be remarkably robust. Despite this, progress on solving the problem has been minimal. . . . [A]t the political level, despite some notable efforts, overall progress has been less than impressive. In the global arena, several weak agreements have been made but then broken. At national levels, even those nations who express the strongest commitment to tackling climate change publicly—such as the European Union, United Kingdom, and New Zealand—have had difficulties in restricting their emissions.

Plainly, there is a mismatch between the apparent seriousness of the problem and our collective institutional response. What accounts for this? . . .

[O]ne root of the problem lies in [its] inter-generational structure. The basic idea can be illustrated (in a simplistic way) as follows. Imagine a sequence of groups occupying the same territory at different times. Suppose, for the sake of simplicity, that each group is temporally distinct: no member of one group exists at the same time as any member of another group.... Call each group so conceived a "generation." Suppose then that the preferences of the members of each group are "generation-relative": they concern only things that happen within the time frame of that group's existence. Finally, suppose that there are such things as temporally extended goods: goods that have benefits and costs that accrue in more than one generation. For current purposes, let us distinguish two kinds of such goods: those that have benefits on one generation and costs in later generations can be called "front-loaded goods"; those that have costs in one generation and benefits in later generations can be called "back-loaded goods."

Other things being equal, in such a situation we would expect that if each group does exactly as it pleases, it will consume as many front-loaded goods as possible and eschew all back-loaded goods. But this observation appears to raise a basic problem of intergenerational fairness. Surely, the thought goes, there are situations in which a given group ought to forgo at least some front-loaded goods and invest in some back-loaded goods; presumably there are at least some constraints (of fairness, or justice, or some other such notion) on legitimate intergenerational behaviour. If this thought is correct, then we might expect that groups that are unaware of, or do not recognize, such constraints will tend to over-consume front-loaded goods and under-consume back-loaded goods.... Hence, one can expect a systematic bias in overall decision-making across generations. Let us call this "the problem of intergenerational buck passing" (PIBP)....

Elsewhere, I have argued that the PIBP is manifest in the case of climate change. On the gradualist paradigm, this looks especially likely.[27] But might the abrupt paradigm help? Initially, it appears so, since the potential proximity of the relevant thresholds appears to undercut the intergenerational aspect of climate change. Consider, for example, the following statement by the (now former) British prime minister Tony Blair:

> What is now plain is that the emission of greenhouse gases, associated with industrialization and strong economic growth from a world population that has increased sixfold in 200 years, is causing global warming at a rate that began as significant, has become alarming and is simply unsustainable in the long-term. I mean within the lifetime of my children certainly; and possibly within my own. And by unsustainable I do not mean a phenomenon causing problems of adjustment. I mean a challenge so far-reaching in its impact and irreversible in its destructive power, that it alters radically human existence.[28]

Blair's main claim appears to be that the impacts of climate change are both extremely serious and coming relatively soon.... If this is right, it seems to give current people powerful reasons to act. Again, the abrupt paradigm appears to extinguish a major source of political inertia.

II. Against Undermining

I shall now argue that . . . this appearance is deceptive. Instead, it is plausible to think that the possibility of abrupt climate change will actually make the . . . problem worse, rather than better. . . .

Let us begin with the intergenerational problem. Blair suggests that some impacts of climate change are serious enough to "[alter] radically human existence," "within the lifetime of my children certainly; and possibly within my own."[29] A rough calculation suggests that this means possibly within the next 26 years and certainly within the next 75 (or 58) years. At first glance, such claims do seem to undermine the usual intergenerational analysis. But this is too hasty. For the notion of proximity is made complicated in the climate change case by the considerable time lags involved—the same lags that give rise to the PIBP.

Consider the following. First, the atmospheric lifetime of a typical molecule of the main anthropogenic greenhouse gas, carbon dioxide, is often said to be around 200 to 300 years. This introduces a significant lagging effect in itself but obscures the fact that around 25 per cent remains for more than 1000. Moreover, many of the basic processes set in motion by the greenhouse effect continue to play out for more than 1000 years. Second, these facts have implications for the shape of the climate change problem. For one thing, the problem is resilient. Once the emissions necessary to cause serious climate change have been released, it is difficult—and perhaps impossible—to reverse the process. For another, the problem is seriously back-loaded. At any given time, the current impacts of anthropogenic climate change do not reflect the full consequences of emissions made up to that point. Finally, this implies that the full effects of current emissions are substantially deferred. Even if we are to reap some of what we sow, we will not reap all of it.

These points suggest that it is worth distinguishing two kinds of proximity: temporal and causal. When Blair claims that the impacts of

climate change are coming soon, he means to speak of temporal proximity: the impacts are near to us in time. But claims about causal proximity are different. Here the claim is that the point at which we effectively commit the Earth to an abrupt change by our actions is close at hand. Given the presence of resilience, serious back-loading, and substantial deferral, temporal proximity does not always imply causal proximity, and vice versa. This fact has important implications, as we shall now see.

Consider first a scenario where we are in a position to commit humanity (and other species) to a catastrophic abrupt impact, but we ourselves will not suffer that impact because it will be visited on future generations. In other words, there is causal, but not temporal, proximity. (Call this scenario "Domino Effect.") Several of the most worrying impacts currently envisioned seem to fit this scenario. For example, even very rapid ice sheet disintegration is presumed to take place over centuries such that its impacts are intergenerational; similarly, the limited work that has been done on deposits of methane hydrate in the oceans suggests that the associated impacts would not arise for several centuries, if not millennia. Hence, the real concern in these cases is with causal proximity. The worry is that by our actions we may commit future generations to catastrophic climate changes. However, such a scenario clearly raises, rather than undermines, the intergenerational analysis. So we will have to look elsewhere for a challenge to the PIBP.

A second kind of scenario would involve temporal but not causal proximity. Suppose, for example, that we are already only a few years from crossing a major climate threshold and that at this point we are already committed to doing so. The most obvious reason why this might be the case would be because, given the time lags, our past emissions make breaching the threshold literally inevitable. But it might also be that we are already committed because there are emissions that we are morally no longer going to be able to avoid, for example, because avoiding them would impose intolerable costs on current people and their immediate descendants. (Call this scenario "On the Cards.")

If it turned out that On the Cards characterized our situation, and if we knew that it did so, then the implications of the abrupt paradigm for political inertia would be more mixed than the basic objection to the intergenerational analysis suggests. First, and most obviously, On the Cards might simply reinforce inertia. Suppose, for example, that a given generation knew that it would be hit with a catastrophic abrupt change no matter what it did. Might it not be inclined to fatalism? If so, then the temporal proximity of abrupt change would actually enhance political inertia, rather than undercut it. (Why bother?)

Second, and less obviously, On the Cards may provoke action of the wrong kind. For example, assume, for simplicity, that the two main policy responses for climate change are mitigation of future impacts through reducing the emissions that cause them and adaptation to minimize the adverse effects of those impacts that can or will not be avoided. Then, the following may turn out to be true of On the Cards. On the one hand, the incentives for the current generation to engage in mitigation may at least be weakened and might disappear altogether. This is because if a given abrupt change is, practically speaking, inevitable, then it appears to provide no incentive to a current generation with purely generation-relative motivations for limiting its emissions. Perhaps the current generation will still have reasons to engage in some mitigation, since this might help it to avoid further impacts (including abrupt impacts) after the

given abrupt change. But the given abrupt change does no motivational work of its own. Hence, its presence does not help future generations. On the other hand, the incentives for the current generation to engage in adaptation might be substantially improved. If big changes are coming, then it makes sense to prepare for them. In itself, this appears to be good news for both current and future people. But there are complications. For it remains possible that the current generation's adaptation efforts may be unfair to the future. This point is important, so it is worth spending some time on it.

Let us consider [how] the improved motivation for adaptation provided by On the Cards may come into conflict with intergenerational concerns. . . . [C]onsidering only its generation-relative preferences, a current generation aware of an impending abrupt change will have an incentive to over-invest in adaptation relative to mitigation (and other intergenerational projects). That is, given the opportunity, such a generation will prefer to put resources into adaptation (from which it expects benefits) rather than mitigation (which tends to benefit the future). Moreover, even within the category of adaptation, the current generation will have an incentive to prioritize projects and strategies that are more beneficial to it (e.g., temporary "quick fixes") over those that seem best from an intergenerational point of view. . . .

[More] importantly, the proximity of the abrupt change may actually provide an incentive for increasing current emissions above the amount that even a completely self-interested generation would normally choose. What I have in mind is this. Suppose that a generation could increase its own ability to cope with an impending abrupt change by increasing its emissions beyond their existing level. (For example, suppose that it could boost economic output to enhance adaptation efforts by relaxing existing emissions standards.)

Then, it would have a generation-relative reason to do so, and it would have this even if the net costs of the additional emissions to future generations far exceed the short-term benefits. Given this, it is conceivable that the impending presence of a given abrupt change may actually exacerbate the PIBP, leaving future generations worse off than under the gradualist paradigm (or than they would be if the earlier generation had not discovered the falsity of that paradigm).

Furthermore, just like the original PIBP, this problem can become iterated. That is, if the increased indulgence in emissions by earlier generations intent on adapting to a specific abrupt climate change worsens the situation for a subsequent generation (e.g., by causing a further threshold to be breached), then the later generation may also be motivated to engage in extra emissions, and so on. In short, under the On the Cards scenario, we may see the structural equivalent of an Intergenerational Arms Race surrounding greenhouse gas emissions. Abrupt climate change may make life for a particular generation hard enough that it is motivated to increase its emissions substantially in order to cope. This may then increase the impact on subsequent generations, with the same result. And so it goes on. . . .

On The Cards shows that it is possible for abrupt change to make matters worse. But perhaps that scenario is too pessimistic. Hopefully, even though there is a sense in which the climate thresholds are close, it is not true that we are already committed to crossing one. Interestingly, this thought reveals a tension in the proximity claim that is supposed to undermine the intergenerational problem: to be successful, the threatened abrupt change must be temporally close enough to motivate the current generation but distant enough so as not yet to be "on the cards." This tension suggests that the argument against the intergenerational analysis presupposes a very specific scenario: that there is an abrupt change that would affect the current generation to which the planet is not yet committed but to which it will become committed unless the current generation takes evasive action very soon. (Call this scenario "Open Window.")

Several issues arise about Open Window. The first, obviously enough, is whether there is such a window, and, if so, how big it is. These are empirical questions on which our information is sketchy. Still, the preliminary estimates are not particularly encouraging. First, two parameters loom large. At the present time, scientists often say that there is a further temperature rise beyond that which the Earth has yet experienced but which is "already in the system." Estimates of this commitment typically range from 0.5 to 1.0°C, suggesting that a fair amount of climate change is already literally "on the cards." The vital issue then becomes how much more is on the cards, since we cannot stop the world economy (and so the current trajectory and level of global emissions) on a dime. Barring a sudden technological miracle, the answer to this question would also seem to be "a substantial amount." These facts suggest that we are already committed to any abrupt changes likely to arise in the short to medium term. Thus, On the Cards has substantial relevance.

Second, preliminary calculations suggest that our ability to avoid a more substantial commitment is limited. Consider, for example, the European Union's call for limiting the global temperature rise to 2°C in order to avoid "dangerous climate change." The origins of this target are a little unclear, but according to one recent analysis, its policy implications are sobering. At the very least, strong action appears to be needed very quickly:

[T]o have a high probability of keeping the temperature increase below 2C, the total global 21st-century carbon budget must be limited to about 400 Gigatonnes.... A budget of 400 Gt is very small. To stay within this budget, global emissions would almost certainly have to peak before 2020 and decline fairly rapidly thereafter. If emissions were to continue to grow past 2020, so much of the 400 Gt budget would he rapidly used up that holding the 2C line would ultimately require extraordinary rates of emission reduction, rates corresponding to such large and historically unprecedented rates of accelerated capital-stock turnover that, frankly, it's difficult to imagine them occurring by virtue of any normal, orderly economic process. Time, in other words, is running out.[30]

But given the current substantial growth in global emissions, stabilization in less than 15 years is a very ambitious target. According to the authors of the analysis, the 400 Gt budget is so tight that even if the developed nations were to reduce their emissions to zero by 2028, it would require serious reductions by developing countries starting in 2030. Obviously, absent some technological miracle, the antecedent of this claim is politically (and morally) impossible. But the consequent is almost as implausible, given that projections indicate that in 2030 the developing nations will still be quite poor.

It is thus unclear whether the 2°C target is feasible. Hence, if meeting that target is really necessary for avoiding any catastrophic abrupt impacts for the current people—and some, of course, believe that 2°C is too high—then the prospects for motivating action on those grounds appear slim. Much would then depend on how many other impacts that are still caus-

ally proximate are temporally close enough to have notable effects on the current generation. This remains an empirical question. But the projections suggest that we are now dealing with a very limited subset of the impacts of climate change. In short, for a generation interested only in impacts that affect its own concerns, the window may be closed, or at best, only slightly ajar.

Hopefully, the projections just cited will turn out to be unduly pessimistic. Hence, it is worth making some observations about the importance of the PIBP even if there is an Open Window for the current generation. Our second issue then is whether, if the window is open, this undermines the relevance of the PIBP. One concern is that generations might care less about the end-of-life abrupt climate change than earlier-in-life ones. Another is that even an Open Window severely restricts the relevance of future people's concerns. For Open Window to be effective, there have to be enough effects of present emissions that accrue within the window to justify the current generation's action on a generation-relative basis. But this ignores all the other effects of present emissions—that is, all those that accrue to other generations. So, the PIBP remains. Moreover, in light of the PIBP, there is a realistic concern that solutions that avoid a particular abrupt climate change will be judged purely on how they enable a present generation to avoid that change arising during their lifetime, not on their wider ramifications. In other words, each generation will be motivated simply to delay any given abrupt climate change until after it is dead. So it may endorse policies that merely postpone such a change, making it inevitable for a future generation. Finally, sequential concerns may arise even under the Open Window scenario. Considering the PIBP, it would be predictable that earlier generations tend to use up most of any safety margin left to them. Given this, it may turn out that some later generation cannot help pushing over a given threshold and using up most of the safety margin for the next. . . .

Does the abrupt paradigm impose some limit on how bad climate change can get? Perhaps. But again the intergenerational problem rears its ugly head. If climate change is resilient and seriously back-loaded, the effects on a present generation that experiences an abrupt change and knows these facts are unclear. If further bad impacts are already on the cards, or if the Open Window is only slightly ajar, then, if the present generation is guided by its generation-relative preferences, we may still expect substantial intergenerational buck-passing and so more climate change. Experience of abrupt impacts may not teach and motivate, precisely because for such a generation, the time for teaching and motivating has already passed—at least as far as its own concerns are implicated. Moreover, we should expect other factors to intervene. If a generation experiences a severe abrupt change, we might expect long-term concerns (such as with mitigation) to be crowded out in the finite pool of worry by more immediate concerns. We might also expect such a generation to be

morally justified in ignoring those concerns to at least some extent. In short, we might expect something akin to the beginnings of an intergenerational arms race. . . .

III. Concluding Remarks

I conclude that we should not look to the disasters of abrupt change—either the actual experience of them or increasing scientific evidence that they are coming—to save us. One implication of this is that we should not waste precious time waiting for that to happen. If severe abrupt climate change is a real threat, the time for action is now, when many actions are likely to be prudentially and morally easier than in the future. Still, how effectively to motivate such action remains a very large practical problem about which the psychologists have much to teach us. In my view, if we are to solve this problem, we will need to look beyond people's generation-relative preferences. Moreover, the prevalence of the intergenerational problem suggests that one set of motivations that we need to think hard about engaging is that connected to moral beliefs about our obligations to those only recently, or not yet, born. . . .

Gardiner points our attention to an alarming fact. Current climatological observations are leading scientists to believe that the climate is changing even faster than the IPCC predicted. That is, evidence that the climate system is entering a phase of non-linear adaptation to human-induced forcing is mounting. The key general component of non-linear change is the **positive feedback**. This is any effect of a process that intensifies the process itself. For example, increased air temperatures are leading to a melting of the permafrost in northern Canada and Siberia. Permafrost is partly decomposed organic material rich in methane that lies below the frozen tundra. As it melts, bacterial decomposers get to work on it, a process that causes it to give up its methane (a super-powerful greenhouse gas), thus causing more warming of surface temperatures, which causes the permafrost to melt further, releasing more of its methane. And so on. It is this non-linearity that has scientists worried, because it could trigger the sort of catastrophic climate events we have mentioned. Consider the following two further examples of non-linear changes in the climate system.

The first concerns the amount of Arctic Ocean area covered by ice in summertime. Between 1980 and 2007, this area went from 10 million to 4 million square kilometres. At this pace, the Arctic Ocean will be ice-free in the summer within 15 years. The IPCC predicted this would not happen before 2050. No more ice up there means that every last drop of sunlight that falls on that vast ocean will be absorbed by it rather than reflected back. This will dramatically accelerate the heating process. Second, there are new satellite observations of ocean areas that show a dramatic decline in the population of ocean algae. The dead area comes up a deep blue in the photos rather than a bluey-green. This area has increased in size by 15 per cent over the past nine years. The reason is that the surface of the ocean is warmer than previously, which means that the nutrient-rich cooler water from below has a harder time mixing with it. Why is this dangerous? The growth of algae in the oceans cools the planet by pumping down carbon dioxide. Less algae means more carbon stays in the atmosphere.[31] Again, the heating process is not just advanced but *accelerated*.

Still, even if these events are evidence that the gradualist paradigm is no longer tenable and that we may be visited by climate catastrophes sooner than we thought, Gardiner is sceptical that the latter will undermine the current state of political inertia. In fact, his arguments suggest just the opposite: that the prospect of such catastrophes could reinforce political inertia. Suppose we were convinced that catastrophe was on the cards. Would this make us change our ways? Not likely, according to Gardiner, because it may mean that we simply give up on **mitigation** and focus exclusively on **adaptation**.

This is a key distinction. Efforts at mitigation are aimed at reducing our carbon footprint through conservation and efficiency improvements. Cuts to emissions are the most obvious form mitigation takes. Adaptation involves reacting to climate changes as they occur or preparing for those changes considered likely. It usually involves targeted technological and infrastructural innovation. Extending the height of sea walls is an example. Gardiner argues that to the extent we believe climate catastrophe is inevitable, we will abandon efforts at mitigation and focus exclusively on adaptation. Moreover, even though some adaptations will benefit future generations, most will likely be the sort of quick fixes that are meant to secure only the interests of the present. And each generation will burn as much fossil fuel as it deems necessary to cope with its own problems, thus exacerbating conditions for every future generation. This is the essence of the intergenerational arms race.

Gardiner's is a deeply pessimistic conclusion, but it is hard to deny that things do indeed look pretty dire on the climate front. Setting aside Canada's woeful record on this issue, the mitigation efforts of the world's major players are clearly not going to help us avoid catastrophic climate change. The overwhelming majority of scientists believe that if we are to avoid dangerous climate interference we need to ensure that global average temperatures rise no higher than 2°C above pre-industrial levels. This translates into capping greenhouse gas concentrations in the atmosphere at 450 ppm, then gradually reducing them to 350 ppm. Currently, as we have seen, they are just above 400 ppm and are rising at roughly 2 to 3 ppm per year. Nor, despite their rhetoric, are the giants of the developing world ready to act meaningfully. China, now the world's largest total emitter, has agreed to reduce the amount of carbon per unit of GDP by 5 per cent by 2020. Such "intensity" reductions might sound impressive, but in an economy with a growth rate like China's they are hollow promises. The increases in fossil fuel–based economic growth will vastly outstrip any reductions in carbon intensity, the result of which will be a net rise in greenhouse gas emissions. This is an instance of the general proposition that efficiency gains in the production of a good are very often swamped by *greater* increases in the consumption of that good.

ⓔ Conclusion

But if we are not to abandon hope in the face of this reality, what is the most defensible ethical stance to take? The key to any solution to climate change lies in engaging intergenerational motivations, as Gardiner puts it. And one way to do this is to imagine the potential effects of abrupt change on future people. Let's begin this act of imagination by returning briefly to the tragedy of the commons. It might be suggested that one of the reasons we have gotten ourselves into a state of such deep political inertia on this issue is that we cannot divorce our thinking about what we should do from what we think others have done or are going to do. The herdsman thinks obsessively about what the other herdsmen will do before deciding what he should do. But why not instead decide what sorts of people we really want to be and behave accordingly no matter what we think others are likely to do? In short, we need to develop a virtue ethical standpoint for climate change. Here's how philosopher Dale Jamieson puts the point:

> Instead of looking to moral mathematics for practical solutions to large-scale collective action problems, we should focus instead on non-calculative generators of behaviour: character traits, dispositions, emotions.... When faced with global environmental change, our general policy should be to try to reduce our contribution *regardless of the behaviour of others*, and we are more likely to succeed in doing this by developing and inculcating the right virtues than by improving our calculative abilities.[32]

It's a slender reed of hope, but it might just work. Think again of an image from the film *The Age of Stupid*, discussed in Schönfeld's article: an old man watching footage from our times and wondering what sort of people would have allowed this catastrophe to unfold. This is one way to engage those intergenerational motivations Gardiner is talking about. Speaking specifically of Canada here's how the challenge has recently been put:

> Think of a future person whose world has been blighted by climate change and put yourself for a moment into his or her shoes. Suppose that person could somehow address members of our generation. "Do you really mean to tell me," this bewildered person might say to a Canadian of the early 21st-century, "that you refused to take emissions reduction measures that would have reduced your GDP by 3.2% over a ten year period relative to business as usual because such measures were

deemed too *costly*? Even though your economy would still have grown 23% in this period? What mind-boggling greed!"

If this image makes us uncomfortable, it is because it casts us as vicious people. Are we? We certainly don't think so. For example, in December 2009 evidence emerged that Canadian forces stationed in Afghanistan had been transferring prisoners to Afghani authorities, whereupon many of these prisoners were tortured. It was revealed that the Canadians knew this would likely happen but did nothing to protect the prisoners. The government then tried to conceal the relevant evidence. Canadians were outraged that their government appeared to be colluding in torture. Fast forward again to our future person. She might ask us why we reacted this way. Our answer would likely be that those prisoners have moral standing on a par with every other person.

We might add that no just or benevolent or mindful person would countenance this sort of treatment. "Really?" our future critic might respond, "And what about us? In spite of, or perhaps because of, some of your economists' attempts to 'discount' our interests, here we are—with monumental existential challenges. Do we lack some morally essential property? What could it be? Perhaps it is the fact that at the time you made the decision to abandon efforts at mitigation we did not *yet* exist? But is this a morally relevant property, given the fact that your actions clearly had the power to bring into being a world that was better than this one? In condemning your government's actions on the Afghani detainee issue you refused to take the geographical remoteness of the prisoners—not to mention their dark skin or foreign habits—as reasons to deny their moral standing. You effectively claimed joint membership with them in a moral community. And yet in virtue of the causal connection between your actions and the state of our world, *we too were members of your moral community* when you made your fateful choices. We are genuinely baffled, to put it mildly, that you could not see this."[33]

We need to think seriously about our historical legacy, how future generations will judge us. Thought experiments like this one, suitably expanded, can cause us to feel shame about our actions, and this *can be* a very potent motivator. Jamieson, for his part, talks in a more positive vein about some dispositions and virtues we may want to cultivate so as to act responsibly in the face of climate change: humility, temperance, mindfulness, moral courage. Properly developed, these virtues would build a commitment to sustainability directly into our characters. In this case, we would attempt to do right by future generations as a matter of conscience, regardless of what others were doing. The game might already be lost—think again of Gardiner's "on the cards" scenario—but if we make ourselves into this sort of person, those future people might at least not despise us. That thought might provide us with both a small measure of comfort and a call to political action.

CASE STUDY

Canada, Climate Change, and Bullshit

Source: © Ashley Cooper/Alamy

The Canadian economy runs largely on fossil fuels, but climatologists believe we need to decarbonize the global economy as quickly as possible to avert climate catastrophe. Is Canada doing its part?

The Conference of Parties (COP) is the main decision-making body of the United Nations Framework Convention on Climate Change (UNFCCC). Among those around the world working hard for a meaningful global treaty on climate change, hopes were high that the 2009 COP meetings in Copenhagen would be fruitful. As it happened, the meetings were a total flop, no significant deal to curb emissions having been agreed upon. There were multiple reasons for this failure, but in trying to account for it, most observers point to the behaviour of the Canadian government led by Stephen Harper. Canada came away from the conference with a dubious honour: Fossil of the Year. This "award" is given to the country that has "done the most to block progress towards a fair, robust and binding climate treaty." Here's the statement from the awards committee:

> Fossil of the Year goes to Canada, for bringing a totally unacceptable position into Copenhagen and refusing to strengthen it one bit. Canada's 2020 target is among the worst in the industrialized world, and leaked cabinet documents revealed that the government is contemplating a cap-and-trade plan so weak that it would put even that target out of reach. Canada has made zero progress here on financing, offering nothing for the short

Continued

term or the long term beyond vague platitudes. And in last night's high-level segment, Canada's environment minister gave a speech so lame that it didn't include a single target, number or reference to the science. Canada's performance here in Copenhagen builds on two years of delay, obstruction and total inaction. This government thinks there's a choice between environment and economy, and for them, tar sands beats climate every time. Canada's emissions are headed nowhere but up. For all this and more, we name Canada the Colossal Fossil.

In the teeth of this assessment, the Canadian minister of the environment, Leona Aglukkaq, has claimed more recently that "our government is taking action to address climate change." She then trotted out the pledge to cut emissions by 17 per cent from 2005 levels by 2020, an impossible ideal given that emissions have risen 7 per cent since the Conservatives first took office. But as Jeffrey Simpson argues, "no one—not senior civil servants, not foreign diplomats, not academics, not even people in the oil and gas industry—believes Canada will bring down its emissions by 24 per cent (17 per cent plus 7 per cent) in the next eight years. Canada struts on the world stage, naked as a newt, and can't fool those who know what's really going on." So what *is* really going on? Is the government simply lying? For some enlightenment, let's take a quick look at the nature and ethics of lying.

To lie to someone is intentionally to deceive that person about some state of affairs. This requires the deceiver to "track the truth" about that state of affairs. I'll come back to this important point in a moment. First, consider a different question: What exactly is wrong with lying? When you lie to someone you disrespect that person, and we generally think this is an immoral stance to take toward other rational agents. So is it *ever* permissible to lie? Most of us can think of circumstances in which it is. For example, you might think it permissible to lie to (a) a child about the Tooth Fairy; (b) an axe murderer who wants to know where his next victim is hiding; or (c) an emotionally frail grandfather about his terminal heart condition. Such examples indicate that our default position appears to be that lying is wrong most of the time but not always. Because it can be motivated by justified paternalism (the child), the duty to protect third parties (the axe murderer), or compassion (the grandfather) lying has this essentially ambiguous character as regards its permissibility.

Now come back to the point about the liar tracking the truth. Philosopher Harry Frankfurt has argued that this points to an important distinction between lying and bullshitting:

> Both in lying and in telling the truth people are guided by their beliefs concerning the way things are. These guide them as they endeavor either to describe the world correctly or to describe it deceitfully. For this reason, telling lies does not tend to unfit a person for telling the truth in the same way that bullshitting tends to. . . . . The bullshitter ignores these demands altogether. He does not reject the authority of the truth, as the liar does, and oppose himself to it. He pays no attention to it at all. By virtue of this, bullshit is a greater enemy of the truth than lies are.

Unlike the bullshitter, the liar cares about the truth enough to track it in her beliefs. Paradoxically, the liar might help to keep the truth alive because of this. But in this respect, at least when it comes to climate change, Canadian officials are quintessential bullshitters. The *truth* about the dangers of climate change, the extreme dirtiness of Canadian tar sands, the pathetic inadequacy of the government's emissions reduction targets, and, most importantly, the impossibility of reaching these targets under current policies, seems weirdly irrelevant to these officials. When Simpson says that the government is "naked as a newt" on this issue, he means that the insupportability of the Canadian position is entirely *obvious* to anyone willing to look. But the government is shameless about its deception, a hallmark of the bullshitter (if they are caught, liars can typically be shamed about their deception). The belief seems to be that if you can repeat a claim often enough it's as good as true, and so the impossible emissions targets are repeated *ad nauseam*. Note, finally, that there is also no moral ambiguity in bullshit as there is in the lie. Bullshit is *always* designed to manipulate and disrespect its target audience. As Frankfurt suggests, what is most worrisome here is the possibility that we are all becoming "unfit" for truth because we are awash in bullshit. How might this affect the functioning of our democracy?

Sources: Fossil of the Day website, at: http://climateactionnetwork.ca/2013/11/22/canada-wins-lifetime-unachievement-fossil-award-at-warsaw-climate-talks/; Harry Frankfurt, "On Bullshit," in *The Importance of What We Care About* (Cambridge: Cambridge University Press, 1988), 117–33; Jeffrey Simpson, "Ottawa Denies Its Own Emissions Stats," *Globe and Mail*, 1 November 2013, at: www.theglobeandmail.com/globe-debate/ottawa-denies-its-own-emissions-stats/article15190539/; Jeffrey Simpson, "Canada's Message. The World and Its Climate Be Damned," *Globe and Mail*, 10 September 2012, at: www.theglobeandmail.com/globe-debate/canadas-message-the-world-and-its-climate-be-damned/article4403806/.

Study Questions

1. Explain how climate change is a tragedy of the commons. Do you think that this is the best way to conceptualize it? Why or why not?

2. Schönfeld argues that the United States is the chief culprit in the climate crisis. Why does he say this, and is his argument well supported?

3. Do you agree with Gardiner that we will not necessarily be "saved by disaster"? How is this argument supported by the concept of an intergenerational arms race?

4. Do you believe we should focus our efforts on adaptation to climate change and effectively abandon the attempt to mitigate our emissions? What effects would this have on poor countries and future generations?

Further Reading

Tim Flannery. 2005. *The Weather Makers*. Toronto: HarperCollins.

Stephen Gardiner. 2011. *A Perfect Moral Storm: The Ethical Tragedy of Climate Change*. Oxford: Oxford University Press.

James Garvey. 2008. *The Ethics of Climate Change: Right and Wrong in a Warming World*. London: Continuum.

Dale Jamieson. 2014. *Reason in a Dark Time: Why the Struggle against Climate Change Failed and What It Means for Our Future.* Oxford: Oxford University Press.

Henry Shue. 2014. *Climate Justice: Vulnerability and Protection.* Oxford: Oxford University Press.

Jeffrey Simpson, Mark Jaccard, and Nic Rivers. 2008. *Hot Air: Meeting Canada's Climate Change Challenge.* Toronto: Emblem.

Byron Williston. 2015. *The Anthropocene Project: Virtue in the Age of Climate Change.* Oxford: Oxford University Press.

CHAPTER 12

Population and Consumption

Ecological Intuition Pump

Can societies develop economically without placing inordinate burdens on the natural environment? Are the goals of development and environmental sustainability *inevitably* at odds? One reason for thinking that they might be is that we tend to think of development and population increase as going hand in hand. And it is difficult for us to imagine an indefinitely expanding population that does not place significant, perhaps overwhelming, burdens on the natural environment. Lessons drawn from ecology generally support this way of thinking. In the rest of nature, species populations do not expand exponentially over the course of thousands of years. But this is precisely what we have done. The populations of other species are for the most part tightly constrained by local environmental realities: availability of the appropriate amount of water, enough prey, not too many predators, just the right climate, and so on. When one or more of these background conditions changes radically, so too, generally, do the numbers of the species in question. Are we different from the rest of nature? Does our technology effectively equip us to defy natural constraints? We behave as though we believe that it does. Is this a dangerous illusion or an accurate description of a species that just is, in some sense, super-natural? But isn't our belief that we are apart from nature the source of our environmental crisis? Should we rethink the role technology plays in our lives? What would an environmentally responsible attitude toward technology look like?

Ⓐ Introduction

In 2009, the Indian automobile company Tato Motors launched its newest product, the Nano, otherwise known as "The People's Car." At a cost of less than 100,000 rupees (about Can$2500), the tiny Nano is likely to add up to 300 million new drivers on India's roads by 2015. But as Canadian economist Jeff Rubin notes, Tato Motors is not alone in the push to service an ever-growing market of car buyers in the developing world. In 2007, China's auto sales increased by 20 per cent, Russia's by 60 per cent, and Brazil's by 30 per cent. Meanwhile, many major automobile companies—Nissan, Ford, and Volkswagen among them—are rushing to produce cars that sell for under $10,000.[1]

The current international boom in car sales is a nice illustration of this chapter's focus. Two forces are at work here: a rising human population and a corresponding overall increase in that population's consumption. Together, these two forces are placing dramatic strains on the coping mechanisms of the natural environment. More cars on the road—especially where the figures are as high as they are projected to be—obviously entails more consumption of fossil fuels. But the environmental effects of these developments extend well beyond this. More people consuming more resources will inevitably affect, possibly to the breaking point, all kinds of natural goods and services: the soil, the climate, the water supply, wilderness spaces, the viability of other species, and the quality of forests and air.

Most of us can intuitively appreciate the idea that we humans rely on the natural world for basic sustenance but that there are limits to what nature can provide. Our numbers surely cannot grow indefinitely. This intuition hints at the ecological concept of **carrying capacity**. For ecologists, carrying capacity refers to the maximum number of individuals of a species that can be sustained indefinitely in a given physical space. To determine this number, we need to look at factors like the species' ability to resist diseases and parasites, its reproductive rate, its ability to adapt to environmental change and to migrate and live in other habitats, the adequacy of its food and energy supplies, and so on.[2] These factors place a strict limit on the species' ability to expand its numbers.

Ⓑ Malthus's Challenge

The concept of carrying capacity raises many important questions. Most obvious, perhaps, is the question about what precise limit nature places on the numbers of *our* species and what the consequences might be of our exceeding that limit. Have we already exceeded the Earth's carrying capacity, or are we very close to doing so? As we will see, experts disagree fundamentally on this. On one side stand those who believe that we humans are close to or already in excess of Earth's carrying capacity. For simplicity, we can refer to this group as defending the "ecological" perspective on population. On the other side are those who believe that we are nowhere near such a limit. Indeed, many in this latter group think that the whole ecologically based notion of carrying capacity, whatever value it has as applied to non-human populations, is misleading when applied to our species.

Since this is the view of many defenders of free-market capitalist solutions to our environmental problems, let's refer to it, again for the sake of simplicity, as the "economic" perspective (more specifically, it is the view of neo-liberal economics, examined in Chapter 4). This is by no means to suggest that all ecologists and all economists espouse every feature of the views being attributed to them here, but by and large the labels fit quite well. Each view is articulated in one of the two essays of this chapter. Let's begin with the ecological view as expressed more than 200 years ago by Reverend Thomas Robert Malthus, the figure some point to as the original inspiration for the notion of carrying capacity.[3]

AN ESSAY ON THE PRINCIPLE OF POPULATION

Reverend Thomas Robert Malthus

I have read some of the speculations on the perfectibility of man and of society with great pleasure. I have been warmed and delighted with the enchanting picture which they hold forth. I ardently wish for such happy improvements. But I see great, and, to my understanding, unconquerable difficulties in the way to them. These difficulties it is my present purpose to state, declaring, at the same time, that so far from exulting in them, as a cause of triumph over the friends of innovation, nothing would give me greater pleasure than to see them completely removed.

The most important argument that I shall adduce is certainly not new. The principles on which it depends have been explained in part by Hume, and more at large by Dr Adam Smith. It has been advanced and applied to the present subject, though not with its proper weight, or in the most forcible point of view, by Mr Wallace, and it may probably have been stated by many writers that I have never met with. I should certainly therefore not think of advancing it again, though I mean to place it in a point of view in some degree different from any that I have hitherto seen, if it had ever been fairly and satisfactorily answered.

I think I may fairly make two postulata. First, That food is necessary to the existence of man. Secondly, That the passion between the sexes is necessary and will remain nearly in its present state. These two laws, ever since we have had any knowledge of mankind, appear to have been fixed laws of our nature, and, as we have not hitherto seen any alteration in them, we have no right to conclude that they will ever cease to be what they now are, without an immediate act

of power in that Being who first arranged the system of the universe, and for the advantage of his creatures, still executes, according to fixed laws, all its various operations.

I do not know that any writer has supposed that on this earth man will ultimately be able to live without food. But Mr Godwin has conjectured that the passion between the sexes may in time be extinguished. As, however, he calls this part of his work a deviation into the land of conjecture, I will not dwell longer upon it at present than to say that the best arguments for the perfectibility of man are drawn from a contemplation of the great progress that he has already made from the savage state and the difficulty of saying where he is to stop. But towards the extinction of the passion between sexes, no progress whatever has hitherto been made. It appears to exist in as much force at present as it did two thousand or four thousand years ago. There are individual exceptions now as there always have been. But, as these exceptions do not appear to increase in number, it would surely be a very unphilosophical mode of arguing, to infer merely from the existence of an exception that the exception would, in time, become the rule, and the rule the exception.

Assuming then, my postulata as granted, I say, that the power of population is indefinitely greater than the power in the earth to produce subsistence for man. Population, when unchecked, increases in a geometrical ratio. Subsistence increases only in an arithmetical ratio. A slight acquaintance with numbers will show the immensity of the first power in comparison of the second. By that law of our nature which makes food necessary to the life of man, the effects of these two unequal powers must be kept equal. This implies a strong and constantly operating check on population from the difficulty of substance. This difficulty must fall some where and must necessarily be severely felt by a large portion of mankind.

Through the animal and vegetable kingdoms, nature has scattered the seeds of life abroad with the most profuse and liberal hand. She has been comparatively sparing in the room and the nourishment necessary to rear them. The germs of existence contained in this spot of earth, with ample food, and ample room to expand in, would fill millions of worlds in the course of a few thousand years. Necessity, that imperious all pervading law of nature, restrains them within the prescribed bounds. The race of plants, and race of animals shrink under this great restrictive law. And the race of man cannot, by any efforts of reason, escape from it. Among plants and animals its effects are waste of seed, sickness, and premature death. Among mankind misery and vice. The former, misery, is an absolutely necessary consequence of it. Vice is a highly probable consequence, and we therefore see it abundantly prevail, but it ought not, perhaps, to be called an absolutely necessary consequence. The ordeal of virtue is to resist all temptation to evil.

This natural inequality of the two powers of population and of production in the earth and that great law of our nature which must constantly keep their effects equal form the great difficulty that to me appears insurmountable in the way to the perfectibility of society. All other arguments are of slight and subordinate consideration in comparison of this. I see no way by which man can escape from the weight of this law which pervades all animated nature. No fancied equality, no agrarian regulations in their utmost extent, could remove the pressure of it even for a single century. And it appears, therefore, to be decisive against the possible existence of a society, all the members of which should live in ease, happiness, and comparative leisure; and feel no anxiety about providing the means of subsistence for themselves and families.

Consequently, if the premises are just, the argument is conclusive against the perfectibility of the mass of mankind. I have thus sketched the general outline of the argument, but I will examine it more particularly, and I think it will be found that experience, the true source and foundation of all knowledge, invariably confirms its truth. I said that population, when unchecked, increased in a geometrical ratio, and subsistence for man in an arithmetical ratio.

Let us examine whether this position be just. I think it will be allowed, that no state has hitherto existed (at least that we have any account of) where the manners were so pure and simple, and the means of subsistence so abundant, that no check whatever has existed to early marriages among the lower classes, from a fear of not providing well for their families, or among the higher classes, from a fear of lowering their condition in life. Consequently in no state that we have yet known has the power of population been left to exert itself with perfect freedom.

Let us now take any spot on earth, this Island for instance, and see in what ratio the subsistence it affords can be supposed to increase. We will begin with it under its present state of cultivation. If I allow that by the best possible policy, by breaking up more land and by great encouragements to agriculture, the produce of this Island may be doubled in the first twenty-five years, I think it will be allowing as much as any person can well demand.

In the next twenty-five years, it is impossible to suppose that the produce could be quadrupled. It would be contrary to all our knowledge of the qualities of land. The very utmost that we can conceive, is, that the increase in the second twenty-five years might equal the present produce. Let us then take this for our rule, though certainly far beyond the truth, and allow that by great exertion, the whole produce of the Island might be increased every twenty-five years, by a quantity of subsistence equal to what it at present produces. The most enthusiastic speculator cannot suppose a greater increase than this. In a few centuries it would make every acre of land in the Island like a garden. Yet this ratio of increase is evidently arithmetical. It may be fairly said, therefore, that the means of subsistence increase in an arithmetical ratio. Let us now bring the effects of these two ratios together.

The population of the Island is computed to be about seven millions, and we will suppose the present produce equal to the support of such a number. In the first twenty-five years the population would be fourteen millions, and the food being also doubled, the means of subsistence would be equal to this increase. In the next twenty-five years the population would be twenty-eight millions, and the means of subsistence only equal to the support of twenty-one millions. In the next period, the population would be fifty-six millions, and the means of subsistence just sufficient for half that number. And at the conclusion of the first century the population would be one hundred and twelve millions and the means of subsistence only equal to the support of thirty-five millions, which would leave a population of seventy-seven millions totally unprovided for.

A great emigration necessarily implies unhappiness of some kind or other in the country that is deserted. For few persons will leave their families, connections, friends, and native land, to seek a settlement in untried foreign climes, without some strong subsisting causes of uneasiness where they are, or the hope of some great advantages in the place to which they are going.

But to make the argument more general and less interrupted by the partial view of emigration, let us take the whole earth, instead of one spot, and suppose that the restraints to population were universally removed. If the subsistence for the man that the earth affords was to be increased every twenty-five years by a quantity equal to what the whole world at present produces, this would allow the power of production in the earth to be absolutely unlimited, and its ratio of increase much greater than we can conceive that any possible exertions of mankind could make it.

Taking the population of the world at any number, a thousand millions, for instance, the human species would increase in the ratio of—1, 2, 4, 8, 16, 32, 64, 128, 256, 512, &c. and subsistence as—1, 2, 3, 4, 5, 6, 7, 8, 9, 10, &c. In two centuries and a quarter, the population would be to the means of subsistence as 512 to 10; in three centuries at 4096 to 13, and in two thousand years the difference would be almost incalculable, though the produce in that time would have increased to an immense extent.

No limits whatever are placed to the productions of the earth; they may increase for ever and be greater than any assignable quantity; yet still the power of population being a power of a superior order, the increase of the human species can only be kept commensurate to the increase of the means of subsistence, by the constant operation of the strong law of necessity acting as a check upon the greater power.

The theory on which the truth of this position depends appears to me so extremely clear that I feel at a loss to conjecture what part of it can be denied. That population cannot increase without the means of subsistence is a proposition so evident that it needs no illustration. That population does invariably increase where there are the means of subsistence, the history of every people that have ever existed will abundantly prove. And that the superior power of population cannot be checked without producing misery or vice, the ample portion of these too bitter ingredients in the cup of human life and continuance of the physical causes that seem to have produced them bear too convincing a testimony. The only true criterion of a real and permanent increase in the population of any country is the increase of the means of subsistence....

The happiness of a country does not depend, absolutely, upon its poverty or its riches, upon its youth or its age, upon its being thinly or fully inhabited, but upon the rapidity with which it is increasing, upon the degree in which the yearly increase of food approaches to the yearly increase of an unrestricted population. This approximation is always the nearest in new colonies, where the knowledge and industry of an old State, operate on the fertile unappropriated land of a new one. In other cases, the youth or the age of a State is not in this respect of very great importance. It is probable, that the food

of Great Britain is divided in as great plenty to the inhabitants, at the present period, as it was two thousand, three thousand, or four thousand years ago. And there is reason to believe that the poor and thinly inhabited tracts of the Scotch Highlands, are as much distressed by an overcharged population, as the rich and populous province of Flanders....

Famine seems to be the last, the most dreadful resource of nature. The power of population is so superior to the power in the earth to produce subsistence for man, that premature death must in some shape or other visit the human race. The vices of mankind are active and able ministers of depopulation. They are the precursors in the great army of destruction; and often finish the dreadful work themselves. But should they fail in this war of extermination, sickly seasons, epidemics, pestilence, and plague, advance in terrific array, and sweep off their thousands and ten thousands. Should success be still incomplete, gigantic inevitable famine stalks in the rear, and with one mighty blow, levels the population with the food of the world.

This is a bleak, but wonderfully lucid, picture of the human enterprise. Malthus's thoughts can be summarized in these three claims:

1. Technological advances cause an increase in the means of subsistence, which causes an increase in population.
2. This population increase vastly outstrips the increase in subsistence that causes it.
3. The only effective "checks" on such expanding population are "misery and vice."

What are we to make of this set of claims? Claim (1) is surely indisputable. Since the advent of agriculture some 10,000 years ago, the human population has increased dramatically. At that point in our history, humans gained the ability—mainly through the domestication of plants and animals and improved food storage technologies—to increase our food supply dramatically. As a direct result, our numbers soared. However, Malthus's main contribution to our understanding of these issues lies with (2) and (3), the ideas that population increases outstrip increases in subsistence and that the only effective checks on such population increases are misery and vice. The first of these claims has to do with the effect in total numbers of **geometrical growth**. This kind of growth is contrasted with **arithmetical growth**. The latter works by adding units of 1 to the (new) total: 1, 2, 3, 4, 5, and so on. Geometrical growth works by doubling this total: 1, 2, 4, 8, 16, 32, and so on. Another way to put this claim is to say that geometrical growth (population) outstrips arithmetical growth (subsistence), even though the latter is a major cause of the former. Indeed, as you can see if you run the two number streams further, population growth will quickly explode beyond the point at which it can continue given the relatively meagre corresponding expansion in the food supply. Once this kind of population explosion has occurred, Malthus claims, the only way to achieve a new balance between the population and its means of subsistence is through "misery and vice." Of these two forces, misery is by far the more impressive:

> [S]ickly seasons, epidemics, pestilence, and plague, advance in terrific array, and sweep off their thousands and ten thousands. Should success still be incomplete, gigantic inevitable famine stalks in the rear, and with one mighty blow, levels the population with the food of the world.

And thus it inevitably goes throughout history. Humanity may achieve a state of relative equilibrium with its means of subsistence, but this can only be temporary. Technological advances upset this harmony, causing the population to outstrip its means of subsistence. Then the terrible forces Malthus cites kick in to generate a new equilibrium.

An apparent example of Malthus's claims is currently playing out in Pakistan. Although ravaged by floods over much of its territory in the summer of 2010, Pakistan is in fact facing severe long-term water scarcity. From 1981 to 2010, the amount of water available per capita dropped roughly 50 per cent, from nearly 3000 cubic metres to just under 1500, and is projected to fall below 1000 cubic metres per capita by 2030. The trouble is in the basin of the huge Indus River, an arid region that has over the years seen huge amounts of water used in the expansion of irrigation projects. The spread of irrigation technology has allowed for a huge increase in the region's population. The area of irrigated lands has nearly doubled since independence in 1947, a period that also saw a fivefold increase in population. Now, however, the desert is taking back the territory because there is not enough water left in the Indus to support all the people that live there. Indeed, a recent World Bank report pointed out that the volume of water that reaches the sea from the Indus has dropped to virtually zero in some years. Here we see the key elements of the Malthusian analysis: a disproportionate increase in population fuelled by technological advance and a consequent increase in the means of subsistence. The last element—the strife—may be just around the corner, because population growth in the area has now clearly exceeded what the means of subsistence can support. According to Abdul Salam, deputy head of Badin district, "the next war will be over water."[14]

For much of the twentieth century, Malthus's fears seemed warranted. Even he would no doubt have been shocked at the expansion in the numbers of our species since his time. Since the nineteenth century, the population has expanded almost sevenfold, from 1 billion to roughly 6.8 billion. It is projected to reach at least 8.9 billion by 2050, with a whopping 97 per cent of this increase expected to take place in the developing world. In the middle of the 1960s, the human population had gotten so out of control that many thought substantial Malthusian misery was just around the corner. In an influential book, *The Population Bomb*, written in 1968, Paul Ehrlich had this to say:

> The battle to feed all of humanity is over. In the 1970s the world will undergo famines—hundreds of millions of people are going to starve to death in spite of any crash programs embarked upon now. At this late date nothing can prevent a substantial increase in world death rate.[5]

Many of those examining the situation at the time were inclined to think that we had entered a state of severe population emergency. Garrett Hardin, for instance, famously suggested that we should think of the Earth and its resources as a *lifeboat* with a limited seating capacity that cannot be exceeded or expanded. We in the boat are surrounded by all those people struggling to get in and share our resources with us. But the stark fact is that there simply is not enough room for everyone wanting to get in. There are different ways to apply the lifeboat metaphor. For our purposes, we can think of the inhabitants of the boat as the entire present generation and those in the surrounding waters clamouring to get in as those future generations who will be born *at the current population growth rate*. Let them all in, and the boat will surely capsize. So given that it cannot support all those people, those of us in the lifeboat have a difficult moral task. Assuming we do not, with Ehrlich, throw up our hands in despair, we need to act so as to slow dramatically or stabilize population growth.

Ecologists will claim that the main reason we have reached our carrying capacity is that we consume so much. Since 1850, there has been an eightfold increase in per capita consumption of energy, the vast bulk of it in the developed world.[6] And much of the developing world is currently striving to match these consumption levels. In this connection, it is sobering to notice the rise in numbers of what UN Environment Programme consultant Matthew Bentley calls the **consumer class**. This class includes those whose income is equivalent to at least US$7000. The consumer class currently totals around 1.7 billion people worldwide. As a percentage of national population, the United States and Canada lead the world with 85 per cent of their populations in the consumer class, while developing countries as a whole have just 17 per cent in this class.[7]

Nevertheless, as the impending success of automobiles like the Nano suggests, the consumer class is expanding nearly everywhere. Indeed, the world's fastest-growing economies are not members of the Organisation for Economic Co-operation and Development (OECD), the current in-group of developed nations. The so-called BRIC countries (Brazil, Russia, India, and China) together have a population of 2.7 billion people, a number projected to increase to 3.2 billion by 2025. In 2002, they had an average of roughly 26 per cent of their citizens in the consumer class. Although this number looks low compared to the 85 per cent in Canada and the US, two points must be emphasized. First, these countries have an enormous share of the world's population (around 40 per cent). Second, the 2002 number is rising steadily. To come back to car sales again, while mature markets like those in Western Europe and the US have growth rates in this sector of 2 to 3 per cent, rates are 10 to 20 times that number in the BRIC countries.[8]

Those who find the Malthusian, or ecological, analysis of population and consumption compelling will suggest that given these numbers, some terrible calamity surely awaits us. Perhaps it does, and at the end of this chapter we will return to the implications of this expansion in the consumer class. But before we go any further with our analysis, we must confront the possibility that—in spite of arresting examples like that of Pakistan's water crisis—*Malthus was demonstrably wrong.* This assessment rests on the undeniable empirical observation that although human population has indeed expanded geometrically (especially since the Industrial Revolution), the miseries cited by Malthus—the ones that are supposed to bring us to new states of balance with the means of subsistence through the brutal mechanism of massive die-off—have not occurred. The 1970s famines predicted by Ehrlich did not occur on anything like the scale he imagined. What's more, the rate of population growth is now *declining* in spite of the absence of such extreme events, a fact that directly contradicts claim (6) in Malthus's argument (see above). Between 1963 and 2004, the world's annual population growth has been cut virtually in half, going from 2.2 per cent to 1.25 per cent. What exactly has gone wrong—or right!—here?

Ⓒ The Economist's Retort

Some critics of the ecological view suggest that its proponents didn't just happen to make a few bad predictions. Instead, the bad predictions were the inevitable product of their reliance on a faulty assumption, the one contained in the metaphor of a lifeboat and the associated notion of carrying capacity. The key error with these ideas is that they entail that resources are essentially finite. Perhaps this is the case for other species, and such species must therefore maintain a more or less constant population over time. If they do not, they will have big population crashes. But humans are different, so the ecological analysis does not fit us. Because of our ingenuity, we can use technology to manipulate the physical world to such a degree that there is little sense in the idea that the resources we require are fixed in quantity or quality. We will *always* learn how to find or make more of what we need, and the more brainpower around to help us find it or make it—that is, the greater our population—the likelier we are to succeed in the enterprise. Does the evident predictive failure of Malthus and Ehrlich doom the whole ecological perspective on consumption and population? Can we safely expand our population indefinitely? Are resources really infinite? To find answers to these questions, we need to look at what we are calling the economist's perspective, ably and famously defended here by Julian Simon.

CAN THE SUPPLY OF NATURAL RESOURCES REALLY BE INFINITE? YES!

Julian Simon

I. Introduction

Natural resources are not finite. Yes, you read correctly. This [essay] shows that the supply of natural resources is not finite in any economic sense, which is why their cost can continue to fall in the future.

On the face of it, even to inquire whether natural resources are finite seems like nonsense. Everyone "knows" that resources are finite, from C.P. Snow to Isaac Asimov to as many other persons as you have time to read about in the newspaper. And this belief has led many persons to draw far-reaching conclusions about the future of our world economy and civilization. A prominent example is the Limits to Growth group, who open the preface to their 1974 book, a sequel to *Limits to Growth*, as follows.

> Most people acknowledge that the earth is finite. . . . Policy makers generally assume that growth will provide them tomorrow with the resources required to deal with today's problems. . . . Recently, however, concern about the consequences of population growth, increased environmental pollution, and the depletion of fossil fuels has cast doubt upon the belief that continuous growth is either possible or a panacea.[9]

(Note the rhetorical device embedded in the term "acknowledge" in the first sentence of the quotation. That word suggests that the statement is a fact and that anyone who does not "acknowledge" it is simply refusing to accept or admit it.) . . .

The assumption of finiteness is responsible for misleading many scientific forecasters because their conclusions follow inexorably from that assumption. From the Limits to Growth team again, this time on food: "The world model is based on the fundamental assumption that there is an upper limit to the total amount of food that can be produced annually by the world's agricultural system."[10]

II. The Theory of Decreasing Natural-Resource Scarcity

We shall begin with a far-out example to see what contrasting possibilities there are. (Such an analysis of far-out examples is a useful and favourite trick of economists and mathematicians.) If there is just one person, Alpha Crusoe, on an island, with a single copper mine on his island, it will be harder to get raw copper next year if Alpha makes a lot of copper pots and bronze tools this year. And if he continues to use his mine, his son Beta Crusoe will have a tougher time getting copper than did his daddy.

Recycling could change the outcome. If Alpha decides in the second year to make new tools to replace the old tools he made in the first year, it will be easier for him to get the necessary copper than it was the first year because he can reuse the copper from the old tools without much new mining. And if Alpha adds fewer new pots and tools from year to year, the proportion of copper that can come from recycling can rise year by year. This could mean a progressive decrease in the cost of obtaining copper with each successive year for this reason alone, even while the total amount of copper in pots and tools increases.

But let us be "conservative" for the moment and ignore the possibility of recycling. Another scenario: If there are two people on the island, Alpha Crusoe and Gamma Defoe, copper will be more scarce for each of them this year than if Alpha lived there alone, unless by cooperative efforts they can devise a more complex but more efficient mining operation—say, one man on the surface and one in the shaft. Or, if there are two fellows this year instead of one, and if copper is therefore harder to get and more scarce, both Alpha and Gamma may spend considerable time looking for new lodes of copper. And they are likely to be successful in their search. This discovery may lower the cost of copper to them somewhat, but on the average the cost will still be higher than if Alpha lived alone on the island.

Alpha and Gamma may follow still other courses of action. Perhaps they will invent better ways of obtaining copper from a given lode, say a better digging tool, or they may develop new materials to substitute for copper, perhaps iron. The cause of these new discoveries or the cause of applying ideas that were discovered earlier is the "shortage" of copper—that is, the increased cost of getting copper. So a "shortage" of copper causes the creation of its own remedy. This has been the key process in the supply and use of natural resources throughout history.

Discovery of an improved mining method or of a substitute product differs, in a manner that affects future generations, from the discovery of a new lode. Even after the discovery of a new lode, on the average it will still be more costly to obtain copper, that is, more costly than if copper had never been used enough to lead to a "shortage." But discoveries of improved mining methods and of substitute products, caused by the shortage of copper, can lead to lower costs of the services people seek from copper. Let's see how.

The key point is that a discovery of a substitute process or product by Alpha or Gamma can benefit innumerable future generations. Alpha

and Gamma cannot themselves extract nearly the full benefit from their discovery of iron. (You and I still benefit from the discoveries of the uses of iron and methods of processing it that our ancestors made thousands of years ago.) This benefit to later generations is an example of what economists call an "externality" due to Alpha and Gamma's activities, that is, a result of their discovery that does not affect them directly.

So, if the cost of copper to Alpha and Gamma does not increase, they may not be impelled to develop improved methods and substitutes. If the cost of getting copper does rise for them, however, they may then bestir themselves to make a new discovery. The discovery may not immediately lower the cost of copper dramatically, and Alpha and Gamma may still not be as well off as if the cost had never risen. But subsequent generations may be better off because their ancestors suffered from increasing cost and "scarcity."

This sequence of events explains how it can be that people have been using cooking pots for thousands of years, as well as using copper for many other purposes, and yet the cost of a pot today is vastly cheaper by any measure than it was 100 or 1000 or 10,000 years ago.

It is all-important to recognize that discoveries of improved methods and of substitute products are not just luck. They happen in response to "scarcity"—an increase in cost. Even after a discovery is made, there is a good chance that it will not be put into operation until there is need for it due to rising cost. This point is important: Scarcity and technological advance are not two unrelated competitors in a race; rather, each influences the other. . . .

III. Resources as Services

As economists or as consumers, we are interested in the particular services that resources yield, not in the resources themselves. Examples of such services are an ability to conduct electricity, an ability to support weight, energy to fuel autos, energy to fuel electrical generators, and food calories.

The supply of a service will depend upon (a) which raw materials can supply that service with the present technology; (b) the availabilities of these materials at various qualities; (c) the costs of extracting and processing them; (d) the amounts needed at the present level of technology to supply the services that we want; (e) the extent to which the previously extracted materials can be recycled; (f) the cost of recycling; (g) the cost of transporting the raw materials and services; and (h) the social and institutional arrangements in force. What is relevant to us is not whether we can find any lead in existing lead mines but whether we can have the services of lead batteries at a reasonable price; it does not matter to us whether this is accomplished by recycling lead, by making batteries last forever, or by replacing lead batteries with another contraption. Similarly, we want intercontinental telephone and television communication, and as long as we get it, we do not care whether this requires 100,000 tons of copper for cables or just a single quarter-ton communications satellite in space that uses no copper at all.

Let us see how this concept of services is crucial to our understanding of natural resources and the economy. To return to Crusoe's cooking pot, we are interested in a utensil that we can put over the fire and cook with. After iron and aluminum were discovered, quite satisfactory cooking pots, perhaps even better than pots of copper, could be made of these materials. The cost that interests us is the cost of providing the cooking service rather than the cost of copper. If we suppose that copper is used only for pots and that iron is quite satisfactory for the same purpose, as long as we have cheap iron it does not matter if the cost of copper rises sky high.

(But in fact that has not happened. As we have seen, the prices of the minerals themselves, as well as the prices of the services they perform, have fallen over the years.)

IV. Are Natural Resources Finite?

Incredible as it may seem at first, the term "finite" is not only inappropriate but is downright misleading when applied to natural resources, from both the practical and philosophical points of view. As with many of the important arguments in this world, the one about "finiteness" is "just semantic." Yet the semantics of resource scarcity muddle public discussion and bring about wrong-headed policy decisions.

The word "finite" originates in mathematics, in which context we all learn it as schoolchildren. But even in mathematics the word's meaning is far from unambiguous. It can have two principal meanings, sometimes with an apparent contradiction between them. For example, the length of a one-inch line is finite in the sense that it is bounded at both ends. But the line within the endpoints contains an infinite number of points; these points cannot be counted, because they have no defined size. Therefore, the number of points in that one-inch segment is not finite. Similarly, the quantity of copper that will ever be available to us is not finite, because there is no method (even in principle) of making an appropriate count of it, given the problem of the economic definition of "copper," the possibility of creating copper or its economic equivalent from other materials, and thus the lack of boundaries to the sources from which copper might be drawn.

Consider this quote about potential oil and gas from Sheldon Lambert, an energy forecaster. He begins, "It's like trying to guess the number of beans in a jar without knowing how big the jar is." So far so good. But then he adds,

"God is the only one who knows—and even He may not be sure."[11] Of course Lambert is speaking lightly. But the notion that some mind might know the "actual" size of the jar is misleading, because it implies that there is a fixed quantity of standard-sized beans. The quantity of a natural resource that might be available to us—and even more important the quantity of the services that can eventually be rendered to us by that natural resource—can never be known even in principle, just as the number of points in a one-inch line can never be counted even in principle. Even if the "jar" were fixed in size, it might yield ever more "beans." Hence, resources are not "finite" in any meaningful sense.

To restate: A satisfactory operational definition of the quantity of a natural resource, or of the services we now get from it, is the only sort of definition that is of any use in policy decisions. The definition must tell us about the quantities of a resource (or of a particular service) that we can expect to receive in any particular year to come, at each particular price, conditional on other events that we might reasonably expect to know (such as use of the resource in prior years). And there is no reason to believe that at any given moment in the future the available quantity of any natural resource or service at present prices will be much smaller than it is now, or non-existent. Only such one-of-a-kind resources as an Arthur Rubenstein concert or a Julius Erving basketball game, for which there are no close replacements, will disappear in the future and hence are finite in quantity.

Why do we become hypnotized by the word "finite"? That is an interesting question in psychology, education, and philosophy. A first likely reason is that the word "finite" seems to have a precise and unambiguous meaning in any context, even though it does not. Second, we learn the word in the context of mathematics, where all propositions are tautologous def-initions and hence can be shown logically to be true or false (at least in principle). But scientific subjects are empirical rather than definitional, as twentieth-century philosophers have been at great pains to emphasize. Mathematics is not a science in the ordinary sense, because it does not deal with facts other than the stuff of mathematics itself, and hence such terms as "finite" do not have the same meaning elsewhere that they do in mathematics.

Third, much of our daily life about which we need to make decisions is countable and finite—our weekly or monthly salaries, the number of gallons of gas in a full tank, the width of the backyard, the number of greeting cards you sent out last year, or those you will send out next year. Since these quantities are finite, why shouldn't the world's total possible salary in the future, or the gasoline in the possible tanks in the future, or the number of cards you ought to send out also be finite? Though the analogy is appealing, it is not sound. And it is in making this incorrect analogy that we go astray in using the term "finite."

A fourth reason that the term "finite" is not meaningful is that we cannot say with any practical surety where the bounds of a relevant resource system lie, or even if there are any bounds. The bounds for the Crusoes are the shores of their island, and so it was for early man. But then the Crusoes found other islands. Mankind travelled farther and farther in search of resources—finally to the bounds of continents and then to other continents. When America was opened up, the world, which for Europeans had been bounded by Europe and perhaps by Asia too, was suddenly expanded. Each epoch has seen a shift in the bounds of the relevant resource system. Each time the old ideas about "limits," and the calculations of "finite resources" within those bounds, were thereby falsified. Now we have begun to explore the sea, which contains amounts of metallic and other resources that dwarf any deposits we know about on land. And we have begun to explore the moon. Why shouldn't the boundaries of the system from which we derive resources continue to expand in such directions, just as they have expanded in the past? This is one more reason not to regard resources as "finite" in principle.

You may wonder, however, whether "non-renewable" energy resources such as oil, coal, and natural gas differ from the recyclable minerals in such a fashion that the foregoing arguments do not apply—energy is particularly important because it is the "master resource": energy is the key constraint on the availability of all other resources. Even so, our energy supply is non-finite, and oil is an important example. (1) The oil potential of a particular well may be measured and hence is limited (though it is interesting and relevant that as we develop new ways of extracting hard-to-get oil, the economic capacity of a well increases). But the number of wells that will eventually produce oil, and in what quantities, is not known or measurable at present and probably never will be and hence is not meaningfully finite. (2) Even if we make the unrealistic assumption that the number of potential wells in the Earth might be surveyed completely and that we could arrive at a reasonable estimate of the oil that might be obtained with present technology (or even with technology that will be developed in the next 100 years), we still would have to reckon the future possibilities of shale oil and tar sands—a difficult task.

(3) But let us assume that we could reckon the oil potential of shale and tar sands. We would then have to reckon the conversion of coal to oil. That, too, might be done, yet we still could not consider the resulting quantity to be "finite" and "limited." (4) Then there is the oil that we might produce not from fossils but from new crops—palm oil, soybean oil, and so on. Clearly, there is no meaningful limit to this source except the sun's energy. The notion of finiteness does not make sense here either. (5) If we allow for the substitution of nuclear and solar power for oil, since what we really want are the services of oil, not necessarily oil itself, the notion of a limit makes even less sense. (6) Of course, the sun may eventually run down. But even if our sun were not as vast as it is, there may well be other suns elsewhere.

About energy from the sun: The assertion that our resources are ultimately finite seems most relevant to energy but yet is actually more misleading with respect to energy than with respect to other resources. When people say that mineral resources are "finite," they are invariably referring to the Earth as a boundary, the "spaceship Earth," to which we are apparently confined just as astronauts are confined to their spaceship. But the main source of our energy even now is the sun, no matter how you think of the matter. This goes far beyond the fact that the sun was the prior source of the energy locked into the oil and coal we use. The sun is also the source of the energy in the food we eat and in the trees that we use for many purposes. In coming years, solar energy may be used to heat homes and water in many parts of the world. (Much of Israel's hot water has been heated by solar devices for years, even when the price of oil was much lower than it is now.) And if the prices of conventional energy supplies were to rise considerably higher than they now are, solar energy could be called on for much more of our needs, though this price rise seems unlikely given present technology. And even if the Earth were sometime to run out of sources of energy for nuclear processes—a prospect so distant that it is a waste of time to talk about it—there are energy sources on other planets. Hence, the notion that the supply of energy is finite because the Earth's fossil fuels or even its nuclear fuels are limited is sheer nonsense.

Whether there is an "ultimate" end to all this—that is, whether the energy supply really is "finite" after the sun and all the other planets have been exhausted—is a question so hypothetical that it should be compared with other metaphysical entertainments such as calculating the number of angels that can dance on the head of a pin. As long as we continue to draw energy from the sun, any conclusion about whether energy is "ultimately finite" or not has no bearing upon present policy decisions....

V. Summary

A conceptual quantity is not finite or infinite in itself. Rather, it is finite or infinite if you make it so—by your own definitions. If you define the subject of discussion suitably and sufficiently closely so that it can be counted, then it is finite—for example, the money in your wallet or the socks in your top drawer. But without sufficient definition, the subject is not finite—for example, the thoughts in your head, the strength of your wish to go to Turkey, your dog's love for you, the number of points in a one-inch line. You can, of course, develop definitions that will make these quantities finite, but that makes it clear that the finiteness inheres in you and in your definitions rather than in the money, love, or one-inch line themselves. There is no necessity either in logic or in historical trends to suggest that the supply of any given resource is "finite."

Simon's argument, and the general position on resources that it has inspired, rests on two fundamental claims: (1) that humans are themselves resources and (2) that the idea that (non-human) resources are finite is not meaningful. Let's look at these claims in order. As we will see, the two ideas are very closely related.

First, the claim that humans are themselves resources. Those, like Simon, writing from the economic perspective think that the ecologists are working with a flawed conception of the value of humans in the overall economy. Humans, on this mistaken view, are mainly seen as a burden, a force that strains the natural environment, forever drawing down or otherwise degrading its resources. Opposed to this is the idea that human labour and ingenuity are essentially wealth-creating, value-adding forces. This is part of the point of the thought-experiment Simon begins with. We are to imagine Alpha Crusoe on an island with a resource, copper, that is essential to his survival. But he needs to work hard to get at it. The ecologist might see Crusoe's copper-mining activities as burdensome to the natural world: he degrades an otherwise pristine environment while scratching out a living. Moreover, if he *is* absolutely reliant on the resource, and assuming there is not a readily accessible superabundance of it—in other words the resource is *scarce*—he will eventually exhaust it and come to grief.

However, Simon claims that this is not likely to happen. Crusoe will not mindlessly work away at the ever-receding stocks of copper with an ever-diminishing payoff until, exhausted, he expires in the dusty earth. Instead, he will learn how to recycle the copper materials he already has, he will invent more efficient ways to mine the copper, or he will discover or invent a substitute product, one more readily accessible but capable of providing the same service.

In other words, he will create economic value where none existed before. This is what we humans *do*. We are value and wealth *creators*, not just ecological burdens. And since it is our brains and brawn that allow us to do this, the more of us there are, the more likely we will be to overcome all resource scarcity. The introduction of additional characters into Simon's little drama—Beta Crusoe and Gamma Defoe—is thus no accident: we *require* population increase if we are to master nature efficiently.

This brings us to the second claim. Because of the virtually boundless wealth-creating potential of our labour and ingenuity, it is misleading to talk about resources—*any* resources—as finite. In the words of E. Zimmerman, "resources *are* not, they *become*."[12] The reason for this is straightforward. From the economic standpoint, we are not interested in a particular resource per se. Rather, we are interested in the services it provides for us, and those services can very often be provided in different ways as we draw down the readily accessible stocks of the scarce resource. This is why the economist's view of the human prospect tends to look so much more optimistic than the ecologist's. For example, regarding food resources we encounter the following sort of claim:

> [W]orld agricultural resources are capable of providing an adequate diet (2,500 kilocalories a day) as well as fibre, rubber, tobacco, and beverages, for 40 billion people, or eight times the present [1988] number. This . . .would require less than one-fourth—compared with one-ninth today—of the earth's ice-free land. . . . Clearly, better yields and/or use of a larger share of the land area would support over 40 billion persons.[13]

It's pretty difficult to imagine a view more at odds with the pessimistic outlook of a Malthus or an Ehrlich! Moreover, the focus on food resources is instructive. Those impressed by Simon's analysis often point to the two so-called Green Revolutions to bolster their optimistic claims about the non-finite nature of resources. Up until roughly 1950, it had been believed that we were getting just about as much agricultural output from our arable lands as they could provide. But just like Simon's Crusoe, we humans found a way to increase this output dramatically. The Green Revolutions involve three key developments: (1) selectively bred monocultures like rice, corn, and wheat; (2) more efficient irrigation techniques; and (3) heavy application of synthetic pesticides and fertilizers to crops. The first of the two revolutions took place in the developed world (chiefly North America and Western Europe) between 1950 and 1970 and has allowed Canada to become a major food exporter. Even though only about 7 per cent of its terrestrial area is suitable for agriculture, Canada currently exports $21 billion worth of food to 200 different countries. The second revolution began in 1967 in the developing world (mainly Central America, southern South America, and Southeast Asia) and is ongoing. Together, the two Green Revolutions have substantially increased yields per area unit of cropland.[14]

So in 1948 it would have been premature to claim that the human population could expand no more because we had reached the limit of an essentially finite resource—namely, the capacity of the available soil to yield food. Similarly, the argument might go, it would be premature to claim now that the second Green Revolution could not spread even farther than it has. What about central South America, most of sub-Saharan Africa, much of the Middle East, and Eastern Europe? There is no *in principle* limit to the advances we could achieve in either irrigation technology or the genetic modification of seeds to withstand harsh climates—drought-resistant and flood-resistant seeds are already in advanced stages of development—in order to bring increased yields to these and other places. And since Malthus is at least correct in claiming that population inevitably increases where subsistence does—see claim (2), above—our numbers are also capable of significant further expansion. Even if we cannot expand our numbers greatly, we may not have to take aggressive measures to reduce them. Here is what one economist has to say about this slightly softened brand of optimism:

> The optimists have two huge points in their favour. The first is the likelihood that the global population will stabilize this century. Malthus could certainly not anticipate the rise of modern contraception and the ready uptake of contraception in most societies in the world. The second cause for optimism is that technological advancement continues to be rapid and is probably accelerating. The revolutions in computing, data management, materials science (including nanotechnology) and other areas of knowledge all suggest that technologies, at least potentially, can rescue us and the planet yet again. The science and technology can be harnessed. The harder question is whether we will be well enough organized, and cooperative enough on a global scale, to seize the chance.[15]

As noted above, the central point of contention between the two perspectives on population and consumption—the ecological and the economic—has to do with the appropriateness of the concept of carrying capacity as applied to humans. We have seen that the lifeboat metaphor is the central expression of the ecologist's commitment to the idea that the Earth's carrying capacities are fixed for humans. The economist disagrees:

> In Garrett Hardin's metaphor, the lifeboat's capacity is written on its side. The doomsday literature of limits is shot through with the conceit of absolute capacity, which is alien to economics.[16]

Both sides in this dispute have something important to tell us. Is there some way to decide the debate in favour of one or the other or, failing that, to find some sort of compromise position?

Ⓓ Peak Oil and the Tar Sands

To see whether this is possible, let's look closely at what Simon calls the "master resource," energy. There is a connection between energy and food production, of course. As Canadian geologist David Hughes notes, the Green Revolutions have been fuelled by hydrocarbons, especially diesel fuel for farm machinery and natural gas–based fertilizers. But Hughes sees trouble ahead, because he thinks we are quickly running out of fossil fuels:

> The looming issue of fossil fuel depletion will put ever more pressure on food production. As a result there will be higher food prices and greater civil unrest, particularly within the large segment of the world's population that is living at close to subsistence levels. Peak oil and natural gas means peak food, without a rethink of our food production systems.[17]

This looks like Malthus all over again. Hughes's claims rest on the **peak oil hypothesis**, first put forward by the Shell geologist M. King Hubbert in the 1950s. The peak oil claim is simple: we have reached or are fast approaching that point at which we will have tapped half of the world's oil reserves. We are on, or near, the downward side of oil's supply curve, meaning that from now on supply will be unable to keep up with demand, especially given the ever-*increasing* demand we are now placing on it. And because our food production systems are so reliant on these fuels, if we do not massively reorganize them we will likely see large-scale social and political unrest—can famines be far behind?—in the not-too-distant future.

Is this view of our current situation tenable? The economist will surely balk at Hughes's conclusions. After all, even if it is true that reserves of conventional oil are in decline, there are other fossil-fuel sources to exploit. And these resources will inevitably be developed as long as the dwindling supply of conventional oil means that its price remains sufficiently high (but not too high). And Hughes does appear to underestimate the potential for alternative fuel sources such as coal-to-gas, tar sands, and oil shale. This debate is of central concern to Canadians, because Canada has a huge repository of tar sands in Alberta. As the name implies, this is heavy crude oil—otherwise known as bitumen—that is trapped in sand and is consequently extremely viscous.

Since the oil does not readily flow, the process of extracting it from the ground is extremely energy-intensive and environmentally disruptive. The bitumen must be extracted either by strip mining the area or by in situ techniques that involve pumping solvents or steam into the sands to make the oil less viscous. Once extracted, it is equally difficult and environmentally disruptive to refine the bitumen into useful synthetic fuel. For these reasons, it is not cost-effective to develop this resource unless the price of a barrel of oil is relatively high. In the first half of 2008, the price of a barrel of oil reached $140. It dropped dramatically in 2014, to around $50, which seriously eroded investment in the tar sands. Clearly, the price of this commodity is very volatile and difficult to predict. But many believe the tar sands will bounce back and become, once again, an economically viable resource.[18]

By some estimates, *total* fossil-fuel reserves (i.e., resources that are extractable given current technology) would last 130 years at current use rates. Taking into consideration all unconventional oil resources, we may be nowhere near the peak. The exploitation of Alberta's tar sands alone testifies to this: in 2008, the tar sands were producing 1 million barrels per day, a number that could rise to 4 or 5 million barrels per day by 2020 (assuming 2014's price fall is significantly reversed at some point).[19] Again, the 165 million barrels referred to just above are *proven reserves*. So to the extent that they are purely a product of our running out of conventional oil, it looks as though the large-scale problems predicted by Hughes can be avoided. Score one for Simon and the paean to human ingenuity.

However, the obvious problem is the damage that will be done to the environment with the exploitation of the tar sands on this scale (not to mention the even larger degradation caused by development of shale oil and coal-to-gas). Quite apart from the link between the tar sands and climate change (see Chapter 11 for a detailed analysis of this problem), production generates massive quantities of pollutants such as nitrogen oxides, sulphur dioxide, and volatile organic compounds (all of which degrade air quality). With open-pit mining of the bitumen, the entire surface of an area of more than 140,000 square kilometres will be radically upset: rivers diverted, wetlands drained, all surface vegetation (including massive tracts of boreal forest) removed.

Further, tar sands production uses a huge amount of fresh water and groundwater. Production of one cubic metre of synthetic crude oil requires 2 to 4.5 cubic metres of water. This amount increases with the Steam-Assisted Gravity Drainage (SAGD) method of oil extraction. This is the alternative to open-pit mining of bitumen. The technology involves pumping steam through horizontal pipes set deep in the ground. The steam loosens up the oil, which can then be pumped to the surface. This looks better on the surface (less ground is disturbed) but it uses even more fresh water than open-pit mining (as noted) as well as more natural gas to heat the water. Thus, despite appearances and contrary to what industry apologists claim about it, SAGD has a *more* harmful overall impact on the environment than open-pit mining. To meet the need for fresh water, up to 349 million cubic metres of water per year is currently being diverted from the Athabasca River, an amount that could rise as high as 500 million cubic metres. Finally, tar sands extraction processes produce "tailings," fine particles of toxic waste that nobody seems to know what to do with. So they sit in vast "ponds" dotting the landscape north of Fort McMurray. The extent of these ponds is expected to grow to 220 square kilometres by 2020.[20] This catalogue of environmental woe only scratches the surface of the problems associated with the exploitation of this resource. As the Pembina Institute puts it in its 2008 tar sands report card, the technologies involved in tar sands production "make the product among the most environmentally costly sources of transport fuel in the world."[21]

The example of the tar sands illustrates that the development of some resources carries an environmental burden that may be far too costly to warrant such development. Indeed, the damage to the biosphere might rise to a level at which the Earth's capacity to sustain us is severely compromised. It is crucial to bear in mind the fact that *waste disposal* is an unavoidable part of consumption. This is something no good ecologist would overlook, although economists too often do. But there is nothing intrinsic to the economic standpoint that forces its proponents to take on this blind spot. The most hard-headed economist might conclude that although we can, given our technological prowess, exploit a given resource, it is impermissible to do so because of the environmental harms caused by such exploitation.

This recognition opens the possibility for compromise between the economist and the ecologist. Thus, David Hughes, arguing from a broadly ecological point of view, can argue that the chief moral imperative for current generations is to "reduce consumption through conservation and efficiency."[22] In particular, we should move to a renewable energy–based economy as quickly as possible, diverting what precious little conventional fossil-fuel reserves we still possess to the task of building the future non-hydrocarbon infrastructure. And Canadian climate researcher Mark Jaccard, who would agree with Simon about the non-finite nature of fossil-fuel energy resources, nevertheless writes as though conservation and efficiency concerns should place legitimate constraints on the development of resources. We should develop fossil-fuel resources, he says, "provided we do this without causing unacceptable disruption to earth's climate and key biophysical systems."[23] Presumably, if we could not develop them without this result, we should instead engage in the sort of conservation strategies Hughes advocates.

Ⓔ Conclusion

Have we reached Earth's carrying capacity? Should aggressive measures be taken to stabilize population in developing countries? Should we scale back the rate at which developed countries are consuming resources? If so, to what degree and how exactly? There are no easy answers to these questions, but two points are noteworthy.

First, there is ample evidence to suggest that the best way to reduce fertility rates is through economic development, especially through the economic empowerment of women.[24] This goal has multiple applications in the lives of women in the developing world: legal protection (against domestic violence for example), property rights on a par with those of men, microfinance for small business ventures, the ending of discrimination in the workplace, and so on. As Jeffrey D. Sachs points out, with greater empowerment in these facets of social life, women are able to

> shift from quantity to quality in child-rearing because of the much higher opportunity cost of the mother's time. Husbands are also much more likely to agree to fewer children when their wives are money-earners in the labour market. This kind of empowerment is also likely to strengthen the mother's bargaining power vis-à-vis the husband in case of a difference of opinion between the spouses.[25]

But probably the most important factor in reducing fertility rates in the developing world is an emphasis on the education of girls. This move has a number of positive effects. Data show that with female secondary school enrolment comes lower fertility. Sachs again:

> There is the most direct effect: girls in school, notably in secondary school, are likely to remain unmarried until a later age, and, therefore, are likely to begin child-rearing much later than girls without schooling. Of course the content of education matters as well. Girls can and should be educated about sexual and reproductive health, and about the options for contraception. They can learn to analyze the quality/quantity tradeoff in the size of families and thereby overcome existing cultural biases more easily. This is critical since the cultural assumptions may have developed under a set of demographic conditions (for example very high child mortality rates) that are no longer applicable.[26]

Finally, one of the more subtle effects of educating girls, according to Sachs, is that it will also raise the future earnings potential of women in these societies, which will, in turn, raise their market value and reduce the preference for boys:

> When a household is aiming for three sons, it will need to have six children on average, half girls and half boys. If the household decides instead that it wants three children of any gender mix, it can reduce fertility by half and still achieve its objective. As son-preference diminishes, therefore, the overall fertility rate will fall as well.[27]

If these considerations are right, there may not be a stark choice between the goals of economic development and population reduction in the developing world. As long as economic development is rooted in the social and economic empowerment of women, it will likely not lead to catastrophic increases in population.

This brings us to the second noteworthy point: that such development must not proceed in a business-as-usual fashion. The rise in the world's consumer class, analyzed earlier in this chapter, is both an opportunity and a danger. If the wealth and social opportunities being created in the developing world are justly distributed, especially to women, we will likely see a dramatic drop in the rate of population growth in the foreseeable future in many countries (though not all). This provides some grounds for hope. However, if this economic growth is not achieved in an environmentally sustainable manner, its effects on the biosphere could be catastrophic. Nor is this to suggest that the moral burden facing humanity rests entirely with the developing world: the developed world obviously needs to reduce its consumption of resources massively. Learning how to consume in a sustainable manner is a moral challenge for *all* of us. If we fail to meet it, Malthus may yet be proven right.

CASE STUDY

Making Space for Grizzlies

Source: © iStockphoto.com/Stephen Schwartz

Grizzly bears, Prince George, BC. Are humans prepared to share the summit with species needing lots of land to thrive?

In his book *Dominion*, paleontologist Niles Eldridge provides a picture of the forces driving the vast increases in human population since the agricultural revolution some 10,000 years ago. Prominent among these forces has been an ever-increasing inwardness on the part of our species, a tendency to think and behave as though we are dependent for everything we need to thrive only on other humans. In fact, this tendency is both a cause and a consequence of the expansion of our numbers. The result is that we now think of ourselves as a globally interconnected species. This is a thought we take for granted these days, but compared to the situation of other species, it is profoundly weird. Every other species is organized into mutually isolated sub-populations, and each of these is intimately connected to local ecosystems. Even within local ecosystems, members of non-human species do not appear to spend *all* their time "wondering" about the goings-on of other members of their own species (though it is difficult in principle to know what sort of observations could confirm this). But we humans are different. For Eldridge, the fact that we exchange roughly $1 trillion per day among ourselves in the global economy is a "most riveting statistic"; it symbolizes our extreme conspecific inwardness and interdependency. But all of this comes with a danger:

[T]he story that we in the Western world have been telling ourselves for the last 10,000 years or so is only partly correct: we have indeed stepped outside of local ecosystems. But we have not stepped outside the natural world. We merely changed our stance towards that world. Until our population began to reach its current huge proportions our status vis-à-vis the rest of nature was equivocal. But now it has emerged as a full fledged reorientation: we stand foursquare as interactors with all the dynamic elements of the earth's natural system—its atmosphere, hydrosphere (lakes, rivers, and oceans), its soils and its rocks. And its species. We harvest; we alter the physical surroundings—mostly inadvertently as we pursue our inner-directed ways of realizing total control and self-sufficiency.

Our stepping outside of local ecosystems and focusing on ourselves has caused us to neglect our impact on those ecosystems. Let's take a single example to make this more concrete: the grizzly bear. The grizzly has an almost iconic status for Canadians. In its habitat—mainly the Rocky Mountains of Alberta, parts of the Yukon, and much of British Columbia—it is a top predator, apart from humans. But it needs a huge amount of space in order to thrive: anywhere from 26 to 42 square kilometres per animal. Gradually, pressures of human population expansion have been reducing the amount of land available to the grizzly.

Here's how this works. We tend not to think of our ever-increasing numbers—and the expansion in consumption and territory that comes with them—as in any way problematic or up for debate. Economic growth is an unchallengeable imperative, and it is premised on population growth. But we expand at the expense of other species' habitats, which we continually shrink. When the species we threaten are summit predators—which as such pose a threat to us—we begin to see *them* as the problem, we think *their* numbers must be too high. In British Columbia, this way of thinking has resulted in the trophy hunting of grizzlies. The David Suzuki Foundation has discovered that 10,000 grizzlies have been legally hunted in BC since the government began keeping records in the late 1970s. This accounts for 88 per cent of all human-caused grizzly deaths.

Tourists come from all over the world—often from places where they have already hunted their own wild megafauna to extinction—to participate in the slaughter of our grizzlies. Why do we allow this? Part of the answer to this question is that we are not yet ready to question some of our most fundamental assumptions about ourselves. Do we unconsciously resent the notion that the world contains summit predators other than ourselves? Do we want to sit entirely alone on that summit, with, at best, a few nicely domesticated and therefore docile animals to keep us company (sitting, as it were, at our feet on the summit)? Are we that bent on *controlling* wild nature, on achieving absolute self-sufficiency? Or are we just blind to the destruction we cause? Are we, as Eldridge surmises, so wrapped up in the human world that we cannot see that we are not in fact separate from the natural world our numbers and consumption are making less and less wild all the time?

Eldridge, for one, thinks we need to start telling ourselves a different kind of story about ourselves, one in which we are *essentially* integrated into various local ecosystems. As far as the grizzlies are concerned, this reorientation may result in substantial efforts to create geographical zones where these animals—along with other summit predators like polar bears, wolverines,

Continued

and mountain lions—are truly protected. Making space for grizzlies entails relinquishing our own claim to such space, and this can only take the form of reducing our population and consumption. As Ian McTaggart-Cowan suggests:

> In these new areas it will not be enough that the bears, wolves, wolverines and mountain lions can live without threat from man. Here their well-being will be first in all decisions. Road construction, timber harvest, hydro-electric impoundments, forest-fire control, the hunting of large mammals by man—all will be examined and regulated for their impact on the priority species and their prey base.

Currently, the parks system in Canada does not afford this broad measure of protection to such animals. Indeed, the current trophy hunt of grizzlies in BC takes place partly in parks and "protected" areas. True protection for other species will come only with the recognition that the way we consume, and the numbers in which we live, cannot be sustained. If we ever come to the realization that grizzlies need a substantial amount of the wild territory remaining in this country, and act so as to secure it for them, we will have taken a huge step in the new *outward-directed* reorientation required of us.

Sources: Niles Eldridge, *Dominion* (Berkeley: U of California P, 1995); Ian McTaggart-Cowan, "Room at the Top?," in Hummel 1989, 249–66; David Suzuki Foundation, "The Grizzly Truth about the Trophy Hunt," DSF *Newsletter,* Spring 2010.

Study Questions

1. Malthus's analysis resurfaces in many discussions of population and resource scarcity. Do you think his views are still relevant for the crises we face on these fronts? Why or why not?

2. Is Simon correct to claim that resources are infinite? Do you see examples today of people—in government or business, for instance—who think the same way? That is, even if they would not put the point the way Simon did, do their actions imply that they agree with him? Give some examples.

3. How big a problem is overpopulation in your view? If you think it is a big problem, what morally defensible responses can we adopt toward it?

4. What is the connection between Canada's exploitation of the tar sands and its very high per capita levels of wealth? What environmental issues are we overlooking as we develop this resource? Would you be willing to tolerate a reduction in your standard of living in order to constrain our development of the tar sands?

Further Reading

Joel E. Cohen. 1995. *How Many People Can the Earth Support?* New York: Norton.

Gary Cross. 2000. *An All-Consuming Century: Why Commercialism Won in America.* New York: Columbia UP.

Paul Ehrlich. 1968. *The Population Bomb.* New York: Ballantine Books.

Lindsey Grant. 1996. *Juggernaut: Growth on a Finite Planet.* Santa Ana, CA: Seven Locks Press.

Garrett Hardin. 1974. "Lifeboat Ethics." *Psychology Today,* September, 38–40, 123–4, 126.

Thomas Homer-Dixon and Nick Garrison, eds. 2009. *Carbon Shift.* Toronto: Random House Canada.

Bill McKibben. 1999. "A Special Moment in History: The Challenge of Overpopulation and Overconsumption." *The Atlantic Monthly,* May.

CHAPTER 13

The Biodiversity Crisis

Ecological Intuition Pump

Imagine you are alone on the planet with four other species. Just before you die, you wonder whether you should reduce this number to two (suppose you could do so easily), your favourite number. If you decide against this, it could be because you value species diversity as such. But why would you? Some philosophers have argued that the only, or at any rate the best, reason for working to maximize species diversity has to do with the aesthetic appeal of various individual animals and plants. An individual zebra's pattern of stripes is striking, and therefore the species is worth preserving. You might think this approach has some rather obvious shortcomings. Think for example of the humble horseshoe crab: although it has been around for nearly 400 million years, it is decidedly ugly. But surely that's not a *reason* to allow it to become extinct? More generally, there are lots of non-charismatic species that we think should be preserved despite their lack of charm. But perhaps it is a mistake to think that there is just one way in which we value species diversity. In fact, this issue looks like a very clear illustration of the power of value pluralism. It is possible to think that species diversity is valuable for aesthetic, economic, medical, religious, and symbolic reasons. Are these too anthropocentric? If so, return to the thought experiment above. In allowing all four species to survive you have decided to maximize species diversity even though no valuer will be around to appreciate it. Does *this* make sense?

Ⓐ Introduction

In Canada, 13 species have been driven to extinction by humans. Among them are the great auk (a large penguin-like bird), the Labrador duck, the passenger pigeon, the sea mink, and the Dawson caribou.[1] In all five of these cases, the main cause of extinction was over-hunting of the animals by humans. Of course, it is possible to be somewhat underwhelmed by these numbers. After all, Canada is a huge country with a rich variety of plant and animal species. Against that bounty, 13 species does not seem like anything to get alarmed about. To guard against whatever complacency this thought might induce, we need to be reminded of two important facts.

The first has to do with the scale of anthropogenic species loss worldwide. Biologists estimate that there are anywhere between 5 and 100 million extant species, of which we have identified only 1.4 to 1.8 million. The rate of extinctions since the arrival of modern *Homo sapiens* on the scene is 1000 to 10,000 times the rate before our arrival. That is, whereas the rate of species loss was about 0.0001 per cent per year before we arrived (this is sometime called the **background rate of extinction**), it has been somewhere between 0.1 per cent and 1 per cent per year since then. The consensus among biologists is that the total number of species is probably between 10 million and 14 million. At the upper end, we are therefore losing 14,000 species every year, at which rate we will lose 1 million species in 71 years. However, this number assumes that the extinction rate is at the lower end of the 0.1 per cent to 1 per cent scale, and there are good reasons to think this is too conservative an estimate. If we instead take 1 per cent as our measure of the rate of species loss, biologists have estimated that 20 per cent of all species will be lost by 2030 and 50 per cent by 2050.[2]

And there are some biologists who think that even the 1 per cent rate is too conservative, which, if correct, means that the number of lost species will be even more staggering than this. In the process of natural selection, extinctions happen all the time. Indeed, given the long history of life on Earth—more than 3 billion years—99 per cent of all species that have existed have become extinct. However, this history has also been punctuated by relatively abrupt, massive die-offs, usually caused by some sort of natural disaster. This has happened five times in the past 500 million years—for instance, 65 million years ago when an asteroid crashed into the planet and destroyed half the species on Earth, including the dinosaurs. The difference between species losses like that one and the one we are now experiencing is that this time *we* are the natural disaster. So devastating is the imminent loss of flora and fauna that it will take at least 5 million years for the slow process of **speciation** (whereby species evolve through natural selection) to replenish the number of species we will likely remove this century.[3]

The second point to bear in mind is that it can be misleading to look only at rates—or absolute numbers—of *extinctions*. Just as revealing are the numbers of species classified as **extirpated** (local, though not global, extinction), **endangered** (nearing extinct status), **threatened** (nearing endangered status), and of **special concern** (nearing threatened status). If we focus just on endangered species, 34 per cent of saltwater fish species are in this category, 51 per cent of freshwater species, 25 per cent of amphibians, 24 per cent of mammals, 20 per cent of reptiles, 14 per cent of plants, and 12 per cent of birds.[4] In Canada alone, there are 22 extirpated species, 205 endangered species, 136 threatened species, and 153 species of special concern, for a total of 529 species.[5]

Clearly, there is no room for complacency about possible biodiversity loss, here or elsewhere. And no catalogue of such loss would be complete without mentioning two of the world's most important biodiversity hotspots, one for terrestrial life, the other for marine life. The first are the equatorial rainforests of Asia, Africa, and South America. They contain more than half of all species. So rich in biodiversity are they that biologists have reported finding 425 different kinds of trees in a single hectare of Amazonian jungle and 1300 butterfly species in a fragment of Peru's Manu National Park.[6] This cornucopia is vitally dependent on the health of the rainforest ecosystem, and yet the rate of deforestation in these countries is extremely high: between 0.4 per cent and 1.98 per cent of the total forest per year in South American countries like Brazil, Ecuador, and Colombia. Indonesia's rainforests are being stripped at the rate of 2 per cent per year, chiefly in order to provide palm oil to companies like Nestlé, which makes some of its chocolate bars from this product.[7] At this rate, most of these forests—and the species that depend on them—will be gone within decades if aggressive protection measures are not adopted.[8]

If anything, the situation is even more dire in the "rainforests of the sea," the world's coral reefs. Coral reefs are found within 30 degrees latitude north and south of the equator. They are comprised of huge numbers of coral polyps surviving on a limestone bed of calcium carbonate skeletons that millions of previous generations of organisms have laid down.[9] This assemblage of biotic and abiotic material attracts a huge diversity of sea animals. A single reef may contain as many as 3000 species of corals and fish. According to some estimates, fully one-third of all fish species in oceans depend on coral reefs for food, although the reefs make up only 0.1 per cent of total ocean area in the world.[10]

But these wonders of biodiversity are under extreme threat. Over the past 20 years, coral has disappeared at a rate five times greater than the rate of rainforest loss. More than 80 per cent of Southeast Asia's coral reefs are threatened. Globally, about 3000 square kilometres of coral reef is lost each year. There are two deadly forces at work here. The first is **coral bleaching** due to warming waters. When tropical waters warm, the tiny algae that live on the corals and are the key link in the coral food web (they supply food for organisms further up the chain) are expelled, and the corals subsequently die. The second force is increased **acidification** of ocean water as the ocean absorbs more CO_2, producing carbonic acid. This acid releases hydrogen ions, which lower the water's pH level. The problem is that the coral's skeleton cannot form with this much acid in the water, so the coral will simply fail to grow.[11] The proximate cause of these two forces—bleaching and acidification—is anthropogenic climate change (see Chapter 11). Working together, they could lead to the loss of 80 per cent to 100 per cent of the world's coral reefs this century.

What exactly are humans doing that is putting so much stress on other species? Conservation biologists have coined the acronym **HIPPO** to highlight the key anthropogenic causes:[12]

> **H**abitat destruction
> **I**nvasive species
> **P**ollution
> **P**opulation
> **O**verharvesting

There are four points to note about this list. First, some conservation biologists think that these causes are listed in descending order of impact. So, for example, habitat destruction exceeds overharvesting as a cause of species loss worldwide. But second, this is true only at the present time. Previously—for instance, when humans first entered North America, quickly dispatching many of its largest fauna—overharvesting was a more salient cause of extinction than habitat loss. And as we have just seen, climate change—which we can classify under the pollution heading—is currently the primary threat to coral reefs. Third, the causes are in some cases mutually reinforcing. For instance, massive growth in human population leads to habitat loss as forests are cleared to make way for agricultural land. And finally, the HIPPO list is somewhat coarse-grained. If we want to be more precise about assigning causal responsibility for species loss, we would have to include such factors as resource use, climate change, poverty, sale of exotic pet and animal parts, predator and pest control, the absence of full environmental accounting, and so on.[13]

Another way to think about these impacts is to note that humans now appropriate roughly 40 per cent of the biosphere's **primary productivity**, the amount of solar energy converted by plants into potential food. For one species among 14 million, that truly constitutes the lion's share (too bad for the lions). When the distribution of basic and finite resources among species—considered as competitors for these resources—becomes this disproportionate, there is bound to be a significant diminution in the number of competitors. But should we be at all dismayed by these numbers? Is there something morally wrong about the fact that we are appropriating so much of the Earth's primary productivity, thus driving other species to the brink and beyond? What could it be? Is biodiversity as such valuable? Even if we are inclined to answer yes to this question, what things ought to be the main objects of our moral concern?

We can organize these questions into two main groups: (1) those that ask whether our moral concern should be directed at species or the individuals that make them up and (2) those that ask what biodiversity is and why it is valuable. We will look at question (1) in the next section (B), then turn to question (2) in sections C and D.

Ⓑ Species or Individuals?

The view that species ought to be our primary objects of concern has been forcefully expressed by the evolutionary biologist Stephen Jay Gould. However, at first glance it does look rather odd to claim that what is of direct moral concern to us are species. After all, it might be suggested that a species is just a category of thought, not something existing in its own right. Darwin himself once said:

> [W]e shall have to treat species as merely artificial combinations made for convenience. This may not be a cheering prospect; but we shall at least be freed from the vain search for the undiscovered and undiscoverable essence of the term *species*.[14]

Another way to put the worry here is to ask why we care about species. If we look back to the arguments about who or what has moral standing—arguments canvassed in Chapters 1–3—we might wonder how a species could possess such standing. After all, a species has no psychological states, and it is not a teleological centre of life. Recall that these are the reasons for thinking, respectively, that animals or living things have moral standing. Alternatively, it may be that the moral standing of species is on a par with, or similar to, that of ecosystems, but this too is problematic. According to the land ethic, the morally significant characteristic of ecosystems is their integrity, but this is a quality that does not look to be applicable to species. So in virtue of what characteristics should we grant moral standing to species (does a species have interests, and if so, how are they determined)?

For Gould, the moral status of species follows from the fact that they are, as he puts it, "nature's objective packages." That is, he wants to deny the claim, made above by Darwin, that species are merely conceptual or "artificial" entities. There are two criteria used to define a species that show they are real. The first criterion is historical. Speciation usually occurs when populations of a species become geographically isolated from one another and eventually develop distinct gene packages. Although for a short time after this process there will be only minor differences between the original species and the newly isolated one, if the latter survives at all it can quickly become a unique species. Thus, although at any given geologic time 1 per cent of species will be difficult to define in a way that clearly separates them from all other species—these are the populations that have just become geographically isolated— this will not be the case for the other 99 per cent. The point is that the history of speciation breaks up populations of living things into genetically distinct groups, known as species.

The second criterion is functional. Members of a species are capable of interbreeding only with members of their own species. Again, although there are some exceptions to this rule, Gould asserts that it "generally works well." And again, these two criteria—the historical and the functional—confer objective existence on species. By contrast, the higher categories of biological classification are mere contrivances, although no less useful to us for all that. We organize species into genera, genera into phyla, and phyla into kingdoms as a matter of theoretical convenience. But the evolutionary tree—Gould prefers to think of it as a *bush*—is out there, comprised of species and independent of our classificatory designs.

The question is, what follows about our duties? Gould is clear about this:

> By grasping the objective status of species as real units in nature (and by understanding why they are not arbitrary divisions for human convenience), we may better comprehend the moral rationale for their preservation. You can expunge an arbitrary idea by rearranging your conceptual world. But when a species dies an item of natural uniqueness is gone forever.[15]

However, at most this establishes only that there is a difference between expunging ideas and expunging things outside our heads. By itself, the distinction carries no moral implications. Indeed, we can construct examples in which the moral implications are exactly the opposite of what Gould supposes they are. Suppose we had isolated the world's last particle of the human immunodeficiency virus (HIV-1), which we could easily obliterate but could not safely store. Suppose further that we possessed the mathematical formula showing us how to synthesize the cure for cancer but that this formula was contained in the mind of one person and nowhere else. A simple and harmless memory alteration technique would remove the idea from this person's mind forever. Finally, imagine we were told we must destroy one—but only one—of these two things. One is a real piece of the world's furniture, the other a mere idea. But clearly we would be morally obliged to destroy the virus particle rather than the formula.

The key to this thought experiment is that the objective/non-objective distinction, though we can make some sense of it, plays no role whatever in the determination of our duties. Instead, we will, for example, weigh the harms that will likely result from each course of action. In a widely cited article, philosopher Lilly-Marlene Russow argues for a novel approach to these issues, in two crucial respects. First, she claims that what we really value are the individual members of species, not the species as such; second, she claims that such value is chiefly aesthetic. Both points are captured in this passage:

> The reasons ... given for the value of a species are, in fact, reasons for saying that an individual has value. We do not admire the grace and beauty of the species *Panthera tigris*; rather we admire the grace and beauty of the individual Bengal tigers that we may encounter. What we value then is the existence of that individual and the existence (present or future) of individuals like that. The ways in which other individuals should be "like that" will depend on why we value that particular sort of individual: the stripes on a zebra do not matter if we value zebras primarily for the way they are adapted to a certain environment, their unique fitness for a certain sort of life. If, on the other hand, we value zebras because their stripes are aesthetically pleasing, the stripes do matter.[16]

Problems remain with Russow's proposal (are aesthetic considerations *that* important to us?), but on the whole it may be an attractive alternative to the view that species as such are morally considerable. In any case, from the standpoint of environmental policy, nothing much changes here. It remains useful to employ the concept of a species. We can still talk meaningfully about threatened or endangered species, and we can still direct our energy and resources toward protecting them. And we can do so while recognizing that what we ultimately *care* about are the individuals that make up those species.

ⓒ The Nature of Biodiversity

But this brings us to another question: What is the nature of species *diversity*? And why exactly is it valuable? Canadian environmentalist David Suzuki can help us get started on the first of these questions.

THE POWER OF DIVERSITY

David Suzuki

In past decades, the scientific community has undergone a tremendous expansion, and knowledge has increased proportionately. However, too often the accumulation of information is mistaken for knowledge that provides understanding and control. We can't afford to make such an assumption, because it fosters the terrible illusion that we can "manage" wilderness and it has resulted in destructive consequences.

Globally, old-growth forests are being cleared with alarming speed. In the past decades, geneticists have made a surprising finding that foresters should heed. When seemingly homogeneous populations of organisms were analyzed using molecular techniques, they were unexpectedly found to be highly diverse. When looked at from individual to individual, the products of a single gene are found to vary considerably. Geneticists call such variability "genetic polymorphism," and we now know that a characteristic of wild populations of any species is a high degree of genetic polymorphism. Apparently, maximum genetic diversity optimizes the chances that a species can withstand changes in the environment.

When individuals of the same species are compared, their patterns of genetic polymorphism differ from region to region. Thus, whether a tree, fish, or bird, different geographic subgroups exhibit different spectra of variation. So Ronald Reagan was dead wrong—if you see one redwood tree, you haven't seen them all. Stanford ecologist Paul Ehrlich says, "The loss of genetically distinct populations within species is, at the moment, at least as important a problem as the loss of entire species."

The biological value of diversity can also be applied to a collection of species. A forest is more than an assemblage of trees; it is a community of plants, animals, and soil microorganisms that have evolved together. This aggregate of species creates a highly resilient forest with a great capacity to recover from fire, flooding, landslides, disease, selective logging, or storm blowdowns. That's because the diverse species remaining in the surrounding areas can replenish the damaged parts. Clearly, we should try to maximize forest diversity by protecting as many old-growth forests as possible. That's the best way to ensure the maintenance of a broad genetic base on which the future of the forestry industry will depend.

There is a way to illustrate the power of diversity by looking at our bodies. Just as a forest is made up of vast numbers of individuals of different species, we are an aggregate of some 100 trillion cells that vary in size, shape, and function. These different cells are organized at many levels into tissues and organs that all come together in a single integrated whole—a functioning body.

The collective entity that is each of us thus is a mosaic of an immense array of different cell, tissue, and organ types that have enormous resilience and recuperative powers. If we suffer a cut, bruise, or infection, the body has built-in mechanisms to overcome the assault. We even have the ability to regenerate skin, liver, blood, and other body parts and compensate for damage to the brain and circulatory systems. We can function pretty well with the loss of some body parts, such as a digit, tonsils, or teeth. In short, our bodies can absorb considerable trauma and recover well, a tribute to cellular diversity in form and function.

If we amputate large parts of the body, we can still function and survive. Thus, we can live with the loss of limbs, eyes, ears, and other parts, but each loss confers greater dependence on other people and on human technological ingenuity to compensate for lost abilities. With the power of modern science and high technology, we can make artificial substitutes for teeth, bones, skin, and blood, and we have even devised machines to take over for the heart, lungs, and kidneys. In principle, it should be possible for an individual to survive the combined loss of organs that are not absolutely necessary for life and those that can be mimicked by machines. Thus, a blind, deaf, quadruple amputee who is hooked to a heart-lung and kidney machine could live and would still be a person but one with capabilities and resilience radically restricted in comparison with a whole individual. Essentially, such a patient would be a different kind of human being, created by and dependent upon human expertise and technology.

In the same way, a forest bereft of its vast biodiversity and replaced by a limited number of selected species is nothing like the original community. It is an artifact created by human beings who foster a grotesque concept of what a forest is. We know very little about the basic biology of a forest community, yet road-building, clear-cut logging, slash burning, pesticide and herbicide spraying, even artificial fertilization have become parts of silviculture practice. The integrity of the diverse community of species is totally altered by such practices with unexpected consequences—loss of topsoil, death from acidification, weed overgrowth, disease

outbreak, insect infestation, and so on. But now, caught up in the mistaken notion that we have enough knowledge about forests to "manage" them in perpetuity, we end up ricocheting from one contrived Band-Aid solution to another.

Medical doctors today are struggling to readjust their perspective to treat a patient as a whole individual rather than as an aggregate of autonomous organ systems, and eco-psychology recognizes the relevance of our surroundings to our psychological health. A similar perspective has to be gained on forests. The key to development of sustainable forests must reside in the maintenance of maximum genetic diversity both within a species and between the species within an ecosystem. If we begin from this basic assumption, then the current outlook and practices in forestry and logging have to be radically overhauled.

What is biodiversity? How is it measured? When we refer to a crisis of biodiversity, of what are we bemoaning the loss? According to Suzuki, our preservation efforts should be directed at two things: the total number of species and the genetic variability within species. The value of "genetic polymorphism" is that the species as a whole is able to withstand disruptive environmental changes when there is a high degree of it. This makes intuitive sense. Suppose a toxin is introduced into the food supply of a particular species. In virtue of small genetic differences between them, the toxin might be deadly to members of one subpopulation of the species but not to another. So the presence of variability has ensured the survival of the species in the face of this external threat, whereas the species would have been killed off without it. This is sometimes referred to as the **insurance principle of biodiversity**.

Genetic polymorphism within a species is a by-product of natural selection, but it cannot happen if the numbers of individuals in the species fall below a certain minimal threshold. The reason for this is that below that threshold there is too much inbreeding to ensure the development of optimal genetic variability. For example, in Alberta and Saskatchewan the greater sage-grouse has been reduced from numbers in the millions 200 years ago to just 800 now, which has led to a massive loss of genetic diversity in the Canadian population of the bird. This makes it highly vulnerable to environmental disturbances. In an effort to restore the population's genetic diversity, researchers are pondering the translocation of individuals from populations in northern Montana.[17]

Given that species are located in relatively closed ecosystems or habitats, there are two ways to think about species diversity. First, there is **within-habitat complexity**, which refers to the number of species within a habitat. Think of this with respect to **trophic levels**. A trophic level is a particular place, or node, in an ecosystem's energy transfer network. So producers like plants belong to the first trophic level, while herbivores like cattle belong to the second, and carnivores like mountain lions belong to the third. Within-habitat complexity therefore refers to (a) the number of species at each trophic level and (b) the number of trophic levels.[18] Second, there is **between-habitat complexity**, which refers to the lack of similarity between systems. As Bryan Norton points out, between-habitat complexity "must measure and compare dissimilarities in the types of species, the mix of species, and the types of interactions between them, as these exist in different systems." If we combine these two measures of biodiversity, then add genetic variability within species, we get **total complexity**. Total complexity is therefore the "proper object of concern for species preservationists."[19]

Ⓓ The Value of Biodiversity

Notwithstanding the identification of the proper object of concern for preservationists, the question remains as to why total diversity or complexity is of value. Ecologists David Tilman and John Downing conducted a 12-year study in which they attempted to answer this question. They planted 207 discrete plots containing various plant species. Each plot had a different number and mixture of species in it, a mini-ecosystem. The study region then underwent a one-year drought. The results were twofold. First, the more species-diverse the plot, the fewer species it lost to drought; second, the more species-diverse the plot, the quicker it returned to pre-drought conditions (inertia and resilience: see Chapter 3).[20] Such research is crucial to our understanding of how ecosystems work. The insight does not so much replace the main point from Suzuki's article—that biodiversity reduces a species' vulnerability to environmental shock—as widen the focus from species to whole ecosystems. And this is important, because we humans depend on healthy ecosystems as much as other species do.

So what exactly is the connection between biodiversity and healthy ecosystems? Here is how biologist E.O. Wilson describes the fundamental claim:

> In conserving nature, whether for practical or aesthetic reasons, diversity matters. The following rule is now widely accepted by ecologists: the more species that inhabit an ecosystem, such as a forest or lake, the more productive and stable is the ecosystem. By "production," the scientists mean the amount of plant and animal tissue created each hour or year or any other unit of time. By "stability" they mean one or the other or both of two things: first, how narrowly the summed abundances of all species vary through time; and secondly how quickly the ecosystem recovers from fire, drought, and other stresses that perturb it.[21]

The ecological lesson we are slowly learning is that diversity begets diversity. That is, the production of biodiversity is self-enhancing, an upward spiral. Now, most biologists believe there is probably a limit to how much biological diversity is required to keep an ecosystem stable, but there is consensus on two points. First, there is a threshold below which stability is threatened, and second, it is usually difficult to identify *all* the species that are essential to the stability of an ecosystem.

Still, some species clearly do more work than others in the process of building stability and adding diversity. Ecologists refer to these as **foundational species**. The beaver—that Canadian icon—is one such species. In the process of building its dams, it converts streams to ponds and bogs. This activity creates the conditions for many other species to flourish. Many amphibians are attracted to these ponds, as are waterfowl species and riparian plant species. In turn, the activities of these species—niche construction, consumption, death and decay—provide conditions for even more species to thrive. And when beavers abandon their ponds, the latter are gradually drained, and the logs and other organic materials that have been decaying at the bottom of the pond—put there by the beavers—provide rich soils for the formation of beaver meadows. These provide habitat for still more species.[22]

However, since we are focusing in this chapter on the biodiversity *crisis*, we should not overlook the logical corollary of the process just described. This is the downward spiral of biodiversity loss. Anne and Paul Ehrlich have provided us with an elegant analogy to describe the danger here. The story they tell is called the "Rivet Poppers." A person about to board an airplane notices that there is a worker removing some of the rivets holding the wings in place. The concerned passenger questions the worker about the wisdom of this. The worker replies that the rivets can be sold for a modest sum and that the profits from the sale can be used to keep the cost of flights low. And besides, he adds, the procedure is obviously not unsafe, since he and his colleagues have been at it for some time and no wings have yet come off the planes.[23] Obviously, this explanation would, or should, fail to reassure the passenger. And the reason is straightforward: the more rivets the worker removes, the greater the likelihood that the whole wing will come loose.

Species are the rivets of ecosystems. Is the analogy correct? That is, is it the case that the removal of each species makes a catastrophic loss of species more likely? Bryan Norton thinks it is and that in order to appreciate this, we need to make a fundamental distinction between independent and dependent occurrences:

> Two kinds of situation need to be distinguished here. In some situations, such as catastrophic core meltdowns of nuclear reactors, the *individual occurrences are independent*. The absence of a meltdown to date can be considered (very weak) confirmation of a low probability assignment to future meltdowns.

But species reduction is not like this.

> Assuming, as most ecologists would, that there is some minimal number of species such that, if extinctions diminished the stock of species below that minimum an ecological disaster would occur, species extinctions represent *dependent occurrences*. Each species extinction increases, however slightly, the probability that

the next will prove disastrous. Hence the rivet-popping parable exhibits the false assumption that dependent occurrences are independent.[24]

In the case of the rivet-popping, the passenger thinks that her own safety will be best secured if *no rivets* are popped. This is because she assumes that every one of the rivets is doing something vital—namely, bearing a fraction of the total load of the attached wing. That is, she assumes, correctly, that even if there is *some* functional redundancy in the system, it is probably better to play it safe. If every popped rivet is a dependent occurrence, it raises the likelihood of a total system failure.

By analogy, we might construct the following argument:

1. Because species are functionally interconnected in ecosystems, species extinctions are dependent occurrences.
2. Our own survival is intimately tied to the existence of healthy ecosystems.
3. Therefore, it is prudent to avoid the extinction of *any* species as much as possible.

Even if there is some functional redundancy and overlap in ecosystems, it is imprudent to proceed as though the work done by any particular species can be done by some other species such that it is permissible to eliminate it. If, for instance, three species perform some vital function, then the removal of even one of them might disable the system profoundly. And since our knowledge of ecosystems is generally so incomplete, we should probably assume that we rarely know for sure whether or not the removal of one species will fatally undermine the system, sinking a boatload of other species with it. In other words, just like the wing's rivets, the presence of some redundancy in an ecosystem does not imply that extinctions are independent occurrences. Rather, the redundancies are, as we have seen, part of a broader *insurance policy*: they are meant to make the ecosystem more resilient in the face of external perturbations. To erode the insurance policy willfully or through mismanagement is folly.

However, isn't this argument entirely too anthropocentric? Do we really want to insist that the only reason we have for preserving biodiversity is that our own survival depends on it (premise [2], just above)? Is prudence our only guide here? Our next article challenges not only this formulation but any attempt to specify what is valuable about biodiversity.

WHY PUT A VALUE ON BIODIVERSITY?

David Ehrenfeld

In this [paper], I express a point of view in absolute terms to make it more vivid and understandable. There are exceptions to what I have written, but I will let others find them.

That it was considered necessary to have a section in this volume devoted to the value of biological diversity tells us a great deal about why biological diversity is in trouble. Two to three decades ago, the topic would not have been thought worth discussing, because few scientists and fewer laymen believed that biological diversity was—or could be—endangered in its totality. Three or four decades before that, a discussion of the value of biological diversity would probably have been scorned for a different reason. In the early part of this century, that value would have been taken for granted; the diversity of life was considered an integral part of life and one of the nicest parts at that. Valuing diversity would, I suspect, have been thought both presumptuous and a terrible waste of time.

Now ... we have meetings, papers, and entire books devoted to the subject of the value of biological diversity. It has become a kind of academic cottage industry, with dozens of us sitting at home at our word processors churning out economic, philosophical, and scientific reasons for or against keeping diversity. Why?

There are probably many explanations of why we feel compelled to place a value on diversity. One, for example, is that our ability to destroy diversity appears to place us on a plane above it, obliging us to judge and evaluate that which is in our power. A more straightforward explanation is that the dominant economic realities of our time—technological development, consumerism, the increasing size of governmental, industrial, and agricultural enterprises, and the growth of human populations—are responsible for most of the loss of biological diversity. Our lives and futures are dominated by the economic manifestations of these often hidden processes, and survival itself is viewed as a matter of economics (we speak of tax shelters and safety nets), so it is hardly surprising that even we conservationists have begun to justify our efforts on behalf of diversity in economic terms.

It does not occur to us that nothing forces us to confront the process of destruction by using its own uncouth and self-destructive premises and terminology. It does not occur to us that by assigning value to diversity we merely legitimize the process that is wiping it out, the process that says, "The first thing that matters in any important decision is the tangible magnitude of the dollar costs and benefits." People are afraid that if they do not express their fears and concerns in this language, they will be laughed at, they will not be listened to. This may be true (although having philosophies that differ from the established ones is not necessarily inconsistent with political power). But true or not, it is certain that if we persist in this crusade to determine value where value ought to be evident, we will be left with nothing but our greed when the dust finally settles. I should make it clear that I am referring not just to the effort to put an actual price on biological diversity but also to the attempt to rephrase the price in terms of a nebulous survival value.

Two concrete examples that call into question this evaluating process come immediately to mind. The first is one that I first noticed a number of years ago: it was a paper written ... by Clark, an applied mathematician at the University of British Columbia. That paper, which everyone who seeks to put a dollar value on biological diversity ought to read, is about the economics of killing blue whales. The question was whether it was economically advisable to halt the Japanese whaling of this species in order to give blue whales time to recover to the point where they could become a sustained economic resource. Clark demonstrated that in fact it was economically preferable to kill every blue whale left in the oceans as fast as possible and reinvest the profits in growth industries rather than to wait for the species to recover to the point where it could sustain an annual catch. He was not recommending this course—just pointing out a danger of relying heavily on economic justifications for conservation in that case.

Another example concerns the pharmaceutical industry. It used to be said, and to some extent still is, that the myriad plants and animals of the world's remaining tropical moist forests may well contain a great many chemical compounds of potential benefit to human health—everything from safe contraceptives to cures for cancer. I think this is true, and for all I know, the pharmaceutical companies think it is true also, but the point is that this has become irrelevant. Pharmaceutical researchers now believe, rightly or wrongly, that they can get new drugs faster and cheaper by computer modelling of the molecular structures they find promising on theoretical grounds, followed by organic synthesis in the laboratory using a host of new technologies, including genetic engineering. There is no need, they claim, to waste time and money slogging around in the jungle. In a few short years, this so-called value of the tropical rainforest has fallen to the level of used computer printout.

In the long run, basing our conservation strategy on the economic value of diversity will only make things worse, because it keeps us from coping with the root cause of the loss of diversity. It makes us accept as givens the technological/socio-economic premises that make biological impoverishment of the world inevitable. If I were one of the many exploiters and destroyers of biological diversity, I would like nothing better than for my opponents, the conservationists, to be bogged down over the issue of valuing. As shown by the example of the faltering search for new drugs in the tropics, economic criteria of value are shifting, fluid, and utterly opportunistic in their practical application. This is the opposite of the value system needed to conserve biological diversity over the course of decades and centuries.

Value is an intrinsic part of diversity; it does not depend on the properties of the species in question, the uses to which particular species may or may not be put, or their alleged role in the balance of global ecosystems. For biological diversity, value is. Nothing more and nothing less. No cottage industry of expert evaluators is needed to assess this kind of value.

Having said this, I should stop, but I won't, because I would like to say it in a different way.

There are two practical problems with assigning value to biological diversity. The first is a problem for economists: it is not possible to figure out the true economic value of any piece of biological diversity, let alone the value of diversity in the aggregate. We do not know enough about any gene, species, or ecosystem to be able to calculate its ecological and economic worth in the larger scheme of things. Even in relatively closed systems (or in systems that they pretend are closed), economists are poor at describing what is happening and terrible at making even short-term predictions based on available data. How then should ecologists and economists, dealing with huge, open systems, decide on the net present or future worth of any part of diversity? There is not even a way to assign numbers to many of the admittedly most important sources of value in the calculation. For example, we can figure out, more or less, the value of lost revenue in terms of lost fisherman-days when trout streams are destroyed by acid mine drainage, but what sort of value do we assign to the loss to the community when a whole generation of its children can never experience the streams in their environment as amenities or can never experience home as a place where one would like to stay, even after it becomes possible to leave?

Moreover, how do we deal with values of organisms whose very existence escapes our notice? Before we fully appreciated the vital role that mycorrhizal symbiosis plays in the lives of many plants, what kind of value would we have assigned to the tiny, threadlike fungi in the soil that make those relationships possible? Given these realities of life on this infinitely complex planet, it is no wonder that contemporary efforts to assign value to a species or ecosystem so often appear like clumsy rewrites of *The Emperor's New Clothes*.

The second practical problem with assigning value to biological diversity is one for conservationists. In a chapter called "The Conservation Dilemma" in my book *The Arrogance of Humanism*, I discussed the problem of what I call non-resources. The sad fact that few conservationists care to face is that many species, perhaps most, do not seem to have any conventional value at all, even hidden value. True, we cannot be sure which particular species fall into this category, but it is hard to deny that there must be a great many of them. And unfortunately, the species whose members are the fewest in number, the rarest, the most narrowly distributed—in short, the ones most likely to become extinct—are obviously the ones least likely to be missed by the biosphere. Many of these species were never common or ecologically influential; by no stretch of the imagination can we make them out to be vital cogs in the ecological machine. If the California condor disappears forever from the California hills, it will be a tragedy: but don't expect the chaparral to die, the redwoods to wither, the San Andreas fault to open up, or even the California tourist industry to suffer—they won't.

So it is with plants. We do not know how many species are needed to keep the planet green and healthy, but it seems very unlikely to be anywhere near the more than quarter of a million we have now. Even a mighty dominant like the American chestnut, extending over half a continent, all but disappeared without bringing the eastern deciduous forest down with it. And if we turn to the invertebrates, the source of nearly all biological diversity, what biologist is willing to find a value—conventional or ecological—for all 600,000-plus species of beetles?

I am not trying to deny the very real ecological dangers the world is facing; rather, I am pointing out that the danger of declining diversity is in great measure a separate danger, a danger in its own right. Nor am I trying to undermine conservation; in fact, I would like to see it find a sound footing outside the slick terrain of the economists and their philosophical allies.

If conservation is to succeed, the public must come to understand the inherent wrongness of the destruction of biological diversity. This notion of wrongness is a powerful argument with great breadth of appeal to all manner of personal philosophies.

Those who do not believe in God, for example, can still accept the fact that it is wrong to destroy biological diversity. The very existence of diversity is its own warrant for survival. As in law, long-established existence confers a powerful right to a continued existence. And if more human-centred values are still deemed necessary, there are plenty available—for example, the value of the wonder, excitement, and challenge of so many species arising from a few dozen elements of the periodic table.

And to countenance the destruction of diversity is equally wrong for those who believe in God, because it was God who, by whatever mechanism, caused this diversity to appear here in the first place. Diversity is God's property, and we, who bear the relationship to it of strangers and sojourners, have no right to destroy it. There is a much-told story about the great biologist, B.S. Haldane, who was not exactly an apostle of religion. Haldane was asked what his years of studying biology had taught him about the Creator. His rather snide reply was that God seems to have an "inordinate fondness for beetles." Well why not? As God answered Job from the whirlwind in the section of the Bible that is perhaps most relevant to biological diversity, "Where were you when I laid the foundations of the earth?" (Job 38:4). Assigning value to that which we do not own and whose purpose we cannot understand except in the most superficial ways is the ultimate in presumptuous folly.

The great biochemist Erwin Chargaff, one of the founders of modern molecular biology, remarked not too many years ago, "I cannot help thinking of the deplorable fact that when the child has found out how its mechanical toy operates, there is no mechanical toy left." He was referring to the direction taken by modern scientific research, but the problem is a general one, and we can apply it to conservation as well. I cannot help thinking that when we finish assigning values to biological diversity, we will find that we don't have very much biological diversity left.

Let's begin our analysis of this provocative article by focusing on one of the two "practical problems" with assigning value to biodiversity identified by Ehrenfeld. The claim is that our knowledge of wild nature is too limited to warrant assigning value to it. This is a direct challenge to anthropocentrism and more particularly to the economist. Our limitation shows up in three distinct ways: (1) our ignorance of the real economic value of species and ecosystems; (2) the incompleteness of our catalogue of species; and (3) our inability to determine the non-economic value of species. Let's begin by investigating claims (1) and (2). Together, they assert that it is folly to assign a purely economic value to biodiversity because we have identified only a fraction of the species that currently exist and we don't know enough about the ones we *have* identified.

We should emphasize the point about our ignorance of the value of species we have not yet identified. To illustrate it, E.O. Wilson points our attention to the importance of wild species to medicine. As he notes, approximately 40 per cent of all prescriptions issued by pharmacies are derived from substances extracted from plants, micro-organisms, and animals. Even so, only a small portion of nature's bounty has been scientifically tested for its medicinal potential. For example, the control of bacterial disease has been assigned almost exclusively to species of ascomycete fungi. Eighty-five per cent of the antibiotics we now use were extracted from species of this fungi. So they are obviously very useful, but even though we have identified some 30,000 of them, this represents just 10 per cent of the estimated total.

Or think of the flowering plants. Only 3 per cent of all identified species of flowering plants have been tested for alkaloids, a powerful substance in fighting some cancers. As Wilson points out, there is a sound evolutionary logic to the usefulness of wild species in medicine. What nature presents us with right now are victors in the long evolutionary struggle for survival. For millions of years, plant and animal species have developed chemical mechanisms for fighting cancer-producing toxins, among other things. Since many of the threats our bodies face are the same ones their bodies have faced, we do well to investigate them for the secrets of their success. The medicinal potential here is virtually inexhaustible: not just antibiotics but anti-malarial drugs, anesthetics, blood-clotting agents, immunosuppressive agents, antidepressants, sedatives, and so on.[25]

However, even supposing we have identified a particular species and determined what we believe to be its economic value with some accuracy, we cannot know in the present exactly what its value will be in the future. As Ehrenfeld argues, a cautionary tale emerges from our attempt to calculate the value of the blue whale. The numbers of the species plummeted in the twentieth century because the animals were so big and easy to kill, and so the question naturally arose as to whether or not it should simply be hunted to extinction given its clear economic value. A UBC mathematician infamously calculated that it would be best to kill it off and invest the money in "growth industries." Universalized, this reasoning would cause an instant and catastrophic loss of species. But setting aside this obvious objection, Wilson's dismissal of the economic argument is blunt and decisive:

> The dollars-and-cents value of a dead blue whale was based only on the values of an existing market—that is, on the going price per unit weight of whale oil and meat. There are many other values, destined to grow along with our knowledge of living [blue whales] in science, medicine, and aesthetics, in dimensions and magnitudes still unforeseen. What was the value of the blue whale in AD 1000? Close to zero. What will be its value in AD 3000? Essentially limitless, plus the gratitude of the generation then alive to those who, in their wisdom, saved the whale from extinction.[26]

Consideration of points (1) and (2), above, thus leads to two important insights. First, even if we consider only the use-value of species—that is, their value as commodities—prudence suggests that we should exploit them with extreme caution. This is because, as we have seen, species are not isolated in nature but come instead as parts of ecosystems. Like the loss of a single rivet on a plane's wing, the loss of one species can increase the likelihood that others in the ecosystem will be lost. And since we know so little about how many species there are, we might be endangering ourselves—or at least losing the opportunity to better our lot—in the process. Our ignorance suggests that we adopt a **precautionary principle** here: since we are unaware of many of the complex causal links in ecosystems, we should take an extremely conservative approach to managing them, taking care to prevent harm whenever possible.

But, second, we must in any case be much more open to the possibility that the use-value of a species—even if we can determine it accurately—does not exhaust its value. This is the point of claim (3) above and of Wilson's response to the notion that we should simply kill off all the blue whales. Let's agree that a purely economic determination of aesthetic, scientific, and spiritual values—that is, their inclusion in a cost-benefit analysis—will be artificial and distorting.

Ehrenfeld's more general point, however, is that we should not assign value to species anthropocentrically—"conventionally," as he would have it—at all. This view seems based on the notion that anthropocentric valuation inevitably takes the form of economic valuation, which in turn reduces to cost-benefit analysis. But this is false. We can talk intelligibly about anthropocentric value while insisting that it is not possible to quantify such value without distorting it. Moreover, we can do this and conclude—on the basis of the values we identify as threatened by species loss—that the best stance to adopt toward biodiversity is an extremely conservative one. That is, there is no obvious inconsistency in being both an anthropocentrist about the value of biodiversity and an uncompromising preservationist. Since Ehrenfeld also wants maximal preservation of species, the difference between his position and that of the anthropocentrist is therefore merely theoretical.

Ⓔ Species Preservation: The Challenge of Climate Change

These points also bear on Ehrenfeld's other practical problem, the one for conservationists. There are, says Ehrenfeld, so many species whose disappearance would "not be noticed by the biosphere." Losing some of these seems unproblematic *if* we insist on thinking about value in the conventional manner (i.e., instead of adopting the intrinsic value approach). However, the moral of the rivet-popper story is that our knowledge of ecosystems is too spare to license this cavalier an approach to wild nature. Instead, our ignorance dictates the need for the precautionary principle. Ehrenfeld himself admits that we do not know which species might be redundant. What he does not see is that this very consideration suggests we probably ought to act as if they are *all* "vital cogs in the ecological machine." But can we act this way? As our next article suggests, the losses we are in store for because of climate change might make a tragic attitude more appropriate (not that this should be taken as a call to despair).

GLOBAL CLIMATE CHANGE AND SPECIES PRESERVATION[27]

Ronald L. Sandler

Polar bears are in trouble. Sea ice is a crucial component of polar bear habitat. The bears depend upon it as a platform for hunting seals and other marine mammals, which are their primary food source. As the climate has warmed, the sea ice has begun breaking up earlier in the year, so bears have less time to build up the fat reserves they need during the period of food scarcity until the sea ice reforms. They also must swim longer distances between ice platforms, further depleting their energy reserves. The result has been decreases in the average body weight of the bears in some populations. This has, in turn, led to higher mortality rates, lower percentages of bears having litters, and smaller litter sizes. Consequently, bear numbers are declining in those populations. Because climatic change, and not just local factors such as hunting or mining, is driving the decreases in population sizes, local management plans alone are inadequate for protecting bears. Designating and protecting critical habitat areas will not itself preserve the bears in their current locations, since it will not limit increases in surface air temperature.

This is a terrible situation. An amazing and distinctive animal is imperilled, and we are the reason. It is greenhouse gas emissions from our industrial activities that are the primary cause of global climate change. It seems that we have an ethical responsibility to help them. But what should we do? We cannot preserve polar bears by creating wildlife refuges, since this will not prevent the polar ice losses. We could put some of them in zoos—we've done that before—and in this way keep some individuals of the species alive. But this will not preserve what is most important and valuable about polar bears, their unique form of life—that they are up north roaming the Arctic, hunting seals, and swimming in frozen seas. (And it could be detrimental to the animals by preventing them from engaging in species-typical behaviours.) If we want to preserve polar bears in the wild, we are going to have to do something *different*, something *more*. Some have suggested that we ought to deliver food to them to help maintain their body weight. Others have proposed translocating them to Antarctica, where there remains ice for them to roam. Others have proposed freezing tissue samples so that they can be cloned when suitable habitat reforms. Nobody *wants* to do these things. It would be best, all agree, if we could find a way to conserve polar bears as they live now in their co-evolved habitat. But if we cannot do that, and it increasingly appears that we cannot, then we are faced with this choice: allow anthropogenic polar bear extirpation to occur or take more drastic measures.

I. The Species Conservation Dilemma

The polar bear is a charismatic and high-profile case, as well as a particularly difficult and stark one, given its habitat. But it is not atypical. The distinctive features of anthropogenic climate change are the increased magnitude, rate, and uncertainty of climatic and ecological change in comparison to the recent geological past. There has always been ecological change, and

species populations have always had to adapt or else go extinct. However, the greater the rate and magnitude of change the more difficult is adaptation. Many species populations are dependent upon environmental conditions that will no longer obtain in their current and historic ranges as a result of global climate change.[28] For such populations, place-based preservation strategies—creating protected parks and reserves—are not going to be as effective as they have been in the past. It is not possible to preserve coral reefs and the species that depend upon them by designating their locations marine sanctuaries when increases in ocean temperatures due to climate change and ocean acidification due to elevated atmospheric levels of carbon dioxide are the causes of coral declines. It is not possible to preserve American pika populations in the western United States or golden toad populations in Costa Rica by protecting the mountaintops where they live when climatically altered temperature and precipitation patterns, and not local land uses, threaten them. It is not possible to preserve Canada lynx and wolverine populations in greater Yellowstone through local management plans when it is increases in air temperature that are reducing the crucial snowpack. Place-based preservation strategies depend upon the relative stability of background climatic and ecological conditions. Global climate change disrupts that stability. To the extent that it does so in a particular location, place-based preservation strategies for the at-risk species that are there are less viable. They cannot preserve the species' form of life in their ecological context.

How widespread is this effect likely to be? Studies have found that 35 per cent of bird species and 52 per cent of amphibians have traits that put them at increased risk of extinction due to global climate change; that 20 per cent of lizard species are likely to be extinct by 2080 due to global climate change (and that 4 per cent of local populations already are); and that 15 to 37 per cent of species will be committed to extinction by 2050 on mid-level warming scenarios.[29] Non-human species populations that only slowly change their geographical ranges (e.g., that disperse seed only locally or migrate slowly) are less likely to meet the challenge of adaptation than are those that are more mobile. For them, suitable habitat might contract, shift, or otherwise disappear more quickly than they are able to adjust. Mountain, small island, and other geographically constrained populations are also highly vulnerable, given the limits (e.g., mountaintops and coasts) on their capacity to migrate as their environments change. Small and non-diverse populations (phenotypically and genetically) are also more vulnerable, as are populations of species that depend on very particular environmental conditions (or on particular other species) or are otherwise highly sensitive to environmental changes. In addition, populations of species that have fewer offspring and longer developmental periods (e.g., large mammals) are less likely to be able to adapt biologically to changing ecological conditions than are populations of species that reproduce rapidly and abundantly (e.g., weedy plants). Overall, the fourth assessment report of the Intergovernmental Panel on Climate Change (IPCC) concludes:

> There is medium confidence that approximately 20–30% of species assessed so far are likely to be at increased risk of extinction if increases in global average warming exceed 1.5-2.5°C (relative to 1980–1999) [i.e., the low scenario]. As global average temperature increase exceeds about 3.5°C [i.e., the confirmed proposals scenario], model projections suggest significant

extinctions (40–70% of species assessed) around the globe.[30]

Given that the background historical rate of extinctions is one species per million per year (.000001 per cent annually), this constitutes a dramatic increase in extinction rates and may be the highest since the beginning of the fossil record.[31]

Thus, global climate change (in combination with anthropogenic impacts, such as land-use patterns and invasive species) generates a *conservation biology dilemma*.[32] It both dramatically increases the number of species at risk of extinction and dramatically decreases the effectiveness of the predominant and preferred strategies for preserving at-risk species and populations—that is, the creation of parks and reserves and ecological restoration. In response to this situation, many conservation biologists have begun to argue for novel approaches to species conservation and ecosystem management, such as assisted colonization, rewilding, ecosystem engineering, conservation cloning, and de-extinction.

II. Novel Species Conservation Strategies

If the standard species conservation strategies are likely to be less effective for climate-threatened species and we want to try to preserve them outside of zoos or captive breeding programs, then new strategies are needed.[33] Several novel strategies have been proposed.

Assisted colonization is intentionally moving species beyond their historical range and establishing an independent population of them in order to prevent them from going extinct. The idea is that there may be suitable habitat for them in the future, but they cannot reach it without our assistance. Assisted colonization is rapidly gaining proponents and, in some cases, practitioners. In the United Kingdom, two butterfly species have been successfully translocated northward to sites that climate-species models suggest will be more conducive to their long-term survival than their prior ranges. In Canada, scientists have relocated dozens of tree species to locations beyond their recent historical range, and in the United States, an environmental group called the Torreya Guardians has translocated specimens of *Torreya taxifolia,* a threatened conifer, from its present range in Florida to a more northerly location in North Carolina. Assisted colonization has been proposed for a wide variety of other species, from lobster to lynx.

Rewilding involves assisted colonization and relocation of multiple species in order to "re-wild" human-impacted habitats. Advocates of *Pleistocene rewilding* propose translocation of wild (or de-domestication of non-wild) tortoises, camels, cheetahs, horses, elephants, and lions from Asia and Africa, among other places, to expansive parks in the American West and the Great Plains. (It has been proposed for Australia as well.) They believe that the species are appropriate ecological proxies for the large vertebrates that became extinct in North America 13,000 years ago, in part owing to over-hunting by humans. Moreover, many of the proxy species are at risk in their current habitats, so this would contribute to species conservation by providing an intercontinental refuge. There are other, more modest, rewilding projects. For example, in the Netherlands, Konic ponies, Heck cattle, and Galloway cattle have been introduced as proxies for extinct herbivores in two comparatively small and lightly managed reserves. Several similar efforts are being planned throughout Europe.

Rewilding is an instance of *ecosystem engineering*. It involves designing ecosystems as we believe they ought to be, or need to be, in the future for reasons related to species conservation or ecological function. Traditionally, ecosystem engineering for conservation reasons (as opposed to economic or agricultural ones) has been largely restorationist. It has aimed at replicating species and processes that obtained in a space in the past—for example, removing invasive species, reintroducing native ones, and remediating contaminants. It was reasonable to believe that species assemblages that obtained in a place in the past would be well-suited for it and conducive to its ecological integrity in the future. However, with global climate change, the ecological futures of many places are going to be very different from their recent ecological past. Therefore, ecosystem engineering for conservation purposes has begun to take on a more forward-looking, rather than historical, outlook. For example, researchers have begun to study and develop processes for cultivating heat-tolerant coral species, with an eye on engineering climate-change-resilient reef systems.

Cloning is now performed regularly with agricultural and companion animals. *Conservation cloning* refers to cloning animals for preservationist reasons, such as increasing the genetic diversity or population size of a threatened species. It has been done successfully with the guar (a bovine species) and African wildcat, for example. *De-extinction* is cloning species that have already become extinct. The "last" bucardo (a Spanish ibex species) died in 2000, but in 2009 a bucardo clone was born, using DNA from preserved tissue and an interspecific host cell and surrogate. (The animal died after only a few minutes due to a lung abnormality.) There is also research underway to clone the gastric brooding frog, an Australian species that has been extinct for decades; and several "frozen zoos" have been established, repositories of tissue from at-risk animal species that could be used to clone them in the future. Researchers are even exploring the potential for using synthetic genomics to reconstruct the genomes of species that have been extinct for much longer periods of time—for example, thylacine, passenger pigeon, and mammoth—in hopes that they can be "revived."

What these strategies have in common is that they involve engineering and designing organisms and ecological systems. They therefore appear antithetical to the commitments, such as native species prioritization, that have traditionally characterized conservation biology and preservation-oriented ecosystem management. Rather than deferring to where species are now or have been in the past (or what species exist now), the strategies involve putting species where we think they ought to be in the future. The primary justifications offered for this shift in outlook and approach, already alluded to above, are (1) that traditional approaches to species conservation are no longer feasible, (2) that the biotic and abiotic features of ecological systems are already so impacted by human activities that there is no "naturalness" or independence-from-humans left to preserve or to which to defer, and (3) that we must do these things to make up for the harm that we have done. On this view, we have so altered the planet that we must now take responsibility for its management.

III. Three Concerns about Interventionism

Should we embrace the interventionist outlook that, as a matter of expediency and justice, would have us engineer species and design systems in order to forestall extinctions? Should we adopt the view that we are now in the "Anthropocene" and that this means that we must take on the responsibility of actively managing ecological systems?

Critics of this outlook typically emphasize that human interventions into ecological systems are the cause of ecological degradation and the species extinction crisis in the first place. Global climate change is not intentional, but it illustrates that our actions very often have significant and detrimental unintended consequences. Moreover, many of our other ecological interventions, from clearing forests for agriculture to damming rivers for energy, have been both intentional and ecologically destructive. (Habitat destruction from land and resource use is at present the largest contributor to species extinction.) Proponents of interventionism might respond that in this case the interventions will be done for the good of non-human species and the biotic community, rather than for human interests. However, even granting this is so (and critics might be sceptical that it is), challenges remain. After all, we will be choosing which species and functions to prioritize, and even well-intentioned interventions often have unintended impacts. On this view, the interventionist outlook is hubristic: it overestimates our ability to predict and control the consequences of our alterations of complex ecological systems.

Consider, for example, assisted colonization and rewilding. The species most commonly proposed for translocation are those that we care about—charismatic or economically significant ones—and not necessarily those that are most ecologically important. Moreover, assisted colonization and rewilding involve establishing independent non-native populations. Non-native species are often benign, but in some cases become invasive and ecologically (and economically) disruptive. Furthermore, the translocations would occur in the context of rapid and uncertain ecological change, which increases the difficulty of predicting where suitable habitats will be in the future and ensuring that the target species do not become ecologically problematic in the recipient systems. (The species may also have to be moved again in several decades, since elevated rates of change will continue and candidate species often have lower adaptive capacity.)

So, one concern about interventionism is that, particularly under conditions of global climate change, it will often not be successful and/or will further stress ecological systems.

A second concern is that, even if the interventions are successful—that is, species are translocated and established without becoming ecologically problematic—what is most important about the species is not preserved. As illustrated by the polar bear case discussed earlier, many of the types of value that species possess are tied to their ecological and evolutionary situatedness. Moving species outside their historical ranges, engineering new systems, and reintroducing long-extinct species do not maintain or re-establish those relationships. Therefore, when value is tied to ecological and historical properties, such as ecosystem function and independence from human design, interventionist approaches to species conservation might preserve the species without preserving their value.[34] Cheetahs and lions in Kansas, although exciting, would not have the same ecological or intrinsic value as cheetahs and lions on the Serengeti.

Yet a third concern is that these interventionist strategies are something of a distraction. On some widely accepted projections, there will be on the order of tens of thousands of species extinctions per year within a few decades. However, most of these strategies focus on only one or a few species at a time. They cannot scale to the magnitude of the problem that we face. At most, they might enable us to forestall the extinction of some of the species that we most care about for some amount of time. But

we should be concerned about the extinction crises as a whole and the ecological collapse that it might precipitate. What we need are conservation strategies that capture large numbers of species in the way that the creation of parks and reserves has in the past. Furthermore, the interventionist conservation strategies do not address what is causing the species extinctions in the first place—such as climate change, habitat destruction, and over-extraction. They are adaptive, reactive, and might perpetuate the problematic view that we have ways to deal effectively with the extinction crisis, and thereby foster inaction on the causes of extinction. On this view, conservation cloning is an amazing techno-scientific achievement, and de-extinction would be even more so, but these are not the sort of "solutions" that we need for the crisis we face.

IV. What Is the Alternative?

As we have seen, proponents of interventionism are critical of the traditional park and reserve model of ecosystem management on the grounds that its effectiveness for accomplishing species conservation is undermined by global climate change—it is "mismatched to a world that is increasingly dynamic."[35] Perhaps this conclusion is too hasty. Although anthropogenic climate change diminishes the effectiveness of parks and reserves in preserving particular species, species assemblages, and ecosystems, these areas are likely to maintain *comparatively high* ecological (including species preservation) value when measured against non-protected areas. Protected areas and corridors provide some adaptive space (and so more adaptive possibilities) for populations and systems. Moreover, more biodiverse places, often the target of protection, are likely to have more species with sufficient behavioral and evolutionary adaptive potentials to meet the adaptation challenge of global climate change. Therefore, identifying and protecting biologically diverse and rich habitats (including diverse physical environments), crucial or productive wildlife corridors, and ecological gradients, and promoting landscape permeability continue to be well justified under conditions of global climate change. In addition, familiar stressors—e.g., pollution, extraction, and habitat fragmentation—decrease the robustness (e.g., resistance and resilience) of ecosystems and species populations. Reducing or managing such factors can increase the adaptive potential of species and ecosystems, again by removing anthropogenic impediments, rather than by more interventionist activities. Thus, managing in traditional ways for the resilience and protection of biodiverse (and physically diverse) places and corridors increases the capacity of populations and systems to adapt to global climate change. It can lessen the magnitude of the ecosystem management dilemma.

Another reason not to give up on parks and reserves is that, under conditions of rapid ecological change, these protected areas are often conducive to accomplishing non-preservationist goals. For example, light management of less-impacted areas is often an effective approach to protecting ecosystem services and providing instrumental values (e.g., clean water, storm surge protection, carbon sequestration, and option value), particularly when measured against non-protected areas. Moreover, lightly managed spaces will continue to have value as places where ecological and evolutionary processes play out comparatively independently of human intention, design, and manipulation. Therefore, natural value, natural historical value, and the worth of wild organisms continue to be supportive of parks and preserve-based management. Under conditions of rapid eco-

logical change, place-based protection, rather than being valuable for maintaining a space largely as it is, is valuable for the processes of change that occur—for example, human independent adaptation and reconfiguration. This requires changing expectations for what these approaches can (and cannot) accomplish. It also requires shifting management practices appropriately (e.g., de-emphasizing historicity in assisted recovery), as well as refraining from intensive efforts to prop up dwindling populations or communities, when it is associated with climate change–driven ecosystem change.

In addition to revising management goals, expectations, and practices for protected places, global climate change requires adapting our attitudes toward those places and to ecological change more broadly. *Openness* toward the ecological future, *accommodation* of human-independent processes in determining that future, and *appreciation* of new ecosystem arrangements (even if they are partly anthropogenic) are crucial to place-based management, as well as to good ecological engagement more generally, under conditions of rapid ecological change. These attitudes involve cultivating sensitivity and appropriate responsiveness to the value of biotic systems and living things that are the successors or beneficiaries of rapid ecological change—that is, to the species that thrive and species assemblages that emerge—even if they are not the ones we would have preferred or prioritized. The human independent ecological and evolutionary processes that produced what is valued now will continue, and over time will generate new species populations, communities, and systems. The salience of related attitudes, such as *flexibility*, *tolerance*, and *restraint*, are also amplified in place-based management and ecological engagement, given the uncertainties involved with the ecological future and the rate at which ecological change will occur.[36] So, too, is the salience of *patience*, since ecological transitions, the reconfiguration of systems, and the evolution of populations may not occur or abate on the time scale that we might prefer.

For many species and species communities, the most justified response to their inability to meet the challenge of adaptation may not be to engage in highly interventionist activities to preserve them in nature, but to let them go. For this reason, the significance of *reconciliation* is increased under conditions of global climate change. Reconciliation, in environmental contexts, is the disposition to accept and respond appropriately to ecological changes that, though unwanted or undesirable, are not preventable or ought not be actively resisted. Reconciliation has always been relevant to ecological practice. Even independently of global climate change, ecosystems are always dynamic, and individuals, species, and abiotic features are always coming into and going out of existence. Good ecological engagement requires accepting and not resisting too strongly such changes and losses. The increased rate and magnitude of ecological change and loss associated with global climate change make reconciliation still more necessary.

Reconciliation is not indifference. Species are rapidly becoming extinct, ecological relationships are being disrupted, and human activities are the cause. We are responsible for an enormous loss of value in the world. *Recognition* of the magnitude of the loss and *remorsefulness* for our contributions to it are appropriate. The fact that we are now at the point at which we should, not actively seek to prevent the losses or restore or replace what is lost, but instead reconcile ourselves to them, is absolutely *tragic*.

V. Conclusion

When it comes to responding to the imminent species extinction crisis, we are choosing among bad options. On the one hand, interventionist strategies cannot be implemented on a large-enough scale, often do not preserve value, could pose a moral hazard, and could cause further ecological disruption. On the other hand, strategies of restraint require reconciling ourselves to letting many amazing species become extinct, even though we are the cause. For the reasons given above, I believe that the approach of restraint is in general more justified, though heroic measures may be warranted for some highly valued species.

But perhaps the most important point to take away from this discussion is that there is no good strategy for responding to the large-scale species extinctions associated with climate change. In the political discourse about how to respond to global climate change, the following question is often posed: Should we prioritize mitigation or adaptation? When it comes to biodiversity conservation, the answer to this question is absolutely clear: *mitigate, mitigate, mitigate*. There is no adequate adaption option. We need to get on as low an emissions trajectory as possible to reduce the number of species populations that are climate-change threatened.

Environmental problems often come together, sometimes in bunches. This notion is implicit in the HIPPO formula: in many cases all five features work together and become mutually reinforcing. This can be cause for deep worry, and even despair, because it can make our environmental problems look very large. But it can also be a cause for hope because it suggests that if we tackle one problem in a comprehensive way, solutions to some of the other problems might follow or suggest themselves. Thanks to analyses like Sandler's we now see how deeply implicated climate change is—and will increasingly be—in the biodiversity crisis.

Sandler shows us just how threatening climate change is to terrestrial biodiversity. Most of the planet's living things are extremely sensitive to the specific temperature range to which they have become adapted over the long course of their evolution. When their local climate changes, even minutely, they therefore tend to migrate to a place where temperatures are more like what they are used to. The general movement we have seen over the past 50 years or so has been toward the Earth's poles (or to higher altitudes, where available). Species have always migrated like this in response to natural climate change. The difference now is the speed at which climate change is occurring, and the evidence suggests that many species cannot move as fast as they need to. James Hansen, the world's leading climatologist, reports that the average migration rate of species north and south was about four miles per decade in the last half of the twentieth century. But this is not fast enough, because the world's isotherms (temperature zones) have been moving in the same directions at a rate of 35 miles per decade in the same period. Hansen predicts that if we do not take aggressive measures to curb our greenhouse gas emissions, the rate of isotherm movement will double to 70 miles per decade by the end of this century.[37]

In Canada, for example, the boreal forest may move between 700 and 1000 kilometres northward, the southern permafrost border of the Arctic may move 500 kilometres to the north, and the tree line could move 300 kilometres to the north.[38] The effect of these movements of habitat on the huge numbers of species that live in the forest should be obvious. If they find it increasingly difficult to locate a suitable temperature zone, they will simply die, mostly because they will not be able to find the sort of food to which they are adapted. As Hansen notes, while some species will be able to survive all this upheaval, many others, like the polar bear, will be "pushed off the planet."[39] Because of this, Sandler's call to "mitigate, mitigate, mitigate" is surely correct: the most pressing task is to slow our emissions of greenhouse gases. Again, the upside of this observation is that if we manage to do so, we could solve a whole cluster of environmental problems in one fell swoop.

Ⓕ Conclusion

In the meantime we also need to tackle the problem by other means. Let's close the chapter by looking at two broad approaches to protecting species diversity. The first is political legislation aimed explicitly at such protection. In Canada, the most important federal act is the **Species at Risk Act (SARA)**, which became federal law in 2003. Here are three of SARA's more important objectives:

- establish the Committee on the Status of Endangered Wildlife in Canada (COSEWIC), an independent body of experts to identify species at risk;
- devise prohibitions to protect threatened and endangered species and their habitat;
- be consistent with Aboriginal and treaty rights.[40]

Under the legislation, it is a criminal offence to kill, harm, harass, possess, collect, buy, sell, trade, or damage the critical habitat of members of threatened or endangered species. How effective has SARA been at protecting species at risk in this country? Some concerns have been expressed since its implementation. First, in 2004 only 66 per cent of the species at risk identified by the COSEWIC made their way onto SARA's list of endangered species. Second, SARA applies only in areas of federal jurisdiction, amounting only to about 5 per cent of Canada's land area. Finally, many environmental groups have criticized the implementation of SARA. The David Suzuki Foundation, for example, has argued that imperatives of economic growth have slowed the speed at which species at risk, as identified by COSEWIC, have acquired the full legal protection of SARA.[41]

The second device represents a way of thinking of our relation to wild nature that is, in some sense, much more radical than legislation to protect species at risk (as important as this is). It is a new approach to land management known as **agroforestry**, a cluster of agricultural and silvicultural techniques that challenge the dominant agribusiness approach to managing these resources. The most salient feature of large-scale agribusiness is the planting of monocultures, which involves the stripping away of virtually all species from a given area in order to make room for vast quantities of the desired species. Trees are especially anathema. To a farmer bent on extracting as much productivity per hectare from his land as possible, what after all is a tree but an obstacle to increased profit? By contrast, a key goal of agroforestry is to boost the biodiversity of the areas used by humans to grow their food. This means integrating not just trees but *forests* as well as the animals that inhabit them into such practices. There has been some push to develop agroforestry practices in Canada. Agriculture and Agri-Food Canada (AAFC) claims that "agroforestry continues to be high on AAFC's agenda." For example, the Prairie Shelterbelt Program "provides technical services and tree and shrub seedlings at no charge for planting shelterbelts, or for agricultural conservation and land reclamation projects" in the western provinces.[42] Still, these efforts are very small relative to what is required in order to blunt the assault to biodiversity created by an ever-expanding agribusiness in Canada.

CASE STUDY

Biodiversity and the Decline of Pollinators

When human populations expand, their impact on the environment becomes more pronounced because their consumption of biophysical goods and services increases. Paul Ehrlich has developed a formula that helps bring some precision to the process of determining the effect of human population on the environment. It is known as the **IPAT formula**:

$$\text{Impact} = \text{Population} \times \text{Affluence} \times \text{Technology}$$

That is, the human impact (I) on an area is a function of three forces: total population (P), income per person (A), and the technological means required to transform resources into goods and services (T):

Continued

The pollinating activity of bees is vital to the health of many other species. If we care about biodiversity, we must take extraordinary measures to protect species like this. Are we doing so?

> When *T* is high, the kind of technology being used imposes a high environmental burden (for example, extensive use of land or high emissions of greenhouse gases) per unit of Gross National Product (GNP). . . . Clearly, the IPAT relationship signals that a dramatic rise in population and income per person, as we've experienced since 1950 and will experience again until 2050, has a similarly dramatic impact on the environment, unless technology changes in a way to protect the environmental impact.

Relative to its vast territory, Canada is lightly populated. This fact might make us think that the impact we Canadians have on our own environment is negligible. But applied to particular regions of the country, the IPAT formula provides a different picture. Consider, for example, the impact of our agricultural practices on pollinators in southwestern Ontario. By spreading pollen among flowers, pollinators are potent builders of biodiversity. Their simple activity spreads life outwards beyond the plants they help to proliferate, to the many animals that depend on those plants. Here is how the point is put by Stephen Buchman and Gary Nablan:

> The process of pollination keeps the verdant world, that delicate film of life around us known as the biosphere, running with endless cycles. . . . This intrafloral commerce by the birds and the bees is what makes the living world go round on its reproductive cycle.

In Canada, the lion's share of pollination is performed by bees. We are intimately dependent on this service. Fully one-third of all the food we eat—apples, peaches, alfalfa, pumpkins, and much more—depends on pollinators. The economic value of bees' pollination services is hundreds of millions of dollars in Canada alone. However, as E.O. Wilson notes, "the evidence is overwhelming that wild pollinators are declining." This has been happening for many years, but recently there has been a pronounced uptick in the rate of decline, especially in southwestern Ontario. What is to blame for this? Studies are showing that a certain class of pesticides called neonicotinoids—"neonic" pesticides—is the likely culprit. Neonics are used extensively, especially in Ontario's corn and soy crops, to combat wireworm and white grubs, among other pests. The Task Force on Systemic Pesticides, comprised of 50 independent scientists, has concluded that these chemicals are "a key factor in the decline of bees." The losses are huge: in Ontario 40 per cent of beekeepers reported a loss of up to 50 per cent of their bees in 2013, a dramatic worsening of an already alarming trend.

Again, it is at this point unclear what knock-on effects the decline of our bees will have for biodiversity generally, but the impact is likely to be huge, especially since neonics negatively affect not just bees but also other biodiversity-boosters like butterflies and earthworms. If we wish to avoid huge and wide-ranging species losses, we will need to lessen the effect of our agricultural technology in the affected areas. Here's how to think in the abstract about this. In the IPAT formula, it is assumed that the more intensive our technology, the greater its negative impact on the environment. However, it has been suggested that we can turn T on its head and replace it with S if our technology is suitably "green." The equation becomes:

$$I = \frac{P \times A}{S}$$

In this formula, our technology actually *mitigates* the overall environmental impact of our population and affluence combined rather than intensifying it (as with IPAT). The suggestion is that people living in the region need not contemplate a drastic reduction in levels of affluence in order to keep their environmental impact to a minimum. In particular, although they would need to resist the spread of neonic pesticide use (or eliminate it altogether) in this intensively agricultural zone, there is no need to contemplate drastic reductions in human population or affluence there. Is this wishful thinking and a recipe for complacency? Can we solve such problems *just* by altering our technologies?

Sources: Draper and Reed 2009, 380–1; Dearden and Mitchell 2009, 111; David Suzuki Foundation, "Pollinators: What's the Buzz?" at: www.davidsuzuki.org/issues/downloads/pollinator_fact_sheet_final. pdf; "Are Neonic Pesticides Killing Bees?," *Globe and Mail*, 25 June 2014, at: www.theglobeandmail. com/report-on-business/industry-news/controversial-pesticides-linked-to-wave-of-bee-deaths/ article19330959/; David Suzuki Foundation, "Unprecedented Study Confirms Neonic Pesticides Endanger Bees, Butterflies and Earthworms," 25 June 2014, at: www.davidsuzuki.org/media/news/2014/06/ unprecedented-study-confirms-neonic-pesticides-endanger-bees-birds-butterflies-a/.

Study Questions

1. What are "assisted colonization" and "rewilding"? According to Sandler, what problems does global climate change pose for these activities or projects? Do you agree with him?
2. What do you make of the parable of the rivet poppers? Does it apply neatly to the role played by species in ecosystems?
3. How does Suzuki characterize the "power" of biodiversity? Can you think of concrete policies that would enhance this power, or detract from it?
4. Is Ehrenfeld correct to claim that it is difficult or unwise to put a value on biodiversity? Is he perhaps operating with an overly narrow conception of "value"?

Further Reading

Yvonne Baskin. 1997. *The Work of Nature: How the Diversity of Life Sustains Us*. Washington: Island Press.

Elizabeth Kolbert. 2014. *The Sixth Extinction: An Unnatural History*. New York: Henry Holt.

Bryan G. Norton, ed. 1986. *The Preservation of Species*. Princeton, NJ: Princeton UP.

——. 1990. *Why Preserve Natural Variety?* Princeton, NJ: Princeton UP.

E.O. Wilson, ed. 1988. *Biodiversity*. Washington: National Academy Press.

——. 2002. *The Future of Life*. New York: Vintage Books.

CHAPTER 14

Sustainability

Ecological Intuition Pump

Geologists divide history into geological epochs. Some have suggested that we have now exited the Holocene period (which began about 12,000 years ago) and have entered the "Anthropocene" period. As the name implies, the Anthropocene is the age in which humans themselves have begun to affect the biophysical systems of the planet directly. The concept of sustainability can be seen against this backdrop. Some think of sustainability as the effort to scale back our effects on the planet, to remove our fingerprints from the biosphere as much as possible. Others think exactly the opposite: since we are now fated to play God, we might as well try to be good at it. So sustainable living might involve intelligent intervention in the biosphere rather than withdrawal from it. For example, some scientists have suggested that to combat the effects of climate change, we should seriously consider various geo-engineering schemes. One of them would have us mimic the activity of volcanoes by injecting sulphur dioxide into the stratosphere. This stuff can cool the planet by reflecting sunlight back to space. Another idea is to dump huge quantities of iron filings into the oceans. This can stoke algae growth, which draws down atmospheric carbon, thus diminishing the greenhouse effect. This is control of the planet on a massive scale. Are we wise enough to do it responsibly? Can considerations of sustainability help us to determine how much geo-engineering, if any, we should engage in? Or should we really just do less to the environment?

Ⓐ Introduction

The Sydney Tar Ponds are a two-square-kilometre area on the west side of Sydney, Nova Scotia, on Cape Breton Island. It has been referred to by federal officials as "the largest chemical waste site in Canada," containing highly elevated levels of polycyclic aromatic hydrocarbons (PAHs), polychlorinated biphenyls (PCBs), benzenes, ammonium sulphates, and naphthalene.[1] These chemicals, contained mainly in two large ponds, are the by-products of the area's steel mills, operating from 1891 until they were shut down in 2001. That roughly 100-year history is a model case of unsustainable resource extraction in the service of unrestrained industrial development in Canada.

Steel is made from coke—a solid fuel produced by baking coal—which is used to smelt iron ore into the final product. But the basic ingredients used in the Cape Breton operations were of low quality: the coal was high in sulphur, and the iron ore (procured from Wabana, Newfoundland) was loaded with impurities like silica and rock. So more coal had to be baked to produce decent coke, and the iron ore had to be treated with more limestone to remove the impurities. As a result, more waste was produced in the production of this steel than is normally the case, and all of it was simply dumped into nearby Muggah Creek. The Dominion Iron and Steel Company (DISCO), formed in 1899 with a land grant along the Sydney water-front, ran its operations with near total disregard for the long-term health and environmental consequences of what they were doing. Specifically, according to Barlow and May, they failed to test the quality of their ingredients:

> In a rush to begin full operations, they failed to run the most basic tests on their coals and ores. DISCO tapped its first steel on New Year's Eve 1901, and right from the start there were problems.[2]

The most widely cited definition of sustainable development is that of the Brundtland Commission in 1987: "[S]ustainable development meets the needs of the present without compromising the ability of future generations to meet their own needs."[3] According to this definition, we might think that practices like those that produced the tar ponds are uncontroversially unsustainable. While it is true that the steel mills have brought relative economic prosperity to the region, they have clearly done so at a huge cost to future generations and the environment. To focus on the human health consequences, studies have revealed that between 1977 and 1980 the cancer deaths among citizens of Sydney was 347 per 10,000, compared to the national average of 192 per 10,000. Another study discovered that life expectancy for residents of Sydney was five years less than the national average. Finally, it has been revealed that the rates of disabling birth anomalies are higher in Sydney than in the rest of the province.[4] The effects on non-human land and aquatic species are equally pronounced.

According to some, nearly as disturbing as the reckless way in which this became a problem in the first place are the steps that have been taken since 2001 to correct it. The proposed remedy repeats the pattern of disregarding good science in a search for what is best for the bottom line. The idea is to first solidify the waste by mixing it with drying agents like cement, then to cap it with a high-density polyethylene liner, and finally to top it with layers of gravel and soil. The joint press release from the federal and Nova Scotia governments touted the project as a "long-lasting, sustainable and environmentally responsible solution for the people of Cape Breton."[5] Elizabeth May, the current leader of the federal Green Party, disagrees:

> What the province of Nova Scotia has done is steer the cleanup to the very cheapest mechanism possible with no plan for future use of the site and really tenuous science. . . . It's a Mickey Mouse solution. It's not high-tech.[6]

A more emotional response to the proposed solution was recently provided by Alex Gillis:

> Generations of the men in my family helped create the hell on earth at the gargantuan steel plant in Sydney. . . . The sky-high smoke stacks that my great grandfather once patched are long gone, but the same cannot be said for one of Canada's most toxic sites, the Sydney tar ponds. They hold more than 700,000 tonnes of PCBs, dioxins and other pollutants. . . . My relatives in Sydney have often wondered where 700,000 tonnes of hell will go. Well, now they know. After decades of controversial negotiations, the federal and provincial governments are spending $400 million to cap the tar ponds. Intrepid souls will dig more than two kilometres of channels through the ponds and erect immense underground walls, before capping the entire site. . . . When completed the old ponds will become 100 hectares of prime development property. Hell will have been paved over.[7]

Both May and Gillis seem to be questioning the sustainability of the proposal. This sort of case raises immense difficulties for those charged with devising wise and responsible environmental policy. As we have seen in the governmental press release, public officials will inevitably attempt to justify their environmental projects by appeal to the concept of sustainability. But is it possible to assess, in the present, the extent to which any such proposal is in fact likely to be to the benefit of future generations? The case of the tar ponds illustrates that it is one thing to invoke considerations of sustainability to criticize past practices and quite another to do so to justify current or future plans. With hindsight, we can say that the industrial practices of DISCO were clearly unsustainable, but is the same true of the proposal to solidify and cap the toxic sludge?

Moreover, although the Sydney tar ponds are the most egregious case of industrial pollution in Canada, they not the only one. In Squamish, BC, there is heavy-metal pollution from a choralkali plant as well as the abandoned Britannia mine site; in Port Hope, ON, there is waste from a radium and uranium refinery; in Deline, NWT, there is an abandoned uranium mine; in Newcastle, NB, and Transcona, MB, there are abandoned wood preservative plants.[8] All of these sites—and there are many more—pose challenges of responsible decision-making for their local communities and public officials. But is such decision making at all aided by appeals to considerations of sustainability? Is this just an empty catchword used by self-serving politicians and bureaucrats or woolly-headed greens? Or is it, on the contrary, a concept with substantive philosophical content that, properly employed, can serve as a genuine moral constraint on our actions? To get a better handle on what is probably the most ubiquitous, contested, and slippery concept in environmental discourse—sustainability—we'll need to investigate these, and related, questions.

Ⓑ Sustainability and Human Needs

Recently, two researchers at the University of British Columbia, Mathias Wackernagel and William Rees, devised a tool for calculating the impact our way of life is having on the environment. Ever since then, **ecological footprint analysis** has become bound up with the discourse of sustainability. An ecological footprint is the amount of water and land, measured in hectares, required to sustain a given level of consumption and waste disposal (including the greenhouse gases these activities produce). Wackernagel and Rees discovered that availability of land and water was decreasing even as our demand for them was increasing. Although there are 1.8 hectares per capita of productive land and water area—the planet's **biocapacity**—we are currently consuming at a rate of 2.2 hectares each.[9] Of course, this global average masks huge discrepancies among countries. In Canada, for example, with a population of over 35 million and a biocapacity of 14.5 hectares per capita, there is actually a net surplus of about 6.9 hectares per capita because our footprint is 7.6 hectares each. To compare, India has a population of 1.1 billion, a biocapacity of 0.7 hectares per capita, and a footprint of 0.8, for a net deficit of 0.1 hectares per capita.[10] But even the country-by-country breakdown is too coarse-grained for some purposes. In spite of the country's net surplus, there are jurisdictions within Canada where the footprint wildly exceeds the area's biocapacity. For instance, the Lower Fraser Valley in British Columbia, stretching from Vancouver to Hope, depends on an area 19 times its own size to sustain it.[11]

The metaphor of an ecological footprint is helpful because it puts into sharp relief our concerns about the way in which consumption patterns vary from region to region, country to country, and within countries as well. It thus brings us face to face with questions of equity, both among the Earth's current inhabitants and between the latter as a whole and future generations. It has been calculated, for example, that if everyone on the planet consumed like North Americans, we would need the biocapacity of at least two more Earths. However, this is where problems might arise with the metaphor. Although it is crucial to know what the upper limits are on our consumption, the ecological footprint concept cannot by itself tell us how much we ought to be consuming *below that limit*.

Think again of the Brundtland Commission's definition of sustainable development, which talks about meeting our *needs* without compromising the ability of future generations to meet theirs. Clearly, it is crucial to clarify the nature of human needs if we hope to come to a proper understanding of sustainability. However, many are sceptical that we can achieve such clarity. Referring to the Brundtland Commission, Wilfrid Beckerman says:

> But such a criterion [of sustainability] is totally useless since "needs" are a subjective concept. People at different points in time, or in different income levels, or with different cultural or national backgrounds, will differ with respect to what needs they regard as important. Hence the injunction to enable future generations to meet their needs does not provide any clear guidance as to what has to be preserved in order that future generations may do so.[12]

Is this right? Even allowing for all kinds of diversity among people as to what is reckoned of value, can't we say that there are some things—some core values or goods—that are so central to the project of living a decent life that to be without them constitutes a serious deprivation? What are these things? Adequate food, shelter, and medical care, of course. But don't we also need things like access to sound education, the chance to participate in public and political life, meaningful work, the possibility of developing emotional bonds with others (human and non-human), the chance to play, to create and enjoy art, and much more?

When we are pressed up against our natural limits—the planet's biocapacity—we need to ask serious questions about what we need. The questions need to be asked because our answer to them will determine how many people we think the planet can carry in the future. If we are very parsimonious about what we need, then there will be room for more of us, but as our list of needs expands, our numbers must correspondingly shrink or at least stabilize. The point about needing two more Earths if we all want to live like North Americans is frequently made, but it is almost never pointed out that if we were content to live like most Indians currently do, there would be plenty of room on this planet for all of us and many more besides. At this point the calculations and metaphors of ecologists, economists, and demographers fail us. Worse still, we seem to be in a complete muddle about the meaning of sustainability. Without a rigorous understanding of what humans need to flourish, we will have no idea whether or not a current practice or proposal is likely to be sustainable. Has the concept of sustainability therefore become meaningless? In this chapter's first article, Robert F. Litke explains why we should answer this question in the negative.

THE CONCEPT OF "SUSTAINABILITY"
Robert F. Litke

Harsh things have been said about the concept of sustainability; it has been characterized as "deliberately vague,"[13] a "smoke screen for the perpetuation of the status quo,"[14] "muddled, and even self-contradictory."[15] As if that is not bad enough, it has been said that given "hundreds of published definitions and commentaries, [it can mean] anything you like . . . stretched to the snapping point by attempts to . . . fit many agendas."[16] Jennifer Sumner has characterized the conceptual environment of sustainability very well:

> A popular term, *sustainability* has become one of those motherhood concepts that is hard to oppose, but difficult to pin down. Its very popularity hides the contradictions surrounding its use, hampering a clear understanding of the term.[17]

Nevertheless, I want to emphasize another facet of our conceptual situation, namely, that sustainability is also *easy* to understand, a properly conceived notion with a substantial history in English—its roots reaching back through Anglo-Norman to Old French to Latin. Putting it into one sentence, "sustainability" means "the power or ability to sustain something." And this continues to be a useful way to think about a variety of activities and situations. Currently, we are in the midst of an important struggle to flesh out our thinking about sustainability so as to transform it into a viable, intellectually powerful perspective for a specific and new application—the environmental context. I want to suggest that when we look at the motley sustainability literature, we are witnessing birth throes, the messy business of social construction, not fundamental conceptual inadequacy. Thus, my central idea is that the concept of sustainability is, on the one hand, easy to understand, while it is also the focal point of a somewhat chaotic, bewildering social experiment.

To reveal some of its subtler intricacy, I shall begin by describing relevant concepts at a fairly abstract level, where the concepts are so spare of content, so thin, that they are virtually transparent. I doubt that they lend themselves to much controversy at this level. Thus, they may give us some reliable common ground, an agreed-upon set of basic understandings that we can reach back to and touch base with when we find ourselves getting entangled in conflicted and confusing discussions of sustainability so notoriously characteristic of the literature today. My descriptions can be confirmed by consulting any standard English dictionary. (I assume that the same thing would be true for analogous terms in other languages and their dictionaries.)

Power: At its simplest, in its most basic form, power is the ability to do or to act, the capability of doing or accomplishing something. A moment's thought makes it obvious that things in the world at large have a great many abilities; they can do many different kinds of things; they can act in many different kinds of ways; they have a multitude of capabilities. Moreover, abilities develop and decline as time passes and circumstances change. Thus, we cannot know and say everything that is true of power as ability. There is too much to it. It is an infinitely large and complex subject about matters that are always changing. Consequently, discussions of power should always be conducted with an attitude of intellectual humility, since they must always be partial, focused on a few facets, rather than complete.

This is so whether we are thinking about the powers or abilities of things in the natural world, such as a fungus or a watershed, or about the powers or abilities of products of human ingenuity, from pharmaceuticals to political systems. There is no end to discovery and discussion of such things. Now let us turn to one specific kind

of power or ability—sustainability. This can be done in three quick conceptual steps. I think each step pulls into focus certain facets of what is going on below the surface when we speak of something being sustainable.

Functionality: A function is an action or use (or role) for which something is suited, an action that is taken as natural or proper to something; being functional is serving a use or being operational with respect to some purpose; functionality is the capability for being functional.

Adaptability: An adaptation is a change to fit new circumstances; being adaptable is being able to adjust to new conditions; adaptability is the capability to adjust so as to maintain functionality under new circumstances. It is the ability to keep on being functional despite changing conditions.

Sustainability: To sustain (in the relevant sense) is the ability to make something continue to exist; being sustainable is being able to hold up or maintain some level and kind of functioning; sustainability is the ability to keep on keeping on (being functional). It is functionality over the longer term and across a broader spectrum of circumstances.

These terms have a tremendous range of application. Recently, I have seen the concept of sustainability used in the discussion of lifestyles, the financial plan for a women's shelter, the design of our health care system. Also, it is easy to make the case that racism and sexism are unsustainable in our contemporary world (word gets around as never before). In principle, there is no limit to the things, activities, and arrangements concerning which we can raise the issue of their sustainability. This is even more obviously true of the other two concepts—functionality and adaptability.

For the rest of this paper, I shall be focusing on the context of living things. So let us simply agree to direct our attention in the following way: (1) living things typically exercise some of their power so as to preserve their life; (2) one use or purpose to which a function may be directed is the preservation of life; (3) sometimes living things must adapt in order to maintain their existence; and (4) one can raise the issue of sustainability from the point of view of life maintenance. I want to emphasize that the latter three terms reveal some of the ways in which we naturally articulate our thinking about power as ability or capability.

I also want to suggest that functionality, adaptability, and sustainability form a tight, small conceptual circle: they are a lens through which we can view all of life, its species, individual members of each species, and the larger systems in which they dwell. Now I want to move toward the context in which these concepts are given a much more specific application, namely, environmental considerations. To this end, consider the remarks of Paul Ekins and Manfreed Max-Neef, editors of *Real-Life Economics*:

> Sustainability at its simplest refers to the ability of some function or activity to be sustained. The concern over environmental sustainability has arisen over uncertainty as to the ability of the natural environment to maintain its various functions because of the destructive impacts of human activities. Because these and other human activities depend in turn on environmental functions, their sustainability is then brought into question and, thence, that of the way of life of which they are a part.[18]

I take this to be an observation about the recent historical context in which the concept of sustainability has come to be widely used in discussions about environmental matters. This usage stems from the understanding that human functionality is intertwined with the proper functioning of various things in the natural world and the observation that human functioning is interfering with the functionality of the natural world.

Here is another example of the same observation, by Sim van der Ryn and Stuart Cowan in *Ecological Design*:

> We live in two interpenetrating worlds. The first is the living world, which has been forged in an evolutionary crucible over a period of four billion years. The second is the world of roads and cities, farms and artifacts, that people have been designing for themselves over the last few millennia. The condition that threatens both worlds—unsustainability—results from lack of integration between them.[19]

I think we can agree that there is at this time in our intellectual history widespread, if not universal, acceptance of the two points these authors are making. I think of it this way. First, we must be prepared to look in two directions at once—toward the life-preserving functionality of the natural world and, simultaneously, toward the life-preserving functionality of the human world. Second, we must strive to integrate what we see so as to sacrifice the functionality of both kinds as little as possible. Here is a common way of diagramming this consensus position:

Sustainability

Sustainability: Natural World   Sustainability: Human World

Obviously, a complete account of the life-preserving functions constitutive of the natural and the human worlds would be unmanageably complex and indefinitely voluminous. How natural things have worked out their sustainability conditions over millions of years, on the one hand, and over thousands of years, for the case of human culture, is intellectually staggering and mostly unknown. The most we can do is go forward with what we do know and grope in the dark for what we think will be important in the longer run. And as various insights, findings, predictions, alarms, catastrophes, intuitions, premonitions, and specialized concerns are brought to bear on the life-preserving functionality of the natural and human worlds, we have an explosion of diverse material. We have thousands of flowers blooming! One way of coping with this profusion of alternative views is to think of them as located on a spectrum ranging from the most simple to the richly complex. For example, in their first *Sustainability Report*, CH2MHILL directs attention to the simple end of the spectrum this way:

> Sustainability . . . is a yardstick that can be applied to any technical area and problem solving method—and anyone can use it.[20]

The yardstick: a simple tool that anyone can easily learn to use. The most spare account of sustainability that I am aware of is that offered by David Cook's *The Natural Step*. Looking in two directions, it outlines sustainability in three conditions for the natural world and one condition for the human side of things.[21] However, as is well known, many prefer to work in terms of three dimensions—the triple bottom line of environment, economy, and society.[22] But then it has often been pointed out that there is an undeniably important fourth condition—

social and political institutions and forms of governance.[23] And in this way we come to face the fact that there really is no end to adding further important dimensions, up to and including an array of political–moral–ideological goals such as equity, social justice, democracy, and human rights, goals that themselves are subject to considerable controversy and elaboration.[24]

Thus, as accounts of sustainability become more and more richly inclusive of values that pertain to the human world, we lose the simplicity and elegance of the yardstick and replace it with the impression of a smorgasbord banquet, which can have something for everyone's taste. Thus, we obtain a spectrum of accounts ranging from the spare and simple to the luxuriant and complicated. And in this way, agreement about what sustainability actually means is totally eclipsed by the competitive multiplicity of accounts of how various commentators propose to use the term according to their varying agendas. In this way, the underlying consensus that sustainability is essentially some kind of combination of the life-preserving functionality of the human and natural worlds can be lost sight of as our attention is shifted to the dazzle of controversy.

We can all appreciate that it is one thing to pin down an idea or concept at the abstract level and something else to grapple with the multiple uses to which it may actually, perhaps legitimately, be put. It is worth noting that this is a common problem. Certainly it is true of the concept of power in general. Discussions of power issues are notorious for resisting consensus and spawning contradictory accounts. Here is how one seasoned student of power, a political scientist, illustrates the point:

> The "scandal of philosophy," Kant once remarked, was its failure to address and resolve the problem of our knowledge of

the external world. The scandal of social science—or perhaps only one of its several scandals—is that it has so far failed to arrive at a satisfactory understanding of power. This is all the more remarkable when we consider that power is arguably the single most important organizing concept in social and political theory.[25]

It remains difficult to discuss power issues without generating intellectual controversy: indeed, many academics have built careers in the space provided by this feature of "power." This seems to be in the nature of the subject. And the same thing appears to be occurring with regard to "sustainability."

Another way of coping with this confusing multiplicity is offered by Andres R. Edwards in his recent book *The Sustainability Revolution*. He recommends that we think of sustainability as the name for a global movement or perspective, which is spawning thousands of groups and organizations, which is embraced by millions of people worldwide, which is producing a rich ferment of intellectual and ideological activity—a movement that is appropriating and eclipsing its precursor, environmentalism. When seen this way, we can more easily accept the fact that we face uncontrollable diversity, unchecked visionary fervour, sometimes even approaching a religious or world view status.[26] Now, this raises an interesting question: Why has the banner of sustainability received such rapid and widespread allegiance in the past several decades that it can plausibly be called a social revolution or a paradigm shift, as Edwards does?[27]

This takes us back to my central claim that sustainability is both easy to understand and a bewildering social experiment. I want to suggest that this global embrace of sustainability has occurred only because adult humans carry in their lives a tacit and fundamental appreciation of the core concepts: functionality–adaptability–sustainability. Who among us doesn't naturally and spontaneously want to be functional rather than dysfunctional, and adaptable rather than useless or obsolete, and sustainable—that is, able to keep on keeping on being functional—rather than self-defeating or self-terminating? We do not need a complex ideology or an imaginative moral justification to legitimate our having such desires. They naturally arise in our daily lives. They are part of the underlying conceptual structure of what it is for a human being to be competent, developing, and alive. The fact that the impulse to be powerful in these ways can run in so many different directions may be intellectually inconvenient, but it should be no surprise.

It is in this way that we do have a stable and well-conceived standpoint from which to face the multiplicity of interpretations and applications of sustainability. The challenge then is to hold fast to what the relevant core concepts do mean in their spare, content-thin version and sort through content-rich and practicable accounts of sustainability-in-the-real-world, which are inevitably partial, myopic, and inadequate to the complexity of what we are trying to deal with, namely, our current situation where human functionality threatens natural functionality and, thereby, humanity itself. In other words, we can remain clear about the properly conceived conceptual trio—functionality–adaptability–sustainability—and use it to sift through and critically evaluate the diversity of proposals; we can scrutinize them for what really does bear on the relevant forms of life-preserving functionality in the human and natural worlds.

No doubt some of what has been offered in the name of sustainability will fail to meet the test, and it will have to be set aside as not strictly relevant. In this way, we can slowly develop a coherent and effective account of how to pursue genuine sustainability. Through time and careful work, a consensus will probably grow to replace the current welter of sustainability proposals. Thus, sustainability could emerge as a viable and intellectually powerful concept to help us deal with our increasingly dysfunctional relations with the natural world.

In conclusion, I want to confess that some of my hope for progress in these matters comes from reflections in two other contexts, those indicated by "peace" and "health." We all yearn for health and peace, we wish them upon each other in daily life, we respond to what promises to be conducive to them, we invest enormous resources in our pursuit of them—from education, to research, to institutional arrangements—and in the end we each die. There is no shame or chagrin in that. I think "peace" and "health"

are mostly about how we want to live, not much about our wanting to avoid death.

I think the same is true of "sustainability." It is mostly about how we are slowly coming to want to live—more elegantly and more consciously intertwined with the natural world. The challenge is to orchestrate whatever proves to be genuinely conducive to this end. For peace, we have designed everything from the United Nations and various global alliances, through military and police forces, to neighbourhood watch programs. For health, we have created the World Health Organization and centres for disease control, through to health care systems, to local health units and neighbourhood clinics. I predict the day will come when we know enough to establish a World Sustainability Organization down through to a full range of regional and local sustainability infrastructures, all for the sake of one thing: our deeply embedded, perfectly natural desire to be functional rather than self-destructive.

This is a subtle analysis of the current discourse of sustainability. As Litke would have it, although the concept of sustainability has become annoyingly ubiquitous in our culture such that it seems to have lost all substance, we should not abandon it altogether. The concept retains a core meaning that is both clear and valuable in our current environmental predicament. Living sustainably, for Litke, involves intertwining ourselves with the natural world to a greater degree and more consciously than we do now. Analytically—though perhaps not practically—we can distinguish between the human and natural worlds. The key is to look beyond all the contentious definitions to "the underlying consensus that sustainability is essentially some kind of combination of the life-preserving functionality of the human and natural worlds."

Let's examine the "human world" more closely. It is often pointed out that ecological sustainability should not be pursued in the absence of social and economic sustainability, the two major components of the human world. Litke is determined not to sacrifice whatever is of importance in the natural world for the achievement of human ends, but it works the other way too: there are genuinely important human ends that should not be sacrificed in an effort to live as close to the natural bone as we can get. Let's put this another way, a way that has a distinctly Aristotelian ring to it. Social and economic sustainability are important because of the *goods* they are meant to protect.

Social sustainability protects things like the equitable distribution of social benefits and burdens, a respect for human rights, the chance to make and enjoy art, to participate in the political process, to have meaningful work, and so on. Economic sustainability protects goods associated with material prosperity.[28] A life that is severely deprived of any of these goods falls short of what makes us distinctly human. And this way of putting the point immediately sets some *lower limits* on the quest for sustainability, in a general sense. For Aristotle, the full or flourishing human life contains the chance to participate in political processes, to be bound by the rule of law, to help craft and learn from the sundry products of culture, to enjoy an adequate level of material prosperity, and so on. To aspire to a life without these things was, for him, to want to be something literally un-human and therefore unnatural. In other words, social and economic necessities are the goods of civilization.

This is important because if we insist that these goods—and perhaps others—are in fact *constitutive* of a flourishing human life, we will have gone some way toward defining sustainability. Sustainable practices and proposals are those that allow future generations of humans to flourish in the sense just described. But it should be emphasized that this is a standard that requires a fairly high level of material wealth. The goods of culture do not come cheap. If we want these things—if we define social and economic sustainability *in part* as the ability to have these things—then this will put real downward pressure on how many of us there can be. Of course, as we have seen, it is always open to us to make sacrifices in one way or another, to say, for example, that we should let our numbers grow and simply trim back radically on the "luxuries" of culture. But we are, just for the sake of argument, supposing that this is not the best option to pursue. We are instead thinking of the three kinds of sustainability—ecological, social, and economic—as protecting *necessary* aspects of a flourishing human life.

ⓒ Sustainability and Substitutability

To say that a cluster of values or goods is necessary—*all* the items in the cluster—is to claim that you cannot make up for the total loss of one by getting more of the other. That is, they are not mutually substitutable (for more on this concept, see Chapter 4). Suppose, by way of analogy, that on the personal level there are three fundamental goods: happiness, autonomy, and moral agency. Now imagine someone who sacrifices one of these things to achieve another, or more of another than she had before. We might imagine someone who values happiness above all. This person agrees to be hooked up to a computer program—think of the plot of the film *The Matrix*—that fulfills all of her desires and thus makes her happy but robs her of autonomy and moral agency. Or think of the ancient Stoic philosophers who believed that the only thing in life that really mattered was moral virtue. As long as one always acts in a manner that is in accordance with the rationally accessible demands of virtue, one is living the highest possible existence. It does not matter that such a life might bring unhappiness (which it often does). Although these two examples do describe a possible human ideal, most of us think that a life lived in these ways is missing some key ingredients. If we are drawn to this thought, it is because we believe that a large dose of one fundamental or necessary good cannot compensate for the loss of another.[29]

Something analogous is going on with all of the things we are trying to combine in thinking about sustainability. And, of course, the analogy to what makes an individual life go well is not accidental, because in talking about living sustainably, we are trying to determine how to conduct ourselves in such a way as to allow for the lives of other individuals—future people—to go as well as they can. This is not simply to say that we are trying to provide good lives for them. Defining what they need is a way of providing the *conditions* for them to live good lives. Whether, and in what way, they take advantage of such conditions is entirely up to them. In the spirit of this suggestion, Jeffrey Sachs has argued that in this millennium humanity should aspire to these three goals:

> First, we will have to develop and adopt, on a global scale, and in a short period of time the sustainable technologies . . . that can allow us to combine high levels of prosperity with lower environmental impacts. Second, we will have to stabilize the global population, and especially the population in the poorest countries in order to combine economic prosperity with environmental sustainability. And third, we will have to help the poorest countries escape from the poverty trap.[30]

Here, Litke's two worlds are inextricably intertwined. Sachs does not believe we should abandon the goal of bringing "high levels of prosperity" to the whole world and to future generations. But the only way to do this within the constraints imposed on us by the Earth's biocapacity is to stabilize human population. Together, Litke and Sachs—interpreted in the Aristotelian fashion suggested here—bring a good deal of clarity to the concept of sustainability. At the end of the day, sustainability is, as Litke reminds us, about "how we want to live." The notion is disarmingly simple, but that is surely his point.

Perhaps the difficult part comes in universalizing the thought (here we will add a dash of Kant to our Aristotle). Your desire to prosper or flourish is shared by all normal humans, whichever part of the world—or whatever generation—they inhabit. There is nothing morally special about your desire such that it should be pursued at the cost of others' ability to prosper or flourish. As philosophers sometimes put it, the desire being indexed to you adds nothing of moral significance to it. To be consistent, you should recognize that if something is a *fundamental good* for you—that it is *necessary* for your flourishing—then it probably is for everyone else too. This insight places constraints on how you are permitted to act so as to acquire this and other goods. So another way to look at the sustainability of practices and proposals is to ask whether or not they are universalizable in this way. As Kant saw, formulating and acting on the principles revealed in this step of our thinking is what ethics is all about.

The notion that some goods are interchangeable and that acting morally always involves trade-offs among them—the principle of substitutability—is central to many debates in environmental ethics, so let's not leave it just yet. Wilfrid Beckerman has posed the following broad challenge to the concept of sustainability:

> The advocates of sustainable development as a constraint, therefore, face a dilemma. Either they stick to "strong" sustainability, which is logical but requires

subscribing to a morally repugnant and totally impractical objective, or they switch to some welfare-based concept of sustainability, in which case they are advocating a concept that appears to be redundant and unable to qualify as a logical constraint on welfare maximization.[31]

This is a serious challenge: the concept of sustainable development is

(a)   either morally repugnant; or
(b)   logically redundant.

How did Beckerman arrive at such a conclusion, and what can be said in response? The first thing to point out is that we *do* want the idea of sustainability to function as a constraint on our actions. Suppose we have two proposals before us, *A* and *B*, regarding the development of an area rich in some resource. It may be that proposal *A*, which involves mining the region aggressively, maximizes welfare for the local residents. But *A* also degrades the local environment, an ecologically sensitive wetland. Proposal *B* would preserve the wetland but fail to raise the overall welfare of the residents. From the economist's standpoint, proposal *A* is therefore optimal. If the appeal to sustainability is to have significant normative bite, it must be capable of providing an alternative to considerations of welfare maximization in cases like this.

With this requirement—that appeals to sustainable development provide a genuine alternative to welfare-maximizing proposals—in mind, Beckerman introduces a crucial distinction between strong and weak sustainability.

- **Strong sustainability**. The environment must be preserved in perpetuity, just the way we find it now in every aspect.

This would mean, for example, that it is impermissible to allow, or cause, the extinction of any plant or animal species. But doing this would require a huge output of resources. Here is where charge (a) enters the picture:

> Clearly, such a strong conception of "sustainable development" is morally repugnant. Given the acute poverty and environmental degradation in which a large part of the world's population live, one could not justify using up vast resources in an attempt to preserve from extinction, say, every one of the several million species of beetles that exist. For the cost of such a task would be partly, if not wholly, resources that could otherwise have been devoted to more urgent environmental concerns, such as increasing access to clean drinking water or sanitation in the Third World.[32]

This is clearly right, and we should eschew such an absolutist conception of sustainable development. But if we do, we land on the other horn of Beckerman's dilemma, charge (b). The idea here is that we should in some cases allow for the drawing down or even degradation of a local environmental resource, as long as the relevant stakeholders are adequately compensated for the loss through some payment "in kind": the provision of schools, roads, bridges, jobs, and so on. But, clearly, with this move, the idea of welfare maximization has simply replaced that of sustainable development. Explaining why this is the case brings us to the next form of sustainability.

- **Weak sustainability**. We ought to act so as to maximize welfare.

Here is how David Pearce describes this position:

> "Sustainability" therefore implies something about maintaining the level of human well-being so that it might improve but at least never declines (or not more than temporarily, anyway). Interpreted this way, sustainable development becomes equivalent to some requirement that well-being does not decline through time.[33]

As Beckerman points out, there are two implications of a view like this. First, natural capital could in principle be entirely replaced by man-made capital as long as the result was not a net decline in the welfare of affected people. That is, a group of people could decide to degrade their region's stock of natural capital—for example, its fresh water, biodiversity, or carbon-absorbing potential—as long as they were compensated for these losses through the provision of man-made capital. On this conception, natural and man-made capital are fully substitutable. But, second, this means that the concept of sustainable development has lost its logical independence. As Beckerman puts it,

> [I]f the choice between preserving natural capital and adding to or preserving man-made capital depends on which makes the greatest contribution to welfare the concept of sustainable development becomes redundant.[34]

Now, we might not worry too much about this latter result, except that we already expressed a commitment to a notion of sustainability that was able to act as a unique constraint on welfare maximization. That is a difficult commitment to abandon, but Beckerman thinks we should do so. We should, he suggests, jettison appeals to sustainability and sustainable development and instead stick with the economist's tried and true concept of optimality (welfare maximization).

But this solution may be hasty. In criticizing Beckerman, ecological economist Herman Daly has in effect found a version of sustainability that is neither strong nor weak. According to Daly, Beckerman's "strong sustainability" is indeed morally repugnant, but it is better dubbed "absurdly strong sustainability." And, crucially, we can reject it without becoming advocates of weak sustainability. The key for Daly is to reject the principle of *unlimited* substitutability between natural and human-made capital on which Beckerman's critique of sustainable development relies. Daly insists on thinking of the two kinds of capital as "complementary" rather than substitutable. Two things are complementary in this sense when both are required, when it is not possible to compensate for the loss of one by heaping on more of the other:

> If natural and manmade capital were substitutes (weak sustainability) then neither could be a limiting factor. If however they are complements . . . then the one in short supply is limiting. Historically, in the "empty world" economy, manmade capital was limiting and natural capital superabundant. We have now, due to demographic and economic growth, entered the era of the "full world" economy, in which the roles are reversed.[35]

As things currently stand, natural capital—the Earth's biocapacity—is a limiting factor. This, if correct, eliminates Beckerman's dilemma, for we can reject both strong and weak versions of sustainable development. This is moderate sustainability.

- **Moderate sustainability**. This is stronger than weak sustainability because it rejects the principle of *unlimited* substitutability between natural and human-made capital. However, it is weaker than strong sustainability because it recognizes that *sometimes* it will be necessary to provide human-made capital to (impoverished) people at the cost of degrading some aspect of their environment.

Embracing moderate sustainability has real implications for our duties to future generations:

> [Future generations] have ownership claims to as much natural capital as the present—i.e., the rule is to keep natural capital intact. [Moderate] sustainability requires that manmade and natural capital each be maintained intact separately, since they are considered complements: weak sustainability requires that only the sum of the two be maintained intact, since they are presumed to be substitutes. As natural capital more and more becomes the limiting factor, the importance of keeping it separately intact increases.[36]

One way to cash out the notion that our duty to keep the planet's natural capital *increasingly* intact is to say that we ought to employ the precautionary principle in our dealings with the natural world. According to this principle, when our activities threaten to do damage to the environment, "the lack of full scientific certainty shall not be used as a reason for postponing cost-effective measures to prevent environmental deterioration."[37]

Clarifying what we mean by sustainability in these last few sections has been difficult work. Let's summarize what we have so far, with the following four points.

1. There is more than one aspect of our practices that needs to be sustainable. At the very least, our practices must be ecologically, socially, *and* economically sustainable.

2. These kinds of sustainability are not substitutable because we—and this includes members of future generations—need *all* the goods they protect: a healthy environment, basic and wide-ranging equity, and a certain level of material prosperity.

3. We need these things because having them is constitutive of a fully flourishing human life.

4. Focusing on ecological sustainability, because the biocapacity of the Earth is already so stressed, we must treat the stock of natural capital as an especially limiting factor in our decisions. This is why we should adopt the moderate version of sustainability rather than the strong or weak versions.

Ⓓ Sustainability and the City

Now that we have a working, if complex, account of sustainability, let's switch gears somewhat. Very little work has been done by philosophers on sustainable city living. This is not true of ecologists: open any recent ecology textbook, and you will see large sections devoted to this theme. The lack of philosophical attention to the problem of urban existence is strange, since so many of us live in cities. Indeed, we are becoming an increasingly urbanized species. In 1900, 13 per cent of us (220 million) lived in cities, a proportion that rose to 29 per cent (732 million) by 1950 and 49 per cent (3.2 billion) by 2005.[38] Moreover, the twentieth century has seen the emergence of the megacity, defined as an urban centre with a population of 10 million or more. By 2005, there were 20 megacities in the world, led by Tokyo (35.2 million), Mexico City (19.4 million), and New York–Newark (18.7 million).[39] Each of these cities is a true colossus, but most of us still live in smaller cities (under 50,000).[40]

From the standpoint of sustainability, the key fact to bear in mind is that although cities comprise just 21 per cent of the world's surface area, they are responsible for 75 per cent of all resource use.[41] The most important urban environmental concerns have to do with the quality of urban climate and atmosphere, noise pollution, water use, energy use, materials use (including waste disposal), the urbanization of land, and the loss of agricultural land.[42] How do Canadian cities fare in tackling these problems in an environmentally sustainable manner? Obviously, we will get a different answer depending on which issue we look at, but let's take the problem of air quality as representative. Air-quality indicators track the components of smog, mainly ground-level ozone and fine particulate matter. Between 1990 and 2005, ozone levels increased 12 per cent overall, with the biggest increase in southern Ontario (17 per cent). The three biggest problem areas in the country for ozone are the Lower Fraser Valley, the Windsor–Quebec corridor, and the Fundy region of southern New Brunswick and western Nova Scotia.[43] Although substantial progress has been made nationwide in reducing overall smog levels, it is sobering to note that compared to 30 other OECD countries, Canada is dead last on a number of air-quality indicators. For instance, Canada's per capita emissions of sulphur dioxide were, in 2005, three times the OECD average.[44]

Statistics aside, what are the deeper philosophical issues that the fact of increasing urbanization is making salient? For help on this question, we turn to the chapter's second reading.

SUSTAINABILITY AND SENSE OF PLACE

Ingrid Leman Stefanovic

When we speak of "nature" or "environment," we often envision pastoral fields or wilderness areas, untouched by human settlement. The reality is, however, that pressing environmental issues such as global climate change or air and water contamination immediately implicate our cities and the way in which humans inhabit the Earth. To separate "natural" from "urban" issues is increasingly problematic, both empirically and philosophically.

The United Nations reports that half of the world's population already resides in urban areas.[45] Under these circumstances, the field of environmental ethics can no longer assume that concerns about the intrinsic value of nature or the welfare of animals can be debated separately from the global challenges of urbanization.

To be sure, we cannot afford to simply urbanize and industrialize as we have done in the past, precisely because of the enormous environmental impact of our actions. As we develop our cities and our urban regions, we must do so *sustainably*. What does that mean?

According to the World Commission on Environment and Development, it signifies that we must urbanize in ways that meet "the needs of the present without compromising the ability of future generations to meet their own needs."[46] It means that we must rethink, in the most comprehensive way, the manner in which we build our dwelling places.

I. Questioning Calculative Measures of Sustainability

While the concept of sustainability has become increasingly popularized over recent decades, its original formulation by the World Commission on Environment and Development has not been without its critics. Some believe that the notion of sustainable *development* belies an economic growth model and a Western utilitarian ethic that are the very source of environmental problems in the first place. The emphasis on future *human* generations is also seen to imply an unquestioning anthropocentric paradigm at the expense of broader ecological interests.

Much discussion about sustainability indicators is based upon quantitative parameters and statistical measures of individual elements, such as population changes, number of endangered species, percentage of waste water treated, and the like. Yet the question can be raised as to whether full justice is done to the concept if one operates simply within calculative, reductionist parameters. While green roof design and hybrid vehicles may be necessary conditions of a sustainable city, is an inventory of such discrete initiatives sufficient when it comes to capturing a broader, non-calculative vision of sustainability? Does a quantitative, utilitarian assessment of costs and benefits take us far enough in our understanding of sustainable cities?

Some philosophers suggest that in addition to calculative measures, a different kind of originative thinking is required if we are to move toward a healthier planet.[47] Rather than restrict ourselves simply to measuring sustainability by quantifying individual *things* or *entities*, the suggestion is to look more broadly and focus on holistic, interpretive horizons of understanding and the essential relations *between* things.

The National Round Table on the Environment and the Economy in Canada echoes this thought when they write that sustainable development deals with interrelationships and linkages. It means looking at decisions in a holistic way.... More than in anything else, the power of sustainable development lies in its bridging capability—its ability to facilitate integration, synthesis and collaborative approaches to problem solving.[48]

There is a need to move beyond positivist, reductionist paradigms and capture the full, lived experience of sustainability in a different way. In this respect, some insights from environmental phenomenology and post-modernism may prove to be particularly instructive.

II. Moving beyond Modernism

"Continental philosophy" is a term used to describe a variety of philosophical approaches, originally emerging from European countries such as France and Germany. From existentialism to post-modernism, a common tendency has been to critique positivist, "modernist" foundations of philosophical and ethical inquiry. These modernist metaphysical foundations are themselves seen to be grounded in a dualistic framework of subject/object dichotomies that favours either abstract, theoretical speculation or objective, empirical facts. Critics of the modernist tradition argue that such a simplistic dichotomy fails to respect the diversity, complexity, and vibrancy of everyday lived experience.

Much of this analysis of traditional Western metaphysics originated with the philosophy of the German phenomenologist Martin Heidegger, who believed that the modernist, calculative way of thinking had to be supplemented by what he called "meditative thinking."[49] The prevalent calculative world view is one that computes explicit empirical realities,

forgetting to reflect on the broader ontological meanings that sustain them. It is a world view that seeks an understanding of the natural world by identifying linear, deterministic causal relations between discrete atomistic entities and events. It focuses on measurable empirical realities (whether they are explicitly and logically articulated concepts or empirically defined facts) and hopes to draw universally applicable conclusions that are deemed to be "objectively" or scientifically "true."

Generally speaking, Continental philosophers and critics of modernism suggest that such calculation does not begin to capture the whole story when it comes to understanding the complexities of our lived environments. Rather than reducing the richness and density of our lives to simple linear, causal relationships, the tendency among this group of thinkers is to proceed with a broader exploratory attitude, one that is non-positivist in nature and, as open to diverse perspectives, is "non-essentializing."

From the beginning of the phenomenological tradition, the bifurcation of "subjective" and "objective" realities is itself brought into question. We are not seen to be self-contained human subjects, interacting with an objective world that is separate and external to us. On the contrary, to use Heidegger's phrase, we are essentially defined as "in-the-world."[50] The environment is not some-*thing* out there, independent of who we *are* as human beings. On the contrary, we are defined by our integral, ontological belonging to the world in which we find ourselves. To live sustainably means to *be* sustainable as a condition of our comportment within the world.

How does this belonging manifest itself, and why is it important for environmental ethics? It helps us to realize that "the environment" is not a singular entity outside of our ways of being, serving simply as a physical container of human activities. On the contrary, building on the French sense of *"environs,"* the environment surrounds us and defines us as the very incarnation of our existence. Inasmuch as we cannot *be* without an environment, the quality and well-being of human lives is integrally linked to the quality of how and where we are environmentally *implaced*.[51]

As Heidegger pointed out, the Greek root of the word "ethics" is *ethos*, meaning the fundamental comportment of human beings and their bearing (*Haltung*) in life. Drawing on the philosophy of Heraclitus, Heidegger suggests that the pre-Socratic roots of *ethos* meant abode, dwelling place.[52] Ethics itself is fundamentally an expression of how human beings dwell on the Earth. On this reading, ethics is, essentially, environmental ethics.

III. Recovering a Sense of Place

So inasmuch as we exist, we exist somewhere. In the words of Edward Casey, "to exist at all . . . is to have a place—to be implaced. The point is that place, by virtue of its unencompassability by anything other than itself, is at once the limit and the condition of all that exists."[53] Our dwelling places are not something merely incidental to our sense of self, but they are the spatial and temporal locations that reveal who we are and what our values are as well.

Consider a medieval town and how the cathedral visually defines the town's centre, even from afar. That centrality of the church spire tells us something about the importance of Christianity in that time and place. Contrast that town with the centre of the city of Toronto, clearly indicated by the CN Tower and the skyscrapers that house the banking district. Do those core human settlement patterns not reveal something significant about the centrality of telecommunications and financial aspirations when it comes to the modern Canadian city? Our world views, attitudes, and values are embodied in our buildings and dwelling places.

In the words of architectural phenomenologist Christian Norberg-Schulz, "when we identify with a place, we dedicate ourselves to a way of being in the world."[54] The way in which we are implaced is not simply as one physical body within another material entity. We are as ahead-of-ourselves, embodied within the spaces that we inhabit and always interpreting those places. Heidegger gives the example of heading toward the door of a lecture hall. "I am already there," he writes, "and I could not go to it at all if I were not such that I am there. I am never here only, as this encapsulated body; rather, I am there, that is, I already pervade the room and only thus can I go through it."[55] Consider how we may be *meaningfully* closer to the home we are approaching than we are to the sidewalk we physically touch as we walk upon it. Our lives are defined by both our embodied existence and how we interpret the world through a temporal and spatial horizon of understanding. In short, our understanding of our environments is framed not only by positively existing entities but by prethematically intuited relations, contexts, and futural horizons that are not reducible to measurable "things." We may be situated *here*, on the sidewalk, but also, in the words of poet Pierre-Jean Jouve, "nous sommes où nous ne sommes pas" ("we are where we are not").[56]

Some refer to this originary belonging to our lives' spaces as a "sense of place."[57] Casey emphasizes that this relation to place is not simply secondary to our way of being but is the condition of the possibility of comporting ourselves in the world in the first place. In this sense, implacement is a primordial, ontological condition of being human.[58] As architect Juhani Pallasmaa points out, this understanding also suggests that our built settlements are not simply external artifacts, but they define us in the most fundamental way. "I experience myself in the city," he writes, "and the city exists through my embodied experience. The city and my body supplement and define each other. I dwell in the city and the city dwells in me."[59] We are each implaced in urban areas as an embodied presence. "There is no body separate from its domicile in space, and there is no space unrelated to the unconscious image of the perceiving self."[60]

As ontological, implacement is not something that is consciously always recognized or explicitly acknowledged. Sometimes, sense of place reveals itself implicitly: consider the different experience of travelling along a highway rather than meandering along a gravel country road. Time moves differently with the divergent configurations of space. Similarly, when we imagine a basement—dark, damp, and hidden—it is a different sense of place than that revealed in an attic—light, high, and full of cobwebs.[61] The lived experience of place in these instances exceeds modernist, calculative measures. In Gaston Bachelard's words, "inhabited space transcends geometrical space."[62]

One of the most famous examples that illustrate how we are implaced comes from the magnum opus of Martin Heidegger, when he describes the way in which we use tools in our everyday lives. We might imagine the process of hammering a nail into the wall. When the hammer is grasped and put to use, it reveals itself in its "readiness-to-hand" (*Zuhandenheit*). This immediacy stands in stark contrast to the hammer when we view it in terms of its "presence-at-hand" (*Vorhandenheit*)—that is, when we gaze upon the hammer and note its size, its shape, its purpose. When I actually put the hammer to *use*, it is no longer the object of my explicit attention, but it "subordinates itself to the 'in order to'. . . . The less we just stare at the hammer-Thing and the more we seize hold of it

and use it, the more primordial does our relationship to it become."[63]

Similarly, to be "implaced" means that as embodied beings, we do not always explicitly reflect on physical spaces as if they were "present-at-hand" entities. Sometimes we understand our built environments pre-reflectively, spontaneously, immediately: when we cannot find the main entrance to a building, we feel disoriented and unwelcome, even if we are not consciously aware of these intuitions. We are engaged in the world as rational, reflective beings but also as beings who are aware of the world pre-reflectively. As Heidegger reminds us, the world is much more than "something subsequent that we calculate as a result of the sum of all beings. The world comes not afterward but beforehand, in the strict sense of the word. . . . It is so self-evident, so much a matter of course, that we are often completely oblivious to it."[64]

Why is this fact important to discussions of sustainability and sense of place? Consider the experience of owning an automobile. On the one hand, we rationally recognize that reliance upon the car has led to road development that has decreased agricultural lands and to unsustainable subdivision planning and construction. On one level, we rationally *know* that even if we inhabit such a subdivision, the "environmentally responsible" thing to do is to leave the car at home and take public transit instead.

Yet on a deep level, the *experience* of driving can be so liberating! Many people who publicly espouse sustainability principles privately *love* their cars and love driving as well. The phenomenology of automobility reveals that the car encapsulates values, such as freedom, privacy, and individual property rights, that are entrenched in Western political democracies. As planners and ethicists implore citizens to use public transit in order to advance sustainability ideals, how much attention is being paid to intuitive, pre-reflective attitudes and behaviours that affect decision-making? How much attention is paid to the experience and love of driving instead of to the promotion of abstract, rational environmental ideals? It is that pre-reflective level of people's awareness that is often forgotten when it comes to policies and planning initiatives, and yet our taken-for-granted values and perceptions are often key to how we understand and act in the world. Unless we attend to this level of understanding, our sustainability initiatives will amount to little more than Band-Aid solutions and busy-work—and our ethical deliberations, while rational, will not realistically encompass the full range of human behaviour and people's *implicit* value systems and beliefs.

IV. Sustainability, Place, and Environmental Ethics

If the goal of sustainability is to ensure that present-day needs are met without preventing future generations from meeting their own needs, such a task requires a change in our present patterns of urbanization and ways of inhabiting the Earth. Rather than relying simply upon new technologies to solve the problems of global climate change or engaging in a technocratic re-ordering of individual entities and "things" that we do, we need to rethink—in a fundamental way—the way in which we dwell. As economist Bill Rees points out, "for sustainable development . . . the need is more for appropriate philosophy than for appropriate technology."[65]

If we recognize that human behaviour incorporates both explicit, rational ways of acting and implicit, taken-for-granted world views and attitudes, then it becomes evident that ethics itself, as a largely deductive inventory of overt,

individual directives, will need to be rethought. In an effort to ensure that both reflective and pre-reflective world views are elicited, a postmodernist approach to environmental ethics and sustainable values of place will necessarily aim at something more than the articulation and enumeration of discrete, explicitly articulated, universalizing principles and rules.

A post-modern approach will also look to moving beyond the traditional disputes between "anthropocentric" and "ecocentric" extremes in the field of environmental ethics. In the words of phenomenologist Bruce Foltz, "it is possible to approach [current ecological problems] from a different direction altogether, taking our fundamental *relation* to nature, rather than nature alone, as the primary subject of the crisis."[66] In fact, an ethic of place suggests that we must do more than reduce the complexity of the human–nature relationship to either an arbitrary, subjectivist interpretation of the world or an argument in favour of objective rules built upon calculative facts about an external world.[67]

What might such an alternative, non-universalizing ethical approach consist of? From the phenomenology of Werner Marx to the postmodern interpretations of an ethic of place, the responses to this question cover a range of perspectives.[68] Nevertheless, in each of these instances, environmental ethics is no longer seen as a monistic, essentializing enterprise of deducing a single, totalizing theory and then applying it indiscriminately across all cases. Rather than developing a speculative ethical theory and generating universal principles and rules to be followed, the move beyond modernism in the field of environmental ethics recognizes the legitimacy of *diverse environmental narratives* as we move toward a more sustainable future. There is a growing recognition that the complexities of the lived world exceed reductionist parameters and that it is naive to expect that we can construct answers to ethical dilemmas that are universally applicable across all cultures and languages. While good arguments can still be advanced in favour of how we ought to behave and what actions we should pursue, it is not a matter of arguments developing top-down, from theory to practice, but rather it is *praxis* that informs the evolution of understanding in discourse ethics.

How does such an approach to ethical decision-making occur in the lived world of everyday challenges? Parents recognize that wise strategies for the complex processes of child-rearing are not available through a simple "how-to" manual or with some sort of mathematical precision. As we navigate through life's unexpected predicaments, many moral dilemmas similarly evade single, automatic prescriptions for success. No one is invulnerable to error, and as we learn from mistakes and commit ourselves to listening to and engaging with others with genuine goodwill, the goal is to do the best job we can to act as ethical citizens, learning from one another and from those who preceded us. As moral pluralist Christopher Stone reminds us, life is usually not simply a matter of either/or decisions, and consequently, "we simply may not be able to devise a single system of morals that is subject to closure and in which the laws of noncontradiction and excluded middle are in vigilant command."[69] Rather than eliciting a single abstract, monistic moral theory, it may be more discerning to approach ethical problems from different perspectives, to understand the alternatives available to us, and to engage in a collaborative, intersubjective dialogue to jointly evolve what is arguably the best moral alternative under the present circumstances.

If one accepts that ethics is a deliberate, discursive exercise in justifying one's claims rather than an egocentric elucidation of abstract principles, then it becomes clear how a post-modern philosophical approach impacts even upon social science methodologies. From phenomenological to hermeneutic qualitative research methodologies, researchers investigate broader, lived world experiences by conducting interviews and relying upon other less quantitative approaches than traditional surveys or statistical analyses.[70] The field of environmental ethics—and the important changes in behaviour that are intimately linked to the critical investigation of value systems—reminds us that philosophy can no longer afford to be a hypostatizing ivory tower exercise but is intimately related to *praxis*. Interdisciplinary conversations between the humanities, social sciences, and natural sciences are mandated by a broad-based awareness of the relationship between people's values, their actions, and the impact upon the world in which they are implaced. Philosophy is no longer a solitary exercise but is intimately related to social discourse. As Henri Lefebvre reminds us in *Writings on Cities*, "the philosopher no longer has a right to independence vis-à-vis social practice. Philosophy inserts itself into it."[71]

V. Conclusion

The field of environmental ethics is not only important in terms of how we value wilderness preservation but is also intimately related to how we urbanize and build sustainable human settlements. The goal of sustainability is, however, global and comprehensive in nature. It requires a rethinking of how we dwell and how we are fundamentally implaced within our built environments. For this reason, it becomes important to supplement traditional, reductionist calculative ways of thinking with richer, more diverse ways of thinking beyond the modernist framework. Instead of simply debating theoretical principles, we are invited to engage in a description of lived experience in its broadest sense and to engage in moral and ontological reflections that are discursive, non-essentializing, and intimately linked to *praxis*. The complex, multicultural, multi-scale goals of building a sustainable world demand nothing less.

Stefanovic presents us with a difficult challenge: to think of our existence in urban spaces in a way that goes beyond calculative, reductionist, modernist, and purely theoretical approaches. To eschew these forms of understanding and embrace instead a post-modernist notion of "implacement" is essential to grasping how we might live sustainably in our cities. What is implacement? Let's try to cash the idea out with three broad points.

First, underlying this notion is the rejection of an abstract distinction between the subject (us) and the objective world existing outside that subject. We should cease thinking of the environment, urban or otherwise, as something out there, a place we find ourselves in and that provides various services to us (or fails to do so). The very term "environment" invites this kind of misunderstanding, for it characterizes the external world as a thing that is separate from us, to which we must adapt or, perhaps, that we can master and subdue. As philosophers John O'Neill, Alan Holland, and Andrew Light have recently argued, we should for this reason stop thinking of the environment as something we live from—that provides "a living" for us—and think of it instead, or also, as something we both live *in* and *with*. Here is how they describe these terms:

> We live in the world. The environment is not just a physical precondition for human life and productive activity, it is where humans and other species lead their lives.... We live with the world: the physical and natural worlds have histories that stretch out before humans emerged and have futures that will continue beyond the disappearance of the human species.[72]

The main idea is to go beyond the narrow or economic approach to the environment that tends to see the latter as a mere resource base. Because we live *in* environments, we need to take seriously both recreational and aesthetic considerations; because we live *with* environments, we need to take seriously the "accumulation of data provided by the life sciences." This is better than seeing the environment as an external resource, but it probably still does not capture what Stefanovic is after. Maybe we should think about what earthworms do:

> Geophysiology emphasizes that organisms play an important role in the creation of their environments. Organisms thereby play an important role in creating themselves, for the environmental conditions they contribute to forming subsequently exert selective pressures on them and their descendants: on straight evolutionary reasoning, it follows that organisms which create a favourable environment for themselves will tend to be selected for through consequent feedback effects from

the environment. Darwin's understanding of the shaping force of earthworms on their surroundings resonates strongly with this view: he did not take the land as the given background to which worms adapt, but saw it as a medium actively created and maintained, in large part, by these animals themselves.[73]

Earthworms do this by moving huge amounts of soil so that it is fit to retain moisture, retain soluble substances, and be ready for nitrification. What matters for our purposes are not the scientific details but the larger philosophical point that in manufacturing the earth in specific ways, earthworms are in fact "creating themselves." This seems to get closer to Stefanovic's meaning. Recall, for instance, her reference to what the CN Tower, an aspect of the world we Canadians have made for ourselves, says about who we are. The idea is that the environments we build also *make us*. Or better: that in constructing our cities, we are at the same time constructing ourselves. But the point about the earthworms is instructive on another level. The built environment exerts subsequent selective pressures on the population of earthworms: organisms that "create a favourable environment for themselves" will tend, other things being equal, to pass on their genes. Similarly, humans can expect to flourish in built environments that are genuinely sustainable. Thinking this way can help us overcome the subject (human)/ object (built environment) distinction that seems to plague us.

This brings us to the second broad point. The built environment will be sustainable if and only if we remember two things as we go about planning. The first is that many of our values and preferences are pre-reflective. Recall Stefanovic's elegant example of the love of driving: "the *experience* of driving can be so liberating! Many people who publicly espouse sustainability principles privately *love* their cars and love driving as well." This is not to say that because people love driving we cannot challenge the dominance of automobility in our lives if we think it is having deleterious environmental impacts. Rather, it is that we won't make effective—*sustainable*—changes unless we understand and respect the values that underlie the phenomenon of automobility: freedom, privacy, property rights. *If* the car has got to go, we will need to find alternative ways to instantiate these values. Second, building sustainable cities requires attention to the *plurality* of values and ideals that one finds in cities. This means putting an ear to the ground and constructing the "diverse environmental narratives" that are the stuff of city life rather than engaging in monistic, top-down environmental ethical theory (think here of environmental pragmatism, from Chapter 5).

The third broad point is that we should emphasize the fact that cities, though they sometimes get a bad rap from those attracted to more bucolic settings, can in fact be models of environmental sustainability. Here's the bad rap:

> The city is often a cesspool of pollution, a sewer of vice, violence, crime, corruption, poor schools, poverty, unemployment, high taxes, suffering and alienation. Typically dense with the anonymous homeless, panhandlers, muggers and drug addicts, pungent with the smell of decay, the ugly sights of gaudy graffiti and garish advertisements . . . the urban ulcer ubiquitously bombards our senses and crowds out our thoughts, alienating us from ourselves.[74]

From the standpoint of the three necessary aspects of sustainable practice—environmental, social, and economic—this is a pretty damning indictment of city life. It is certainly a far cry from what the worms have done with their little cities of earth, but beyond the hyperbole is it at all correct? Are cities inevitably such wastelands? Not really. Consider, to begin, Kleiber's law, which has it that organisms tend to increase in metabolic efficiency as they get larger. Physicist Geoffrey West says this:

> One of the basic principles of cities is that it is more efficient to bring people together. You need a little bit less of everything per person. It's exactly the same way in biology. As animals get bigger they require less energy to support each unit of tissue.[75]

Compared to their country cousins, people in cities can generate more wealth with a lower ecological footprint. Stewart Brand has argued that this is true even of the huge slums in cities like Mumbai, India: "[W]hat you see up close is not a despondent populace crushed by poverty but a lot of people busy getting out of poverty as fast as they can."[76] Recall a key point from section B, above: that eliminating poverty is a key element of social sustainability. If Brand is right, then cities are where this is happening. Speaking of Mumbai, Suketu Mehta has this to say:

> Why would anyone leave a brick house in the village with its two mango trees and its view of small hills in the East to come here? So that someday the eldest son can buy two rooms in Mira road, at the northern edges of the city. And the younger one can move beyond that, to New Jersey. Discomfort is an investment.[77]

Moreover, the trend in large cities is towards a reduction in fertility rates. We have seen that stabilization of population, especially in the developing world, is crucial to building a sustainable future. But according to development expert Paul Polak, rural life does not foster this goal:

> A one-acre farm family in Bangladesh needs three sons to get ahead—one to help with the farm, one to get a good enough education to land a government job capable of supporting the family . . . and one to get a local job that pays enough to keep his brother . . . in school. But to end up with three sons means having eight babies, two of which are likely to die before the age of five, leaving three boys and three girls.[78]

Things are different in cities, as a 2007 UN report indicates:

> In urban areas, new social aspirations, the empowerment of women, changes in gender relations, higher-quality reproductive health services and better access to them, all favour rapid fertility reduction.[79]

So much for the myth of the city as an environmentally destructive cesspool of despair and alienation. And if this is a myth as applied to the slums of Mumbai, it is surely just as false as applied to cities in the developed world, whatever their other faults.

Ⓔ Conclusion

What we need to emphasize, in conclusion, is that sustainability is a moral ideal, not merely a technical yardstick. Furthermore, it is in many ways the most important concept in environmental ethics. Whether we make them explicit or not, considerations of sustainability are implicated in what we are doing when we decide to increase urban density, rethink our approach to land use, expand public transportation, impose a moratorium on the further development of a particularly polluting resource, build interconnected natural areas in our cities, reduce our energy consumption, consume locally grown and/or organic food, and so on.

Jeffrey Sachs, an economist, has argued that "sustainability has to be a choice, a choice of a global society that thinks ahead and acts in unaccustomed harmony." This is no doubt correct as far as it goes, but it is incomplete in a way that invites misunderstanding. What it implies is that the choice to act sustainably is *criterionless*. The statement—which expresses a sentiment one finds ubiquitously in the non-philosophical literature on sustainability—might incline us to believe that it is fruitless to try to provide a philosophical justification for our choice. But this is false. In fact, deciding whether or not we ought to engage in sustainable environmental practices inevitably takes us right back to the theoretical beginnings of environmental ethics. This is because considerations of sustainability assume that the interests of future generations, not to mention those of non-human plants and animals and ecosystems, *matter morally*.

If we think this is right, however, we need to say why. In virtue of what property or cluster of properties do the entities just mentioned have moral standing? If we do not, or cannot, provide a rigorously defensible answer to this question, then we will have no good reason to act sustainably. Though we are bombarded with calls to live sustainably, we should remember that we cannot dispense with the more difficult task of providing a theoretical foundation for our sustainable actions. Sustainable environmental practice *requires* sound environmental ethics.

CASE STUDY

Nuclear Power: Unsustainable Train to the Future?

For many years, nuclear power was viewed by many as the epitome of an unsustainable approach to meeting our energy needs. Philosophers Richard Routley and Val Plumwood express this idea by asking us to think of our use of nuclear power as akin to placing a container filled with a "highly toxic and explosive gas" on a train, then sending it casually on its way, knowing (a) that it could leak out along the way and (b) that when it arrives at its destination, others will have to deal with it somehow. Replace the spatial metaphor with a temporal

Source: © Bill Brooks/Alamy

CANDU nuclear reactor at Pickering, ON. With climate change an urgent issue, can we afford to wait for renewables?

one, and it is easy to grasp the alleged problem with nuclear power. The radioactive waste material generated by our nuclear facilities can take many thousands of years to degrade, during which time it poses a distinct threat to future generations. Routley and Plumwood therefore characterize nuclear power as a dangerous "train to the future." Just as we think it impermissible to put that package on the train, it is impermissible to saddle future generations with the risk of radioactive contamination:

> Even where there are serious risks and costs to oneself or some group for whom one is concerned one is usually considered not to be entitled to simply transfer the heavy burden of those risks and costs onto other uninvolved parties, especially where they arise from one's own, or one's group's, chosen lifestyle.

Many people in the environmental movement—including many of those who went on to start up environmental NGOs like Greenpeace, the World Wildlife Fund, and the Sierra Club—cut their teeth on the nuclear issue in the 1960s and 1970s. Opposition to nuclear power virtually defined the environmental movement in these early days. But that is changing because of the looming threat of climate change. According to mainstream scientific opinion, humanity needs to stabilize its emissions of greenhouse gases by 2015, with the long-term goal of reducing them by 80 per cent by 2050. All of this is to avoid passing the critical threshold of a 2°C rise in temperature (relative to pre-industrial times), beyond which we will have runaway climate

Continued

change and untold catastrophe (see Chapter 11 for more on climate change). NASA's James Hansen, the world's leading climatologist, has argued that nuclear power is, at this point in our history, by far the lesser of two evils. He thinks that the only responsible energy policy is to place an immediate moratorium on the development of Alberta's oil sands as well as a ban on all coal-fired power generation. He is not alone in this new environmentalist campaign for nuclear. James Lovelock, originator of Gaia Theory, claims that there is no way at present for us to meet our energy needs—our baseload requirement—with renewables. If we are to achieve the required stabilization of greenhouse gases by 2015, we must resort to nuclear power on a massive scale.

But how do proposals like this meet the objection of Routley and Plumwood? There are two possibilities. First, you might simply argue that nuclear power is a deadly train to the future but that passing the 2°C mark is a far deadlier train. Hansen, for example, calls our current heavy use of coal a "gross case of intergenerational injustice." Second, you might argue that it is within our power to substantially reduce the risk of radioactivity to future generations by developing so-called "fast reactors." These reactors allow the neutrons of uranium to move at a higher speed than they do with normal thermal reactors. The result is that they burn 99 per cent of the uranium that is mined, compared to 1 per cent burned by thermal reactors. So they increase efficiency by a factor of about 100 over the older reactors. Moreover, the waste they do produce degrades in a few hundred rather than a few thousand years, and this waste is useless for making bombs. The technology is not exactly ready to go, but neither is it just a pipe dream.

So what should Canada do? In Canada, we produce about 12 to 15 per cent of our electricity from nuclear power, mostly in Ontario. In December 2005, the Ontario Power Authority argued that it would require $83 billion to update its system of nuclear reactors over a 20-year period, and the government of New Brunswick is also looking into doing the same to its Point Lepreau reactor at a cost of $1.4 billion. In Ontario, the case for the upgrades relies on the relative environmental cleanliness of generating power this way. But if this is the argument, then given the fact that we have so little time to wean ourselves from fossil fuels, how can we justify not adopting nuclear power on a much greater and nationwide scale? As long as some of the money we spend is earmarked for research and development of fast reactors and the search for a workable renewables option is enhanced, we may have a duty to do so. It does seem that we are committed to sending a train to the future, a train loaded with something nasty. But the nature of the substance and how much of it there will be depend on choices we make now. When we have a choice between two evils, acting sustainably might mean taking the least unsustainable path.

Sources: Dearden and Mitchell 2009, 471–4; Hansen 2009; Lovelock 2006; Richard Routley and Val Plumwood, "The Nuclear Train to the Future," in Pierce and Van DeVeer 1998a, 449–50.

Study Questions

1. As noted throughout the chapter, "sustainability" is one of the most overused words in environmental discourse. Do you think it is still a useful concept, or is it too easy to abuse it (through, for instance, the practice of "greenwashing")? Can we do without it? What would be a good successor concept?

2. What do you think about cities and the environment? Are cities inevitable cesspools or is there a way to think about and design them that is compatible with environmental stewardship? What would your ideally environmentally friendly city look like?

3. What is the distinction among the three degrees of sustainability: strong, weak, and moderate?

4. Do you agree with the claim that the best approach to ecological sustainability is the moderate one? Why or why not?

Further Reading

Stewart Brand. 2009. *Whole Earth Discipline*. New York: Viking Press.

Edward T. Casey. 1993. *Getting Back into Place: Toward a Renewed Understanding of the Place-World*. Bloomington and Indianapolis: Indiana UP.

David Cook. 2004. *The Natural Step*. Foxhole, Devon: Green Books.

Jeffrey D. Sachs. 2008. *Commonwealth: Economics for a Crowded Planet*. New York: Penguin.

Jennifer Sumner. 2005. *Sustainability and the Civil Commons*. Toronto: U of Toronto P.

Notes

Preface

1. Reported in the *Globe and Mail*, www. theglobeandmail.com/news/world/ canada-dead-last-in-oecd-ranking- for-environmental-protection/ article15484134/.

Introduction

1. The following taxonomy is adapted from Pojman 2002, 9–12.
2. Plato 1992, Book II359a-b, 35.
3. Pojman 2002, 95. The friendship argument, considered next, is adapted from ibid., 93.
4. The following argument is adapted from ibid., 26–8.
5. Quoted in Rachels 1999a, 2.
6. Quoted in Rachels 1999b, 22.
7. The following argument is adapted from Shafer-Landau 2004, 15–18.
8. An excellent discussion of these points is provided in ibid., 38–42.
9. Kant 1988, 254.
10. Ibid., 273.
11. Ibid.
12. Hobbes 1994, 76.
13. Ibid., 23.
14. Aristotle 1947, 1094a (16–27), 394.
15. Ibid., 1102b (5–6), 328.
16. Ibid., 1105b (5–10), 337.
17. Ibid., 1104a (29–36), 336.
18. Ibid., 1106b (18–24), 339–40.

Chapter 1

1. Pierce and Van DeVeer 1998b, 76.
2. Rowlands 2008, 124.
3. *Introduction to the Principles of Morals and Legislation*, ch. XVII.
4. I owe the term "speciesism" to Richard Ryder.
5. Harrison 1964. For an account of farming conditions, see my *Animal Liberation*, 1975.
6. Belshaw 2001, 97.
7. Regan 1983, 243.
8. Belshaw 2001, 105.
9. Ibid., 104.
10. Francione 2008.
11. Regan 1983; Singer 1975.
12. Palmer 2010.
13. Acampora 2004.
14. Donovan and Adams 2007.
15. Nussbaum 2006.
16. Haraway 2007.
17. Svärd 2013.
18. Donaldson and Kymlicka 2011.
19. Smith 2001; O'Sullivan 2011.
20. Valentini 2014; Matarrese 2010.
21. Hadley 2005.
22. Sagoff 1984.
23. Ibid.
24. Dearden and Mitchell 2009, 269–70.
25. Mulrennan 1998, 66.
26. Ibid., 60.
27. See Belshaw 2001, 104
28. Quoted in Valpy 2010, A6.

Chapter 2

1. Dressel and Suzuki 2002, 225.
2. Ibid.
3. Schweitzer 2008, 133.
4. Hegel 1945, 41.
5. McCullers 2005, 150–1.
6. Feinberg, quoted in Goodpaster 2008, 158.
7. Goodpaster 2008, 161.
8. Gribbin and Gribbin 2009, 53.
9. Clark 1977, 112.
10. My criticisms of the dogma of human superiority gain independent support from a carefully reasoned essay by R. Routley and V. Routley showing the many logical weaknesses in

arguments for human-centred theories of environmental ethics: "Against the Inevitability of Human Chauvinism," in Goodpaster and Sayre 1979, 36–59.

11. For this way of distinguishing between merit and inherent worth, I am indebted to Vlastos's "Justice and Equality," 1962, 31–72.

12. Dearden and Mitchell 2009, 295.

13. Ibid., 300.

14. Ibid., 317.

15. Dressel and Suzuki 2002, 226.

16. Ibid., 227.

17. Ibid., 214.

18. Belshaw 2001, 139–40.

19. Ibid., 141.

20. Ibid., 144.

Chapter 3

1. Greenpeace website: www.greenpeace.org/ international/en/news/Blogs/Rex-Weyler/.

2. Leopold 1949, 15.

3. Callicott 2008a, 180.

4. Ibid., 181.

5. Mulrennan 1998, 108.

6. Hackett and Miller 2008, 221.

7. Hummel 1989, 268.

8. Hackett and Miller 2008, 221–2.

9. Parks Canada website: www.pc.gc.ca/eng/ progs/np-pn/ie-ei.aspx.

10. Mulrennan 1998, 105–6.

11. Descartes 1987, 356.

12. Horowitz, quoted in Belshaw 2001, 189.

13. Hettinger and Throop 2008, 188–9.

14. Ibid., 189.

15. Gribbin and Gribbin 2009, 57.

16. Dearden and Mitchell 2009, 93–4.

17. Regan 1983, 362.

18. Callicott 2008a, 183.

19. Ibid.

Chapter 4

1. Suzuki 2004b, 138.

2. See Heath 2001, chs 1–3.

3. Ackerman and Heinzerling 2004, 153–5.

4. Mulrennan 1998, 20.

5. Ibid., 20–2.

6. Ibid., 22–3.

7. Ibid., 23.

8. Jamieson 2006.

9. US Environmental Protection Agency 1989.

10. For discussion, see Sagoff 1984.

11. Krugman 2009.

12. Jamieson 1988b, 1990.

13. See Jamieson 1988a.

14. Borza and Jamieson 1990; Thomas et al. 2004.

15. McKibben 1989.

16. Andrews and Waits 1978; Jamieson 2002, ch. 15.

17. Weiskel 1990.

18. Bacon (1620) 1870; Locke (1690) 1952; Mandeville (1714) 1970; see also Hirschman 1977.

19. Leonard and Zeckhauser 1983.

20. Ackerman and Heinzerling 2004.

21. Kelman 1981.

22. Heath 2001, passim.

23. Hackett and Miller 2008, 609–18.

24. Mulrennan 1998, 27.

25. Sagoff 1981, 23.

26. Singer, Peter, 2010. "One Atmosphere," in Gardiner et al., eds, 2010, 186–7.

Chapter 5

1. Casey 2009, 72.

2. Light and Katz 1996, 5.

3. Weston 1996, 297.

4. Jamieson 2008, 74.

5. Norton 1996, 117.

6. See, for example, Williams 1985; Walzer 1984; and Warren 1991.

7. Stone 1974. G.E. Varner (1987) points out that the creation of new legal rights—as, for example, in the Endangered Species Act—helps expand what W.D. Lament calls our "stock of ethical ideas—the mental capital, so to speak, with which [one] begins the business of living." There is no reason that the law must merely reflect "growth" that has already occurred, as opposed to motivating some growth itself.

8. See Perelman 1982 and Perelman and Olbrechts-Tyteca 1969 for an account of rhetoric that resists the usual Platonic disparagement.

9. See Norton 1986a and Rodman 1983, 89–92. Remember also that Leopold insists that ethics are "products of social evolution" and that "nothing so important as an ethic is ever 'written'"—which again suggests that we ought to rethink the usual reading of Leopold as an environmental-ethical theorist with a grand criterion for ethical action.

10. Leopold 1949, 225.

11. Callicott 1990.

12. On "ecosteries," see Drengson 1990. On "re-inhabitation," a good starting point is Berg 1991.

13. See, for instance, Foreman 1987; Bookchin 1989; and Devall 1987.

14. Berry 1987, 13.

15. Snyder 1990; Birch 1990; Cheney 1987. Snyder also speaks of "grace" as the primary "practice of the wild"; Peacock 2001.

16. Alexander et al. 1977. On windows, see sections 239, 159, and 107; on "site repair," section 104; on water in the city, sections 25, 64, and 71; on "accessible green," sections 51 and 60; and on "holy ground," sections 24, 66, and 70.

17. MacEwen 2009.

18. Weston 1996, 299.

19. For a similar scenario, see Thompson 1996.

20. Thompson 1996, 191.

21. Callicott 2002, 12.

22. See ibid., 13. Norton presented his convergence hypothesis in *Towards Unity among Environmentalists* (1991).

23. Callicott 2002, 13.

24. See, for example, Dewey's "Theory of Valuation," in Dewey 1981, 189–251. Lewis offers a different but related set of reasons for rejecting the intrinsic/instrumental distinction as it is commonly employed in current environmental and applied ethics literature in *An Analysis of Knowledge and Valuation* (1946).

25. In the following discussion of this distinction, I draw upon arguments in my paper "Pragmatism and the Value(s) of Nature" (2009).

26. In what follows, I am drawing upon widely cited discussions by Korsgaard (1996, 249–74) and Kagan (1998, 277–97). Also, compare this with Rabinowicz and Rønnow-Rasmussen 1999, 33–52.

27. Woodland caribou are, for the most part, non-migratory. They live in old-growth forests rich in the lichens on which they feed, whose dense cover, if undisturbed, discourages predators such as wolves.

28. See Simon 1955, 1959. For recent discussions of satisficing versus maximizing or optimizing, see Byron 2004. Norton also employs a version in his *Sustainability: A Philosophy of Adaptive Ecosystem Management* (2005).

29. Others have recast Simon's "aspiration levels" as "stopping rules" and as "thresholds of expected utility." However, since the reason for doing this is to adapt satisficing to a utilitarian outlook, we can pass over these for present purposes. See Schmidtz 1996; Byron 1998; and Pettit 1984.

30. Callicott 2002, 13.

31. As in Katz 1996.

32. Thompson 1996, 203, 205, 206.

33. Mulrennan 1998, 45.

34. Ibid., 52.

35. Ibid., 53.

36. Ibid., 54.

Chapter 6

1. Desjardins 2001, 233–5.

2. Shiva 1998, 277.

3. Saul 2008, 292.

4. Bookchin 1987, ch. 9.

5. It may be that in contemporary Western society, which is so thoroughly structured by categories of gender, race, class, age, and affectional orientation, there simply is no meaningful notion of "value-hierarchical

thinking" that does not function in an oppressive context. For the purposes of this paper, I leave that question open.

6. See, for example, hooks 1984a, 51–2.
7. Lugones 1987, ch. 3.
8. Cheney 1987, 144.
9. Hare 1981.
10. For an ecofeminist discussion of the Chipko Movement, see my "Toward an Ecofeminist Ethic" and Shiva's *Staying Alive* (1988).
11. Cheney 1987, 122.
12. Levin and Levin 2001, 201.
13. Ibid., 200.
14. Ibid., 201.
15. Desjardins 2001, 254.
16. Reed 2005.
17. Desjardins 2001, 254.
18. Shiva 1998, 277.
19. Quoted in Desjardins 2001, 253.
20. From the essay reprinted in this chapter.
21. Merchant 1980; Shiva 1988; Plumwood 1991, 1993.
22. Glazebrook 2001.
23. Glazebrook 2000.
24. Glazebrook and Olusanya 2011.
25. Cf. Finn 1982.
26. From the essay reprinted in this chapter.
27. Shiva 1993, 289.
28. Waring 1988.
29. World Bank Group 2014.
30. Johnson-Odim 1991, 315.
31. Lugones and Spelman 1983; hooks 1984b.
32. Warren 1997.
33. Johnson-Odim 1991.
34. Diamond 1990.
35. Gibbs 1998.
36. SDWW 2013.
37. Social Watch Coalition 2010.
38. Shiva 1993; Kettel 1996.
39. Dwivedi et al. 2001.
40. Camacho 1998; Low and Gleeson 1998; Rechtschaffen and Gauna 2002.
41. Curtin 1999.
42. Rodda 1991; Warren 1992; Reardon 1993; Merchant 1995.

43. Allen 1986, 1990; Braidotti 1994; Shiva 1994; Ruether 1996; Low and Tremayne 2001; Doubiago 1989; Booth and Jacobs 1990.
44. Rao 2002.
45. Curtin 1999.
46. Shiva 1988; Bakker 1994; Venkateswaran 1995.
47. Dwivedi et al. 2001.
48. Rao 2002.
49. Bookchin 2002.
50. Hardin 1974.
51. Glazebrook 2001.
52. Sen 2002.
53. Zimmerman 1987, 34.
54. Salleh 1992, 195.
55. Salleh 1993, 225.
56. Slicer 1995, 151.
57. de Beauvoir 1952, 144.
58. de Beauvoir 1984, 103.
59. Glazebrook and Olusanya 2011.
60. Salleh 1997, 190.
61. Mallory 2009, 2010.
62. Warren 1990.
63. Warren 2000, 103–4.
64. Ibid., 109–13.
65. Ibid., 67
66. Buckingham 2004.
67. Zimmerman 1987, 44.
68. McMahon 2005, 133.
69. Ibid., 139.
70. Plumwood 1994, 252.
71. Ibid., 253.

Chapter 7

1. Moodie 2007, 28.
2. Quoted by Allen Carlson in his contribution to this chapter.
3. Hargrove 1979, repr. in Carlson and Lintott 2008.
4. Callicott 2008b, 106.
5. Hettinger 2005, 76.
6. Conron 2000, 17–18; a classic discussion is Hipple, Jr 1957.

7. The key works include Gilpin 1792; Price 1794; and Knight 1794, 1805; a standard treatment is Hussey 1927.

8. Bell 1913, 30.

9. Ibid., 45.

10. Although traditional aesthetics of nature is historically rooted in the picturesque and formalism, certain aspects of this view are defended in some recent work; for example, see Stecker 1997; Crawford 2007; and Leddy 2005; formal aesthetic appreciation of nature is defended in Zangwill 2001.

11. Callicott 2008b, 106.

12. Leopold, repr. 1966, 179–80.

13. Not all environmental thinkers share this concern; for example, see Brook 2008.

14. Some of these criticisms have been noted since the beginnings of the renewed interest in aesthetics of nature; see for examples Sagoff 1974 and Carlson 1977.

15. I consider seriousness and objectivity to be requirements for the adequacy of aesthetics of nature in general; see Carlson 2007.

16. Rees 1975, 312.

17. Godlovitch 1994, 16, repr. in Berleant and Carlson 2007.

18. Ibid.; Godlovitch's acentrism reflects some of the ideas in Nagel 1986.

19. On wetlands, see Carlson 1999; Rolston III 2000; and Callicott 2003.

20. Saito 1998a, 101, repr. in Carlson and Lintott 2008.

21. Rolston III 1998, 162; repr. in Berleant and Carlson 2007.

22. Callicott 2008a, 108–9.

23. Hepburn 1966, 305, repr. in Berleant and Carlson 2007.

24. It can be argued that formalism underwrites a degree of objectivity of aesthetic value; see Parsons 2008, 41–3.

25. Hettinger 2008, 414.

26. Thompson 1995, 292, repr. in Carlson and Lintott 2008.

27. Ibid., 292.

28. Carroll 1993, 257, repr. in Berleant and Carlson 2007.

29. Rees 1975, 312.

30. Andrews 1989, 59.

31. Matthews 2002, 38, repr. in Carlson and Lintott 2008. Discussions concerning bringing aesthetic appreciation and moral obligation in line with one another include Nassauer 1997; Eaton 1997; and Lintott 2006; all repr. in Carlson and Lintott 2008; on this same topic, although more focused on human environments, is Saito 2007.

32. Other prominent non-cognitive approaches include the arousal model championed by Carroll and the mystery model defended by Godlovitch. Carroll holds that we may appreciate nature simply by opening ourselves to it and being emotionally aroused by it, without invoking any particular knowledge about it; see Carroll 1993. Godlovitch contends that neither knowledge nor emotional attachment yields appropriate appreciation of nature, for nature itself is ultimately alien and unknowable and thus the appropriate experience of it involves a sense of mystery; see Godlovitch 1994; see also Godlovitch 1998, repr. in Carlson and Lintott 2008.

33. See Berleant 1992, esp. ch. 11, "The Aesthetic of Art and Nature," repr. in Berleant and Carlson 2007; and Berleant 1997, 2005.

34. Berleant 2005, 169–70.

35. Saito 1998b.

36. See Carlson 1979, repr. in Berleant and Carlson 2007; Carlson 1998, repr. in Carlson and Lintott 2008; and Carlson 2000.

37. Other cognitive approaches emphasize other kinds of information, such as derived from history, folklore, or mythology; see, for example, Saito 1998b; Sepänmaa 1993; and Heyd 2001, repr. in Berleant and Carlson 2007.

38. For a classic illustration of this difference, see Muir 1894, repr. in Carlson and Lintott 2008.

39. Godlovitch explicitly challenges this claim in "Icebreakers: Environmentalism and Natural Aesthetics."

40. See Carlson 1981; for follow-up, see Parsons 2006.

41. See the discussion of *Bambi* in Eaton 1998, repr. in Berleant and Carlson 2007.

42. Other philosophers also suggest that non-cognitive and cognitive approaches are not necessarily in conflict; for example, in presenting his arousal model, Carroll remarks, "In defending this alternative mode of nature appreciation, I am not offering it in place of Carlson's environmental model [aka scientific cognitivism]. . . . I'm for coexistence" (Carroll 1993, 246).

43. Berleant seemingly would not accept this conclusion, for he apparently holds not only that a cognitive component is not necessary for appropriate aesthetic experience, but also that engagement is both necessary *and sufficient* for such experience. I argue that engagement is not sufficient in Carlson 2006a.

44. Rolston III in Carlson and Lintott 2008, 337; for an overview of Rolston's aesthetics, see Carlson 2006b; another constructive attempt to combine cognitive and non-cognitive approaches is Moore 2008.

45. Some of the points made in this essay are treated in more detail in the introduction to Carlson and Lintott 2008. A longer version of the essay with the title "Contemporary Environmental Aesthetics and the Requirements of Environmentalism" appears in *Environmental Values* 19 (2010): 289–314.

46. Kant, repr. 1952, 157.

47. For an account of the controversy between separatists and non-separatists, see Moore 1995.

48. This aspect of Eaton's defence of value holism appears in Eaton 1989, 168–79; she joins this with a full-scale attack on separatism in Eaton 2001.

49. Brady points out the way in which our language both tracks and reveals the cross-fertilization of aesthetic and moral values in Brady 2003, 255.

50. J. Thompson has argued that aesthetic value can, by itself, provide a justification for the environmental preservation ethic on the ground that the recognition of this value in some natural objects involves a transcending of the individual, subjective perspective, and the acquiescence in a wider objectivity; Thompson 1995; Parsons discusses Thompson's position at some length elsewhere in this chapter.

51. Carlson has argued powerfully in favour of this position, which he calls the Natural Environmental Model of the aesthetic appreciation of nature, in Carlson 2000; Carlson and Parsons 2008; Carlson 2009; and numerous other articles.

52. Carlson 2000, 12–13.

53. Kant, repr. 1952, 56.

54. Rolston III, for instance, maintains that beauty judgments are both too narrow and too weak to provide the moral imperatives needed to ground the ethical imperatives respecting nature unless they are expanded in such a way as to encompass "appropriate respect," a respect that entails duties; Rolston III in Carlson and Lintott 2008, 338.

55. Taylor 2000, 107.

56. If, as some environmental philosophers maintain, the imperatives that count most toward environmental subjects transcend the anthropocentric perspective altogether, our moral duty toward natural objects may be taken to extend beyond, and detach from, all man-made aesthetic judgments. The most forceful exposition of this view is to be found in Godlovitch 1994.

57. Eaton 2001, 111.

58. Brady 1998, 143.

59. Shelley 1891.

60. For a more extended presentation of the notion of growing up morally and aesthetically, see my *Natural Beauty: A Theory of Aesthetics Beyond the Arts* (2008), ch. 9.

61. I defend this view in Moore 2008.

62. Gaita 2002.

63. Rolston III 2002, 127, repr. in Carlson and Lintott 2008.

64. A number of philosophers have defended some version of aesthetic preservation as a sound basis for protecting nature. Rolston himself argues that environmental ethics

needs some concept of aesthetic value in order to be "adequately founded" (Rolston III 2002, 140, repr. in Carlson and Lintott 2008). Leopold included aesthetic considerations in his influential "Land Ethic," claiming that "a thing is right when it tends to preserve the integrity, stability, and beauty of the biotic community" (Leopold, repr. 1966, 262). Carlson defends aesthetic preservation in advocating a version of what he calls "the eyesore argument": certain forms of human intrusion into nature make it aesthetically worse; therefore, we ought not to engage in these forms of intrusion (Carlson 2000, ch. 9). Other philosophers have defended stronger versions of aesthetic preservation. Hargrove's *The Foundations of Environmental Ethics* (1989) gives pride of place to the aesthetic value of nature: according to him, the aesthetic value of nature provides the best motivation we currently have for preserving it. Sober, after considering various other reasons for preserving nature, settles on its aesthetic value as the most defensible (Sober 1986).

65. The objectivity of aesthetic value is maintained by a number of contemporary views on the aesthetics of nature; for discussion, see Carlson (reading in this chapter, section III.4).

66. This case is described in Lee 1995.

67. As another example, think of a rare bird species on the verge of extinction due to some natural cycle of disease or a natural decline in its food supply. If the species were particularly impressive aesthetically, one might argue that we ought to intervene to protect it, by providing food for it, perhaps, or inoculating it against potentially fatal diseases. A similar example is discussed in Godlovitch 1989, 175.

68. This point of view is developed in Godlovitch 1989 and in Lee 1995.

69. Loftis 2003.

70. Ibid., 43.

71. Sober 1986, 194.

72. This example, which involved Hetch Hetchy Valley in Yosemite National Park, is discussed in Saito 1998b, 146.

73. Saito 1984, 42.

74. The recent work of the photographer Edward Burtynsky demonstrates powerfully (and perhaps unintentionally) the potential aesthetic value of industrial landscapes; on Burtynsky's works, see Pauli 2003.

75. Hettinger 2005, 72.

76. This dilemma is a version of Beardsley's "dilemma of aesthetic education," which he presented in "The Aesthetic Point of View" (1970, repr. 1982); the dilemma is also applied to nature in Carlson 2000, ch. 9.

77. Thompson 1995.

78. Ibid., 304.

79. Ibid.

80. Thompson's list raises other potential problems as well. One of these is that it is unclear whether some of the features on it are truly criteria of *aesthetic* value. Consider the fourth factor, for instance: the capacity to put things into perspective. There is no doubt that helping us to put things into perspective is a valuable characteristic, but it does not seem clearly tied to the perceptual appearance of things in the way aesthetic qualities usually are. Plainly, lots of things can have this capacity without being aesthetically good at all: a slap in the face, for example. This raises the worry that Thompson is unduly stretching the concept of the aesthetic in order to bolster the case for the aesthetic superiority of nature.

81. Hettinger 2005, 75.

82. Saito 1984, 45.

83. Hettinger 2005, 75.

84. For further discussion of these and other philosophical issues relating to aesthetic preservation, see Parsons 2008, ch. 7.

85. Moore 2008.

Chapter 8

1. Dickason 2006, 264.

2. Westra 1997, 277.

3. Ibid., 281.

4. Draper and Reed 2009, 25.

5. Berry 1988, 189–90.

6. Sapa (Black Elk) 1975, 26.
7. Remarks by Peter Ochees, Anishinaabe (Ojibwa) holy man, during a time devoted to holy rites. Author's personal files.
8. Brinton 1868, 20.
9. Ibid., 6.
10. Clastres 1974, 28.
11. Brinton 1868, 21.
12. See Cohen 1952.
13. Bachofen 1983, 43.
14. Krech III 1978, 722.
15. Bachofen 1903, 66–7.
16. Lafitau 1974, 69.
17. Ibid., 343–4.
18. Bachofen 1903, 37–8.
19. Brown 1970, 164.
20. Bachofen 1903, 64–5.
21. Georgina Tobac (Athabascan) 1979.
22. Dearden and Mitchell 2009, 19.
23. Callicott 2008b, 181.
24. Krech III 1999, 14–16.
25. See Booth and Jacobs 1990.
26. See Callicott 1989.
27. Hughes 1996, 22.
28. Ibid., 50.
29. Gill 1987, 5.
30. Ibid., 66.
31. Ibid., 14.
32. Rousseau 1964, 151.
33. Lahontan 1703, 245–6.
34. Thwaites 1896.
35. Arneil 1996, 26.
36. Trigger and Washburn 1996.
37. Rousseau 1964, 151.
38. Ibid., 128.
39. Booth and Jacobs 1990.
40. Sioui 1992, 9.
41. Gill 1987, 100–2.
42. Borrows 1997, 429.
43. Johnston 1976.
44. Ibid., 28.
45. Alfred 1999, xiv.
46. Fenton n.d.; Skye 1998; Wallace 1958.
47. Jennings 1971, 1975, 1988.
48. Lovelock 1987.
49. Sapontzis 1984.
50. Levins and Lewontin 1985.
51. Grande 1999, 308.
52. Ibid., 309.
53. Ibid., 312.
54. Wilson 2002, 94–5.
55. Ibid., 94.
56. Turner, Ignace, and Ignace 2000, 1278.
57. Ibid., 1282.
58. Berkes, Colding, and Folke 2000, 1255.
59. Ibid., 1257.
60. Ibid., 1259.
61. Quoted in Atwood 1972, 118.
62. Atwood 1972, 123.
63. Saul 2008, 88.

Chapter 9

1. Ehrlich and Ehrlich 2009, 223.
2. Leopold 1949, 210.
3. Aristotle 1947, 1106b17–24.
4. Moore 1903, 1912.
5. Moore 1959, 95–7.
6. See, for example, Nowell-Smith 1954.
7. Suzuki 2004a, 42.
8. Leopold 1949, 200.
9. Carson 1962, 118.
10. Vaillant 2006, 19.
11. Ibid., 135.
12. Ibid., 136.
13. McKibben 2011, 191.
14. See, for example, Kvanvig 1992; Zagzebski 1996; and Greco and Turri 2012.
15. Ostrom 1990.
16. Jamieson 2007.
17. Ibid.
18. Kovel 2007, 84.
19. See Roberts and Wood 2007, 157–8.
20. Wijkman and Rostrōm 2012, loc. 3747.
21. Rolston III 2005, 68.
22. Sandler 2007, 115.
23. Ibid., 127.
24. Ehrlich and Ehrlich 2009, 243.
25. Sandler 2007, 125.
26. Draper and Reed 2009, 208.
27. Dearden and Mitchell 2009, 360.
28. Draper and Reed 2009, 208.

29. Dearden and Mitchell 2009, 360.
30. Ibid., 361–2.
31. Sandler 2007, 82.

Chapter 10

1. Jackson 2012, xv.
2. Quoted in Bookchin 2005, epigraph.
3. Bookchin 2005, 65.
4. Ibid., 77.
5. Turner, Ignace, and Ignace 2000, 1279.
6. Brody 2000, 255.
7. Ibid., 256.
8. Ibid.
9. Ibid., 117.
10. Bookchin 2005, 129.
11. The legitimacy of violence as a measure is controversial. We differentiate between aggressiveness as a legitimate form of political action and violence that involves instrumental utilization of other people to promote your cause.
12. In previous years, environmental education was more closely related to environmental activism in that environmental educators were activists and the students were urged to take part in activism themselves. It seems that in recent years environmental education is being incorporated into the standard curricula and aims both in the sciences and in the humanities to make students more informed about their own actions.
13. This typology is in line with common differentiations made by various authors, though under different terms. Arne Naess (1995) distinguishes between shallow and deep ecology; Dryzek (2005) differentiates between sustainability and green radicalism; and, similarly, Dobson (2007) discerns between environmentalism and ecologism.
14. Dryzek 2005, 143–80.
15. Characterizing ecologism as radical may be misleading. It does not necessarily imply that ecologism activists will utilize radical activism. Rather, "radical" in this sense means getting down to the root of things, searching for the primary and most essential reasons and points of impact.

16. Dobson 2007; Dryzek 2005, 180–228.
17. For more information on Pinchot's legacy and examples of this type of policy, see the Pinchot Institute for Conservation website: www.pinchot.org/.
18. For more information on Muir's legacy and examples of the types of action undertaken by the Sierra Club, see the club's website: www.sierraclub.org/.
19. This category takes non-human species as a general group. It does not reflect attitudes and justifications for animal rights and neither does the suggested typology as a whole. For a comprehensive introduction to animal rights, see Chapter 1.
20. Leopold 1966, 262.
21. Devall and Sessions 1985, 67–9; Naess 1995, 4.
22. For an elaborate discourse analysis of views that blindly support economic development, see Dryzek 2005, 51–72.
23. For an elaborate discourse analysis of views that blindly reject economic development and technology but have a romantic appreciation of nature, see Dryzek 2005, 183–202.
24. Dobson and Bell 2006; Dobson 2003.
25. Light 2003.
26. Bookchin 1980, 83.
27. de-Shalit 2001.
28. Margalit 2010.
29. Bookchin 2005, 131.
30. Westra 1994.
31. Westra et al. 2000.
32. Ibid., 3.
33. See Westra 1998, ch. 8.
34. See Karr 1981; J.R. Karr (1996) in Schulze 1981.
35. See Shrader-Frechette and McCoy 1994.
36. Bosselmann 2008.
37. Ibid., 32; see also Wackernagel and Rees 1996.
38. Rees and Mickelson 2003.
39. Westra, 1998; see also Westra, Miller, et al. 2000, 33–4.
40. O'Neill 1996, 175.
41. Ibid., 176.

42. O'Neill 1996, 177.
43. Pimentel et al. 2000; Noss and Cooperrider 1994; Westra 1998.
44. O'Neill 1996, 181.
45. Pogge 2001, 15.
46. Gewirth 1982.
47. Ibid., 5.
48. Ibid., 7.
49. Ibid., 5.
50. Beyleveld, Deryck, and Brownsword 2001, 71.
51. Shue 1996.
52. Beyleveld, Deryck, and Brownsword, note 27, p. 70; Gewirth 1982, 54n23.
53. Fluker 2013, 23.
54. Parks Canada, Canadian National Parks Act, SC 2000, c. 32.
55. Fluker 2013.
56. Minister of Canadian Heritage 2001, 1123.
57. Fluker, 2013, 25.
58. Minister of Canadian Heritage 2001, 1426.
59. Canada's Constitution Act 1982, 35 (1), states: "The existing aboriginal and Treaty rights of the aboriginal peoples of Canada are here recognized and defended."
60. Westra 2007, esp. chs 1 and 2.
61. Westra 2011.
62. Frasz 2005.
63. Wolf 1998, 71.
64. Ibid., 72.
65. Ibid., 76.

Chapter 11

1. IPCC 2013, press release, www.ipcc.ch/news_and_events/docs/ar5/press_release_ar5_wgi_en.pdf.
2. Garvey 2008, 28.
3. Archer 2009, 4–3.
4. Friel 2010, 86.
5. Quoted in Friel 2010, 87.
6. Hardin 1986, 1245.
7. This is an update of "Amerigenic Climate Change: An Indictment of Normalcy," which appeared in the first edition of this textbook. Revising this chapter turned out to be so extensive that the result is a largely new essay. For their meticulous research assistance in collecting and comparing the relevant datasets for this update, I would like to thank John Voelpel, Philosophy Department, and Alex Miller, Patel College of Global Sustainability, both at the University of South Florida.
8. Armstrong 2009.
9. Data were posted in 2009 at the old site of the UN Statistics Division (UNSD). These data were annually updated; the official 2009 data were based on the fall 2008 update, which integrated data gathered in 2006. The data in that form are not available anymore, as UNSD redesigned its site. The new site is *UNDATA* at http://data.un.org/Default.aspx.
10. Measured in millions of tons of CO_2 in 2006, China emitted 6103, the US 5975, and Russia 1578 units. For comparison, Germany's 2006 emissions were 880 and Canada's 560 million tons of CO_2.
11. Measured in tons of CO_2 per capita in 2006, Qatar emitted 56, United Arab Emirates 33, and Kuwait 31 units. For comparison, Germany's 2006 per capita emissions were 11 and Canada's 17 units.
12. "China overtakes US in greenhouse gas emissions," *New York Times*, 20 June 2007.
13. Cf. UNSD, note 9 above.
14. The population of the People's Republic of China was 1.331 billion in 2009; cf. Population Reference Bureau, *2009 World Population Data Sheet* (Washington, DC: PRB/USAID, 2009).
15. The US population was 305.8 million in January 2009 and 308.2 million in December 2009; cf. US Census: historical data: vintage 2009: national tables; www.census.gov.
16. In 2002–6, Chinese CO_2 output nearly doubled (from 3.3. billion to 6.1 billion). Cf. UNSD, note 9.
17. den Elzen et al. 2013, table 1, 402.
18. Gore 2006, 250; see also UNSD, note 9.
19. "Obama Affirms Climate Change Goals," *New York Times*, 18 November 2008; "Climate

Change and President Obama's Action Plan," www.whitehouse.gov/climate-change.

20. "House Votes to Deny Climate Science and Ties Pentagon's Hands on Climate Change," *Think Progress,* 22 May 2014.

21. "Solar Market Set to Ignite in 2014, Led by China," *Climate Group,* 27 February 2014.

22. "China Targets 70 GW of Solar by 2017," *pv-Magazine: Photovoltaic Markets and Technology,* 16 May 2014.

23. "Australia Worst Carbon Emitter Per Capita among Major Western Nations: Country Has Failed to Consistently Decrease Its Emissions, Faring Poorly in a Global Climate Report," *The Guardian,* 19 November 2013.

24. Phillips 2007, 220.

25. Garvey 2008, 115.

26. The David Suzuki Foundation released a report in 2009 showing how Canada *could* meet the 2020 target with no significant hit to the economy. The report was ignored by the government. See Bramley et al. 2009.

27. Gardiner 2006.

28. Blair 2004.

29. Ibid.

30. Kartha et al. 2010.

31. Lovelock 2009, 27–9.

32. Jamieson 2007, 167.

33. Williston 2011, 158. The numbers cited in the first paragraph are from the report mentioned in note 26, above.

Chapter 12

1. Rubin 2009, 136.

2. Hackett and Miller 2008, 171.

3. Wolf 2008, 434.

4. Graeme Smith, *Globe and Mail,* 23 July 2010, F5.

5. Ehrlich 1968, Prologue.

6. Hughes 2009, 66.

7. Assadourian, Gardner, and Sarin 2008.

8. Rubin 2009, 136.

9. Meadows et al. 1974, vi.

10. Ibid., 265.

11. Sheldon Lambert, quoted in *Newsweek,* 27 June 1977, 71.

12. Jaccard 2009, 112.

13. Kasun 2008, 406.

14. Hackett and Miller 2008, 285–6.

15. Sachs 2008, 73–4.

16. Kasun 2008, 404.

17. Hughes 2009, 90–1.

18. See Rubin 2009, 150. Rubin seems to have been wrong about where prices were going, but he may turn out to be correct in the long run.

19. Jaccard 2009, 114.

20. Marsden 2009, 159.

21. Draper and Reed 2009, 451–2.

22. Hughes 2009, 94.

23. Jaccard 2009, 110.

24. Willott 2002.

25. Sachs 2008, 188.

26. Ibid., 187.

27. Ibid., 188.

Chapter 13

1. Hackett and Miller 2008, 235–6.

2. Ibid., 237.

3. Ibid., 238.

4. Ibid., 236.

5. Ibid., 248.

6. Wilson 2002, 20.

7. *Greenpeace Magazine* 11, no. 6 (Spring/ Summer 2010): 6–7.

8. Wilson 2002, 60.

9. Dearden and Mitchell 2009, 248–9.

10. Draper and Reed 2009, 93.

11. Dearden and Mitchell 2009, 248–9.

12. Wilson 2002, 50–1.

13. Hackett and Miller 2008, 240.

14. In Gould 1992.

15. Ibid.

16. Ibid., 110.

17. Nepstad 2007.

18. Hackett and Miller 2008, 166.

19. Norton 1986b, 112–13.

20. Tilman and Downing 1994.

21. Wilson 2002, 108.

22. Hackett and Miller 2008, 155.

23. Ehrlich and Ehrlich 1982, xi–xvi.

24. Norton 1986b, 122.
25. Wilson 2002, 119.
26. Ibid., 113.
27. Parts of this essay are adapted from Sandler 2012.
28. IPCC 2014.
29. Foden et al. 2008; Sinervo et al. 2010; Thomas et al. 2004.
30. IPCC 2007, 54.
31. Baillie et al. 2004; Magurran and Dornelas 2010.
32. Sandler 2012.
33. Minteer and Collins 2010.
34. Sandler 2010, 2013.
35. Camacho et al. 2010, 21.
36. Sandler 2007.
37. Hansen 2009, 146.
38. Deardon and Mitchell 2009, 221.
39. Hansen 2009, 146.
40. Government of Canada.
41. Hackett and Miller 2008, 248–9.
42. AAFC 2015.

Chapter 14

1. Dearden and Mitchell 2009, 11.
2. Quoted in Dearden and Mitchell 2009, 10.
3. World Commission on Environment and Development 1987, ix.
4. Dearden and Mitchell 2009, 12.
5. Ibid., 14.
6. Quoted in Dearden and Mitchell 2009, 15.
7. Quoted in Draper and Reed 2009, 562.
8. Dearden and Mitchell 2009, 17.
9. Draper and Reed 2009, 21.
10. Ibid., 22.
11. Ibid., 23.
12. Beckerman 1994, 194.
13. O'Rioden 1988, 37.
14. Buttel 1988, 261.
15. Holland 2003, 392.
16. Costanza, Daly, and Prugh 2000, 3–4.
17. Sumner 2005, 76.
18. Ekins and Max-Neef 1992, 402.
19. van der Ryn and Cowan 1996, 58.
20. CH2MHILL 2005, 15.

21. Cook 2004, 14.
22. For examples, see the following: Fraser Basin Council 2004 and State of Washington 2003.
23. For a brief discussion of this, see Keiner 2004, 382–3.
24. See Keiner 2004; Cook 2004, ch. 6; and Edwards 2005. For an account of social justice, in terms beautifully compatible with my account of sustainability, which illustrates the complex and controversial nature of the matter, see "the capabilities approach" to a "basic social minimum" being developed in Nussbaum 2006, 69ff.
25. Ball 1992, 14.
26. See Sumner 2005, ch. 4; Edwards 2005, chs 2 and 6.
27. The subtitle of his book: *Portrait of a Paradigm Shift*.
28. Draper and Reed 2009, 26.
29. See Kazez 2007.
30. Sachs 2008, 32.
31. Beckerman 1994, 203.
32. Ibid., 195.
33. Pearce 1993, 48.
34. Beckerman 1994, 196.
35. Daly 1995, 51.
36. Ibid., 54.
37. Government of Canada, quoted in Draper and Reed 2009, 26.
38. Draper and Reed 2009, 525.
39. Ibid., 527.
40. Ibid., 528.
41. Ibid.
42. Ibid., 528–48.
43. Ibid., 528–31.
44. Ibid., 531.
45. United Nations 2009.
46. World Commission on Environment and Development 1987.
47. Stefanovic 2000; Brown and Toadvine 2003.
48. Hawke Baxter et al. 1995.
49. Heidegger 1966.
50. Ibid.
51. Casey 1993.
52. Heidegger 1971.
53. Casey 1993, 103.

54. Norberg-Schulz 1984, 12.
55. Heidegger 1971, 156–7.
56. Cited in Bachelard 1964, 211.
57. Relph 1976; Mugerauer and Seamon 1989; Stefanovic 2000.
58. Casey 1993.
59. Pallasmaa 2005, 40.
60. Ibid., 40.
61. Bachelard 1964.
62. Ibid., 47.
63. Heidegger 1962, 98.
64. Heidegger 1982, 165.
65. Cited in Stefanovic 2000, 6.
66. Foltz 1995, 4.
67. Stefanovic 2000.
68. Marx 1992; Oelschlaeger 1995; Smith 2001.
69. Cited in Stefanovic 2000, 120.
70. Stefanovic 2000, 149ff.
71. Cited in ibid. 2000, 153.
72. Holland, Light, and O'Neill 2008, 2–3.
73. Crist, quoted in Gribbin and Gribbin 2009, 186–7.
74. Pojman 2008, 725.
75. Quoted in Brand 2009, 32.
76. Brand 2009, 36.
77. Quoted in Brand 2009, 37.
78. Ibid., 59.
79. Ibid.

Glossary

acidification When ocean water absorbs CO_2, carbonic acid is produced. This acid releases hydrogen ions, which lower the water's pH level. One effect of high levels of carbonic acid in the oceans is that coral reefs are threatened because they cannot build their calcium skeletons in this water. Acidification is increasing in the world's oceans because of anthropogenic climate change.

act-utilitarianism A version of utilitarianism according to which an act is right if and only if it produces the best consequences (or maximizes utility) for all those affected by it.

adaptation A response to the effects of climate change. The focus is on making changes—to our social relations, our infrastructural capacities, and so on—in an effort to deal with the challenges that we face. Example: building higher sea walls in face of the threat of rising sea levels.

agroforestry A cluster of agricultural techniques that challenge the dominant agribusiness approach to growing food. Example: allowing trees to grow among agricultural crops, an approach that runs counter to large-scale monocultural agriculture.

animal welfarism The view that the primary bearers of value are individual animals (rather than species or ecosystems).

anthropocentrism The view that we need look only to what humans value to discover all our duties.

anthropogenic climate change Climate change caused by human activities, especially the burning of fossil fuels and deforestation.

applied ethics The effort to apply the insights of meta-ethics and normative ethics to concrete questions in our business, environmental, medical, or other professional practices.

arithmetical growth Growth that proceeds incrementally. For example, the sequence $1, 2, 3, 4 \ldots n$ describes an arithmetical growth trajectory.

autonomy The capacity to act on the basis of self-legislated reasons.

background rate of extinction The rate at which species become extinct on a natural basis—i.e., without significant human interference in the process.

between-habitat complexity The lack of similarity among the populations of organisms in contiguous ecosystems or habitats.

bioaccumulation The build-up of chemical substances in organisms in ever greater quantities as they are passed up the food chain. Also known as "biomagnification."

biocapacity The amount of available productive land and water area per capita.

biocentric egalitarianism The view that the interests of all livings things have equal positive moral weight.

biomimesis Taking successful processes and patterns in nature as models for solving problems arising in human systems.

biocentrism The view that being alive is necessary and sufficient for moral standing.

biosphere The zone of the Earth containing life. The totality of interconnected ecosystems on Earth.

carrying capacity The maximum number of individuals of a species that can be sustained indefinitely in a given physical space.

categorical imperative A command that binds an agent absolutely, or with no exceptions. Example: "always pay back your debts." A key feature of Kant's ethics, in which it appears in three distinct, but related, versions. Contrast "hypothetical imperative."

consequentialism The view that what matters morally are the outcomes, or consequences, of our actions, rather than the actions per se or the "mental state" of the agent performing them.

conservationism The view that we ought to keep wild spaces relatively pristine but that this goal is compatible with some use by humans of such spaces for economic gain.

consumer class The class of people whose income is equivalent to at least US$7000. The consumer class currently totals around 1.7 billion people worldwide.

contractualism The view that a right action is one that conforms to the terms of a duly constituted agreement among agents who view one another in terms of mutual respect.

coral bleaching When water temperature is too high, corals are killed or "bleached." Anthropogenic climate change is leading to increased bleaching of corals in the world's tropical and subtropical oceans.

cost-benefit analysis (CBA) A procedure that weighs the costs of a project against its benefits to determine whether the project should go forward. Usually this is done entirely in monetary terms.

deep ecology A development of ecocentrism according to which all natural things and systems are worthy of respect. A key element of the view is that humans can achieve self-realization by identifying with nature.

deontology The view that a right action is one that conforms to the dictates of duty. As with its major philosophical proponent, Immanuel Kant, it is usually a morality of principles.

descriptive claim A claim about the way the world is.

discounting An economic theory concerning the value of future goods as determined by the amount we should invest in them in the present. The idea is to invest only so much in future goods as is required to produce the sum we need at the appropriate date, given the amount we can expect to get in interest on our investment.

dynamic equilibrium The characteristic of a system that is neither purely chaotic nor in full stasis (equilibrium).

ecocentrism The view that ecosystems, and perhaps the totality of ecosystems—the biosphere—possess moral standing.

ecological economics An approach to the discipline of economics that emphasizes full-cost pricing and denies that economies must perpetually grow.

ecological footprint analysis An attempt to measure the impact humans have on a specific bioregion by determining the amount of land and water required to sustain their level of consumption and waste disposal (including the greenhouse gases these activities produce). The analysis can be run for individuals or groups.

ecosystem A collection of biotic (living) and abiotic (non-living) things existing and causally interacting in a certain region. The borders of an ecosystem can be difficult to define because its internal causal interactions are *both* relatively closed from those of contiguous ecosystems *and* causally connected to the latter.

ecosystem-based management (EBM) A way of managing ecosystems that focuses on species diversity and the integrity of the system as a whole rather than on this or that resource within the system.

endangered A species nearing extinct status.

environmental pragmatism The view that philosophers need to focus less on the attempt to solve theoretical issues in environmental ethics and think instead about the best ways to solve real and pressing environmental problems.

environmental racism The attempt to discriminate against a racial minority through the environment—for example, by locating toxic waste sites near their property.

ethical egoism The view that, for each of us, it is right to perform those actions that fulfill our own desires and interests regardless of the effect such actions have on others.

ethical relativism The view that moral principles are made true by the fact that they are the product of a particular culture or society. On this view, there can be no transcultural moral standards, and we all have a duty to tolerate the moral beliefs and practices of other cultures.

eutrophication A process by which a body of water is over-nourished with fertilizers, thus causing an explosion of algae-growth. The problem is that when the algae dies, its bacterial decomposers consume an inordinate amount of the water's oxygen, thus killing other oxygen-dependent species. This can create "dead zones" in lakes and oceans.

exemplar theory of moral education In virtue ethics, the idea that the best way to learn how to become a moral person is to "apprentice" with someone who lives a virtuous life.

extirpated A species suffering local, though not global, extinction.

extrinsic value The value possessed by an entity relative only to the valuations of other entities. Derivative value.

foundational species Organisms that play a pivotal role in building and maintaining the diversity and stability of ecosystems.

full-cost pricing The idea of ensuring that prices really do reflect costs, including all environmental costs.

Gaia Theory The view, propounded by James Lovelock, that the Earth is a self-regulating super-organism.

genetically modified (GM) crops Crops—such as rice, wheat, and corn—that have been altered at the level of their genes to make them resistant to things like drought and pests.

geometrical growth Growth that proceeds non-incrementally. For example, the sequence 1, 2, 4, 8, 16 . . . *n* describes a geometrical growth trajectory. Also known as "exponential growth."

HIPPO A way of highlighting the key anthropogenic causes of biodiversity loss. The five key factors are: **H**abitat destruction; **I**nvasive species; **P**ollution; **P**opulation; **O**verharvesting.

hypothetical imperative A command that binds an agent only contingently. Example: "If you want to gain the favour of the powerful, do whatever they say." Insofar as the "if-clause" fails to apply to the agent, the command is not binding on him or her. Contrast "categorical imperative."

impermissible That class of actions (or omissions) that are morally disallowed, or wrong.

inertia Resistance to change within a system. Ecosystems can display varying degrees of inertia.

insurance principle of biodiversity The idea that some redundancy among an ecosystem's functions—i.e., biodiversity within the system—is a good thing because it makes the system as a whole better able to cope with disturbances. That is, it increases the system's inertia and resilience.

Intergovernmental Panel on Climate Change (IPCC) A body created by the United Nations Environment Programme and the World Meteorological Organization in 1988. The body is comprised of scientists and politicians from around the world whose job it is to assess the nature, extent, and impact of anthropogenic climate change. They conduct no research of their own, just gather and synthesize the literature already out there. The IPCC has released five reports: in 1991, 1995, 2001, 2007 and 2013.

intrinsic value The value possessed by an entity independently of the valuations of other entities. Non-derivative value.

IPAT formula A way of determining the human impact on the environment. It is: **I**mpact = **P**opulation × **A**ffluence × **T**echnology.

land ethic The view, first formulated by Aldo Leopold, that an action is right to the extent that it protects or promotes the stability, integrity, and beauty of the "land" (by which Leopold meant all of nature's organisms and processes). The inspiration for ecocentrism.

market failure An outcome in a market exchange in which an item's price fails to be a true reflection of its cost. In a perfect system of transfers, prices always reflect costs precisely. Prominent market failures are monopolies (which lead to under-production of goods) and negative externalities (which lead to overproduction of goods).

maxim In Kant's ethics, a principle of action that has the form of a universal moral law. Example: "Whenever anyone is strapped for cash, he or she should borrow money with no intention of paying it back."

meta-ethics The branch of philosophical ethics that studies abstract questions. These include the nature of ethical assessment, the meaning of key words like "good" and "right," challenges to morality from ethical relativism, ethical egoism, and more.

mitigation Responding to the problem of climate change by attempting to lessen the severity of some of its effects. Example: reducing carbon emissions.

moderate sustainability A position midway between weak and strong sustainability. It is stronger than weak sustainability because it rejects the principle of unlimited substitutability between natural and man-made capital. However, it is weaker than strong sustainability because it allows for the possibility that sometimes it will be necessary to provide man-made capital to people at the cost of degrading some aspect of their environment.

moral scepticism The view that there are no moral truths, either because all moral sentences are false or because there is nothing for them to be "about."

moral standing A status that confers positive moral weight on an entity's existence and fundamental interests. To have positive moral weight means that other agents have duties to constrain themselves in specific ways when dealing with such entities.

naturism The view that nature is there to be dominated by humans.

negative externality A market failure that occurs when the production of a good imposes costs on society that are not reflected in the price of the good. For example, if the production of a certain good causes pollution as a by-product and the pollution causes people to get sick, the cost of their sickness—to themselves and to the health care system—is generally not reflected in the price of the good. The result is that the good is cheaper than it should be and the market will therefore overproduce it.

negative right The right not to be interfered with in particular ways by other individuals or governments. Example: the right to free speech.

neo-liberal economics The view that there should be as few governmentally-imposed constraints on the free market as possible. Regarding environmental issues, the view implies that (a) natural entities like trees have value only as economic resources and (b) the free market determines the real and precise value of such resources (see entry for "cost-benefit analysis"). Most neo-liberal economists are also committed to the "substitutability principle" (see entry for this term).

non-zero-sum game A "game" in which there can be many winners. Example: a good conversation.

normative claim A claim about the way the world ought to be.

normative ethics The branch of philosophical ethics that deals with the nature of our duties. Prominent examples of normative theories are utilitarianism, deontology, contractarianism, and virtue ethics.

original position A hypothetical view of the social contract put forward by the philosopher John Rawls. He asks us to reflect on what principles we, as members of modern liberal-democratic societies, would choose as the foundation of our political institutions. The choice is made by imagining what agreement we could come to if we were all deprived of the kinds of knowledge of our position in society that impede impartial judgment (see entry for the "veil of ignorance").

Pareto Optimality The state of affairs that is preferred to a feasible alternative by at least one person in a circumscribed group, while nobody in that group prefers the alternative to it.

peak oil hypothesis The claim that we have reached or are near the middle point of the world's oil resources and that we are therefore about to embark on an era in which there is an ever-diminishing supply of these resources.

permissible That class of actions (or omissions) that are morally allowed, or right.

phronesis In Aristotle's ethics, "practical wisdom" or "prudence." This is the master virtue, the possession of which gives an agent the ability to discern the correct way to achieve the human end (*eudaimonia*).

picturesque A theory of natural beauty claiming that natural scenes are beautiful to the extent that they are like pictures, especially landscape paintings. As historian John Conron puts it, the picturesque emphasizes aspects of nature that are "complex and eccentric, varied and irregular, rich and forceful, vibrant with energy."

positive feedback An effect that amplifies its cause.

positive right The right to be provided with certain goods, usually by governments. Example: free education.

precautionary principle When our activities threaten to do damage to the environment, "the lack of full scientific certainty shall not be used as a reason for postponing cost-effective measures to prevent environmental deterioration."

preservationism The view that wild spaces are intrinsically valuable and ought to be kept in as pristine a condition as possible. This usually precludes the use of these spaces by humans for economic gain.

primary productivity The amount of solar energy converted by plants into potential food.

psychocentrism The view that in order to have moral standing, an entity must possess, or be capable of possessing, some kind of psychological state (e.g., sentience, awareness).

publicity argument The claim that a necessary feature of a moral system is that, if it is deemed to be correct, we have a duty to make its principles public if we want to justify holding people to account for breaching those principles.

reciprocity thesis The claim that appreciation of the beauty of nature can assist us in appreciating beautiful art objects and vice versa.

resilience An ecosystem's ability, following a disturbance, to regain the pre-disturbance state.

rule-utilitarianism A version of utilitarianism according to which an act is right if and only if it conforms to a rule that, if followed by everyone, will produce the best consequences (or maximize utility).

social contract theory The view that a right action is one that conforms to the terms of a duly constituted agreement among self-interested agents, each of whom is seeking only his or her own advantage.

social ecology A response to the recognition that our environmental woes are in large part a result of the failure to organize the rest of society in a just or rational way.

special concern A species that is nearing threatened status.

speciation The process by which a species comes into being by evolving from some other species.

Species at Risk Act (SARA) A federal law enacted in 2003, the purpose of which is to monitor endangered species in Canada.

speciesism The view that the values or interests of one species have greater weight than those of other species.

strong sustainability The idea that the biosphere, including all of its species, must be preserved in perpetuity, just the way we find it now. This view allows for *no* substitution of man-made capital for natural capital.

substitutability principle The claim that no good is "priceless." That is, there is no good whose loss cannot in principle be rectified by swapping—*substituting*—an appropriate amount of some other good for

it. A prominent feature of the neo-liberal economic approach to environmental resource management.

supererogatory Actions that go "above and beyond the call of duty." Agents can be praised for performing them but cannot be blamed for failing to do so.

threatened A species that is nearing endangered status.

total complexity Between-habitat complexity plus within-habitat complexity plus genetic variability within species.

Traditional Ecological Knowledge (TEK) The knowledge of their local ecosystems possessed by the world's Aboriginal cultures, including the sustainable practices these cultures have developed.

tragedy of the commons Picture a pasture open to the free use of numerous herdsmen. Every herdsman will, in rational pursuit of his own interest, try to graze as many cows on the commons as he can. At any point each herdsman might ask: Should I add one more animal to the commons (this is a decision "at the margin")? It would seem that the answer will always be yes. The addition of one animal increases the wealth of the herdsman incrementally and does not ruin the land. More specifically, the herdsman will reason that the cost of whatever damage is done to the land will be shared by all the herdsmen, while he alone will reap the benefits of the extra cow. Simple cost-benefit analysis indicates that he should graze that cow. However, since each herdsman makes the same calculation, the result is that the commons is overgrazed. What looks rational from an individual standpoint is irrational from the collective point of view.

trophic level A particular rung on or node in an ecosystem's energy transfer network. Producers like plants belong to the first trophic level, while herbivores like cattle belong to the second, and carnivores like mountain lions belong to the third.

unity of the virtues thesis In virtue ethics, the claim that in order to have any one of the virtues an agent must have them all.

universalizability In Kant's ethics, the capacity of a maxim to become a law for everyone, without contradiction. This is a test of the morality of a maxim and thus of the action that flows from it. For Kant, any maxim that allowed for lying would generate a contradiction. Thus, lying is impermissible.

utilitarianism A version of consequentialism according to which a right action is that which maximizes the utility of all parties affected by the action.

value coherentism The view that values can be justified by reference to other values in a system of values, rather than by reference to some single foundational value or set of such values.

value foundationalism The view that there is one value, or set of values, that anchors all the others in the sense that they can ultimately be justified only by appeal to the foundational elements. When Aristotle claims that there is one thing—"happiness"—that is chosen for its own sake and that all others are chosen for the sake of something else, he is expressing a kind of value foundationalism.

value monism The claim that there is just one fundamental kind of value.

value pluralism The claim that there are many kinds of value and that none of them is fundamental.

veil of ignorance A device employed by the philosopher John Rawls to describe the epistemic situation of the agents trying to come to an agreement on principles of justice. Such agents are said to choose in ignorance of things like their gender, race, class, and so on. The "veil" is meant to enable such agents to choose principles in a relatively impartial way (see entry for "original position").

virtue ethics The view that a right action is one that is the product of the proper character or disposition possessed by an agent.

weak sustainability The idea that we ought to act so as to maximize the welfare of future generations. This view allows for the unlimited substitution of man-made capital for natural capital.

within-habitat complexity The number of species within a habitat.

zero-sum game A "game" in which there is one loser and one winner. If winning receives a value of +1 and losing a value of −1, then every such game results in a "zero sum." Example: a hockey game.

Bibliography

AAFC (Agriculture and Agri-Food Canada). 2015. www. agr.gc.ca/eng/science-and-innovation/agricultural-practices/agroforestry/?id=1177431400694. Web.

Acampora, R. 2004. "*Oikos* and *Domus*: On Constructive Co-Habitation with Other Creatures." *Philosophy & Geography* 7: 219–35. Print.

Ackerman, Frank, and Lisa Heinzerling. 2004. *Priceless: On Knowing the Price of Everything and the Value of Nothing*. New York: The New Press. Print.

Alexander, Christopher, et al. 1977. *A Pattern Language*. New York: Oxford UP. Print.

Alfred, Taiaiake (Gerald). 1999. *Peace, Power, Righteousness: An Indigenous Manifesto*. Toronto: Oxford UP. Print.

Allen, Paula Gunn. 1986. *The Sacred Hoop: Recovering the Feminine in American Indian Tradition*. Boston: Beacon Press. Print.

———. 1990. "The Woman I Love Is a Planet; The Planet I Love Is a Tree." In *Reweaving the World: The Emergence of Ecofeminism*. Ed. Irene Diamond and Gloria Feman Orenstein. San Francisco: Sierra Club Books. Print.

Andrews, Malcolm. 1989. *The Search for the Picturesque*. Stanford, CA: Stanford UP. Print.

Andrews, R., and Mary Jo Waits. 1978. *Environmental Values in Public Decisions: A Research Agenda*. Ann Arbor: U of Michigan, School of Natural Resources. Print.

Archer, David. 2009. *The Long Thaw: How Humans Are Changing the Next 100,000 Years of Earth's Climate*. Princeton, NJ: Princeton UP. Print.

Aristotle. 1947. *Nichomachean Ethics, Introduction to Aristotle*. Ed. Richard McKeon. New York: McGraw-Hill. Print.

Armstrong, Franny, director. 2009. *The Age of Stupid*. Spanner Films. Video film.

Arneil, Barbara. 1996. *John Locke and America: The Defence of English Colonialism*. Oxford: Clarendon Press. Print.

Assadourian, Erik, Gary Gardner, and Radhika Sarin. 2008. "The State of Consumption Today." In *Environmental Ethics: Readings in Theory and Application*. Ed. Louis Pojman and Paul Pojman. Belmont, CA: Thomason Wadsworth. 415–16. Print.

Atwood, Margaret. 1972. *Survival: A Thematic Guide to Canadian Literature*. Toronto: McClelland & Stewart. Print.

Bachelard, Gaston. 1964. *The Poetics of Space*. Boston: Beacon Press. Print.

Bachofen, Johann Jakob. 1903. *Le droit de la mère dans l'antiquité*. Paris: Groupe français d'études feministes. Print.

———. 1983. *Du règne de la mère au patriarcat*. Paris: Éditions de l'Aire. Print.

Bacon, F. 1870 [1620]. *Works*. Ed. James Spedding, Robert Leslie Ellis, and Douglas Devon Heath. London: Longmans Green.

Baillie, J.E.M., et al. 2004. *A Global Species Assessment*. UK: IUCN. http://data.iucn.org/dbtw-wpd/html/Red%20 List%202004/completed/cover.html. Web.

Bakker, Isabella. 1994. *The Strategic Silence: Gender and Economic Policy*. London: Zed Books. Print.

Ball, Terrence. 1992. "New Faces of Power." In *Rethinking Power*. Ed. Thomas E. Wartenberg. Albany: State U of New York P. Print.

Beardsley, Monroe. 1970. "The Aesthetic Point of View." In *Perspectives in Education, Religion, and the Arts*. Ed. H. Kiefer and M. Munitz. Albany: State U of New York P. Repr. in *The Aesthetic Point of View: Selected Essays of Monroe C. Beardsley*. Ed. M.J. Wreen and D.M. Callen. Ithaca, NY: Cornell UP. Print.

Beckerman, Wilfrid. 1994. "Sustainable Development: Is It a Useful Concept?" *Environmental Values* 3: 191–209. Print.

Bell, Clive. 1958 [1913]. *Art*. New York: G.P. Putnam's Sons. Print.

Belshaw, Christopher. 2001. *Environmental Philosophy: Reason, Nature and Human Concern*. Montreal and Kingston: McGill-Queen's UP. Print.

Berg, Peter. 1991. "What Is Bioregionalism?" *The Trumpeter* 1: 6–12. Print.

Berkes, Fikret, Johan Colding, and Carl Folke. 2000. "Rediscovery of Traditional Ecological Knowledge as Adaptive Management." *Ecological Applications* 10, no. 5: 1251–62. Print.

Berleant, Arnold. 1992. *The Aesthetics of Environment*. Philadelphia: Temple UP. Print.

———. 1997. *Living in the Landscape: Toward an Aesthetics of Environment*. Lawrence: UP of Kansas. Print.

———. 2005. *Aesthetics and Environment: Variations on a Theme*. Aldershot, UK: Ashgate. Print.

Berleant, Arnold, and Allen Carlson, eds. 2007. *The Aesthetics of Human Environments*. Peterborough, ON: Broadview Press. Print.

Berry, Thomas. 1988. *The Dream of the Earth*. San Francisco: Sierra Club Books. Print.

Berry, Wendell. 1987. "Getting along with Nature." In *Home Economics*. San Francisco: North Point Press. Print.

Beyleveld, Deryck, and Roger Brownsword. 2001. *Human Dignity in Bioethics and Biolaw*. Oxford: Oxford UP.

Birch, Tom. 1990. "Universal Consideration." Paper presented at the International Society for Environmental Ethics, American Philosophical Association, 27 December.

Blair, Tony. 2004. "Climate Change Speech." www .number-10.gov.uk/output.page6333.asp. Web.

Bookchin, Murray. 1980. *Toward an Ecological Society*. Montreal; Buffalo: Black Rose Books.

———. 1987. "Social Ecology versus 'Deep Ecology.'" *Green Perspectives: Newsletter of the Green Program Project*, 4–5. Print.

———. 1989. "Which Way for the US Greens?" *New Politics* 2: 71–83. Print.

———. 2002. "Social Ecology versus Deep Ecology." In *Environmental Ethics: What Really Matters, What Really Works*. Ed. David Schmidtz and Elizabeth Willott. New York: Oxford UP. 126–36. Print.

———. 2005. *The Ecology of Freedom: The Emergence and Dissolution of Hierarchy*. New York: AK Press.

Booth, Annie, and Harvey M. Jacobs. 1990. "Ties That Bind: Native American Beliefs as a Foundation for Environmental Consciousness." *Environmental Ethics* 12, no. 1: 27–43. Print.

Borrows, John. 1997. "Living between Water and Rocks: First Nations, Environmental Planning and Democracy." *University of Toronto Law Journal* 47: 417–68. Print.

Borza, K., and D. Jamieson. 1990. *Global Change and Biodiversity Loss: Some Impediments to Response*. Boulder, CO: U of Colorado, Center for Space and Geoscience Policy.

Bosselmann, K. 2008. *The Principle of Sustainability*. Aldershot, UK: Ashgate.

Brady, Emily. 1998. "Imagination and the Aesthetic Appreciation of Nature." *Journal of Aesthetics and Art Criticism* 56, no. 2: 139–47. Print.

———. 2003. *Aesthetics of the Natural Environment*. Tuscaloosa: U of Alabama P. Print.

Braidotti, Rosi, ed. 1994. *Women, the Environment and Sustainable Development: Towards a Theoretical Synthesis*. London: Zed Books. Print.

Bramley, M., P. Sadik, and D. Marshall. 2009. *Climate Leadership, Economic Prosperity: Final Report on an Economic Study of Greenhouse Gas Targets and Policies for Canada*. Vancouver, BC: David Suzuki Foundation and Pembina Institute. Print.

Brand, Stewart. 2009. *Whole Earth Discipline*. New York: Viking Press. Print.

Brinton, Daniel Garrison. 1868. *The Myths of the New World: A Treatise on the Symbolism and Mythology of the Red Race of America*. New York: Leopold and Holt. Print.

Brody, Hugh. 2000. *The Other Side of Eden: Hunters, Farmers and the Shaping of the World*. Vancouver: Douglas and McIntyre. Print.

Brook, Isis. 2008. "Wildness in the English Garden Tradition: A Reassessment of the Picturesque from Environmental Philosophy." *Ethics and the Environment* 13: 105–19. Print.

Brown, Charles S., and Ted Toadvine, eds. 2003. *Eco-Phenomenology: Back to the Earth Itself*. Albany: State U of New York P. Print.

Brown, Judith K. 1970. "Economic Organization and the Position of Women among the Iroquois." *Ethnohistory* 17, no. 4: 164. Print.

Buckingham, Susan. 2004. "Ecofeminism in the Twenty-First Century." *The Geographical Journal* 170: 146–54. Print.

Buttel, Frederick H. 1988. "Some Observations on States, World Orders, and the Politics of Sustainability." *Organization and Environment* 11, no. 3: 261–86. Print.

Byron, Michael. 1998. "Satisficing and Optimality." *Ethics* 109: 67–93. Print.

———, ed. 2004. *Satisficing and Maximizing: Moral Theorists on Practical Reason*. Cambridge: Cambridge UP. Print.

Callicott, J. Baird. 1989. *In Defense of the Land Ethic: Essays in Environmental Philosophy*. Suny Series in Philosophy and Biology. Albany: State U of New York P. Print.

———. 1990. "The Case against Moral Pluralism." *Environmental Ethics* 12: 99–124. Print.

———. 2002. "The Pragmatic Power and Promise of Theoretical Environmental Ethics." *Environmental Values* 11: 3–25. Print.

———. 2003. "Wetland Gloom and Wetland Glory." *Philosophy and Geography* 6: 33–45. Print.

———. 2008a. "The Conceptual Foundations of the Land Ethic." In *Environmental Ethics: Readings in Theory and Application*. Ed. Louis Pojman and Paul Pojman. Belmont, CA: Thomason Wadsworth. 173–85. Print.

———. 2008b. "Leopold's Land Aesthetic." In *Nature, Aesthetics, and Environmentalism: From Beauty to Duty*. Ed. Allen Carlson and Sheila Lintott. New York: Columbia UP. Print.

Camacho, A.E., et al. 2010. "Reassessing Conservation Goals in a Changing Climate." *Issues in Science Technology* 26: 21–6.

Camacho, David, ed. 1998. *Environmental Injustices, Political Struggles: Race, Class, and the Environment*. Durham, NC: Duke UP. Print.

Carlson, Allen. 1977. "On the Possibility of Quantifying Scenic Beauty." *Landscape Planning* 4: 131–72. Print.

———. 1979. "Appreciation and the Natural Environment." *Journal of Aesthetics and Art Criticism* 37: 267–76. Print.

———. 1981. "Nature, Aesthetic Judgment, and Objectivity." *Journal of Aesthetics and Art Criticism* 40: 15–27. Print.

———. 1998. "Aesthetic Appreciation of the Natural Environment." In *Environmental Ethics: Divergence and Convergence*, 2nd edn. Ed. S.J. Armstrong and R.G. Botzler. Boston: McGraw-Hill. Print.

———. 1999. "Admiring Mirelands: The Difficult Beauty of Wetlands." In *Suo on Kaunis*. Ed. Liisa Heikkila-Palo. Helsinki: Maahenki Oy. Print.

———. 2000. *Aesthetics and the Environment: The Appreciation of Nature, Art and Architecture*. London: Routledge. Print.

——. 2006a. "Critical Notice: Aesthetics and Environment." *British Journal of Aesthetics* 46: 416–27. Print.

——. 2006b. "'We See Beauty Now Where We Could Not See It Before': Rolston's Aesthetics of Nature." In *Nature, Value, Duty: Life on Earth with Holmes Rolston, III.* Ed. W. Ouderkirk and C. Preston. Dordrecht: Springer. Print.

——. 2007. "The Requirements for an Adequate Aesthetics of Nature." *Environmental Philosophy* 4: 1–12. Print.

——. 2009. *Nature and Landscape: An Introduction to Environmental Aesthetics.* New York: Columbia UP. Print.

——. 2015. "Aesthetic Appreciation of Nature and Environmentalism." (This volume, Chapter 7.)

Carlson, Allen, and Sheila Lintott. 2008. "Introduction." In *Nature, Aesthetics, and Environmentalism: From Beauty to Duty.* Ed. Allen Carlson and Sheila Lintott. New York: Columbia UP. Print.

——, eds. 2008. *Nature, Aesthetics, and Environmentalism: From Beauty to Duty.* New York: Columbia UP. Print.

Carlson, Allen, and Glenn Parsons. 2008. *Functional Beauty.* Oxford: Oxford UP. Print.

Carroll, Noël. 1993. "On Being Moved by Nature: Between Religion and Natural History." In *Landscape, Natural Beauty and the Arts.* Ed. I. Gaskell and S. Kemal. Cambridge: Cambridge UP. Print.

Carson, Rachel. 1962. *Silent Spring.* New York: Fawcett World Library. Print.

Casey, Allan. 2009. *Lakeland: Journeys into the Soul of Canada.* Vancouver: Greystone Books. Print.

Casey, Edward T. 1993. *Getting Back into Place: Toward a Renewed Understanding of the Place-World.* Bloomington and Indianapolis: Indiana UP. Print.

CH2MHILL. 2005. *Engineering a Sustainable Future: 2003/2004 Sustainability Report.* Englewood, CO. Print.

Cheney, Jim. 1987. "Eco-feminism and Deep Ecology." *Environmental Ethics* 9, no. 2: 115–45. Print.

Clark, Stephen R.L. 1977. *The Moral Status of Animals.* Oxford: Clarendon Press. Print.

Clastres, Pierre. 1974. *La société contre l'état.* Paris: Éditions de Minuit. Print.

Cohen, Felix S. 1952. "Americanizing the White Man." *The American Scholar* 20, no. 2: 177–90. Print.

Conron, John. 2000. *American Picturesque.* University Park, PA: Pennsylvania State UP. Print.

Cook, David. 2004. *The Natural Step.* Foxhole, Devon: Green Books. Print.

Costanza, Robert, Herman Daly, and Thomas Prugh. 2000. *The Local Politics of Global Sustainability.* Washington: Island Press. Print.

Crawford, Donald W. 2007. "Scenery and the Aesthetics of Nature." In *The Aesthetics of Natural Environments.* Ed. A. Berleant and A. Carlson. Peterborough, ON: Broadview Press. Print.

Curtin, Deane. 1999. "Recognizing Women's Environmental Expertise." In *Chinnagrounder's Challenge: The Question of Ecological Citizenship.* Bloomington, IN: Indiana UP. Print.

Daly, Herman E. 1995. "On Wilfrid Beckerman's Critique of Sustainable Development." *Environmental Values* 4: 49–55. Print.

Dearden, Philip, and Bruce Mitchell. 2009. *Environmental Change and Challenge.* Toronto: Oxford UP. Print.

de Beauvoir, Simone. 1952. *The Second Sex.* Trans. H.M. Parshley. New York: Vintage Books. Print.

——. 1984. *After the Second Sex: Interviews with Simone de Beauvoir.* Ed. Alice Schwarzer. New York: Pantheon Books. Print.

den Elzen, Michel G.J., et al. 2013. "Countries' Contributions to Climate Change: Effect of Accounting for All Greenhouse Gases, Recent Trends, Basic Needs and Technological Progress." *Climatic Change* 121: 397–412.

Descartes, René. 1987. "Passions of the Soul." Article 80. *The Philosophical Writings of Descartes*, vol. 1. Cambridge: Cambridge UP. Print.

de-Shalit, Avner. 2001. *The Environment between Theory and Practice.* Oxford: Oxford University Press.

Desjardins, Joseph. 2001. *Environmental Ethics: An Introduction to Environmental Philosophy.* Belmont, CA: Wadsworth. Print.

Devall, Bill. 1987. "Deep Ecology and Its Critics." *Earth First!* 22 December. Print.

Devall, B., and G. Sessions. 1985. *Deep Ecology.* Salt Lake City, UT: G.M. Smith. Print.

Dewey, John. 1981. *The Later Works*, vol. 13. Ed. Jo Ann Boydston. Carbondale: Southern Illinois UP. Print.

Diamond, Irene. 1990. "Babies, Heroic Experts, and a Poisoned Earth." In *Reweaving the World: The Emergence of Ecofeminism.* Ed. Irene Diamond and Gloria Feman Orenstein. San Francisco: Sierra Club Books. Print.

Dickason, Olive Patricia. 2006. *A Concise History of Canada's First Nations.* Toronto: Oxford UP. Print.

Dobson, A. 2003. *Citizenship and the Environment.* Oxford, New York: Oxford UP. Print.

——. 2007. *Green Political Thought*, 4th ed. Routledge. Print.

Dobson, A., and D. Bell. 2006. *Environmental Citizenship.* London; Cambridge, MA: MIT Press. Print.

Donaldson, S., and W. Kymlicka. 2011. *Zoopolis: A Political Theory of Animal Rights.* Oxford: Oxford UP. Print.

Donovan, J., and C. Adams, eds. 2007. *The Feminist Care Tradition in Animal Ethics.* New York: Columbia UP. Print.

Doubiago, Sharon. 1989. "Mama Coyote Talks to the Boys." In *Healing the Wounds: The Promise of Ecofeminism.* Ed. Judith Plant. Santa Cruz: New Society Publishers. Print.

Draper, Dianne, and Maureen G. Reed. 2009. *Our Environment: A Canadian Perspective.* Toronto: Nelson. Print.

Drengson, Alan. 1990. "The Ecostery Foundation of North America: Statement of Philosophy." *The Trumpeter* 7, no. 1: 12–16. Print.

Dressel, Holly, and David Suzuki. 2002. *Good News for a Change: How Everyday People Are Changing the Planet.* Vancouver: Greystone. Print.

Dryzek, J.S. 2005. *The Politics of the Earth: Environmental Discourses,* 2nd edn. Oxford; New York: Oxford University Press.

Dwivedi, O.P., J.P. Kyba, P. Stoett, and R. Tiessen. 2001. *Sustainable Development and Canada: National and International Perspectives.* Peterborough, ON: Broadview Press. Print.

Eaton, Marcia. 1989. *Aesthetics and the Good Life.* New York: Fairleigh Dickenson UP. Print.

——. 1997. "The Beauty That Requires Health." *Placing Nature: Culture and Landscape Ecology.* Ed. J.I. Nassauer. Washington: Island Press. Print.

——. 1998. "Fact and Fiction in the Aesthetic Appreciation of Nature." *Journal of Aesthetics and Art Criticism* 56: 149–56. Print.

——. 2001. *Merit: Aesthetic and Ethical.* Oxford: Oxford UP. Print.

Edwards, Andres R. 2005. *The Sustainability Revolution.* Gabriola Island, BC: New Society Publishers. Print.

Ehrlich, Anne H., and Paul R. Ehrlich. 1982. *Extinction.* New York: Ballantine Books. Print.

——. 2009. *The Dominant Animal: Human Evolution and the Environment.* Washington: Island Press. Print.

Ehrlich, Paul. 1968. *The Population Bomb.* New York: Ballantine Books. Print.

Ekins, Paul, and Manfred Max-Neef. 1992. *Real-Life Economics.* London: Routledge. Print.

Fenton, William N. N.d. *Structural Relationships of Kin, Age, Rank, Society, Space and Continuity Found in the Deganiwidah Epic.* Ed. American Philosophical Society. Philadelphia. Print.

Finn, Geraldine. 1982. "On the Oppression of Women in Philosophy—Or, Whatever Happened to Objectivity?" In *Feminism in Canada: From Pressure to Politics.* Montreal: Black Rose Books. Print.

Fluker, Shaun. 2013. "Environmental Norms in the Courtroom (The Case of Ecological Integrity in Canada's National Parks)." In *Confronting Ecological and Economic Collapse.* Ed. L. Westra, P. Taylor, and Agnes Michelot. 21–31. London: Routledge/Earthscan. Print.

Foden, W., et al. 2008. "Species Susceptibility to Climate Change Impacts." In *The 2008 Review of the IUCN Red List of Threatened Species.* https://portals.iucn.org/library/efiles/documents/RL-2009-001.pdf#page=101. Web.

Foltz, Bruce. 1995. *Inhabiting the Earth: Heidegger, Environmental Ethics and the Metaphysics of Nature.* Amherst, NY: Prometheus Books. Print.

Foreman, Dave. 1987. "Reinhabitation, Biocentrism, and Self-Defense." *Earth First!* 1 (August). Print.

Francione, G. 2008. *Animals as Persons: Essays on the Abolition of Animal Exploitation.* New York: Columbia UP. Print.

Fraser Basin Council. 2004. "Sustainability Snapshot 2." In *State of the Fraser Basin Report.* Vancouver: Fraser Basin Council. Print.

Frasz, Geoffrey. 2005. "Benevolence as an Environmental Virtue." In *Environmental Virtue Ethics.* Ed. Ronald Sandler and Philip Cafaro. Lanham, MD: Rowman and Littlefield. 121–34. Print.

Freeman III, A. Myrick. 1998. "The Ethical Basis of the Economic View of the Environment." In *The Environmental Ethics and Policy Book.* Ed. Christine Pierce and Donald Van DeVeer. Belmont, CA: Wadsworth. Print.

Friel, Howard. 2010. *The Lomborg Deception.* New Haven: Yale UP. Print.

Gaita, Raimond. 2002. *The Philosopher's Dog.* New York: Random House. Print.

Gardiner, Stephen. 2006. "A Perfect Moral Storm: Climate Change, Intergenerational Ethics and the Problem of Moral Corruption." *Environmental Values* 15 (August): 397–413. Print.

Gardiner, S., et al. 2010. *Climate Ethics: The Essential Readings.* Oxford: Oxford University Press. Print.

Garvey, James. 2008. *The Ethics of Climate Change: Right and Wrong in a Warming World.* London: Continuum Press. Print.

Gewirth, A. 1982. *Human Rights: Essays on Justification and Applications.* Chicago: U of Chicago P.

Gibbs, Lois Marie. 1998. *Love Canal: The Story Continues…* Gabriola Island, BC: New Society Publishers. Print.

Gill, Sam D. 1987. *Mother Earth: An American Story.* Chicago: U of Chicago P. Print.

Gilpin, William. 1792. *Three Essays: On Picturesque Beauty, On Picturesque Travel, and On Sketching Landscape.* London: R. Blamire. Print.

Glazebrook, Trish. 2000. "From *Physis* to Nature, *Technê* to Technology: Heidegger on Aristotle, Galileo and Newton." *Southern Journal of Philosophy* 38, no. 1: 95–118. Print.

——. 2001. "Heidegger and Ecofeminism." In *Re-Reading the Canon: Feminist Interpretations of Heidegger.* Ed. Nancy Holland and Patricia Huntington. University Park, PA: Pennsylvania State UP. 221–51. Print.

Glazebrook, Trish, and Anthony Kola Olusanya. 2011. "Justice, Conflict, Capital, and Care: Oil in the Niger Delta." *Environmental Ethics* 33, no. 2: 163–84. Print.

Godlovitch, Stan. 1989. "Aesthetic Protectionism." *Journal of Applied Philosophy* 6: 171–80. Print.

——. 1994. "Icebreakers: Environmentalism and Natural Aesthetics." *Journal of Applied Philosophy* 11: 15–30. Print.

———. 1998. "Valuing Nature and the Autonomy of Natural Aesthetics." *British Journal of Aesthetics* 38: 180–97. Print.

Goodpaster, Kenneth. 2008. "On Being Morally Considerable." *Environmental Ethics: Readings in Theory and Application*. Ed. Louis Pojman and Paul Pojman. Belmont, CA: Thomason Wadsworth. 154–63. Print.

Goodpaster, Kenneth E., and K.M. Sayre. 1979. *Ethics and Problems of the Twenty-first Century*. Notre Dame, IN: U of Notre Dame P. Print.

Gore, A. 2006. *An Inconvenient Truth: The Planetary Emergency of Global Warming and What We Can Do about It*. New York: Rodale. Print.

Gould, Stephen Jay. 1992. "What Is a Species?" *Discover Magazine*, December. Print.

Grande, Sandy Marie Anglas. 1999. "Beyond the Ecologically Noble Savage: Deconstructing the White Man's Noble Savage." *Environmental Ethics* 21, no. 3: 307–20. Print.

Greco, J., and J. Turri. 2012. *Virtue Epistemology: Contemporary Readings*. Cambridge MA: MIT Press. Print.

Greenpeace Magazine 11, no. 6 (Spring/Summer 2010): 6–7. Print.

Gribbin, John, and Mary Gribbin. 2009. *He Knew He Was Right: The Irrepressible Life of James Lovelock*. London: Penguin. Print.

Hackett, David, and G. Tyler Miller, Jr. 2008. *Living in the Environment*, 1st Cdn edn. Toronto: Nelson. Print.

Hadley, J. 2005. "Nonhuman Animal Property: Reconciling Environmentalism and Animal Rights." *Journal of Social Philosophy* 36: 305–15.

Hansen, James. 2009. *Storms of My Grandchildren*. New York: Bloomsbury. Print.

Haraway, D. 2007. *When Species Meet*. Minneapolis: U of Minnesota P. Print.

Hardin, Garrett. 1974. "Lifeboat Ethics: The Case against the Poor." *Psychology Today*. Print.

———. 1986. "The Tragedy of the Commons." *Science* 162: 1245. Print.

Hare, Nathan. 1981. "Black Ecology." In *Environmental Ethics*. Ed. Kristin Shrader-Frechette. Pacific Grove, CA: Boxwood Press. 229–36. Print.

Hargrove, Eugene C. 1979. "The Historical Foundations of American Environmental Attitudes." *Environmental Ethics* 1: 209–40. Print.

———. 1989. *The Foundations of Environmental Ethics*. Englewood Cliffs, NJ: Prentice-Hall. Print.

Harrison, Ruth. 1964. *Animal Machines*. London: Stuart. Print.

Hawke Baxter, Kelly, et al., eds. 1995. *Pathways to Sustainability: Assessing Our Progress*. National Round Table Series on Sustainable Development. Ottawa: National Round Table on the Environment and the Economy. Print.

Heath, Joseph. 2001. *The Efficient Society: Why Canada Is as Close to Utopia as It Gets*. Toronto: Penguin. Print.

Hegel, G. 1945. *Hegel's Philosophy of Right*. Oxford: Oxford UP. Print.

Heidegger, Martin. 1962. *Being and Time*. Trans. John Macquarrie and Edward Robinson. New York: Harper and Row. Print.

———. 1966. *Discourse in Thinking*. New York: Harper and Row. Print.

———. 1971. "Building Dwelling Thinking." In *Poetry, Language, Thought*. Trans. Albert Hofstadter. New York: Harper and Row. Print.

———. 1982. *Basic Problems of Phenomenology*. Bloomington: Indiana UP. Print.

Hepburn, Ronald. 1966. "Contemporary Aesthetics and the Neglect of Natural Beauty." In *British Analytical Philosophy*. Ed. A. Montefiore and B. Williams. London: Routledge–Kegan Paul. 305. Print.

Hessing, Melody, Rebecca Raglon, and Catriona Sandilands, eds. 2005. *This Elusive Land: Women and the Canadian Environment*. Vancouver: UBC Press. Print.

Hettinger, Ned. 2005. "Allen Carlson's Environmental Aesthetics and Protection of the Environment." *Environmental Ethics* 27: 57–76. Print.

———. 2008. "Objectivity in Environmental Aesthetics and Environmental Protection." In *Nature, Aesthetics, and Environmentalism: From Beauty to Duty*. Ed. A. Carlson and S. Lintott. New York: Columbia UP. Print.

Hettinger, Ned, and Bill Throop. 2008. "Refocusing Ecocentrism: De-emphasizing Stability and Defending Wildness." In *Environmental Ethics: Readings in Theory and Application*. Ed. Louis Pojman and Paul Pojman. Belmont, CA: Thomason Wadsworth. 186–99. Print.

Heyd, Thomas. 2001. "Aesthetic Appreciation and the Many Stories about Nature." *British Journal of Aesthetics* 41: 125–37. Print.

Hipple, Jr, W.J. 1957. *The Beautiful, the Sublime and the Picturesque in Eighteenth-Century British Aesthetic Theory*. Carbondale: Southern Illinois UP. Print.

Hirschman, A.O. 1977. *The Passions and the Interests: Arguments for Capitalism before Its Triumph*. Princeton: Princeton University Press. Print.

Hobbes, Thomas. 1994. *Leviathan*. Indianapolis: Hackett Publishing. Print.

Holland, Alan. 2003. "Sustainability." In *A Companion to Environmental Philosophy*. Ed. Dale Jamieson. Oxford: Blackwell. Print.

Holland, Alan, Andrew Light, and John O'Neill. 2008. *Environmental Values*. London: Routledge. Print.

Homer-Dixon, Thomas, and Nick Garrison, eds. 2009. *Carbon Shift*. Toronto: Random House Canada. Print.

hooks, bell. 1984a. "Black Women: Shaping Feminist Theory." In *Feminist Theory: From Margin to Center.* Boston, MA: South End Press. Print.

——. 1984b. *Feminist Theory: From Margin to Center.* Boston: South End Press. Print.

Hughes, David. 2009. "The Energy Issue." In *Carbon Shift.* Ed. Thomas Homer-Dixon and Nick Garrison. Toronto: Random House Canada. 59–96. Print.

Hughes, J. Donald. 1996. *North American Indian Ecology,* 2nd edn. El Paso: Texas Western Press. Print.

Hummel, Monte. 1989. *Endangered Spaces.* Toronto: Key Porter Books. Print.

Hussey, Christopher. 1927. *The Picturesque: Studies in a Point of View.* London: G.P. Putnam's Sons. Print.

IPCC (Intergovernmental Panel on Climate Change). 2007. *Summary for Policymakers.* Cambridge: Cambridge UP. Print.

——. 2013. Press release. www.ipcc.ch/news_and_events/docs/ar5/press_release_ar5_wgi_en.pd. Web.

——. 2014. *Climate Change 2014: Impacts, Adaptation and Vulnerability.* UNEP/WMO: IPCC. Print.

Jaccard, Mark. 2009. "Peak Oil and Market Feedbacks." In *Carbon Shift.* Ed. Thomas Homer-Dixon and Nick Garrison. Toronto: Random House Canada. 97–132. Print.

Jackson, Ross. 2012. *Occupy World Street: A Global Roadmap for Radical Economic and Political Reform.* London: Chelsea Green Publishers. Print.

Jamieson, Dale. 1988a. "The Artificial Heart: Reevaluating the Investment." In *Organ Substitution Technology.* Ed. D. Mathieu. Boulder, CO: Westview Press. 277–96. Print.

——. 1988b. "Grappling for a Glimpse of the Future." In *Societal Responses to Regional Climatic Change: Forecasting Analogy.* Ed. Michael H. Glantz. Boulder, CO: Westview Press. 73–93. Print.

——. 1990. "Managing the Future: Public Policy, Scientific Uncertainty and Global Warming." In *Upstream/Downstream: New Essays in Environmental Ethics.* Ed. D. Scherer. Philadelphia, PA: Temple University Press. 67–89. Print.

——. 2002. "Is there Progress in Morality?" *Utilitas* (14), 318–38. Print.

——. 2006. "An American Paradox". *Climatic Change* 77(1-2), 97–102. Print.

——. 2007. "When Utilitarians Should Be Virtue Theorists." *Utilitas* 19, no. 2: 167. Print.

——. 2008. *Ethics and the Environment.* Cambridge: Cambridge UP. Print.

Jennings, Francis. 1971. "The Ambiguous Iroquois Empire and the Constitutional Evolution of the Covenant Chain." *Proceedings of the American Philosophical Society* 115, no. 2: 88–96. Print.

——. 1975. *The Invasion of America: Indians, Colonialism, and the Cant of Conquest.* Published for the Institute of Early American History and Culture (Williamsburg, VA). The Covenant Chain. Chapel Hill: U of North Carolina P. Print.

——. 1988. *Empire of Fortune: Crowns, Colonies, and Tribes in the Seven Years War in America.* The Covenant Chain. New York: W.W. Norton. Print.

Johnson-Odim, Cheryl. 1991. "Common Themes, Different Contexts: Third World Women and Feminism." In *Third World Women and the Politics of Feminism.* Ed. Chandra Mohanty et al. Bloomington: Indiana UP. 314–27. Print.

Johnston, Basil. 1976. *Ojibway Heritage.* Toronto: McClelland & Stewart. Print.

Kagan, Shelly. 1998. "Rethinking Intrinsic Value." *Journal of Ethics* 2: 277–97. Print.

Kant, Immanuel. 1952. *The Critique of Judgement.* Trans. J.C. Meredith. Oxford: Oxford UP. Print.

——. 1988. *Foundations of the Metaphysics of Morals.* Trans. Lewis White Beck. New York: Macmillan. Print.

Karr, J.R. 1981. "Assessment of Biotic Integrity Using Fish Communities." *Fisheries* 6, no. 6: 21–7. Print.

——. 1996. "Ecological Integrity and Ecological Health Are Not the Same." In *Engineering within Ecological Constraints.* Ed. P. Schulze. Washington, DC: National Academy Press. 97–109. Print.

Kartha, Sivan, Tom Athanasiou, et al. 2010. "Cutting the Knot: Climate Protection, Political Realism, and Equity as Requirements of a Post-Kyoto Regime." http://www.ecoequity.org/docs/CuttingTheKnot.pdf. Web.

Kasun, Jacqueline. 2008. "The Unjust War against Population." In *Environmental Ethics: Readings in Theory and Application.* Ed. Louis Pojman and Paul Pojman. Belmont, CA: Thomason Wadsworth. Print.

Katz, Eric. 1996. "Searching for Intrinsic Value: Pragmatism and Despair in Environmental Ethics." In *Environmental Pragmatism.* Ed. Andrew Light and Eric Katz. London: Routledge. 307–18. Print.

Kazez, Jean. 2007. *The Weight of Things: Philosophy and the Good Life.* Malden, MA: Blackwell. Print.

Keiner, Marco. 2004. "Re-emphasizing Sustainable Development—The Concept of Evolutionability." *Environment, Development and Sustainability* 6: 382–3. Print.

Kelman, Steven. 1981. "Cost-Benefit Analysis: An Ethical Critique." *AEI Journal on Government and Society Regulation.* January–February, 33–40. Print.

Kettel, B. 1996. "Putting Women and the Environment First: Poverty Alleviation and Sustainable Development." In *Achieving Sustainable Development.* Ed. A. Dale and J.B. Robinson. Vancouver: UBC Press. Print.

Knight, Richard Payne. 1794. *The Landscape: A Didactic Poem.* London: Printed by W. Bulmer and Co. for G. Nicol. Print.

——. 1805. *Analytical Inquiry into the Principles of Taste.* London: Printed by L. Hansard and Sons for T. Payne and J. White. Print.

Korsgaard, Christine. 1996. *Creating the Kingdom of Ends.* Cambridge: Cambridge UP. Print.

Kovel, J. 2007. *The Enemy of Nature*. London: Zed Books. Print.

Krech III, Shepard. 1978. "Disease, Starvation and Social Organization of the Northern Athabaskan." *American Ethnologist* 5, no. 4: 722. Print.

———. 1999. *The Ecological Indian: Myth and History*. New York: W.W. Norton. Print.

Krugman, Paul. 2009. "How Did Economists Get It So Wrong?" *New York Times*, 2 September. www.nytimes.com/2009/09/06/magazine/06Economic-t.html?pagewanted=all&_r=0. Web.

Kvanvig, J.L. 1992. *The Intellectual Virtues and the Life of the Mind: On the Place of the Virtues in Epistemology*. Lexington: Rowman and Littlefield. Print.

Lafitau, Joseph François. 1974. *Customs of the American Indian Compared with the Customs of Primitive Times*. Toronto: The Champlain Society. Print.

Lahontan, Baron de. 1703. *New Voyages to North America*. Burt Franklin: Research & Source Works Series 675—American Classics in History and Social Sciences 177. Series ed. Reuben Gold Thwaites. New York: Lenox Hill. (Reprinted from the English edition of 1703, with facsimiles of original title pages, maps, illustrations—in the Ohio State University library. London: Printed for H. Benwicke in St Paul's Churchyard; T. Goodwin, M. Wotton, B. Tooke, in Fleetstreet; and S. Manship in Cornhil.)

Leddy, Thomas. 2005. "A Defense of Arts-Based Appreciation of Nature." *Environmental Ethics* 27: 299–315. Print.

Lee, Keekok. 1995. "Beauty for Ever?" *Environmental Values* 4: 213–25. Print.

Leonard, Herman B., and Richard J. Zeckhauser. 1983. "Cost-Benefit Analysis Defended." Report from the Centre for Philosophy and Public Policy. Print.

Leopold, Aldo. 1949. *A Sand County Almanac*. New York: Oxford UP. Print.

———. 1966 [1949, 1952]. *A Sand County Almanac with Essays on Conservation from Round River*. Oxford: Oxford UP. Print.

Levin, Margarita Garcia, and Michael Levin. 2001. "A Critique of Ecofeminism." In *Environmental Ethics: Readings in Theory and Application*, 3rd edn. Ed. Louis Pojman and Paul Pojman. Belmont, CA: Wadsworth. 199–204. Print.

Levins, Richard, and Richard Lewontin. 1985. *The Dialectical Biologist*. Cambridge, MA: Harvard UP. Print.

Lewis, C.I. 1946. *An Analysis of Knowledge and Valuation*. La Salle, IL: Open Court. Print.

Light, A. 2003. "Urban Ecological Citizenship." *Journal of Social Philosophy* 34, no. 1: 44–63. Print.

Light, Andrew, and Eric Katz. 1996. "Introduction." In *Environmental Pragmatism*. Ed. Andrew Light and Eric Katz. London: Routledge. 5. Print.

Lintott, Sheila. 2006. "Toward Eco-friendly Aesthetics." *Environmental Ethics* 28: 57–76. Print.

Locke, J. 1952 [1690]. *The Second Treatise of Government*. Indianapolis, IN: Bobbs-Merrill. Print.

Loftis, R.J. 2003. "Three Problems for the Aesthetic Foundations of Environmental Ethics." *Philosophy in the Contemporary World* 10: 41–5. Print.

Lovelock, James. 1987. *Gaia*. Oxford: Oxford UP. Print.

———. 2006. *The Revenge of Gaia: Why the Earth Is Fighting Back and How We Can Still Save Humanity*. London: Allen Lane. Print.

———. 2009. *The Vanishing Face of Gaia: A Final Warning*. London: Allen Lane. Print.

Low, Alaine, and Soraya Tremayne, eds. 2001. *Sacred Custodians of the Earth? Women, Spirituality and the Environment*. New York: Berghahn Books. Print.

Low, Nicholas, and Brendan Gleeson. 1998. *Justice, Society, and Nature: An Exploration of Political Ecology*. New York: Routledge. Print.

Lugones, Maria. 1987. "Playfulness, "World-Travelling," and "Loving Perception." *Hypatia* 2, no. 2: 3. Print.

Lugones, M., and E. Spelman. 1983. "Have We Got a Theory for You! Feminist Theory, Cultural Imperialism and the Demand for 'The Women's Voice.'" *Women's Studies International Forum* 6, no. 6: 573–81. Print.

McCullers, Carson. 1951. *The Ballad of Sad Café*. New York: Bantam Books. Print.

MacEwen, Gwendolyn. 2009. "Dark Pines under Water." In *Open Wide a Wilderness: Canadian Nature Poems*. Ed. Nancy Holmes. Waterloo, ON: Wilfrid Laurier UP. 284. Print.

McKibben, Bill. 1989. *The End of Nature*. New York: Knopf. Print.

———. 2011. *The Global Warming Reader*. New York: Penguin Books. Print.

McMahon, Martha. 2005. "People for Pigs in Pleasant Land: Small-Scale Women Farmers." In *This Elusive Land: Women and the Canadian Environment*. Ed. Melody Hessing, Rebecca Raglon, and Catriona Sandilands. Vancouver: UBC Press. 128–41. Print.

Magurran, A.E., and M. Dornelas. 2010. "Biological Diversity in a Changing World." *Philosophical Transactions of the Royal Society B* 365, no. 1558: 3593–7.

Mallory, Chaone. 2009. "Ecofeminism and the Green Public Sphere." In *Advances in Ecopolitics: The Transition to Sustainable Living and Practice*, vol. 1. Ed. Liam Leonard and John Q. Barry. UK: Emerald Group Publishing. Print.

———. 2010. "What Is Ecofeminist Political Philosophy? Gender, Nature and the Political." *Environmental Ethics* 32: 305–22. Print.

Mandeville, B. 1970 [1714]. *The Fable of the Bees*. Trans. P. Harth. Hammersmith, England: Penguin. Print.

Margalit, A. 2010. *On Compromise and Rotten Compromises*. Princeton: Princeton University Press. Print.

Marsden, William. 2009. "The Perfect Moment." In *Carbon Shift*. Ed. Thomas Homer-Dixon and Nick Garrison. Toronto: Random House Canada. 153–76. Print.

Marx, Werner. 1992. *Toward a Phenomenological Ethics: Ethos and the Lifeworld*. Albany: State U of New York P. Print.

Matarrese, A. 2010. "The Boundaries of Democracy and the Case of Non-Humans." *In-Spire Journal of Law, Politics and Societies* 5. Print.

Matthews, Patricia. 2002. "Scientific Knowledge and the Aesthetic Appreciation of Nature." *Journal of Aesthetics and Art Criticism* 60: 37–48. Print.

Meadows, Dennis L., William W. Behrens III, Donella H. Meadows, Roger F. Naill, Jorgen Randers, and Erich K.O. Zahn. 1974. *Dynamics of Growth in a Finite World*. Cambridge, MA: Wright-Allen. Print.

Merchant, Carolyn. 1980. *The Death of Nature: Women, Ecology, and the Scientific Revolution*. San Francisco: Harper and Row. Print.

——. 1995. *Earthcare: Women and the Environment*. New York: Routledge. Print.

Minister of Canadian Heritage. 2001. *Mikisew Cree First Nation v. Canada*. FCT 1426.

Minteer, B., and J. Collins. 2010. "Move It or Lose It? The Ecological Ethics of Relocating Species under Climate Change." *Ecological Applications* 20: 1801–4. Web.

Moodie, Susanna. 2007. *Roughing It in the Bush*. Toronto: McClelland & Stewart. Print.

Moore, G.E. 1903. *Principia Ethica*. Cambridge: Cambridge UP. Print.

——. 1959. "Is Goodness a Quality?" In *Philosophical Papers*. London: George Allen and Unwin. Print.

Moore, Ronald. 1995. "The Aesthetic and the Moral." *Journal of Aesthetic Education* 2, no. 29: 18–22. Print.

——. 2008. *Natural Beauty: A Theory of Aesthetics beyond the Arts*. Peterborough, ON: Broadview Press. Print.

Mugerauer, Robert, and David Seamon, eds. 1989. *Dwelling, Place and Environment: Toward a Phenomenology of Person and World*. New York: Columbia UP. Print.

Muir, John. 1894. "A View of the High Sierra." In *The Mountains of California*. New York: Century Company. Print.

Mulrennan, Monica E. 1998. *A Casebook of Environmental Issues in Canada*. Toronto: Wiley. Print.

Naess, A. 1995. "The Shallow and the Deep, Long-Range Ecology Movement: A Summary." In *The Deep Ecology Movement: An Introductory Anthology*. Ed. A. Drengson and I. Yuichi. Berkeley: North Atlantic Books. 3–10. Print.

Nagel, Thomas. 1986. *The View from Nowhere*. Oxford: Oxford UP. Print.

Nassauer, Jane Iverson. 1997. "Cultural Sustainability: Aligning Aesthetics and Ecology." In *Placing Nature: Culture and Landscape Ecology*. Ed. J.I. Nassauer. Washington: Island Press. Print.

Nepstad, D. 2007. "The Amazon's Vicious Cycles: Drought and Fire in the Greenhouse." In *Report to the World Wide Fund for Nature (WWF)*. 4. Print.

Norberg-Schulz, Christian. 1984. *The Concept of Dwelling: On the Way to Figurative Architecture*. New York: Electa/Rizzoli. Print.

Norton, Bryan G. 1986a. "Conservation and Preservation: A Conceptual Rehabilitation." *Environmental Ethics* 8: 195–220. Print.

——. 1986b. "On the Inherent Danger of Undervaluing Species." In *The Preservation of Species: The Value of Biological Diversity*. Ed. Bryan G. Norton. Princeton, NJ: Princeton UP. 110–37. Print.

——. 1991. *Towards Unity among Environmentalists*. New York: Oxford UP. Print.

——. 1996. "Integration or Reduction: Two Approaches to Environmental Values." In *Environmental Pragmatism*. Ed. Andrew Light and Eric Katz. London: Routledge. 105–38. Print.

——. 2005. *Sustainability: A Philosophy of Adaptive Ecosystem Management*. Chicago: U of Chicago P. Print.

Noss, R.F., and A.Y. Cooperrider. 1994. *Saving Natures Legacy*. Washington, DC: Island Press. Print.

Nowell-Smith, P.H. 1954. *Ethics*. New York: Penguin. Print.

Nussbaum, Martha. 2006. *Frontiers of Justice*. Cambridge, MA: Harvard UP. Print.

Oelschlaeger, Max, ed. 1995. *Postmodern Environmental Ethics*. Albany: State U of New York P. Print.

O'Neill, Onora. 1996. *Towards Justice and Virtue: A Constructive Account of Practical Reasoning*. Cambridge: Cambridge University Press.

O'Rioden, Timothy. 1988. "Politics of Sustainability." In *Sustainable Environmental Management: Principles and Practice*. Ed. R. Kerry Turner. New York: Belhaven Press. Print.

Ostrom, Elinor. 1990. *Governing the Commons: The Evolution of Institutions for Collective Action*. Cambridge: Cambridge University Press. Print.

O'Sullivan, S. 2011. *Animals, Equality and Democracy*. New York: Palgrave. Print.

Pallasmaa, Juhani. 2005. *The Eyes of the Skin: Architecture and the Senses*. Chichester, UK: John Wiley and Sons. Print.

Palmer, C. 2010. *Animal Ethics in Context*. New York: Columbia UP.

Parsons, Glenn. 2006. "Freedom and Objectivity in the Aesthetic Appreciation of Nature." *British Journal of Aesthetics* 46: 17–37. Print.

——. 2008. *Aesthetics and Nature*. London: Continuum Press. Print.

Pauli, L. 2003. *Manufactured Landscapes: The Photographs of Edward Burtynsky*. Ottawa: National Gallery of Canada and Yale UP. Print.

Peacock, Doug. 2001. *The Grizzly Years*. New York: Owl Books. Print.

Pearce, David. 1993. *Economic Values and the Natural World*. London: Earthscan. Print.

Perelman, Chaim. 1982. *The Realm of Rhetoric*. Notre Dame, IN: U of Notre Dame P. Print.

Perelman, Chaim, and L. Olbrechts-Tyteca. 1969. *The New Rhetoric*. Notre Dame, IN: U of Notre Dame P. Print.

Pettit, Philip. 1984. "Satisficing Consequentialism." *Proceedings of the Aristotelian Society* 58: 165–76. Print.

Phillips, Kevin. 2007. *American Theocracy: The Peril and Politics of Radical Religion, Oil, and Borrowed Money in the 21st Century*. London: Penguin. Print.

Pierce, Christine, and Donald Van DeVeer, eds. 1998a. *The Environmental Ethics and Policy Book*. Belmont, CA: Wadsworth. Print.

——. 1998b. "The Other Animals: Preview." In *The Environmental Ethics and Policy Book*. Ed. Christine Pierce and Donald Van DeVeer. Belmont, CA: Wadsworth. 76. Print.

Pimentel et al. 2000. *Ecological Integrity: Integrating Environment, Conservation and Health*. Washington, DC: Island Press. Print.

Plato. 1992. *The Republic*. Trans. G.M.A. Grube. Indianapolis: Hackett. Print.

Plumwood, Val. 1991. "Nature, Self and Gender: Feminism, Environmental Philosophy and the Critique of Rationalism." *Hypatia* 6, no. 1: 3–27. Print.

——. 1993. *Feminism and the Mastery of Nature*. New York: Routledge. Print.

——. 1994. *Feminism and the Mastery of Nature*. London: Routledge. Print.

——. 1998. "Nature, Self, and Gender: Feminism, Environmental Philosophy, and the Critique of Rationalism." In *The Environmental Ethics and Policy Book*. Ed. Christine Pierce and Donald Van DeVeer. Belmont, CA: Wadsworth. 241–57. Print.

Pogge, Thomas. 2001. *World Power and Human Rights*. London: Polity Press. Print.

——. 2008. *World Poverty and Human Rights*. London: Polity Press. Print.

Pojman, Louis. 2002. *Ethics: Discovering Right and Wrong*, 4th edn. Belmont, CA: Wadsworth. Print.

——. 2008. "The Challenge of the Future: Private Property, the City, the Globe, and a Sustainable City." In *Environmental Ethics: Readings in Theory and Application*. Ed. Louis Pojman and Paul Pojman. Belmont, CA: Thomason Wadsworth. 721–32. Print.

Population Reference Bureau. 2009. *2009 World Population Data Sheet*. Washington, DC: PRB/USAID. PRINT.

Price, Uvedale. 1794. *An Essay on the Picturesque, as Compared with the Sublime and the Beautiful; and on the Use of Studying Pictures, for the Purpose of Improving Real Landscape*. London: J. Robson. Print.

Rabinowicz, Wlodek, and Toni Rønnow-Rasmussen. 1999. "A Distinction in Value: Intrinsic and for Its Own Sake." *Proceedings of the Aristotelian Society* 100: 33–52. Print.

Rachels, James. 1999a. *The Right Thing to Do: Basic Readings in Moral Philosophy*, 2nd edn. Boston: McGraw-Hill. Print.

——. 1999b. *The Elements of Moral Philosophy*, 3rd edn. Boston: McGraw-Hill. Print.

Rao, V. Rukmini. 2002. "Women Farmers of India's Deccan Plateau: Ecofeminists Challenge World Elites." In *Environmental Ethics: What Really Matters, What Really Works*. Ed. David Schmidtz and Elizabeth Willott. New York: Oxford UP. 255–62. Print.

Reardon, Geraldine. 1993. *Women and the Environment*. Oxford: Oxfam. Print.

Rechtschaffen, Clifford, and Eileen Gauna. 2002. *Environmental Justice: Law, Policy, and Regulation*. Durham, NC: Carolina Academic Press. Print.

Reed, Maureen. 2005. "Working at the Margins of Forestry: The Gender of Labour Practices on British Columbia's West Coast." In *This Elusive Land: Women and the Canadian Environment*. Ed. Melody Hessing, Rebecca Raglon, and Catriona Sandilands. Vancouver: UBC Press. 102–27. Print.

Rees, Ronald. 1975. "The Taste for Mountain Scenery." *History Today* 25: 305–12. Print.

Rees, W.E., and K. Mickelson. 2003. "The Environment: Ecological and Ethical Dimensions." In *Environmental Law and Policy*, 3rd edn. Ed. E. Hughes, A.R. Lucas, and W.A. Tilleman. Toronto: Montgomery Publications. 1–40. Print.

Regan, Tom. 1983. *The Case for Animal Rights*. Berkeley, CA: U of California P. Print.

Relph, Edward. 1976. *Place and Placelessness*. London: Pion Books. Print.

Roberts, Robert C., and W. Jay Wood. 2007. *Intellectual Virtue: An Essay in Regulative Epistemology*. Oxford: Clarendon Press. Print.

Rodda, Annabel, ed. 1991. *Women and the Environment*. London: Zed Books. Print.

Rodman, John. 1983. "Four Forms of Ecological Consciousness Reconsidered." In *Ethics and Environment*. Ed. Thomas Attig and Donald Scherer. Englewood Cliffs, NJ: Prentice-Hall. 89–92. Print.

Rolston III, Holmes. 1998. "Aesthetic Experience in Forests." *Journal of Aesthetics and Art Criticism* 56: 157–66. Print.

——. 2000. "Aesthetics in the Swamps." *Perspectives in Biology and Medicine* 43: 584–97. Print.

——. 2002. "From Beauty to Duty: Aesthetics of Nature and Environmental Ethics." In *Environment and the Arts: Perspective on Environmental Aesthetics*. Ed. Arnold Berleant. Aldershot, UK: Ashgate. 127–41. Print.

——. 2005. "Environmental Virtue Ethics: Half the Truth but Dangerous as a Whole." In *Environmental Virtue Ethics*. Ed. Ronald Sandler and Philip Cafaro. Lanham, MD: Rowman and Littlefield. Print.

Rousseau, Jean-Jacques. 1964. "Discourse on the Origin and Foundations of Inequality." In *The First and Second Discourses: Jean Jacques Rousseau*. Ed. Roger D. Master. New York: St Martin's Press. Print.

Routley, R., and V. Routley. 1979. "Against the Inevitability of Human Chauvinism." In *Ethics and Problems of the 21st Century*. Ed. K.E. Goodpaster and K.M. Sayre. Notre Dame, IN: U of Notre Dame P. 36–59. Print.

Rowlands, Mark. 2008. *The Philosopher and the Wolf: Lessons from the Wild on Love, Death and Happiness*. London: Granta. Print.

Rubin, Jeff. 2009. "Demand Shift." In *Carbon Shift*. Ed. Thomas Homer-Dixon and Nick Garrison. Toronto: Random House Canada. 136–52. Print.

Ruether, Rosemary Radford. 1996. *Women Healing Earth: Third World Women on Ecology, Feminism and Religion*. Maryknoll, NY: Orbis. Print.

Sachs, Jeffrey D. 2008. *Commonwealth: Economics for a Crowded Planet*. New York: Penguin. Print.

Sagoff, Mark. 1974. "On Preserving the Natural Environment." *Yale Law Journal* 84: 205–67. Print.

——. 1981. "At the Shrine of Our Lady of Fatima, or Why Political Questions Are Not All Economic." *Arizona Law Review* 23: 1281–98. Print.

——. 1984. *Price, Principle and the Environment*. New York: Cambridge University Press. Print.

——. 2004. *Price, Principle, and the Environment*. Cambridge: Cambridge UP. Print.

Saito, Yuriko. 1984. "Is There a Correct Aesthetic Appreciation of Nature?" *Journal of Aesthetic Education* 18: 35–46. Print.

——. 1998a. "The Aesthetics of Unscenic Nature." *Journal of Aesthetics and Art Criticism* 56: 101–11. Print.

——. 1998b. "Appreciating Nature on Its Own Terms." *Environmental Ethics* 20: 135–49. Print.

——. 2007. "The Role of Aesthetics in Civic Environmentalism." In *The Aesthetics of Human Environments*. Ed. A. Berleant and A. Carlson. Peterborough, ON: Broadview Press. Print.

Salleh, Ariel. 1992. "The Ecofeminist/Deep Ecology Debate: A Reply to Patriarchal Reason." *Environmental Ethics* 14, no. 3: 195–216. Print.

——. 1993. "Class, Race, and Gender Discourse in the Ecofeminism/Deep Ecology Debate." *Environmental Ethics* 15, no. 3: 225–44. Print.

——. 1997. *Ecofeminism as Politics: Nature, Marx and the Postmodern*. London: Zed Books. Print.

——. 2003. "Ecofeminism as Sociology. *Capitalism, Nature, Socialism* 14, no. 1: 61–74. Print.

Sandler, Ronald. 2007. *Character and Environment*. New York: Columbia UP. Print.

——. 2010. "The Value of Species and the Ethical Foundations of Assisted Colonization." *Conservation Biology* 24: 424–31.

——. 2012. *The Ethics of Species*. Cambridge, UK: Cambridge University Press. Print.

Sapa, Hehaka (Black Elk). 1975. *Les rites secrets des Indiens Sioux*. Ed. Joseph Epes Brown. Paris: Payot. Print.

Sapontzis, Steve F. 1984. "Predation." *Ethics and Animals* 5, no. 2: 27–38. Print.

Saul, John Ralston. 2008. *A Fair Country: Telling Truths about Canada*. Toronto: Viking Canada. Print.

Schmidtz, David. 1996. *Rational Choice and Moral Agency*. Princeton, NJ: Princeton UP. Print.

Schulze, P.C., ed. 1981. *Engineering within Ecological Constraints*. Washington, DC: National Academy Press. Print.

Schweitzer, Albert. 2008. "Reverence for Life." In *Environmental Ethics: Readings in Theory and Application*. Ed. Louis Pojman and Paul Pojman. Belmont, CA: Thomason Wadsworth. 131–8. Print.

SDWW. 2013. "Women and Sustainable Food Security." Prepared by the Women in Development Service, Food and Agriculture Organization, Women and Population Division, United Nations. www.fao.org/sd/fsdirect/fbdirect/fsp001.htm. Web.

Sen, Gita. 2002. "Women, Poverty, and Population: Issues for the Concerned Environmentalist." In *Environmental Ethics: What Really Matters, What Really Works*. Ed. David Schmidtz and Elizabeth Willott. New York: Oxford UP. 248–54. Print.

Sepänmaa, Yrjö. 1993. *The Beauty of Environment: A General Model for Environmental Aesthetics*, 2nd edn. Denton, TX: Environmental Ethics Books. Print.

Shafer-Landau, Russ. 2004. *Whatever Happened to Good and Evil?* New York: Oxford UP. Print.

Shelley, Percy Bysshe. 1891. *The Defense of Poetry*. Boston: Ginn and Company. Print.

Shiva, Vandana. 1988. *Staying Alive: Women, Ecology, and Survival in India*. London: Zed Books. Print.

——. 1990. "Development as a New Project of Western Patriarchy." In *Reweaving the World: The Emergence of Ecofeminism*. Ed. I. Diamond and G. Feman Orenstein. San Francisco: Sierra Club Books. 189–200. Print.

——. 1993. "The Impoverishment of the Environment: Women and Children Last." In *Ecofeminism*. Ed. Maria Mies and Vandana Shiva. Atlantic Highlands, NJ: Zed Books. Print.

——, ed. 1994. *Close to Home: Women Reconnect Ecology, Health, and Development Worldwide*. Gabriola Island, BC: New Society Publishers. Print.

——. 1998. "Development, Ecology and Women." In *The Environmental Ethics and Policy Book*. Ed. Christine Pierce and Donald Van DeVeer. Belmont, CA: Wadsworth. 271–7. Print.

Shrader-Frechette, K., and E.D. McCoy. 1994. *Method in Ecology*. New York: Cambridge University Press. Print.

Shue, Henry. 1980. *Basic Rights: Subsistence, Affluence and U.S. Foreign Policy*. Princeton NJ: Princeton UP. Print.

———. 1996. *Basic Rights: Subsistence, Affluence and US Foreign Policy*. Princeton: Princeton University Press. Print.

Simon, Herbert. 1955. "A Behavioral Model of Rational Choice." *Quarterly Journal of Economics* 69: 99–118. Print.

———. 1959. "Theories of Decision Making and Economics in Behavioural Science". *American Economic Review* (49), 253–83. Print.

Sinervo, B., et al. 2010. "Erosion of Lizard Diversity by Climate Change and Altered Thermal Niches." *Science* 328, no. 5980: 894–9. Print.

Singer, Peter. 1975. *Animal Liberation*. New York: HarperCollins. Print.

———. 2010. "One Atmosphere." In Gardiner et al., eds, 186–7. Print.

Sioui, Georges E. 1992. *For an Amerindian Autohistory*. Trans. Sheila Fischman. Montreal and Kingston: McGill-Queen's UP. Print.

Skye, Raymond. 1998. *The Great Peace: The Gathering of Good Minds*. Brantford, ON: Working World Training Centre. CD-ROM.

Slicer, Deborah. 1995. "Is There an Ecofeminism/Deep Ecology Debate?" *Environmental Ethics* 17, no. 2: 151–69. Print.

Smith, Mick. 2001. *An Ethics of Place: Radical Ecology, Postmodernity and Social Theory*. Albany: State U of New York P. Print.

Snyder, Gary. 1990. "Good, Wild, Sacred." In *The Practice of the Wild*. San Francisco: North Point Press. Print.

Sober, Eliot. 1986. "Philosophical Problems for Environmentalism." In *The Preservation of Species: The Value of Biological Diversity*. Ed. B.G. Norton. Princeton, NJ: Princeton UP. 173–94. Print.

Social Watch Coalition. 2010. "MDGs Remain Elusive. Social Watch: Poverty Eradication and Gender Justice." www.socialwatch.org/node/12082. Web.

State of Washington. 2003. *A New Path Forward: Action Plan for a Sustainable Washington*. State of Washington. Print.

Stecker, Robert. 1997. "The Correct and the Appropriate in the Appreciation of Nature." *British Journal of Aesthetics* 37: 393–402. Print.

Stefanovic, Ingrid Leman. 2000. *Safeguarding Our Common Future: Rethinking Sustainable Development*. Albany: State U of New York P. Print.

Stone, Christopher. 1974. *Should Trees Have Standing? Toward Legal Rights for Natural Objects*. Los Altos, CA: William Kaufmann. Print.

Sumner, Jennifer. 2005. *Sustainability and the Civil Commons*. Toronto: U of Toronto P. Print.

Suzuki, David. 2004a. "How Little We Know." In *The David Suzuki Reader*. Vancouver: Greystone Books. Print.

———. 2004b. "The True Price of a Tree." In *The David Suzuki Reader*. Vancouver: Greystone Books. Print.

Svärd, P. 2013. "Animal National Liberation?" *Journal of Animal Ethics* 3: 201–13.

Taylor, Paul W. 2000. "The Ethics of Respect for Nature." In *Earth Ethics*, 2nd edn. Ed. James Sterba. Upper Saddle River, NJ: Prentice-Hall. Print.

Thomas, C.D., et al. 2004. "Extinction Risk from Climate Change." *Nature* 427: 145–8. Print.

Thompson, Janna. 1995. "Aesthetics and the Value of Nature." *Environmental Ethics* 17: 291–305. Print.

Thompson, Paul B. 1996. "The Case of Water." In *Environmental Pragmatism*. Ed. Andrew Light and Eric Katz. London: Routledge. 187–208. Print.

Thwaites, Reuben G. 1896. *The Jesuit Relations and Allied Documents: Travels and Explorations of the Jesuit Missionaries in New France 1610–1791*, vol. 1. Trans. Reuben G. Thwaites. Cleveland: Burrows Brothers Co. Print.

Tilman, David, and John Downing. 1994. "Biodiversity and Stability in Grasslands." *Nature* 367: 363–5. Print.

Tobac, Georgina (Athabascan). 1979. Speaking in a film by René Fumoleau, Dene Nation, National Film Board of Canada, 1979. Video film.

Trigger, Bruce G., and Wilcomb E. Washburn. 1996. *The Cambridge History of the Native Peoples of the Americas*. Vol. 1: *North America*, Part 1. New York: Cambridge UP. Print.

Turner, Nancy J., Maryanne Boelscher Ignace, and Ronald Ignace. 2000. "Traditional Ecological Knowledge and Wisdom of Aboriginal Peoples in British Columbia." *Ecological Applications* 10, no. 5: 1275–87. Print.

United Nations. 2009. *World Urbanization Prospects: 2009 Edition*. http://esa.un.org/unpd/wup/index.htm. Web.

US Environmental Protection Agency. 1989. *Policy Options for Stabilizing Global Climate Change. Draft Report to Congress*. Ed. D. Lashof and D.A. Tirpak. Washington: GPO. Print.

Vaillant, Peter. 2006. *The Golden Spruce*. Toronto: Vintage Canada. Print.

Valpy, Michael. 2010. "The Seal Hunt as a Matter of Morals." *Globe and Mail*, 8 February, A6. Print.

van der Ryn, Sim, and Stuart Cowan. 1996. *Ecological Design*. Washington: Island Press. Print.

Varner, G.E. 1987. "Do Species Have Standing?" *Environmental Ethics* 9: 57–72. Print.

Venkateswaran, Sandhya. 1995. *Environment, Development and the Gender Gap*. New Delhi: Sage Publications. Print.

Vlastos, Gregory. 1962. "Justice and Equality." In *Social Justice*. Ed. R. Brandt. Englewood Cliffs, NJ: Prentice-Hall. 31–72. Print.

Wackernagel, M., and W. Rees. 1996. *Our Ecological Footprint*. Gabriola, BC: New Society Publishers. Print.

Wallace, Anthony F.C. 1958. "The Deganawidah Myth Analysed as the Record of a Revitalization Myth." *Ethnohistory* 5, no. 2: 118–30. Print.

Walzer, Michael. 1984. *Spheres of Justice*. New York: Basic Books. Print.

Waring, Marilyn. 1988. *If Women Counted: A New Feminist Economics*. New York: Harper & Row. Print.

Warren, Karen. 1990. "The Power and Promise of Ecological Feminism." *Environmental Ethics* 12, no. 2: 125–46. Print. Revised, extended, and reprinted in Karen Warren, ed., *Ecological Feminist Philosophies*. Bloomington: Indiana University Press, 19–42. Print. Citations refer to the extended paper.

——. 1991. "Feminism and the Environment: An Overview of the Issues." *APA Newsletter on Feminism and Philosophy*, Fall. Print.

——. 1997. "Taking Empirical Data Seriously: An Ecofeminist Philosophical Perspective." In *Ecofeminism: Women, Culture, Nature*. Ed. K.J. Warren. Bloomington, IN: Indiana UP. 3–20. Print.

——. 2000. *Ecofeminist Philosophy: A Western Perspective on What It Is and Why It Matters*. Lanham, MD: Rowman and Littlefield. Print.

Warren, Sarah, ed. 1992. *Gender and Environment: Lessons from Social Forestry and Natural Resource Management*. Toronto: Aga Khan Foundation. Print.

Weiskel, Timothy. 1990. "Cultural Values and Their Environmental Implications: An Essay on Knowledge, Belief and Global Survival." Paper presented at the American Association for the Advancement of Science. New Orleans, LA.

Welchman, Jennifer. 2009. "Pragmatism and the Value(s) of Nature" (unpublished manuscript). Print.

Weston, Anthony. 1996. "Beyond Intrinsic Value: Pragmatism in Environmental Ethics." In *Environmental Pragmatism*. Ed. Andrew Light and Eric Katz. London: Routledge. Print.

Westra, Laura. 1994. *The Principle of Integrity*. Lanham MD: Rowman and Littlefield. Print.

——. 1997. "Terrorism at Oka: Environmental Racism and the First Nations of Canada." In *Canadian Issues in Environmental Ethics*. Ed. Wesley Cragg, Allan Greenbaum, and Alex Wellington. Peterborough, ON: Broadview Press. 274–91. Print.

——. 1998. *Living in Integrity*, Lanham, MD: Rowman and Littlefield. Print.

——. 2007. *Environmental Justice and the Rights of Indigenes Peoples*. London: Earthscan. Print.

——. 2011. *Human Rights: The Commons and the Collective*. Vancouver: UBC Press. Print.

Westra, L., et al., eds. 2000. *Ecological Integrity: Integrating Environment, Conservation and Health*. Washington: Island Press. Print.

Wijkman, Anders, and Johan Rockström. 2012. *Bankrupting Nature: Denying Our Planetary Boundaries*. London: Routledge. Print.

Williams, Bernard. 1985. *Ethics and the Limits of Philosophy*. Cambridge, MA: Harvard UP. Print.

Williston, Byron. 2011. "Moral Progress and Canada's Climate Failure." *Journal of Global Ethics* 7, no. 2 (August): 149–60. Print.

Willott, Elizabeth. 2002. "Recent Population Trends." In *Environmental Ethics, What Really Matters, What Really Works*. Ed. David Schmidtz and Elizabeth Willott. Oxford: Oxford UP. 274–83. Print.

Wilson, Edward O. 2002. *The Future of Life*. New York: Vintage Books. Print.

Wolf, Clark. 2007. "Population and the Environment." In *Environmental Ethics: Readings in Theory and Application*. Ed. Louis Pojman and Paul Pojman. Belmont, CA: Thomason Wadsworth. Print.

——. 2008. "Population and the Environment." *Environmental Ethics: Readings in Theory and Application*. Ed. Louis Pojman and Paul Pojman. Belmont, CA: Thomson Wadsworth. Print.

Wolf, Robert Paul. 1998. *In Defence of Anarchism*. Berkeley: U of California P. Print.

World Bank Group. 2014. "Poverty Overview." www.worldbank.org/en/topic/poverty/overview. Web.

World Commission on Environment and Development. 1987. *Our Common Future*. Oxford: Oxford UP. Print.

Valentini, L. 2014. "Canine Justice: An Associative Account." *Political Studies* 62, no. 1: 37–52. Print.

Zagzebski, Linda. 1996. *Virtues of the Mind: An Inquiry into the Nature of Virtue and the Ethical Foundations of Knowledge*. Cambridge: Cambridge University Press. Print.

Zangwill, Nick. 2001. "Formal Natural Beauty." *Proceedings of the Aristotelian Society* 101: 209–24. Print.

Zimmerman, Michael. 1987. "Feminism, Deep Ecology, and Environmental Ethics." *Environmental Ethics* 9, no. 1: 21–44. Print.

Credits

Archer, David. *Long Thaw*. Reproduced with permission of Princeton University Press in the format Book via Copyright Clearance Center.

Baxter, William F. Republished with permission of Columbia University Press. *People or Penguins: The Case for Optimal Pollution*, 1974. Permission conveyed through Copyright Clearance Center, Inc.

Berry, Thomas. Copyright © 1988 by Thomas Berry from *The Dream of the Earth*. Extract reprinted by permission of Counterpoint.

Bookchin, Murray. *The Ecology of Freedom: The Emergence and Dissolution of Hierarchy* (Oakland: AK Press, 2005).

Ehrenfeld, David. "Why Put a Value on Biodiversity?" Reprinted with permission from *Biodiversity*. Ed. E.O. Wilson, 1988, by the National Academy of Sciences, Courtesy of the National Academies Press, Washington, DC.

Gardiner, Stephen M. "Saved By Disaster? Abrupt Climate Change, Political Inertia and the Possibility of an Intergenerational Arms Race." *Journal of Social Philosophy* 40(2), Summer 2009, 140–162. © 2009 Blackwell Publishing, Inc. Reprinted by permission of John Wiley & Sons, Inc.

Gillis, Alex. Quote from "Burying Hell." © Alex Gillis. Reprinted by permission of the author.

Hill, Thomas E., Jr. "Ideals of Human Excellence and Preserving Natural Environments." *Environmental Ethics* 5, 1983, 211–224. Reprinted by permission of the author.

Leopold, Aldo. "The Land Ethic." *A Sand County Almanac*. New York: Oxford University Press, 1949. Reprinted by permission of Oxford University Press US.

MacEwen, Gwendolyn. "Dark Pines Under Water." *Open Wide a Wilderness: Canadian Nature Poems*. Ed. Nancy Holmes. Waterloo, ON: Wilfrid Laurier University Press, 2009. 284. Reprinted with permission of the author's family.

Naess, Arne. "Identification as a Source of Deep Ecological Attitudes." © Arne Naess. Reprinted with permission.

Newlove, John. "The Pride." From *A Long Continual Argument: The Selected Poems of John Newlove*, ed. Robert McTavish. Ottawa: Chaudiere Books, 2007.

Simon, Julian L. "Can the Supply of Natural Resources Really Be Infinite? Yes!" *The Ultimate Resource 2*. Princeton, NJ: Princeton University Press. 1996. Reproduced with permission of Princeton University Press in the format Book via Copyright Clearance Center.

Singer, Peter. "All Animals Are Equal." Part of this essay appeared in the *New York Review of Books* (5 Apr. 1973). This version is an abridged form of an essay which was first published in *Philosophic Exchange* 1, 5 (Summer 1974) Copyright © Peter Singer, 1973, 1974.

Sioui, Georges. "The Sacred Circle of Life." *For an Amerindian Autohistory*. Montreal: MQUP, 1995. Print.

Stone, Christopher D. Excerpt, "Should Trees Have Standing? Toward Legal Rights for Natural Objects," V45: 2 *Southern California Law Review* (pp. 450–470) (1972), reprinted with the permission of the Southern California Law Review.

Suzuki, David. "The Power of Biodiversity." Excerpt from the book *The Sacred Balance: Rediscovering Our Place in Nature, Updated and Expanded by David Suzuki*, published in 2007 by Greystone Books Ltd. Reprinted with permission from the publisher.

Taylor, Paul W. "The Ethics of Respect for Nature." *Earth Ethics*. 2nd ed. Ed. James Sterba. Upper Saddle River, NJ: Prentice-Hall, 2000.

Warren, Karen J. "The Power and the Promise of Ecological Feminism." *Environmental Ethics* 12 2, 1990, 125–146. © Karen J. Warren.

Weston, Anthony. From "Before Environmental Ethics" in *Environmental Pragmatism*, edited by Eric Katz and Andrew Light. Copyright © 1996 Routledge, reproduced by permission of Taylor & Francis Books UK.

Wikler, Ben. "Climate Shame as Canada Named 'Collossal Fossil.'"

Index